To

Sushilarani D. Butani
Late (Mrs) Kusum Srivastava
(24.11.1940 – 03.01.2000)

Butanis'

DICTIONARY OF ENTOMOLOGY
(General and Applied)

DR. DHAMO K. BUTANI
DR. K.P. SRIVASTAVA
LAUREL, U.S.A

Studium Press (India) Pvt. Limited
4735/22, 2nd Floor, Prakash Deep Building
Ansari Road, Darya Ganj, New Delhi-110002

Published by
STUDIUM PRESS (INDIA) PVT. LTD.
4735, 2nd Floor, Prakash Deep Building (Near PNB)
Ansari Road, Darya Ganj, New Delhi-110002
Tel.: +91-11-23240257, 65150445, 65150447
Fax: +91-11-23240273
E-mail: studiumpress@gmail.com

First Edition: 2009

Printed at: Rajkamal Electric Press, B-35/9, G.T. Karnal Road, Delhi.

M.S. SWAMINATHAN RESEARCH FOUNDATION

M.S. SWAMINATHAN
Chairman

FOREWORD

I am glad Dr. Dhamo K. Butani and Dr. K.P. Srivastava have compiled this illustrated Dictionary of Entomology. This is no doubt a comprehensive compilation of many important aspects of both basic and applied entomology. The detailed classification contained in the appendices of the dictionary would be of immense benefit to research scholars and students. The enormous genetic diversity of insects and their multiple roles, both as friends and foes of agriculture, make a study of entomology absolutely essential for sustainable agriculture. Studies on host–plant relationships and of beneficial insects are important for developing integrated pest management strategies. I therefore hope, the dictionary will find widespread use.

(M.S. Swaminathan)

3rd CROSS STREET, TARAMANI, INSTITUTIONAL AREA,
CHENNAI - 600113. Phone: +91-44-22541698, 22542790
Fax: +91-44-22541319
E-mail: chairman@mssrf.res.in, swami@mssrf.res.in

PROLOGUE

ENTOMOLOGY (Gr. *entom* = insect + *logos* = discourse or science) - the science of insects, a branch of zoology that deals with the origin of insects, their systematic positions, structure and bodily functions, habits and habitat, behaviours, development, relation to one another as also to the environments and their economic importance. In wider sense it also covers study of losses caused by insect pests to our crops (Agricultural entomology), their prevention and control (Plant protection), insect pests vis-a-vis forest trees (Forest entomology), domestic animals (Veterinary entomology), health and hygiene (Medical entomology) as also commercial products accruing from certain insects (Industrial entomology).

Origin of insects can be traced back to about 380 million years; though fossil records show that a few collembolan-like insects (*Rhyniella praecursor, Rhyniognatha hirsti* etc.) existed during Lower Devonian period (410 to 405 million years ago). Most of the palaeozoics are by now extincts, but the blattids - the oldest living insects - appear to be inhabiting the Earth since the Carboniferous period. The best known insect epoch was no doubt the Lower Permian and entire Triassic period (280 to 225 million years ago), when the insects increased both in number of species and individuals. As compared to this, man has arrived on this Earth only a short while ago (1500000 years).

Knowledge of insects is as old as the human race. In India, study of insects was initiated much before the word 'Entomology' was coined. In our *Vedas* and ancient Sanskrit literature there is mention of honey which is nothing but predigested food, stored by honeybees for rearing their young ones. House of lac in Mahabharat (*Dwapar yuga*) is another example of our knowledge regarding lac-insect. However, earliest available record appears to be an illustration of stick-lac (lac-insect) drawn by Garcia in 1563 at Goa, that was published by Clusius in Antwerp (1567). Regular work on Indian insects was carried out by J.G. Koeing (student of C.A. Linnaeus), who studied the termites of Tanjore district (Tamil Nadu) and published in **Beschaftigungen der Berlinischen Gesellaschaft Naturforschender Freunde,** Volume 4 (I779). C. Kerremans and W. Roxburgh published their papers on lac-insects in 1782 and 1790 respectively, while J. Anderson's monograph on cochineal insects appeared in 1791. Carolus A. Linnaeus has listed 12 Indian species of insects in 10th edition of **Systema Nature** (1758), followed by Johann Christian Fabricius, who included more than 1000 Indian species in **Entomologia Systematica** volumes (1772-98). For the next 100 years, work on entomology in India was confined, mainly to collection of insects as a hobby by British Colonials and sending the same to British Museum (Natural History), London, for identification and retention. The Indian Museum at Calcutta came into being in 1875, wherein an insect collection section was added in 1885, under the direction of Colonel A. Alcock who also started publication of **Indian Museum Notes** (sex volumes, 1889 to 1904). During this period several volumes of **Fauna of British India including Myanmar (Burma and Sri Lanka (Ceylon)** were published from London, giving illustrated and classified account of almost all the Indian species described till then. Government of India appointed Lionel de Niceville as the first Entomologist in 1901 who started working at Indian Museum Calcutta, but unfortunately died after a year. Maxwell H. Lefroy took over in 1902 but worked at Surat on cotton pests till 1905 when he was shifted to Pusa (Bihar) as Imperial Entomologist and worked there till 1912. The post was subsequently held by T. Bainbrigge Fletcher (1913-34), Hem Singh Pruthi (1934-44), P.V. Issac (1944-45) and Taskhir Ahmed (1945-47). The first three entomologists laid the foundation of economic entomology while the last three gave philip to applied entomology. Teaching of entomology started with opening of six agricultural colleges in 1904-06 at Sabour, Lyallpur, Pune, Coimbatore, Nagpur and Kanpur. Research on entomology was initiated at IARI, in 1905 at pusa (Bihar), shifted after the earthquake in 1934 to New Delhi. Simultaneously studies on Forest entomology were started at Dehradun when Edward Percy Stebbings was appointed as first Forest Entomologist in 1901 and (Imperial) Forest Research Institute came into being in 1906; where E.P. Stebbings was followed by V. Subramaniya (1909-11), A.D. Imms (1911-13) and so on. Studies on Veterinary entomology were initiated at Indian Veterinary Research Institute, Mukteshwar (now at Izatnagar); and later, on Lac-insects at Indian Lac Research Institute, Namkum, Ranchi (1925).

Among the various States, early birds were, Chennai who appointed an Entomologist, as an assistant to Government Botanist in 1904. T.B. Fletcher joined as Government Entomologist in 1909 and worked at Coimbatore, where he was succeeded by E. Ballard in 1914. Mysore State appointed L.C. Coleman as Government Entomologist in 1907; M. Afzal Husain joined as Government Entomologist to Punjab in 1919 followed by Khan A.K. Rahman and Kidar Nath

Trehan (after partition); while P.B. Richards became Government Entomologist to U.P. in 1922 and M.L. Bhatia, State entomologist to Bihar in 1939.

Dictionary of Entomology need not limit or confine itself to basic or fundamental entomology but should also include the terminology used in applied entomology including the common pesticides. An attempt has therefore been made to cover all these aspects of entomology in this dissertation. Besides technical terms, attempt has also been made to describe all animal phyla and their major classes; of class Insecta, superorders, orders, suborders, superfamilies and as many families as possible. Lest we forget the old stalwards of entomology, who are no more with us, short biographical notes on as many as 30 eminent entomologists of yesteryear have also been included and another 44 (living ones) listed at the end.

The authors are highly indebted to Dr. M.S. Swaminathan (Padma Vibhushan), for very kindly writing the foreword. Sincere thanks are due to Dr. Gouri Ganguly (Zoologist), Dr. Prabha Rani (Embryologist), Dr. Shashi Verma, Principal Scientist (Retired), Dr. T.R. Sukhani, Principal Scientist (Retired), Dr. N.P. Agnihotri, Principal Scientist (Retired), Dr. Atma Ram, Principal Scientist (Retired), and Dr. Shashi Shekhar Juneja (Entomologist) for their assistance in compiling this compendium. The help rendered by Ms. Anita R. Butani Mention need also be made of Mr. Arvind Jainji, and Mr. Shray Jain of Studium Press (India) Pvt. Ltd. for publishing the book in its present form. Last, but not the least, the authors wish to express their deep sense of gratitude to their spouces Smt. Sushilarani and late Smt. Kusum and other family members for their cooperation and support in bringing out this dictionary of Entomology a unique publication of Science.

(DHAMOK. BUTANI)
M.S. (Agri.), D.Phil.

(K.P. SRIVASTAVAVA)
M.Sc, Ph.D.

A

Å abbreviation of angstrom; a unit of length, named after Swedish physicist and astronomer Andres Jones Angstrom (1814-74); equivalent to one millionth millimetre (10^{-10} metre or 0.0001 micron); used for expressing wavelengths of light, ultraviolet rays, X-ray etc. Also known as tenthmetre; 10 Å = 1 nm (nanometre, 10^{-9}m)

A- used in composition indicating or signifying on, at not, without.

Ab- Latin prefix denoting away from.

Aback backwards; applied generally to aquatic insects that swim backwards or against the current.

Abactinal in radially symmetrical animals (invertebrates), the surface on the opposite side of the mouth.

Abacus a sliding rule for doing arithmetical calculations manually.

Abandon give up: forsake.

Abdomen hinder part of arthopods' body, posterior to thorax (insects) or cephalothorax (annelids/spiders). In mammals part of body between diaphragm and pelvis. In insects, abdomen mostly comprises of eleven segments and a small terminal telson though several segments are often concealed. In Protura and Collembola the number of segments is much less. In true files (Cyclorrhapha: Diptera), it is difficult to ascertain the exact number of abdominal segments without dissection, as only four or five segments are apparently visible. In Hymenoptera the abdomen is restricted physiologically to the region that commences with the II segment; the I abdominal segment being the propodeum. In adult insect usually there are no appendages attached to abdomen except cerci and genitalia whereas many larvae possess abdominal or false legs. In Coleoptera (adult beetles), four different types of abdomen may be recognised:

— **Adephaged** – hind coxae immovably fused with metathorax and completely divide the I visible abdominal sternite, that is more or less fused with the next two sternites, sulci between these being partly obliterated.

— **Hyplogastrous** – the II abdominal segment exhibits a pleurite and a small lateral plate represent the sternite.

— **Symphiogastrous** - pleurite of II abdominal segment fused to that of III; sternite membranous and not visible externally.

— **Hologastrous** - the II abdominal sternite fully sclerotised and distinct from the III, as in some Cantheriodea (Polyphaga).

Abdominal pertaining to or attached to abdomen.

Abdominal legs legs borne of abdomen; common in insect larvae but not in adults. Also known as false legs.

Abductor (levator) any muscle that draws a part of body away from the body axis. *Cf.* adductor.

Abiocoen a nonbiotic habitat.

Abiogenesis autogenesis: spontaneous generation as proponed by T.H. Huxley; the obsolete conception that plant and animal life arise from non-living organic matter. *Cf.* biogenesis.

Abiology science of non-living matter.

Abiotic non-biotic: absence of living organisms.

Abiotic environment chemical and physical factors that influence the living organisms.

Abiotic substance any fundamental chemical element or compound in the environment.

Abnormal not normal: functioing irregularly.

Abnormality state of being abnormal or diseased: malformation.

Aboral away from the mouth: the other end.

Aboriginal earliest: indigenous: primitive.

Abortive imperfectly formed: not well developed.

Abound to be in plenty: to overflow.

Abovo from the egg: from the beginning.

Abrade injure by rubbing: scrape off.

Abræinæ subfamily of Histeridæ (Histeroidea: Polyphaga); comprising of steel beetles distinguished by having 5-5-4 segmented tarsi, first and second hind tarsal segment fused and hind coxae widely separated.

Abrasion rubbing or scrapping off.

Absolute completely independent: free from conditions, limits or restrictions.

Absolute alcohol water free alcohol: 99 per cent pure ethnol (CH_3CH_2OH) by weight; obtained by digesting rectified spirit over quick-lime and redistilling the same.

Absolute humidity amount of water vapour present in atmosphere or in a given amount of air.

Absolute temperature temperature as recorded on Kalvin scale (°K); though measured in degrees Celsius (centigrade); it starts from absolute zero instead of 0° Celsius.

Absolute threshold the minimum stimulus that an organism can detect.

Absolute zero starting point of Kelvin thermometer: temperature at which the particles lose their kinetic energy - 0°K = –273.15°C or –459.67°F.

Absorb to suck in: imbibe: swallow.

Absorbent capable of absorbing.

Abyssoplegic organism found living in ocean at the depth of 2000 metres or more.

Acanthoceridæ family of Scarabæoidea (Polyphaga: Coleoptera); beetles differ from Trogidæ in being capable of rolling up into a ball; segments of antennal club with three closely co-adapted lamellae; mandibles small, five visible abdominal stermites; common in forests, associated with rotten wood - 120 tropical species.

Acanthopsyche common and generic name of various begworms (Pyschidæ: Lepidoptera) that cause severe damage to forest trees in South Africa - *Acanthopsyche junodi* Heylaerts on black acacia.

Acanthosomatidæ family of Pentatomoiea (Geocorisæ: Heteroptera), earlier considered as subfamily of Pentatomidæ; predominantly polyphagous bugs having antennae 5-segmented; scutellum not reaching end of abdomen - *Cyphostethurs* species, *Elasmucha grisea* et al.

Acanthous spiny.

Acariasis disease caused by mites.

Acaricide pesticide that can kill Acarida (mites and ticks).

Acarida (Acari, Acarina, Acariformes) order of class Arachnida (Arthropoda); tiny (0.5 mm) ovate creatures having compressed body consisting of cephalothorax and abdomen, mouthparts adapted for biting, sawing, piercing and sucking; nymphs and adults have four pairs of legs while larvae have only three pairs (600 species) - mites and ticks.

Acariæ Ewing Nesbith family of mites (Acarida: Arachnida) that feed on organic matter and are serious pests of stored products of high moisture content, such as grains, cheese, cured and raw hides, preserved meat, mushrooms etc. - grain mite (*Acarus siro*), bulb mite (*Rhizoglyphus echinopus*) etc.

Acaridomantium (pl. **acaridomatia**) mite-web (colony).

Acariformes old and obsolete name for group (order) or mites, now place under order Acarida (Acarina).

Acarina (Acari, Acarida) order of Arachnida (Arthropoda).

Acarologist person having sound knowledge of mites.

Acarology branch of zoology that deals with mites.

Acarophily symbiotic association of plants with mites.

Acarus (pl. **acari**) mites(s).

Acceptable daily intake (ADI) quantity of chemical that can be taken daily during entire life time, without appreciable risk of injury; expressed usually as mg/day.

Accessary species a species comprising 25 to 50 per cent of a community.

Accessorius any muscle that reinforces the action of another muscle.

Accessory chromosome sex chromosome.

Accessory glands pair of glands that in female insects open on ventor of IX abdominal segment and secrete adhesive substance that forms a cover or case for eggs: in male insects mucous glands opening into ejaculatory duct.

Accessory heart(s) pulsatory organs that comprise of ampullae with valves and are present near the base of anternnæ, legs or wings of some insects.

Accidental species a species which constitutes less than one-fourth (25%) of the population of a community.

Acclimation (accliatisation) adaptation of a species or population to a change environment over several generations.

Accommodation a population's location within a habitat.

Acellular not composed of cells.

Acephalous apparently without head: applies to such larvae, whose head is much reduced and not separatable from throrax - house fly (*Musca domestica* Linnaeus).

Acephate $C_4H_{10}NO_3PS$; phosphoramidothioate insecticide that is white solid, relatively stable; melting point 85°C and specific gravity 1.35; acute oral and dermal LD_{50} to rat 866-945 and 1494 mg/kg respectively - Lancer®, Orthene® etc.

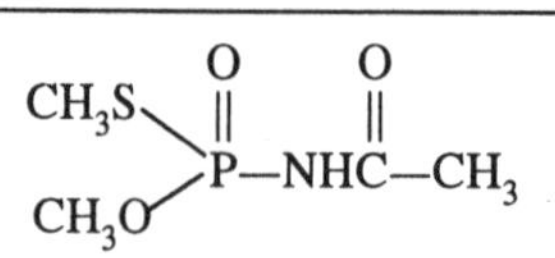

Acephate

Acerentomidæ Berlese 1909 a small diverse family of Acerentomoidea (Protura), primitive apterygote insects, < 2 mm long, having no trachæ and no spiracles; only first one or two pairs of abdominal appendages 2-segmented; about 60 species found in damp places and peat feeding on organic matter – *Acertentulus breviunguis* Conde (on tea in Keralka), *Acerella donica, Acerentomum doderoi* et al.

Acerentulus breviunguis Conde

Acerentomoidea superfamily of primitive wingless, proturan insets, less than 2 mm long, each abdominal trite with two transverse rows of setae; abdominal combs on VIII segment well developed; embraces only two families – Acerentomidæ and Protentomidæ. Acerentulus

Acerous without antennae (all proturan insects): without horns or breviunguis tentacles.

Acetabulum (pl. **acetabola**) coxal cavity on an insect body in which one end of the leg (coaxa) is set for articulation.

Acherontiinæ small subfamily of Sphingidæ (Sphingoidea: Ditrysia) having very short and thick proboscis; antennae thick at the base. Caterpillar thick, tapering to head; thoracic somites retractile; horn recurved at the tip and tuberculate. Pupa without external sheath for proboscis – *Acherontia styx* Westwood and allied species.

Achilidæ a family of Fulgoroidea (Fulgoromorpha: Auchenorrhyncha) homopteran bugs have antennal flagelium not segmented; ocelli outside the lateral carinae of head; apex of hind tarsus with row of small spines and base of abdomen without any appendages. Nymphs found under the bark or in cavities in dead wood. About 250 species having World-wide distribution especially in Tropical regions.

Achilixiiæ a very small family of Fulgoroidea (Fulgoromorpha: auchenorrhyncha) comprising of only 12 species, segregated from the family Coxiidæ and distinguished by the base of abdomen having one or two short appendages bearing three hemispheroidal depressions.

Achrestogonime(s) a caste or form of termites (Isoptera). When swarming occurs, normally all alate imagines leave the termitarium, but occasionally a small number remains behind; these shed their wings, remain xylophagous, retain their protozoal fauna and though these individuals (achrestogonimes) take no part in maintenance of the termitarium, yet are tolerated by the other inmates of the colony.

Achromic colourless: lacking normal pigmentation.

Acicular needle-shaped.

Acid sharp, corrosive substance, having sour taste; turns litmus red; releases hydrogen ions (H^+) in water. *Cf.* alkali.

Acid gland principal poison gland of honeybee (Apoidea: Apocrita), secreting a complex mixture of histamine-liberating enzymes an proteins of low molecular weight. Also known as apitoxin.

Acidaliidæ Guenee 1844 (= Scopulidæ, Sterrhidæ) a small widely distributed family of Geometroidea (Ditrysia: Lepidoptera); small whitish, grey or brown moths with delicate wavy markings on the wings; forewing with areole present and vein R_1 often fused with areole and not with Sc – herbarium moth *Eois ptelearia* et al.

Acidaliniinæ (Acidaliidæ) subfamily of Geometroidæ (Geometroidea: ditrysia); palpi slender hardly reaching beyond frons; forewing with vein 5 from middle of discocellulars; vein 10 stalked with 7, 8 and 9; hind-wing with vein 5 fully developed; usually from just above the middle; vein 8 bent down and touching the cell near the base then rapidly diverging – *Anisodes ignorata* Walker, *Craspedia modesta* Moore, et al.

Acidophil any substance, tissue or organism having an affinity for acid stains. An organism having a preference for an acidic environment.

Acidotrophic having an acid mutrient requirement.

Aclerdiedæ small family of subterranean scale insects (Coccoidea: Homoptera) – *Aderda distorta* Green, *A. japonica* Newstead and *A. japonica inermis* Green infest sugarcane (*Saccharum officinarum* Linnæus) in India.

Acoelomata subdivision of kingdom Animalia characterised by presence of three germ layers (Triploblastica) and no true body cavity; includes three phyla-Platyhelminthes, Gnathosotomula and Nemertinea = Rhynchocoela).

Acoelomate lower animals(s) having no coelom or body cavity – flatworms, ribbon-worms, round-worms etc.

Acoelous lacking a true body cavity or coelom.

Acoene eye compound eye with crystalline cones poorly developed or absent – craneflies (Tripulidæ: Diptera).

Acontiilnæ subfamily of Noctuidæ (Noctuoidæ: Ditrysia); small-sized moths living amongst low herbage; proboscis fully developed; eyes naked; tibia without spines; hind-wing with vein 5 slender, depressed at origin and the outer margin not produced at its termination - *Acontaia malvæ* Esper, *Xanthoptera spicea* Guenee, *Eublemma angulifera* Moore, *Zagira divisa* Walker, et al.

Acoustic pertaining to sence of hearing.

Acoustic nerve auditory nerve.

Acoustics science of hearing.

Acræidæ a family of Papilionoidea that is considered by some entomologists as synonym of family Danaidæ (Papilionoidea: Ditrysia) while many regard Acræinæ and Danainæ as subfamilies of Nymphalidæ.

Acræin protective and distasteful secretion of certain butterflies (Lepidoptera).

Acræinæ subfamily of Nymphalidæ (Papilionoidea: Diptera) often regarded as an independent family acræidæ; characterised by elongated wings that are sparsely scaled or more less diaphanous; the adults readily exude a nauscous fluid. Essentially African, though a few species occur in orient and South America also - *Acræa* spp. et al.

Acrantia (Protochordata) phylum of kingdom Animalia (now included in Chordata) comprising of primitive chordaes having no distinct cranium; divided into three subphylla - Cephalochordata (lancelets), Tunicata or Urochordata (sea-squirts) and Hemichordat (tongue-worms).

Acrididæ Brunner 1900 family of orthopteran insects (Acridoidea: Cælifera) having short antennæ; pronotum extending backwards and down on sides of thorax; forewings narrow and leathery; auditory organs one on either side of I abdominal segment; ovipositor short and stout; all herbivorous - locusts (*Locusta migratoria* Linnæus, *Schistoceraca gregaria* Forskal etc.), short-horned grashoppers [*Atractomorpha acrenulata* (Fabricius), *Poekilecerus pictues* (Fabricius), *Chrotogonus* spp. etc.]

Schistocerca gregaria Forskal

Acridoidea Burmeister 1829 superfamily of Cælifera (Orthoptera), head bent at right angle to the body; antennæ setiform; tarsi 3-segmeted; forewings horny and thickened (tegmina); ovipositor inconspicuou; heterometabolous.

Acroceridæ (= **Acroderatidæ Leach 1815, 1819)** a family of orthorrhaphan flies (Brachycera: Diptera); having head low and small as compared to hunchbacked thorax; calypteres inflated and concealing the halteres; empodium pesent. Mostly parasitic in spiders.

Acroceridæ old name of Crytidæ - a small family of bristleless flies (Brachycera: Diptera).

Acron unsegmented head of ancestral arthropod: non -segmented preoral portion of an arthropod embryo.

Acrosternite narrow precostal marginal flange, pesent usually in abdominal sterna, not in thoracic sterna.

Acrostichal bristles minute bristles arranged in longitudinal rows along the back of thorax of certain flies between the two row of stronger dorsocentrals.

Acrotergite anterior precostal part of secondary tergite.

Acrydiidæ equivalent name for family Tetrigidæ (Acridoidea; Cælifera) distinguished from other acridoids by the process of pronotum backwardly directed covering the abdomen and concealing the hind-wings - grouse locusts.

Actaletidæ Handirsh 1925 small monogeneric family of apterygote insects (Actaletoidæ: Arthroplea), body compact, head hypognathous, IV to VI abdominal segments fused together, tracheal system present and opens by a pair of cephalic spiracles - *Actaletes neptuni* and allied species reported from France.

Actaletoidea Schroder, 1925 suborder of Collembola that includes single family Actaletidæ.

Actalionidæ very small family of tree-hoppers (Cicadelloidea: Auchenorrhyncha), allied to Membracidæ, comprising of about 50 species.

Activator any chemical or substance added to a pesticide for increasing the toxicity of the pesticide.

Active ingredient (a.i.) toxic component of a formulated pesticide. Also called Technical grade.

Active transport passage of a substance through a cell membrane against a concentration gradient, a process requiring consumption of energy by the cell.

Aculea (pl. **aculeus**) sting: microtrichium: microscopic hair found on wings of certain butterflies and other insects.

Aculeata old and obsolete term for a group of stinging insects (Hymenoptera), now placed under superfamily Scolioidea.

Aculeate bearing a sting or homologous organ.

Aculognathidæ a small family of colepteran insects (Cucujoidea: Polyphaga); small, ant-sucking beetles having flat and broad body with short 11-segmented antennæ and 5-segmented tarsi.

Acute toxicity rapid poisoning produced by a single exposure or a few repeated exposures to excessive amounts of toxic material.

Adaptation an inherent capacity to change: useful or beneficial changes that help an insect or organism to live/grow well and successfully in its environment are passed on from generation to generation. It may be short-term or long-term adoption.

— **Short-term adaptation** - æstivation or hibernation, wherein all the activities are suspend or maintained at a very low metabolic rate untill favourable environmental conditions return.

— **Long-term adaptation** - legs modified for clinging, digging, jumping, swimming, walking etc. Form or structure of claw and beak in case of birds.

Adaption accumulation of characteristics through the course of evolution which make organisms better suited to their environments: any physical physiological or bio-chemical process by which an organism adjusts itself to cope with the particular environment.

Adecticous pupa a pupa having its mandibles non-articulated and often reduced; such mandibules are not used for escaping from the cocoon.

Adekocera common and generic name of various slims, short-legged click beetles (Elateridæ: Polyphaga) that are able to jump several centimetres off the ground by means of spring-like mechanism under their thorax.

Adelgidæ Mordvilko 1935 (Chermidæ Koch 1857, Chermesidæ Passeini 1862, 1867); family of pine aphids (Aphidoidea: Homoptera); very small fragile, oval, indistinctly segmented apterous and alate (venation reduced) forms; roof-like over the abdomen, when at rest; infests only conifers, single life-cycle is completed on two different hosts (both conifers), though gall formation is only on primary host - *Adelges cooleyi, A. tsugæ, Dreyfusia nordmannianæ, Pineus lævis* et al.

Adelidæ a microlepdopterous family (Incurvcariod: Monotrysia); very small moths, usually with bright metallic markings and conspicuously long antennæ; wing having minute needle-like spicules all over the surface and wing venation more or less complete. Caterpillar first miner then casebearer. - *Nemotois* spp., et al.

Adelognatha group of weevils having a short rostrum with temporary mandibles that are later cast off; mostly live in soil, the larvæ feeding on roots.

Adeloneuridæ a fossil family of Protorthoptera (Orthopteroidea: Polyneoptera) from the Middle Upper Carboniferous period of France. Head frees, more or less vertical with strong mandibles and heteronomous wings that could be flexed backwards.

Adephlobiidæ a family of fossil insects (Protoblattoidea: Polyneoptera) from the Middle Upper Carboniferous period of Ilinois (USA). Head free, rounded with strongmandibles and long multi-articulate antennæ; abdomen with cerci and with or without gonapophyses - *Adiphlebia lacoana* et al.

Adelphoparasitism self parasitism, one sex devlops parasitically in the body of its opposite sex - Aphelinidæ; males are obligate parasites of females.

Adenopoda Verhoff 1904 equivalent and obsolete term for insect order Embioptera Shipley.

Adenosine diphoshate (ADP) substance formed when adenosine triphosphate (ATP) is utilised by a cell to supply energy. ATP $\rightarrow$ ADP + phosphate + energy.

Adenosine triphosphate (ATP) primary source of usable energy in cells: an organic compound composed of adenine, ribose and three phosphate groups; formed from ADP and phosphate during cellular respiration.

Adenotrophic viviparity phenomenon found in certain flies (Muscidæ: Diptera) wherein the larvæ after hatching are retained and nourished in female body till the larvæ mature, then these come out and pupate - tsetse flies (*Glossina* spp.).

Adephaga Emery 1885 suborder of Coleoptera that includes beetles, characterised by hypopharynx without sclerome, labrum fused to head capsule; notopleural suture present in prothorax; hind coxa immovably fixed to metasternum; majority predaceous, a few secondarily herbivorous, comprises of only one superfamily - Caraboidea (including Gyrinoidea and Paussoidea).

Aderidæ (Xylophilidæ) family of small coleopteran beetles (Cucujoidea: Heteromera); head deflexed and constricted into a neck; eyes large, tarsi with penultimate segment very small, hidden within the lobes of preceeding segment that is strongly bilobed; trochanters not of heteromerous type; found in dead wood and valuable refuse (700 species) - *Aderus (Xylophila)* species, et al.

Adfronatal areas narrow sclerites along the epicranial suture in the head of caterpillar (Lepidoptera).

Adiabatic any physical change wherein there in no loss or gain of heat.

Adimeridæ old and obsolete name for coleopteran family Colydiidæ (Cucujoidea: Polyphaga).

Adipoleucocytes(s) hæmocytes laden with only droplets of wax as in certain pyrrhocorid bugs (Heteroptera).

Adipose fatty.

Adjuvant(s) spray additives to improve physical or chemical properties of pesticides - deactivators, suspending agents, stabilisers, strickers etc.

Admiral nickname of various brush-footed butterflies (Nymphalidæ: Lepidoptera); medium to large sized, found all over the World. Red admiral (*Vanessa atalanta*) having black wings with red and white markings is a well known migrant butterfly while white admiral (*Ladoga camilla*) ha black or dark brown wings with white or pale yellowish markings.

Adults grown up or matures: final stage in the development of an insect in which all the external characteristics have been fully attained and the internal systems, including reproductive, fully matured. Grown up person.

Adulterated any pesticide whose strength or purity fails below the quality stated on its label. A food, feed or product that contains illegal pesticide residues.

Advolution development or growth with increasing similarities; growth towards the opposite of evolutions.

Ædeagus copulatory organ or penis of male insect: diatal part of phallus.

Ægeriidæ Stephens 1829 old name for lepidopteran family Sesiidæ (Yponomutoidea: Ditrysia) that includes clearwings or clearwing moths.

Ægialiidæ very small family of Scarabæoidea (Polyphaga: Coleoptera); minute, rare beetles characterised by 9-segmented lamellate antennæ, segments of antennal club thin and leaf-like; mandibles exposed (not covered by clypeus).

Ægialitidæ equivalent name for family Salpingidæ (Tenebrionoidea: Coleoptera), that includes small, narrow-waisted bark beetles.

Æolothriidæ Uzel 1895 small family of broad-winged thrips (Æolothripoidea: Terebrantia); body not depressed, antennæ 9-segmeted; forewings broad, rounded apically with two longitudinal veins and front margin without fringe of long hair; ovipositor upcurved - *Aeolothris collaris* Priesner, polyphagous pest attacking garlic, mango, onion, sugarcane etc.

Æolothripoidea Hood 1915 superfamily of Terebrantia (Thysanoptera); thrips having ovipositor curved upwards; forewings relatively broad, rounded at apex, front margin without fringe of hairs and bearing two longitudinal and several cross-veins - Æolothripidæ, Orothripidæ, Melanothripidæ, Franklinothripidæ and Dactuliothripidæ.

Æpophilidæ Lethierry and Severin 1896 very small family of amphibicorisan bugs (Saloidea: Gymnocerata), oftern included in family Gerridæ, consists of surface water bugs having small eyes, no ocelli rostrum 3-segmented; hemelytra reduced and hind coxæ hinged.

Æridinae a subfamily of Acridiade (Acridoidea; Caelifera), includes short-horned grasshoppers that are primarly grass-feeders.

Æridinæ Burmeister 1839 family of Æshnoidea (Anisoptera: Odonata); large dragonflies characterised by eyes broadly contiguous dorsally; third branch of radius (vein) strongly arched just below pterostigma; labial palpi 2-segmented; females with complete ovipositor - *Æschna cærulea* Burmeister (Europe), giant Californian dragonfly, *A. walsinghmi* McLachlan is the biggest, being 100 to 110 mm long with wing span of 120 to 130 mm; Emperor dragonfly (*A. imperator*) is a close second being about 95 mm long with wing expanse of 100 mm.

Æshnoidea Selys 1840 superfamily of Anisoptera (Odonata); cross-veins between costa and radius incomplete, only two thick primary antenodals (Ax_i and Ax_2) present; triangles of fore- and hind-wings not distinctly different in shape and position -Æshnidæ, Gomphidæ, Cordulegasteridæ and Petaluridæ.

Æstivation dormancy or periods of inactivity during hot and dry season; summer sleep or summer hibernation.

Æthalionidæ homopteran family of Cicadoidea (Cicadomorpha: Auchenorrhyncha).

Aeon age of the universe: immeasurable period: thousand million (10^9) years.

Aerial of, in or belonging to the air or atmosphere: conducted by aircraft.

Aerial agriculture application of agriculural chemicals (pesticides and fertilisers) through aircrafts. It ranges from a Superfortress spraying insecticide over USA to the Beaver spraying phosphate in New Zealand. It covers the Piper-cub treating vast areas of wheat in Turkey and the helicopter flying across a patch of couple of hectares in England or spraying cotton and sugarcane plantations in India.

Aerial application applying chemicals with help of small æroplanes. Chemicals have been sprayed from aircraft since 1919 when experiments were carried out in USSR spraying liquids through simple boom and nozzle equipment attached to military biplane. In USA lead arsenate was dusted to control sphinx *Ceratomia catalapœ* (Ditrysia: Lepidoptera), at Ohio in 1921 while spraying operations started from 1929 onwards. The real boost to aerial application came in 1944 following the success of large scale aerial spraying for the control of insect-borne diseases during the world war in Tropical areas. In India, ærial application has been used for pest control since 1951.

Aerial sound ranging process of locating an aircraft by means of sound emitted by it.

Aero pertaining to air.

Aerobe (aerobiont) organism that requires air or free oxygen to maintain its life processes.

Aerobic living or active only in the presence of oxygen.

Aerobic respiration respiratory process requiring oxygen.

Aerobiologist person having sound knowledge of aerobiology.

Aerobiology branch of biology that deals with the study of air-borne micro-organisms, especially as agents of infection including insect vectors.

Aerobiosis life existing in air or in presence of oxygen.

Aerobiotic requiring free oxygen for respiration.

Aerodynamics study of air or other gases in motion: study of motion and control of soild bodies such as aircrafts, missiles and rockets.

Aerosol fine mist of solid or liquid particles suspend in air. A type of pesticide formulation, applied as mist or fog that is dispersed or propelled by a gas under pressure.

Aerotaxis the movement of aerobic organisms (including minute insects) with reference to the direction of air or oxygen.

Afferent bringing inwards: nerve that conducts from perifery towards nerve centre. *Cf.* efferent.

Affinity relationship of a group of animals/plants to the other group based on differences and similarities of taxonomic characters.

Agamic species or generation that reproduces without mating.

Agaontidæ Walker 1846 (nec. **Aganoidæ, Blastophafidæ)** family of fig-wasps (Chalcidopidea: Apocrita); exhibits great sexual diamorphism.; females small (2-3 mm long), black in colour having oblong head; antennæ 9 to 13-segmented with long scape; wings stout, ovipositor long and prominent; males very small (< 2 mm) apterous, queer looking with 3 to 9-segmetned antennæ and abdomen broadly sessile, long and tubular; females move from tree to tree (*Ficus* spp.) and pollinate the flowers while males remain and perish in fig fruits – *Blastophaga breeviventirs* Meyr, *Scycobia bethyloides* Walker, *Sycoscapta insignis* Saunders.

Agaptidæ old and equivalent name for lepidopteran family Satyridæ (Papilionoidea: Ditrysia), that is cosmopolitan in distribution and incudes such butterflies as, Graylings, Heaths, Marbled-whites and Meadow-browns.

Agaridstidæ (= Phalænoididæ) a small family of Noctuoidea (Ditrysia: Lepidoptera), moths with vivid colouration resemble those of Arctiinæ; antennæ with shaft dilated distally; hind-wing having $Sc + R_1$ separate from Rs, anastomosing with cell near the base. About 350 species found mostly in North America and the Orient; many diurnal in habit, some produce clicking sound during flight – *Ægocera bimacula, A. tripartita, A. venulia, Eusemia adlatrix* et al.

Aggression intraspecific defence: defensive behaviours by one individual species towards anther individual of same species.

Agitator mechanical device in spray tank to ensure uniform distribution of solution and prevent sedimentation.

Aglycyderidæ small family of Curculionoidea (Polyphaga: Coleotera), that differs from others in having filiform atennæ and 3-segmetned tarsi (120 species) - *Aglycyderes* species (Canary Isaland) et al.

Agnatha class of Chordata that includes primitive-most fossil and living fish-like vertebrates; long slender body; skin smooth and slimming, no scales; jaws wanting, instead a sucker-like

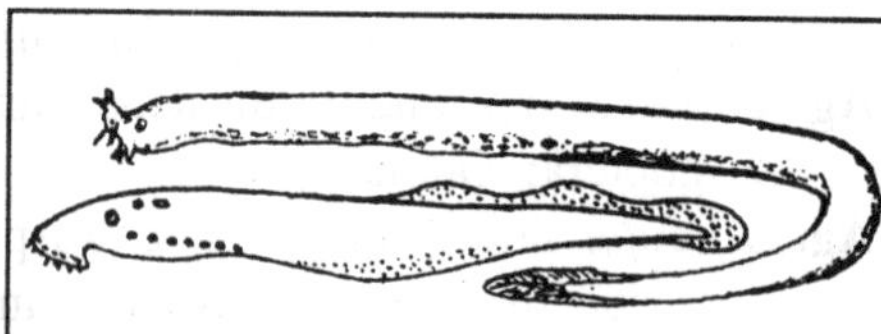

Hagfish Lamprey

buccal funnel at anterior end. Found in sea, feeding on dead, weak and dying fishes - hagfishes, lampreys etc.

Agonoxenidæ a microlepidopterous family often place under family Oecophridæ (Gelechioidea: Ditrysia) - *Aonopterix heractiana* sp. web together flower heads and seeds of parsnip (*Pastinaca sativa*) and allied plants in Europe and North America.

Agricultural pertaining to agriculture.

Agricultural ant s harvestor ant: common name for various species of hebivorous and (Formicoidea: Apocrita); that build large mounds of loose soils or sand and clear the surrounding ground of all vegetation.

Agricultural Entomologist person having sound knowledge of insect pests of various crops and their control.

Agricultural entomology study of insects vis-a-vis crop production, including crop protection and post harvest technology.

Agriculture art and practice of cultivating land; it involves breeding and management of crops (crop husbandry) and animals (animal husbandry) useful to man.

Agriidæ Tillyrad 1926 (Calopterygidæ) large family of dominant and highly specialised damselfies (Agrioidea: Zygoptera); wings indistinctly petiolate at the base, without a distinct angle on hind margin; enire wing densely reticulate; discoidal cell longer and traversed by numerous cross-veins; pterostigma indistinct or obsolete in male - *Hetærina americana* Fabricius (North America), *Vestalis* species (Indo-Malaysia).

Agrioidea Tillyard 1926 superfamily of damselflies (Zygoptera: Odonata), characterised by wings having five or more antinodal cross-veins; arculus equidistant or further away from the nods than the base of wings; includes seven families - Amphipterygidæ, Agriidæ, Polythoridæ, Epallagidæ, Chlorocyphidæ, Pseudolestidæ and Heliocharitidæ.

Agrionidæ a family of Coenagrioidea (Zygoptera: Odonata) comprising of large, broad-wingled damselflies with base of wings devoid of stalks - *Coenagrion* spp., *Ischunra* spp. et al.

Agrionoidea old and equivalent name for superfamily Agrioidea (Zygoptera: Odonata).

Agriotypidæ family of parasitic wasps (Ichneumonoidea: Parasitica: Apocrita) that prey mostly upon immature stages of other insects.

Agro-pesticide(s) chemicals classified as pesticides, adjuvants, curative agents and other chemicals used for improving agricultural production by protecting crops or controlling pests (including pathogens and weeds) and correcting diseased conditions of crop plants.

Agrochemicals chemicals used for raising or boosting agricultural produce – fertilisers, pesticides, plant nutrients etc.

Agroecosystem ecosystem composed of cultivated land with plants growing thereon and the animals associated with these plants.

Agromyzidæ Biglow 1852 large family of small dipteran flies (Agromyzoidea: Schizophora); having vertex without inter frontal bristles, postvertical bristles divergent, vein Sc reduced and female having oviscapt. Maggots are generally either leaf-miners or stem-borers – *Melanagromyrza obtusa* Malloch, *M. phaseoli* Cocquillett, *Ophiomyia phaseoi* (Tryon), *Chromatomyia (Phytomyza) horticola (Goureau)*.

Ophiomyia phaseoli (Tryon)

Agromyzoidea superfamily of dipteran flies (Schizophora: Cyclorrhapa), second antennal segment without distinct external groove ocellar and vertical bristles present; costal vein broken at R_1 or Sc, subcosta usually reduced; squamæ small or vestigial; comprises of large assemblage of families.

Agrotidæ old and equivalent name for family Noctuidæ (Ditrysia;: Lepidoptera) that includes night flying owlet moths.

Air gaseous mixture surrounding the Earth; in addition to atmospheric gases, it contains water vapour, hydrogen carbons, hydrogen peroxide, sulphur compounds as also small dust particles.

Air sac enlargement of tracheæl tube, present in almost all pterygote insects but absent in Apterygota and in holometabolous larvæ. In birds, thin-walled, air-filled extensions of lugs, lying in thorax and abdomen that help in respiration.

***Ak* grasshpper** *Poekilocerus picturs* Fabricius (Acrididæ: Orthoptera); both nymphs and adults stout and conspicuously coloured, yellowish with orange and black stipes and dots all over their bodies; antennæ bluish-black with yellow rings; adults have forewings bluish-green with yellow veins and reticulations; hid-wings hyaline. Polyophagous and voracious leaf defoliators, *ak* or madar (*Calotropis gigantea*) and allied species being the main host.

Akinesea absence of loss of power of motions.

Akinesis motionless state of insect resembling sleep (unconsciousness), particularly after loss of, or damage to sensory organs like antennæ etc.

Akinetic without movement.

Ala (pl. **alæ**) wing or wing-like process.

Alæ cordia wing-like bands of fibrous tissue spending the heart from pericardium in insects and other arthropods.

Alar(y) polymorphism pertaining to wing.

Alary muscles diaphragm muscles or wing muscles of heart; typically fan-shaped groups of fibrous, spreading from their point of origin on tergal plates to their insertions in heart.

Alary polymorphism presence of two or more forms of wings in same species, with or without any relationship to sex.

Alate winged: having wings.

Alcohol distillate obtained by fermentation of sugar or other sacchrine matter, constituting intoxicating element in liquors. Common alcohol (C_2H_5OH) is ethyl alcohol in which a hydroxyl group is substituted by an atom of hydrogen. Methyl alcohol ($CH_3.OH$) is unpalatable.

Alder frogphopper *Aphrophora alni* (Cercopidæ: Cercopoidea); distinguished by two white patches on margins of forewings; beside having only two ocelli and smooth tibia; found on alder and sallow trees specially in British Isles during May to October.

Alder wood-wasp hymenopterous insect, *Xiphydria* spp. (Xiphydriidæ: Symphyta); about 20 mm long, black in colour with red and white markings; females having needle-like ovipositer that looks like sting; lays eggs under the bark of alder and willow trees; larvæ bore inside the wood.

Alderfly (flies) small brownish to black insect(s) (Sialidæ: Megaloptera), feeble flier, found resting in large number on waterside vegetation; having prognathous head, bitting mouthparts and many segmented antennæ; eggs minute, elongated, stone-grey in colour, found in rows attached to water plants; one female can lay several hundred eggs. Larvæ brownish in colour, aquatic, active carnivorous lie burried in bottom of stream for upto two years and then pupate in

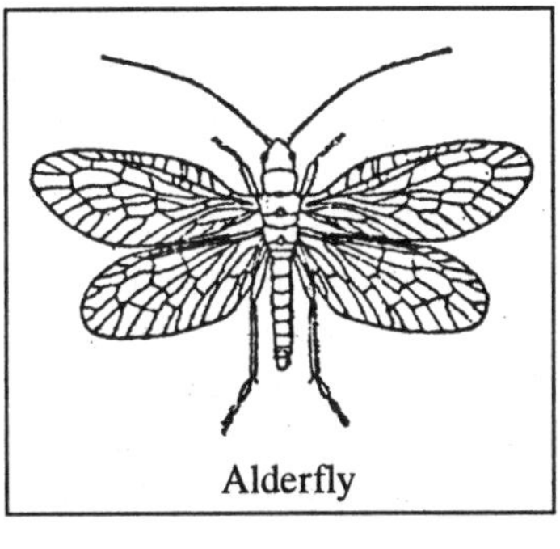
Alderfly

wet soil just above water line - *Ascanthacorydalis asiatica* Woodmason, *Sialis lutaria* et al.

Alderfly, Smoky *Sialis infumata* Newman (sialidæ: Megaloptera), dark brown to black in colour with wing span of about 25 mm, found in eastern USA, inhabiting lakes, ponds and streams.

Aldicarb $C_7H_{14}N_2O_2S$; carbamate compound; crystalline white solid having slightly sulfurous smell, barely soluble in water but readily, dissolves in organic solvents; used as systemic acaricide, insecticide and nematocide - Temik®.

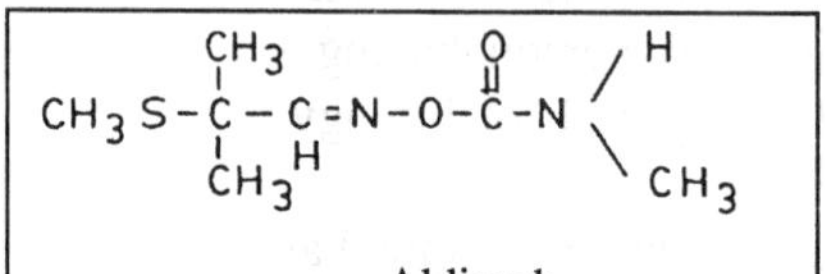

Aldicarb

Aldrin $C_{12}H_8Cl_2$; non-systemic, non-phytotoxic organo-chloride with high contact and stomach activity, insoluble in water but soluble in solvents like acetone, benzene, ethyl alcohol, xylene etc. and stable in strong alkali and mild acids; compatible with most pesticides and fertilisers. Effective against most soil-borne insects, specially termites and beetle grubs. Named after German chemist Kurt Alder - Aldrex.

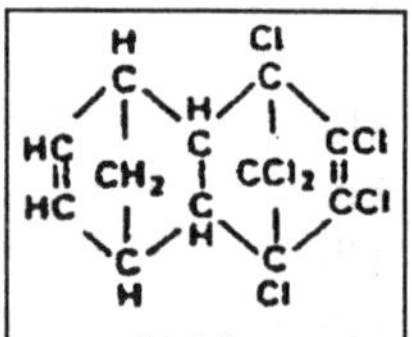

Aldrin

Aleyrodidæ (= Aleuroididæ) Westwood 1840 family of whiteflies (Aleyrodoidea: Homoptera), very small insects (1 to 3 mm long, with wing-span of 3 to 4 white powdery wax; mouthparts sucking type; wings opaque broadly rounded with highly reduced ventation, always held horizontally over the body when at rest; both nymphs and adults exude honeydew; includes many polyphahous pests of economic crops, some also transmit virus diseases - *Bemisia tabaci* (Gennadius), *Dialeurodea citri* Riley and Howard and *Trialeurodes vaporariorum* (Westwood) are most notorious pests found all over the World; *B. tabaci* also transmits leaf curl disease in cotton, okra, tobacco, tomato etc.

Aleyrodina suborder of Homoptera in certain system of classification equivalent to superfamily Aleyrodoidea of suborder Sternorrhyncha (Homoptera) comparising of single family Aleyrodidæ.

Aleyrodoidea Handlirsch 1903 superfamily of Homoptera, very small insects dusted fully with mealy white powdery wax; having two ocelli and two reniform compound eyes; both pairs of wings of similar membranous consistency; some clouded or mottled with spots or bands; honeydew excreted through anus by both nymphs and adults. Only one family - Aleyrodidæ (whiteflies).

Alga (pl. **algæ**) sea-weed(s) (marine and freshwater), usually green in colour, some brown or red; simple photosynthetic, unicellular organisms belonging to kingdom Protista (20000 spectes) - Schizophyta, Euglenophyta, Chlorophyta, Bacillariophyta, Rhodophyta and Phæophyta.

Algicide substance (toxin) that can kill algæ.

Algologist (phycologist) person having sound knowledge of algæ.

Algology (phycology) branch of biology that deals with the study of algæ.

Alien species organism that has invaded and is growing in a new region.

Alienidæ small hymenopteran family of Chalcioidea (Parasitica: Apocrita) that includes small parasitic wasps.

Aliform wing-like.

Aliment food: nourishment.

Alimentary pertaining to food/nourishment.

Alimentary canal food tube or digestive tract, traversing from mouth to anus; comprising of stomodæum, mesentron and proctodæum.

Alimentary castration sterility of worker and soldier castes of some social insects (Hymenoptera) due to diet deficiency.

Alimentation providing mourishment by feeding.

Alinotum wing-bearing plate or dorsum of mesothorax/metathorax of pterygote (winged) insects.

Alitrunk fused thorax and propodeum as fond in Apocrita (Hymenoptera): wing-bearing segments usually meso- and meta-thorax in hymenopterous insects.

Alkali substance that turns red litmus blue; releases hydroxyl ions (OH^-) in water. *Cf.* acid.

Alkaline gland (Dufour's gland) gland found in the sting of bees (Apoidea: Apocrita: Hymenoptera); it produces oily alkaline secretion that is lubricatory.

Allantoid sausage-shaped.

Alleculidæ (Cistelidæ) small tropical family of comb-claw beetles (Tenebrionoidea: Polyphaga); characterised by tarsal claws pectinate; head prolonged into short but distinct blunt rostrum and antennæ long (longer in males) - species of *Allecula, Cistela, Cistrlomorpha, Pseudocistella* etc. *Omophlus lepturoides* damage potatoes (*Solanum tuberosum* Linnæus).

Allelochemic non-nutritional chemical produced by an individual of one species that affects growth, behaviour or biology or another species.

Allelochemicals chemicals significant to the members of a species different from their source, for reasons other than feed as such. These can be allomones, kairomones or synomones, depending upon which organism benefits in the interaction; the emitter (allomones), the receiver (kairomones) or both (synomones).

Allethrin first synthetic pyrothroid produced in 1949 by La Forge; same toxic activity a pyrethrum but low photostability; acute oral LD_{50} to rat 920 mg/kg.

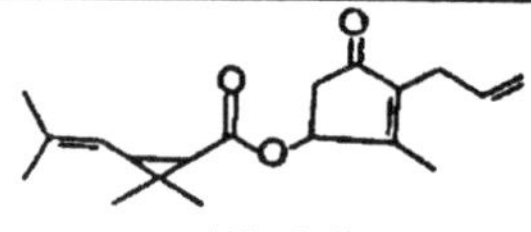

Allethrin

Alometric (heterogonic) growth disproportionate growth of certain body-parts as compared to the entire organism – male stag beetles (Lucanidae: Scarabæoidea); as the beetles grow, their mandibles go on becoming relatively much bigger.

Allomone allelochemic that induces response in individual of another species but is beneficial to emitting organism and not to the receiver – plant scent that attracts bees and thereby facilitate pollination. *Cf.* kairone.

Allopatric species population of two or more clearly related species that occupy different, though often adjacent, geographical regions; these do not interbreed and thereby maintain their reproductive isolation.

Allosomatic resemblance Betasian mimicry: resemblance of a harmless animal to a poisonous or distasteful one; serves as protection against the predators – resemblance of clearwing moth (Lepidoptera) to a wasps (Hymenoptera).

Allotype paratype which is opposite in sex to holotype.

Alpha individual most dominent individual in a dominance hierarchy or peck-order.

Alpha taxonomy the level of taxonomic study concerning characterisning and naming the species.

Alternate host one or other of the two unlike hosts.

Altruism regard for others as a principle of action.

Altruistic behaviour a phenomenon exhibited by certain (social) insects whereby individuals sacrifice themselves in the interest of the population as a whole, common among honeybees, termites etc.

Alucitidæ old name for family Orneodidæ comprising of many plume-moths (Ditrysia: Lepidoptera).

Alucitoidæ old name for Copromorphoidea (Ditrysia: Lepidoptera).

Alula (pl. **alulæ**) a free membranous lobe near the base of inner or posterior margin of the wing as in Coleoptera and Diptera: bastard-wing.

Alveary beehive. Hollow of external ear (pinna) where wax collects.

Alveolus (pl. **alveoli**) pit or small depression. Hair-socket. Tooth-socket.

Alydidæ family of heteropteran bugs (Coreoidea: Pentatomomorpha), earlier considered as subfamily of Coreidæ; it differs in being more elongated; 4th antennal segment curved and longer that third segment; head as broad as pronotum - *Alydus calcaratus, Leptocorisa acuta* Thunberg (*variconis* Fabricius) (rice gundhi bug) et al.

Alysiidæ a small family of Icheneumonoidea (Parasitica: Apocrita); allied to Braconidæ but differs by having a peculiar attachment of mandibles, the apices of the latter being directed outwards and not meeting when closed; parasitic upon dipteran maggots - *Alysia manducator* et al.

Alysonidæ small family of Sphecoidea (Aculeata: Apocrita) that includes beneficial wasps as these are predaceous on other insects and spiders.

Amatidæ (=Syntomidæ) family of tropical day-flying moths (Noctuoidæ: Ditrysia); having forewing with M_2 basally approximated to M_3 and Sc + R_1 fused with Rs. (2000 species). In India, *Amata pasialis* (Fabricius) is pest of cowpeas and *Euchromia polymena* (Linnæus) of sweet potato.

Amatidæ (=Syntomidæ) equivalent names of family Ctenuchidæ (Sphingoidea: Ditrysia).

Ambience environment: surroundings: atmosphere.

Amblycera Kellogg 1896 suborder of Mallophaga (biting lice) hemipteroid insects having short 4- or 5-segmeted antennæ, capitate, concealed in grooves at the side of head; mandibles with dorsal and ventral articulations; maxillary palp 2- or 5-segmented, labial palp absent or single segmented; meso- and metathorax separate; ectoparasites of birds and mammals; comprises of six families - Menoponidæ, Boopidæ, Læmobothriidæ, Rincinidæ, Trimenoponidæ and Gyropidæ.

Ambulatoria Westwood 1859 alternative and equivalent name for insect order Phasmida Leach 1815, the later has precedence.

American blight nickname of wooly aphid, *Erisoma lanigerum* (Hausmann) (Eriosomatidæ: Aphidoidea); native to America, cosmopolitan in distribution, most destructive pest of apple trees in the World, attacks both aerial parts and roots, primarly a bark feeder causing large and small exerescences on the bark.

Amercan cockroach *Periplaneta americana* (Linnæus) (Blattaria: Dictyoptera); largest among the roaches usually 30 to 55 mm long (maximum 50 mm); indigenous to Mexico but found throughout the World.

American cow-killer ant not a true ant but a cuckoo-wasp or valvet and *Dasymutitlla accidentalis* (Linnæus) (Mutillidæ: Vespoidea), 18 to 30 mm long, densely covered with hair having crossbands of contrasting colours; characterised by antennæ 12-segmented and curled in female while 13-segmented and straight in males, male winged, female wingless but very active; found in entire South-eastern USA.

American water-strider(s) common North American water striders *Rhagovelia obesa* Uhler and allied species (Gerridæ: Amphibicoriseæ); apterous insects having short and raptorial forelegs, long middle and hind legs, arrangement of tarsi and claws is such as to give both buoyancy and traction on surface of water, where there are able to walk, run or skate; prey upon such small insects as fall into water.

Ametabola primitive or no metamorphosis in which no distinct external changes are evident, except in size; exhibited by insects belonging to orders Thysanura, Diplura, Protura, Collemnbola and Microcoryphia, as also by certain apterous species of Phasmida and Anoplura. *Cf.* metabola.

Ametabolous metamorhosis growth style of certain insects that show regular increase in size without any external changes or metamorphosis.

Ametoecious parasite that remains in/on same host.

Ametropodidæ Bengtsson 1913 Holarctic family of Heptagenioidea (Ephermeroptera); paleopteran insects having primitve wings with two pairs of intercalaries between Cu_1 andCu_2, the anterior pair being longer; hind tarsi with five free segments. Nymphs having sub-cylingdrical body with long claws on mid and hind leg – *Ametropus* species.

Ammochæta stiff bristles found in bunches on head of an ant; used for cleaning its legs.

Amnion cellular, membranous, inner thin lining of embryonic cavity.

Amniotic cavity space between amnion and embryo.

Amoeboid not having a cell-wall and changing in form like an amoeba.

Amorphic without any form or shape.

Amorphous having no definite form or shape.

Amphibia class of Chordata; cold-blooded, oviparous vertebrates that live partly only land and partly in freshwater; gill-breathing in larval stage and skin or lung breathing a adults; skin moist and grandular without hair or scales; no fins; toes of hind legs webbed together - frog, newt, toad etc.

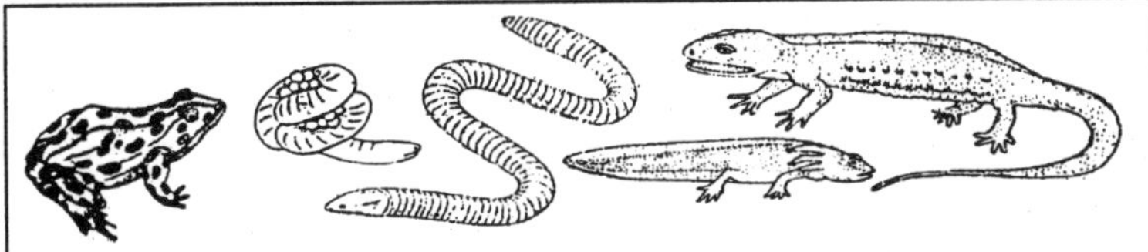

Frog Vermiform Blind-worm Tadpole Newt

Amphibian (amphibious) any member of class Amphibia; capable of living both on dry or moist land and in water - Frog, Salamandar etc.

Amphibicorisæ Durfour 1833 subdividion (series) of Heteropetera, comprising of surface water bugs, characterised b having exposed antennæ; includes single superfamily - Gerroidea, that embraces Gerridæ, Hebriadæ, Hydrometridæ, Mesovellidæ and Vellidæ (Micro-vellidæ).

Amphibiotic aquatic during larval sage and terrestrial in adult stage.

Amphientomidæ Enderlein 1903 family of Trochtomorpha (Psocoptera); body convex, meso- and metanota separate, fore-femora with saw-like row of longitudinally oriented teeth; body and wings scaled, areola postica free.

Amphigean an organism that is native to both Old World and New World.

Amphimixis union of egg and sperm in sexual reproduction.

Amphipneustic having only two pairs of functional spiracles, one anterior and other posterior - maggot of housefly (Diptera).

Amphipodous having both walking and swimming legs.

Amphipsyilidæ small family of fleas (Siphonaptera) having metanotum with small marginal spines, arch of tentorium visible in front of ocellus which is often reduced or vestigial; interantennal suture present but very weak.

Amphipterygidæ family of Agrioidea (Zygoptera: Odonata) that includes damselflies characteristed by wing petiolate at the base, distinct angle on hind margin at the junction of petiole and rest of the wing; antenodal cross-veins very few.

Amphisocidæ a small family of Psocomorpha (Psocoptera: Psocopteroidea); related to family Cæciliidæ but differing in having incomplete cross-vein from the pterostigma towards Rs. About 40 species - *Kolbea quisquiliarum* et al.

Amphitheridæ small family of moths (Ditrysia: Lepidoptera) confined to Australia.

Amphitokay parthenogenetic production of both males and females.

Amphitropical distribution occurrence of insects in temperate regions, discontinuous between northern and southern hemispheres.

Amphizoidæ LeConte 1862 very small family of trout stream beetles (Caraboidea: Adephaga), metasternum not prolonged between posterior coxæ; anterior coxæ globular; tibia and tarsi without swimming hairs, as such these beetles are not adapted for swimming.

Amulicidæ small hymenopteran family of predaceous wasps (Sphecoidea: Apocrita), prothorax narrow and elongate; mesosternum produced into a forked process posteriorly; mesonotum with parapsidial furrows; base of abdomen constricted to form a short pseudo-petiole. Provision their nests with spiders and various kind of destructive insects which they consume and are therefore considered as beneficial.

Amydriidæ family of Tinæoidea (Ditrysia: Lepidoptera); moths having mouthparst well developed, labial palpi bristled; vertex with short flattened hair; wings linear-lanceolate with sharply pointed apices.

Amylase digestive enzyme that initiates attack in starch molecules.

Anabolism constructive phase of metabolism involving building of the tissues. *Cf.* catabolism.

Anacharitiedæ family of parasitic bees (Cynipoidea: Apocrita), abdomen distinctly petiolate, second tergite longer than third; forewings without stigma.

Anacharitinæ a hymenopteran subfamily of Figitidæ (Cynipoidea: Parastica), often regarded as an independent family the small wasps are parasitic on neuropteran insects.

Anaerobe organism which does not require free oxygen for living.

Anæsthesis loss of feeling : insensibility.

Anæsthetise immobilise an insect by exposing it to carbon dioxide, chloroform etc.

Anal relating to or located near anus.

Anal gland(s) ectodermal glands opening close to anus.

Anal papillæ thin walled outgrowths in anal region of some aquatic insects (larvæ); may be respiratory in function, also act as organs for absorption of chloride and other ions from water - mosquito larvæ (Culicidæ: Diptera).

Anal vein(s) most posterior veins in insect wing, between cubitus and jugal region; post-cubitus: vannal.

Analogous organs similar in function but not necessarily alike in genetic relationship - wings of birds and insects.

Analogy agreement in function, as distinguished from homology or agreement in origin. *Cf.* homology.

Anamerentoma Prell 1912 name suggested for insect order Protura Silvestri 1907, but no more in usage, as the later has precedence.

Anamorphosis (anamorphism) type of metamorphosis, a rare phenomenon found only in proturan (ametabolours) insects, wherein during larval development three new segments are added in between the last (VIII) abdominal segment and telsonic lobe. Graudal increase in complexity of form and function during evolution of a group of animals or plants.

Anapleurite eupleuron: dorsal supracoxal sceleroptisation of thoracic pleuron.

Anatomy determination of structural facts: study of structure as revealed by dissection.

Anatomy, Gross science of anatomy that deals with naked-eye appearance of the tissue of plants and animals (extinct and extant) and seeks understanding of life. *Cf.* microscopic anatomy.

Anatomy, Microscopic branch of science of anatomy (biology); it embraces all the knowledge as can be gained through microscope; involves analysis and resynthesis of information concerning the finer structure of the organs and parts of living organisms; often includes embryology as well.

Andrenidæ (Latreill 1802) Samouelle 1819 family of solitary mining bees (Apoidea: Apocrita); characterised by presence of two subantennal sutures (grooves) below each attennal socket on the head; glossa flattened and shorter than mentum; labrum small and mostly concealed by clypeas; forewing with three cubital cells; females dig burrows upto 600 mm deep wherein they lay eggs and also keep balls of pollen and honey for the developing larvæ - *Andrena bradleyi* Veirck, *A. nasoni* Robertson, *A. winkleyi* Viereck, *Perdita* spp., *Panurgus* spp. et. al.

Androconia specialised scales (plumules) found in various shapes on dorsal surface of wins in various male butterflies; these are scent scales that serve as outlets for odoriferous glands; the scent secreted excites the females of the species.

Androgens male sex hormones.

Anecdysis long passive period between two moults of an insect or other arthropod during which there is no preparation for the next moult.

Anellus sclerotised dinner wall of phallotheca forming ring or tube from which ædeagus protrudes.

Anemotaxis tendency of certain insects to orient themselves in relation to wind direction.

Anemotropism orientation response of sessile organism to air currents and wind.

Anepimeron pteropleuron: upper part of epimeron or posterior sclerite of subcoxa at the base of an insect's legs. *Cf.* anepimeron.

Aner male insect specially ant (Formicoidea : Apocrita); sexual individuals with well developed sensory and male reproductive organ; head small and round; mandibles reduced; antennæ long and slender. Relatively large individuals are called **macraners,** dwarfs as **micraners** and those resembling worker ants as **ergataners** while those that look like fertile females are known as **gynæcaners**.

Angle-shades (moths) common British moth, *Phlogohora meticulosa* (Noctuidæ: Lepidoptera) having light brown forewings with dark brown triangular markings; fly by night; when at rest, during day, look like a shrivelled leaf.

Angle-wing nickname for various Amercian butterflies, *Polygonia* sp. (Nymphalidæ: Lepidoptera), having wing margin irregularly notched and angled - *Polyonia c-album* commonly called Comma is European species.

Angoumis grain moth *Sitotroga cerealella* (Olivier) (Gelechiidæ: Lepidoptera), native of Europe, named after a French province (Angoumois); adult is small moth with wing span of about 12 mm, buff-coloured, slightly speckled with black, hind-wings pointed and pale grey with wide fringe of hairs; tiny caterpillars fed on stored products; a serious pest of stored grains (barley, oats, maize, wheat etc.) and dry fruits (apricot, almond etc.) all over the World. Caterpillars and pupæ always found inside the grains.

Animal organised being, having life, sensation and voluntary motion; belonging to kingdom Animalia and vividly distinguished from plants, algæ, bacteria and fungi.

Animal community aggregation of animal species held together in a continuous or discontinuous geographical area.

Animal ecology branch of zoology that deals with study of animals vis-a-vis their environment.

Animalcule extremely small animal (microscopic) that cannot be seen with naked eyes.

Animalia animal kingdom; a major division of living organisms, that includes all animals. Recently, unicellular animals (Protozoa) have been shifted and placed under kingdom Protista (along with algæ), leaving only multicellular animals (Metazoa) under this kingdom. There are about 1.4 million described and several thousand undescribed species of animals in the World.

Anisembiidæ Ross 1940 New World family of Embioptera, small fragile insects found in South-west USA thru Argentina; wings with R_{4+5} simple; X abdominal tergite completely divided to form hemitergites that are separated by a membranous area: cerci in males asymmetrical - *Saussurembia* species and *Chelicera* species.

Anisopodidæ (= Rhyphidæ) only family of Anisopodoidea (Nematocera: Diptera); ocelli disinctly present; eyes not connected by dorsal bridge of facets, pronotum without long bristles; wins without network of folds, fourth posterior cell wide open; coxa elongate, pulvilli small or absent - wood gnats (*Anisopus* spp. et al).

Anisopodoidea monofamilial superfamily of suborder Nematocera (Diptera: Oligoneoptera); antennæ 16-segmentd and as long as thorax; eyes in males heloptic and discal cell present. Maggots whitish in appearance, about 10 mm long, amphipneoustic and saprophagous, live in decaying vegetable matter and manure - Anisopodidæ.

Anisoptera Selys 1834 one of the three suborders of Odonata; eyes large often contiguous; fore- and hind-wings dissimilar in shape and size and not petiolate; held horizontally during reposes; male with two superior and one inferior anal appendages and a jointed penis. Nymphs with rectal gills - Æschnoidea and Libelluoidea.

Anisotomidæ equivalent name for Leioidæ (Staphylinoidea: Polyphaga: Coleoptera) that includes carrion beetles.

Anisozygoptera Handlirsch 1906 monofamilial suborder of Odonata; paleopteran insects having combine characteristics of Anisoptera and Zygoptera; nymphs and adults resemble anisopteran forms; includes many extinct Mesozoic species and two extant species placed under family Epiophlebiidæ - *Epiophelbia* species, one each from India and Japan.

Annelida (Annulata) phylum of metamerically segmented worms (Coelmata: Triploblastica); animals having ring like structure with definite body cavity and food tube. About 15000 species divided into five classes - Polychæta (nereis), Oligochæta (earthworms), Hirudinea (leeches), Gephyrea and Archiannelida.

Annulipalpa suborder of Trichoptera; oligoneopteran insects, having last segment of maxillary palpi annulated, flexible and much longer than other; dorsal part of tentorium not developed. Larva having anal blood-gills, no tracheal gills, live in silken non-portable retreat - Hydropsychoidea.

Anobiidæ Suckard 1840 family of Bostrichoidea (Polyphaga: Coleoptera); tarsi 5-jointed, posterior coxæ sulcate for reception of femora, onychium small - furniture beetle (*Stegobium pancieum* Linnæus), tobacco beetle (*Lasioderma serricorne* Fabricius) et al.

Anomologidæ small and unimportant family of moths (Tortricoidea: Ditrysia); maxillary palpi vestigial or wanting; hind-wings as wide as forewings.

Anomosetidæ very small family of Hepialoidea (Monotrysia: Lepidoptera); that includes moths having maxillary palpi well developed and wins coupled by jugum.

Anoplura Lucas 1840 ((Siphunculata Latreille 1825) order of Psocopteroidea (Paraneoptera: Pterygota); secondary wingless insects; body dorso-ventrally flattened, head conical and pointed; antennæ 3- to 5-segmented; eyes vestigial, no ocelli; mothparts retractile, piercing and sucking; tarsi single segmented having single powerful claw, cerci absent; metamorphosis practically absent or gradual (incomplete); ectoprarasites of mammals (400 species) - Pediculidæ, Hæmatopinidæ, Hæmatopinoididæ, Echinophthiriidæ, Phthiridæ, Neolinognathidæ, Linognathidæ and Hopkopleuridæ e.g., Head Pouse (*Pediculus humanus capitis) Pediculus humanus humanus*.

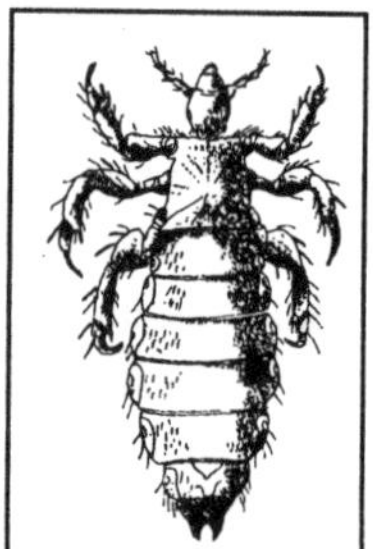
Body louse

Ant(s) small social insects (Formicoidea: Hymenotptera); usually shades of yellow, brown, red or black in colour, having elbowed antennæ and very narrow abdominal constriction (petiole or waist); live in highly organised complex colonies (no solitary form); comprising of several castes - queen, reproductive males, soldiers and workers (incomplete females) etc. - characterised by distinctly marked division of labour; only the workers cause all the damage and nuisance.

Ant, acrobat *Cremastogaster lineolata* (Say) Myrmicinæ: Formicidæ) and its numerous subspecies; occur throughout North America.

Ant, Argentine native to South America, *Iridomyrmex humilis* Mar (Dolichoderinæ: Formicidæ); common household and garden pest in USA, brown in colour, very small in size (queen 6 mm, male 3 mm, workers 2 mm); prolific breeder; colonies are exceedingly populous; eats almost any food article - raw and cooked - does not sting, but bites freely, visits aphids, scales etc., for want of honeydew: can travel 50 meters in an hour.

Ant, Army driver ants or legionary ants: carnivorous ants: *Eciton hamatum* (Fabricius) (Dorylineæ: Formicidæ) of Tropical America; live in temporary nests and migrate at night or occasionally on dull day, moving in long files like an army.

Ant, Australian jumper *Myrmecia nigrocincta* Smith (Ponerineæ: Formicidæ); active jumper and fierce biter; reddish-yellow in colour, builds mounds in dry forests (of Australia).

Ant, Australian sugar household pest (in Australia) that feeds on sugar and other sweets - *Camponotus nigripes* Smith (Formicinæ: Formicidæ).

Ant, Black *Monomorium minimum* Buckley and allied species (Myrmicineæ: Formcididæ) a common household pest, cosmopolitan in distribution.

Ant, Blood red *Formica sanguinea* (Formicinæ: Formicoidæ), a species similar to the wood-ant but bright red in colour, strictly carnivorous feeding almost entirely on other insects; does not build large domed nests, raids nests of smaller ants (*F. fusca* and its allies) carrying off cocoons and keeping its capatives as slaves.

Ant, Bulldog ***Myremcia gulosa*** (Fabricius) (Ponerinæ: Formicidæ); large ferocious, stinging and biting ants of Australia; queen 25 mm long and workers 20 mm, red, yellow or black in colour.

Ant, Cocoa *Azteca chartifex* Forel (Dolichoderinaæ: Formicidæ); is inimical to cocoa culture in tropical America, as it protects the coccids and other homopterous insects injurious to cocoa trees.

Ant, Dark brown *Lasius flavus* de Geer and *L. niger* Linnæus (Formicinæ: Formincidæ); are common European species, while *L. n. americanus* Emery is found in North America.

Ant, Fire *Solenopsis geminata* (Fabricius), *S. aurea* Wheeler, *S. xyloni* McCook, *S.molesta* Say, *S. texana* Emery (Myrmicinæ: Formicidæ) are comparatively bigger in size and sting severely, often fatal in case of new born wild animals and young poultry.

Ant, Harvester some ants of subfamily Myrcicnæ (Formicidæ: Aculeata: Apocrita) that store seeds of grasses, small millets, wheat etc. and include, *Holycomyrmex scabriceps* Mayr, *Messor barbarus* Linnæus, *Pheidologeton offinis* Jeden and *Pheidole grayi* Forel.

Ant, Honey pot some ants have an additional special caste – repletes – which have the crop capable of being greatly distended and serve as reservoir. Repletes hand from ceilings of the galleries of their nest, receive honeydew brought by the workers and store the same in their crop to provide food for the colony during drought periods; common in arid regions – *Camponotus inflatus* Lubbock, *Leptomyrmx varian* Emery (Queensland), *Myrmecocystus hortideorum* (America) etc.

Ant, Leaf cutting *Atta texana* Buckley and allied species (Myrmicinaæ: Formicidæ); cut leaves of the trees and drop them to ground, parasol-ants carry the cut leaves to their subterranean nests where fungi are cultured on these chewed leaves as food for the ants.

Ant, Meadow *Lasius flavus* de Geer (Formicinæ: Formicidæ); the dark brown ant, a subterranean species responsible for grass-covered mounds often seen in uncultivated fields and orchards.

Ant, Negro black, fast-moving medium-sized ant of British Isles and forests of Europe, *Formica fusca* (Formicinæ: Fomicidæ); nests under stones and in old stumps.

Ant, Odourous house *Tapinoma sessile* (Say), native to USA; dark brown to black in colour; workers when crushed give off a sickening sweetish smell; lives in shallow underground nests.

Ant, Pharaoh's *Monomorium pharaonis* (Linnæus) (Myrmicinæ: Formicidæ); very small, yellowish-red species of Old World, now common in Tropical regions of the whole World, frequent inhabitant of bakeries, canteens etc., as it is floors, sills etc.

Ant, Red *Dorylus indicus* Westwood, *D. labiatus* Shuckard and *D. orientalis* Westwood (Dorylinæ: Formicidæ) are common Indian species, living in burrows in soil around the host plants and have plenty of nuisance value.

Dorylus orientalis Westwood

Ant, Robber slave making ant; *Formica sanguinea* Latreille (Formicinæ: Fomicidæ); makes slaves of workers of other species, specially those of *F. fusca* Latreille and *F. lemani*. Band of robber-ants raid the nest of its prey and make off with the worker pupæ, some of these are eaten by these ants and the rest hatch out and become workers for their captors.

Ant, Sauba *Atta cephalotes* Fabricius (Myrmicinæ: Formicidæ); a very large species of ant from Amazon forest, commonly eaten by the natives.

Ant, Tree *Cremastogaster dohrni* Mayr and allied species (Myricinæ: Formicidæ); make globular paper-like nests round the stems or live in hollows of tree trunks.

Ant, Wood (horse ant) *Formica rufa* Linnæus (Formicinæ: Formicidæ); builds large mounds in forests of Europe and North America; these ants have no sting, but can squirt formic acid at is enemy; also give painful nips with their powerful jaws. Pupæ (cocoons) are collected and sold as ant-eggs to be used as bird and fish food.

Ant's nest simple dwelling unit as compared to bee's or wasp's nest; series of galleries and chambers excavated in earth or wood, no separate cells for eggs or grubs; if disturbed, carry away the eggs, grubs and pupæ to new lodging place; when a colony becomes too big, fresh queens each with a group of workers moves off to start fresh colony.

Ant-bear common name of aadvark, *Orycteropus afer* (Tubulidentata: Mammalia), the ant-eater of South Africa.

Ant-beetle(s) small European beetles, *Thanasimus formicarius* (Linnæus) and related species (Cleridæ: Polyphaga) that resemble ant in shape and appearance; both grubs and adults are predaceous on eggs and grubs of bark beetles of coniferous trees.

Ant-cow aphid (Aphididæ: Homoptera) kept and tended by ants for honeydew excreted by the aphid.

Ant-eater(s) mammals (*Tamandua* spp., Edentata) having no teeth but long, slender, sticky tongue for licking ants and termites. Three species are found in Central and South America – giant ant eater is over two metres long with cylindrical snout and large bushy tail whereas silky ant-eater and two-toed ant-eater are smaller in size, arboreal having prehensile tail.

Ant-eater, Banded not an ant but ant- and termite-eating marsupial (mammal); commonly called numbat; 40 to 50 cm long including 18 to 20 cm busy tail; brownish fur with white bands around body; prominent ears and 52 teeth; found in South-west Australia.

Ant-eater, Scaly pangolin, *Manis* spp. (Pholiodota: Mammalia), body covered with overlapping horny plates, elogated snout, no teeth, thin long tongue, short legs with long claws and long tail; lives in forests, feeding on ants and termites.

Ant-eater, Spiny echidna: primitive egg-laying, burrowing mammal or monotremes; body covered with spines (dorsally) and hair (ventrally); toothless with long projectile tongue. Found in rocky regions of Australia feeding on ants and termites.

Ant-eggs pupae of certain species of ants (Formicidæ: Apocrita) commonly (erroneously) called eggs; used for feeding fishes and small birds.

Ant-guest(s) myrmecophiles; insects other than ants found visiting or living in nests of ants.

Ant-hill (ant-heap) small hillock type structure raised by certain species of ants (or termites) as their nest: mound of earth deposited around the entrance to ants' (or termites') nest in the ground.

Ant-lion holokmetabolous carnivorous insect (Myrmeleontidæ: Neuroptera); having narrow dragonfly-like body, knobbed antennæ, chewing mouthparts; two pairs of long narrow wings with numerous longitudinal and cross-veins but no pterstigma - *Myrmelon formicarius, M. contractus* et al.

Ant-nest beetle(s) small rove-beetles (Staphylinidæ: Polyphaga) that live in ants' nest, as their grubs resemble ants and are accepted as members of their colony; both gurbs and adult beetles feed on food of the ants and even prey on ants' larvæ - *Lomechusa strumosa* lives in the nest of red ant, *Formica sanguinea* Latreille (slave-maker ant); it stimulates an ant worker to regurgitate food by tapping its head with its forelegs in evident imitation to another ant.

Ant-wasp (valvet ants) apterous female wasps of family Mutlillidæ (Scolioidea: Apocrita) that resemble ants but without abdominal nodes.

Antarctoperlaria suborder of Plecoptera, according to recent classification; confined to southern hemisphere; consists of four families - Eustheniidæ, Diamphipnoidæ, Anstroperlidæ and Gripopterygidæ.

Antecosta transverse chitinous ridge with which longitudinal muscles are attached on the inside of each cuticular plate (tergum or sternum).

Antenna (pl. **antennæ**) feelers or sensory organs: mobile, often many segmented appendages, found always in pairs on head of certain arthropods - Crustacea, Insecta, Onychophora and Myriapoda. Crustaceans have two pairs of antennæ while all others have only one pair. Insect antennæ show great variation in shape and size and include the following forms:

— **Acuminate** - ending in a point.

— **Bipectinate** - Fringed (pectinate) on both sides (♂ Lymantriidæ).

— **Capitate** - ending in capitulumor knob (*khapra* beetle).

— **Clavate** - club-shaped, gradually clubbed (Curuculionidæ).

— **Elbowed (Geniculate)** - abruptly bent (Formicoidea).

— **Knobbed** - swollen tipped; abruptly clubbed (Papilionidæ).

— **Lamellate** - expanded in flattened plates (Melolonthidæ).

— **Monoliform** - like a string of beads (temites, Isoptera).

— **Mucronate** - terminating in a sharp point or spine.

— **Pectinate** - toothed like a comb (♂ Lasicampidæ and Saturnidæ).

— **Plumose** - feather-like (♂ mosquito).

— **Serrate** - notched like a saw (Elateridæ).

— **Setaceous** - bristle-like (locust, grasshopper).

Antenna-comb row of bristles forming comb-like structure at proximal end of anterior tarsus in bees; used for cleaning antennæ (removing pollens etc.).

Antennule(s) first antenna in Crustacea; small paired, many-jointed appendages found on second segment of head in front of antennæ (third segment); usually sensory.

Anthelidæ small family of noctuid moths (Ditrysia: Lepidoptera).

Antheridium (pl. **antheridia**) male sex organ in the fungi.

Anthicidæ family of small, ant-like, flower beetles (Polyphaga: Coleoptera); head strongly deflexed and constricted into narrow neck, antennæ filiform and 11-segmented, tarsi with penultimate segments lobed beneath; middle cosæ with trochantins; predaceous on aphids and other small insects - *Anthicus santali* Haliday, *Formicomus sulcipes* Picter *Hylobœnus indicus* Pictet, et al.

Anthoboscidæ small and unimportant family of wasps (Vespoidea: Hymenoptera); predaceous on other smaller insects and spiders.

Anthrocorid common name of very small flower-bugs belonging to heteropteran family Anthrocoridae (Cimicoidea: Gymnocerata).

Anthocoridæ (Amyot and Serville 1843) family of small flower-bugs (Gymnocerata: Heteroptera); elongate, oval and flattened with first two antennal segments very short, last two long and pilose and the one in between swollen at the base; ocelli present; rostrum

3-segmented - *Orius indicus* is predaceous on thrips while *O. tantilus* preys on aphids and thrips.

Anthogenesis parthenogenetic reproduction of both males andfemales as found incertain aphids (Aphidoidea: Homoptera).

Anthomyiidæ Loew 1862 family of dipteran, root-maggot flies or flower-flies (Calyptratæ: Schizophora); often included under Muscidæ, from which it differs in having lowest calypter not longer than upper one; maggots mainly phytophagous (pests or root crops) - *Delia* (*Hylemya*) *antiqua* (Meigen), *Aritochæta excisa* Thomson, *Phorbia brassicæ* Bouche et al.

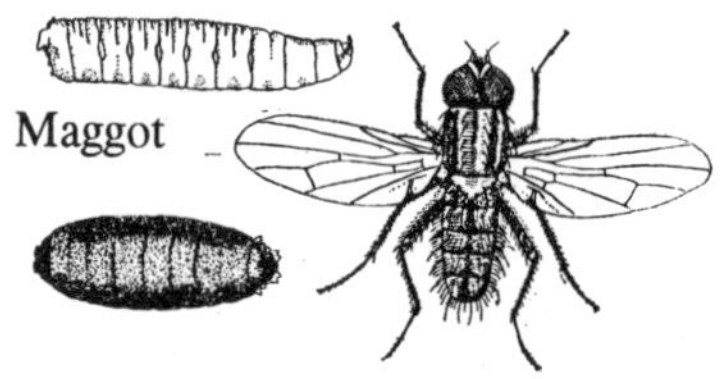

Delia brassicae (*Bouché*)

Anthomyzidæ family of small dipteran flies (Anthomyzoidea: Acalyptratæ) having vibrissæ and fore femur with strong spine-like bristle; Sc vestigial or absent, cell Cu_2 well developed.

Anthomyzoidea superfamily of Acalyptratæ (Schizophora: Diptera), small flies having second antennal segment with no cleft, ocellar bristles well developed, costa usually unbroken; earlier comprised of five families - Heleomyzidæ, Anthomyzidæ, Opomzidæ, Asteiidæ and Sphæroceridæ - to which other small families Rhinotoridæ, Borboropsidæ, Trixoscelidæ, Chyromyidæ and Aulacigastridæ have been added.

Anthonomus common and generic name of cotton boll weevil, *A. grandis* Boheman; native to Mexico, it is one of the most destructive pest of cotton in Central America and Southern USA.

Anthophilous feeding on flowers

Anthophoridæ Kirby 1837 (= Podalirildæ) large family of stout, solitary digger bees, hairly-footed bees etc. (Apoidea: Aporcita); antennæ short, ocelli distinct, eyes reaching the bases of mandibles; clypeus convex tongue greatly enlarged; wings hyaline or clouded, forewing with three cubital cells and radial cell neither long nor narrow - *Diadasia california* Timberlake, *Anthopora parietina* Fabricius et al.

Anthribidæ (= Platystomidæ) large Tropical (Indo-Malayan region) family of Curculionoidea (Polyphaga: Coleoptera); small dull coloured beetles, body clothed with pubescence; antennæ clavate, rostrum short, broad and flat; labrum distinct and not fully free; prothorax elongated elytra covering pygidium; found in old wood and fungi (2400 species) - *Acœcerus carpenemus* Jorden, *A. suturalis* Boheman, *Basitropis hamata* Jorden , *Phloeobis pallipes* Jorden.

Anthroceridæ old and equivalent name of Zygænidæ (Ditrysa: Lepidoptera); antennæ filiform, proboscis well developed; labial plapi short; wings translucent; some moths resemble the butterflies (Batesian mimicry).

Anti- against.

Antibiosis antagonistic relation between associated organisms; resistance in plants that affects development and survival of insects.

Antibiotic substance produced by a micro-organism that inhibits the growth of another micro-organism.

Antibody defensive substance produced in an organism in response to action of a foreign body.

Anticoagulant chemical that induces uncontrolable internal bleeding.

Antidote treatment given to counteract the effect of a poison (insecticide).

Antifeedant substance usually a protein, that is foreign to the host and stimulates production of antibodies.

Antigenicity ability to induce immune response.

Antiliata one of the classis (order) of Insecta suggested by Linnæus (1758); it included sucking insects and mites, now belonging to orders Diptera, Mallophaga and Anoplura as also Acarina and some Crustacea.

Antimone a substance produced or acquired by an organism that when it contacts an individual of another species in the natural context, evokes in the receiver a behavioural or physiological reaction that is maladaptive to both the giver (emitter) and receiver.

Anting introduction of live ants by birds into their plumage as a pleasurable means of cleaning plumage and skin.

Antisepsis destruction or inhibition of growth of bacteria.

Antiseptic substance used for destroying or stopping the growth of micro-orangisms in living tissues.

Antitoxin chemical that can neutralise poison or toxin - injection of atropin on exposure to insecticides.

Antixenosis non-preference.

Antlia (pl. **antliæ**) tubular suctorial proboscis of lepidopteran insects: dilated part of post-pharynx in some flies (Diptera).

Anus posterior opening of alimentary (digestive) canal for voiding undigested residues.

Aorta large artery that carries blood away from the heart; in insects, a slender anterior prolongation of dorsal blood vessel that extends forward through the thorax to terminate in the heart, near the brain; functions as principle artery of the blood.

Apachyidæ Verhioeff 1902 family of brightly coloured earwigs (Apachypoidea: Forficulina); free-living, body elongate flattened or strongly depressed; pronotum elliptical; tegmina smooth; wings long, squamopygidium well developed, empodium present, forceps in both sexes sickle shaped and without teeth; found under bark of dead trees – *Apachyus feæ Bormans,* et al.

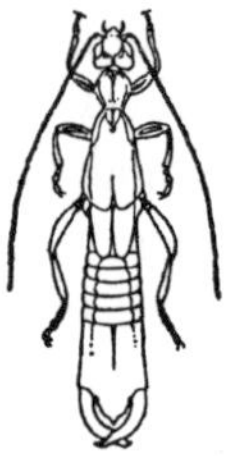

Apachys feæ
(Apachyidae)

Apachyoidea monofamilial superfamily of Forficulina (Dermaptera), characterised by flattened body and sickle-shaped similar forceps in both sexes, XI abdominal tergum and telson fused together with X segment, male genitalia double – Apachyidæ.

Aperture orifice or opening.

Aphaninæ subfamily of Lygæidæ (Heteroptera); characterised by 4-segmented rostrum; short legs, forelegs raptorial and tarsi 3-segmented.

Aphaniptera Kirby 1826 old and obsolete term for oligoneopteran order Siphonaptera (Neoptera: Pterygota) – fleas, e.g. *Pulex* spp., *Xenopsylla cheopis*.

Aphelinidæ (Thunberg 1876) Viereck 1916 family of Chalcidoidea (Apocrita: Hymenoptera); very small sunflies, also called aphelinids or scale parasites; antennæ 8-segmented; notaulices distinct; tarsi 4- or 5-segmented, primary parasites of aleurodes, aphids and coccids; adults feed upon honeydew excreted by thee hosts – *Apheliinus semiflavus* Howard, *Coccophagus flavescens* Howard, *Prospaltella smithi* Silvestri, et al.

Aphelocheirinæ subfamily of Naucoridæ (Hydrocorisæ: Heteroptera), now raised to family; comprises of brachypterous bugs with naked venter; abdomen without terminal respiratory appendages; takes up dissolved oxygen directly from water – *Aphalocheirus* spp. common in China and Japan.

Aphelosetiidæ old and equivalent term for lepidopteran family Elachistidæ (Elachistoidea: Ditrysia).

Aphicide (aphidicid) substance (toxin) that can kill aphids.

Aphid (Aphis) plant louse or greenfly; small (1 to 5 mm long), soft-bodied, pear-shaped sucking insect (Aphidoidea: Homoptera); both alate and apterous forms having relatively long legs and antennæ; a pair of cornicles projecting from posterior part of abdomen. Life-cycle complicated, reproduction sexual exude honeydew through anus that attracts ants, coccids and fungi. Mostly polyphagous and serious pests of various economic crops; some also act as vectors for transmitting plant diseases - banana aphid transmits bunchy top disease of banana while peach green aphid is responsible for mosaic disease of various crops.

Aphid forms based on climate, host plants and type of reproduction include:

— **Fundatrites or stem mother** - first female of the season from overwintering eggs on primary host; mostly apterous and oviviparous.

— **Fundatrigeniæ** - parthenogenetic female offsprings of fundatrices (several generations), may be apterous or alate.

— **Migrantes** - migratory apterous females produced parthenogenetically from second or third generations of fundatrigeniæ on primary host; disperse either to other similar hosts or to secondary hosts, where these reproduce apterous or alate forms.

— **Aleinicolæ or seconds** - parthenogenetic ovoviviparous females, several generations of which are born on secondary hosts during Spring and Summer.

— **Sexuapræ or gynoparæ** - parathenogenetic female offsprings of alienicolæ that give birth (during Autumn) to winged sexuals, either on secondary hosts or after migrating back to primary hosts.

— **Sexuals** - normal sexual males and females, progeny of sexuparæ.

Aphididæ (Buckton, 1881 as Aphidinæ) family (subfamily) of aphids or plant- lice (Aphidoidea: Homoptera); large group of very small, delicate, pear-shaped sucking insects, found feeding on all plant parts (roots to fruits); cornciles and cauda present, sexual forms oviparous, parthenogenetic females viviparous; mostly polyphagous, some act as vectors of virus diseases. Some of the common species are :

Apple green aphid - *Dorsalis pomi* (de Geer)
Banana aphid - *Pentalonia nigrinervosa* (Coquillett)
Been aphid - *Aphis craccivora* (Koch)
Cherry aphid - *Myzus cerasi* (Fabricius)

Citrus black aphid - *Toxoptera aurantii* (Boyer de Fonscolombe)
Citrus brown aphid - *T. citricidus* (Kirkaldy)
Corn (maize) aphid - *Rhopalosiphum maidis* Fitch
Melon (cotton) aphid - *Aphis gossypii* Glover
Pea aphid - *Acyrthosiphon pisum* (Harris)
Peach black aphid - *Pterochlorus persicæ* (Cholodkorsky)
Peach green aphid - *Myzus persicae* (Sulzer)
Peach leaf curl aphid - *Brachycaudus helichrysi* (Kaltenbach)
Peach mealy aphid - *Hyalopterus pruni* (Geoffrey)
Potato aphid - *Aulacortum solani* (Kaltenbach)
Ragi root aphids - *Tetraneura nigriandominalis* (Sasaki)
Sugarcane aphid - *Aphis sacchari* (Zehntner)
Sugarcane root aphids - *Geoica spatulata* Theobald) *Tetraneura javensis* (de Geer)
Walnut green aphid - *Chromaphis juglandicola* (Kaltenbach)
Wheat aphid - *Schizaphis graminum* (Rondani)
Wooly (apple) aphid - *Eriosma lanigerum* (Hausmann)

Aphidiidæ old and equivalent term for hymenopteran family Incubidæ (Inchneumonoidea: Parasitica), formerly a subfamily (Aphidinæ) of Braconidæ - the tiny wasps are parasites of aphids. Not to be confused with Aphididæ (Aphidoidea: Sternorrhyncha: Homoptera).

Aphidina suborder of Homoptera in certain system of classification, equivalent to superfamily Aphidoidea of suborder Sternorrhyncha (Homoptera).

Aphidodea superfamily of Strnorrhyncha (Homoptera) that includes aphids or plant-lice; small (maximum 8 mm), sort-bodied sucking insects, both apterous and alate form, latter with two pairs of transparent membranous wings; paired dorsal processes (cornicles) on V abdominal segment; reproduction sexual as also parthenogenetic, oviparous and viviparous; highly developed polymorphism. Divided into three families - large Aphididæ and two relatively small Adelgidæ and Phylloxeridæ.

Aphis generic name of certain aphids; common name for all aphids.

Aphis-lion nickname of some lacewings (Chrysopidæ: Neuroptera); fragile, gray, yellowish or green insects; larvæ and adults are voracious feeders; they stalk their prey, mainly aphids but also psyllids, whiteflies, mites and spiders.

Aphis-wolf small brownish lacewings (Henmerobiidæ: Neuroptera); devour aphids, mealy bugs, psyllids, whiteflies and other small homopterous insects as also mites.

Aphodiidæ (Leach 1835) family of small dung-beetles (Scarabæoidea: Polyphaga); small, smooth, straited and often hairy; dark brown to black in colour, head hypognathus, antennæ 9-segmented, mandibles concealed below clypeus; breed abundantly in dung – *Aphodius finetarius* (Linnæus), *A. moestus* breed abundantly in dung – *Aphodius finetarius* (Linnæus), *A. moestus* Fabricius, *Rhyssemus germanus* Linnæus, et al.

Aphrodisiac male attractant: chemical (scent) produced by female insects to entice males: that which excites sexually.

Apian pertaining to bees.

Apiarian relating to beehives or bee-keeping.

Apiarist bee-keeper: person who maintains apirary (or apiaries), may be for commercial purpose, hobby or to stud the bees.

Apiary place for maintaining stock of beehives for collecting honey and beeswax.

Apical referes to apex (apices)or tip(s).

Apical plate group of cells at the anterior end of some trochophore larvæ having nervous and sensory functions.

Apicoceridæ (Apioceratidæ) small family of long elogate, thickly pilose dipteran flies (Asiloidea: Brachycera), antennæ with or without short style; palpis spatulate; proboscis non-piercing; two pad-like pulvilli present and vein M_1 terminating before apex of wing (100 species) – *Apicocera maritima* (Australia).

Honeybee
[facsimile of the earliest authentic drawing by Frederico Cesi (1585-1630) published in *Apiarium*, Rome, 1625]

Apiculate ending abruptly in short, sharp point or tip.

Apiculture bee-keeping.

Apiculturist person practicing apiculture.

Apidæ (Latreille 1802, Leach 1815) Samouelle 1819 a hymenopteran family of Apoidea (Aculeata: Apocrita); both social and solitary forms; social bees live in colonies and construct vertical nest (Beehive), solitary ones nests in ground and in hollow stems etc; three subfamilies – Apinæ (honey bees), Bombinæ (bumble bees) and Meliponinæ (Mosquito bees), while Anthophorinæ (digger bees) and Xylocopinæ (carpenter bees) are now considered as independent families.

Apimorophidæ family of coccids (Coccoidea: Homoptera); abdomen narrowed apically or produced into anal tube; pygidium absent; confined to Australasia, infesting *Eucalyptus* trees forming galls on leaves and stems - *Apiomopha duplex* Schrader.

Apinæ subfamily of apidæ (Apoidea: apocrita); bee (honeybees) that are valuable pollinators, besides producing honey and wax - Indian bee (*Apis indica* Fabricius), rockbee (*A. dorsata* Fabricius) et al.

Apioceridæ (Apioceratidæ) small family of long elongate, thickly pilose dipteran flies (Asiloidea: Brachycera); antennæ with or without short style; palpi spatualte; proboscis non-piercing; two pad-like pulvilli present and vein M_1 terminating before apex of wing (100 species) - *Apiocera maritima* (Australia).

Apiologist person well-versed in apiology.

Apiology branch of entomology that deals with scientific study of bees specially honeybees.

Apionidæ small family of Curculionoidea (Polyphaga: coleoptera); tiny weevils with marked sexual differences, having antennæ straight (clavate) rarely geniculate; trochanter elongate and proventriculus not strongly sclerotised; cosmopolitan in distribution (1000 species), majority pests of economic crops - sweet potato weevil (*Cylas formicarius* Fabricius), jute weevil (*Apion corchori* Marshall)(, *A. gagatinum* Motschulsky, *A. strobilanthi* Desbrocher) et al.

Apitoxin poison of bees' sting; it contains histamine producing enzymes and proteins of low moleclar weight.

Apivorous feeding on bees.

Apneumone(s) substances emitted by a non-living material that evoke behavioural or physiological reaction adaptively favourable to the receiving organism, but determinatal to the organism of another species which may be found in or on the non-living material - Chemicals emitted from oatmeal (flour of *Avena sativa*) that attracts the ichneumonid parasite *Venturia canescens* and aid in its search for the hosts: chemicals of meat to which the parasites *Alysia manducator* and *Nasonia vitripennis* are attracted.

Apochrysidæ Brues and Melander 1910 a very small family of fragile lacewings (Hemerobiidæ: Planipennia) separated from Chrysopidæ, confined to Australia, Asia and East Indies; antennæ filiform; hind-wings distinctly larger and costal area having more than 40 cross-veins - *Oligochrysa gracilis* occurs in Eastern Australia; *Apochrysa* species is found in Japan and Papua.

Apocrita Handirsch 1906 (Clistogastra Konow 1905, Petiolata Bingham 1897) larger of the two suborders of Hymenoptera, characterised by having propdodeum fused with metathorax; second segment constricted into slender petiole or pedicel; trochanter 1- or 2-jointed. Larvæ apodous. - ants, bees and wasps.

Apodeme any internal circular process of body wall.

Apodous (apodal) animal (insect) without feet (legs).

Apogamy omission of sexual fusion in reproduction: asexual: parthenogenesis.

Apoidea Ashmead 1899 superfamily of social and solitary bees (Apocrita: Hymenoptera); pubescence of head and thorax plumose; hind-tibia and basal joint of tarsus dilated and densely pubescent for carrying pollens; mouthparts of workers modified and lengthened for gathering pollens; sexual dimorphism common; males smaller than queen, often holoptic but not polliniferous and without sting. Excellent pollinators besides yielding honey and wax; a few are parasitic - Colletidæ (primitive or yellow-faced bees), Helictidæ (mining bees), Andrenidæ, Melittidæ, Megachilidæ (leaf-cutting bees) and Apidæ.

Apomixis parthenogenetic development of sex cells; apogamy.

Apophysis any elongate or tubercular process of body wall (exoskeleton) external or internal.

Aposematic having warning colouration; indicating that the insect is dangerous so as to be avoided; serves as defensive protection for the insect against its predators.

Appendage portion of body that projects out and has a free end - paired jointed limbs of insects like antennæ, legs, cerci etc.

Apple gaint caterpillar *Actias selence* Hubner (Saturniidæ: Lepidoptera); large pinkish moth with wing expanse of 130 to 150 mm (♂) and 140 to 180 mm (♀); having light greenish wings, whitish at base and pale yellowish near margins, hind-wings exhibiting a pinkish tail.

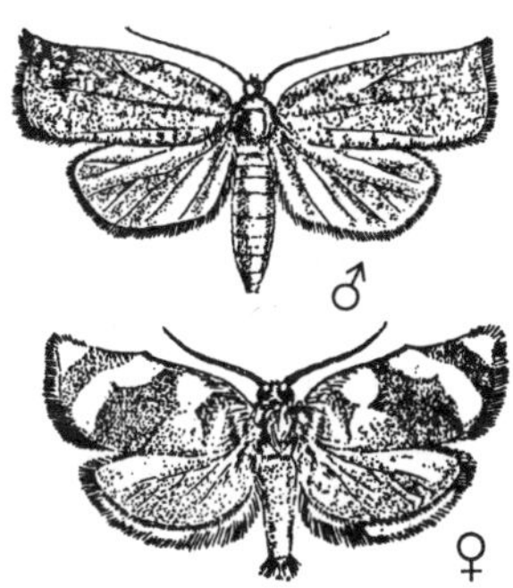

Archips termias (Meyrick)

Apple leaf-roller *Archips termias* (Meyrick) (Torticidæ: Torticoidea); caterpillars feed on foliage of temperate fruit trees and also bore into flower buds and fruits therby affecting adversely the fruit setting capacity and fruit yield of the trees.

Apple root borer *Dorysthenes hungeli* Redtenbacker (Cerambycidæ: Coleoptera); 80 to 110 mm long, chestnut coloured beetle having head and thorax darker than elytra. Grubs bore inside the roots and feed within; young trees die away, older ones become weak and fall down with strong winds.

Apple stem borer *Apricona cinera* Chevrolet (Cerambycidæ: Coleoptera); 350 to 50 mm long, ash-grey beetle with numerous black tubercles at base of elytra; antennæ concolourous and longer than body. Grubs bore through shots and make circutous galleris downwards, leading to main stem and trunk of the tree; adults feed on bark as well cut leaves and young shoots.

Apposition eye compound eye that forms a mosaic image - an errect image composed of large number of dots or points of light, as each crystalline cone (ommatidium) or the eye is surrounded by a sheath of pigment which does not allow the light to pass from one cone to another - common in diurnal insects and crustaceans.

Aproctous having no anus.

Aptera (apterous) without wings, includes both primitively and secondarily winglessness.

Aptera Linnæus 1735 an old order that included all primitive apterous insects (Aptera Linnæus 1759), termites, biting and chewing lice, fleas, as also Crustacea, Arcahnida, Chilopoda, Diplopoda and certain Annualat.

Aptera Linnæus 1759 an old order comprising of primitive apterous species; subsequently splitted into four orders - Thysanura Latrille 1796, Collembola Lubbock 1870, Protura Silvestri 1907 and Entotropha Grassi 1889 (Diplura Borner 1904) and grouped under subclass Apterygota Brauer 1885.

Apterogynidæ hymenopteran family of large hunting wasps (Vespoidea: Aculeata: Apocrita).

Apteropanorpidæ a small Australian family of Eumecoptera (Mecoptera: Mecopteroidea) close to family Panorpidæ, differing only in wing venation.

Apterous without wings.

Aptrygogenea old and obsolete term, proponed by Brauer (1885) for subclass Apterygota.

Apterygota Brauer 1885 subclass of Insecta Linnæus 1758 (Hexapoda Latreille 1825); primitively wingless insects that show many similarlities to crustraceans from which they have probably descended; 0.5 to 50 mm long; characterised by having ventral plate-like abdominal appendage which are vestiges of true abdominal legs; mouthparts either exposed and chewing type or concealed and piercing-sucking type; mostly scavengers feeding on decaying vegetable matter, prefer humid environment; comprises of two infraclasses - Ectortopha and Entotropha.

Aquatic having a water habitat: living or growing in, on or very near the water.

Aqutic caterpillar(s) larvæ of certain pyralid moths (*Nymphula* spp.); eggs are laid on water plants (pond-weed, water-lily etc.); on hatching, the larvæ burrow into the leaf, remain there, breathing dissolved oxygen either through skin [*N. nymphæata* (Linnæus) or by means of tracheal gills (*N. stratiotata*), pupate in white silken cocoons attached inbetween two leaves, from where the adults emerge.

Aquatic insects insects that live in or on water. Nymphs of mayflies, dragonflies, damselflies, stoneflies, alderflies live under water and breath by means of gills and so do the larvæ or mosquitoes. But all adult insects require atmospheric oxygen for breathing as they do not have any gills etc., so there are no true aquatic insects; among the superficially aquatic forms are some heteropterid bugs (back-swimmers or water-boatman, water crickets, water-scorpions etc) and diving beetles (Dytiscidæ), water-beetle (Hydrophilidæ) and whirling beetles (Grynidæ). Water-striders or pond-skaters that glide, skate or walk on water surface by using its surface tension and the tiny bluish-gray springtails (*Podura aquatica*) that are seen in clusters at the edges of lakes and ponds, feeding on algæ and plant detritus are considered as semi-aquatic insects.

Arachcidivoros feeding on arachnids, specially spiders.

Arachnid any animal belonging to class Arachnida (Chelicerata: Arthropoda) - mite, scorpion, spider, tick etc.

Arachnida class of subphylum Chelicerata (Arthropoda: Coelomata) - marine and terrestrial - body comprising of 21 segments forming cephalothorax (head and thorax fused together) and abdomen; cephalothorax bears two pairs of prehensile appendages (pincers) and abdomen; cephalothorax bears two pairs of prehensibe appendages (pincers) and four pairs of walking legs; antennæ wanting. Various orders include Acariformes (mites, ticks), Amblypygi (whip-spiders), Araneida (spiders), Opilioacriformes (daddy-long-legs, harvestmen), Ricinulcida (podogona), Scorpionida (scorpions), Solpugida (Sun-spiders), Uropygi (whip-scorpions), Netostigmata (leathery mites) and Schizomida.

Aradidæ (Spinola 1837) family of hereopteran bugs (Aradoidea: Pentatomorpha); dark brown flattened body with head produced into two pointed antennal tubercles; antennæ projecting freely; eyes prominent; trochanters fused with femora; found under loose bark of dead trees, feeding on fungi; give off offensive odour, to ward off the predators 9400 species) - flat bugs.

Aradoidea superfamily of Pentatomorpa (Geocorisæ: Heteroptera); unusally flat insects, ocelli lacking, stylets very long, tarsi 2-segmented and pretarsal claws without pulvilli; hemelytra, if present, narrower than abdomen with clavus not meeting that of opposite side behind the scutellum; bugs mainly subcorticolous or living in fungi; embraces two families - Aradidæ and Termitaphididæ.

Aræopidæ old and equivalent name for family Delphacidæ (Fulgoroidea: Auchenorrhyncha) - leaf-hoppers.

Araneidivorous spider-eating.

Arboreal animals adapted for living in trees.

Archæognathata suborder of Thysanura (Apterygota: Insecta) in an old system of classification; head reduced, ocelli present, compound eyes large, mouthparts ectognathous, mandibles with single articulation, rolling action; hypopharynx 3-lobbed; maxillary palpi large; abdomen 11-segmented with styli; tergum of XI segment prolonged into articulated median caudal filament and flanked by two shorter articulated cercal filaments; moulting continues after maturity. Only one order (Microcoryphia) with one superfamily (Machiloidea) and two families (Machiliæ and Meinertellidæ) - bristletails.

Archedictyon (Archidictyon) primitive meshwork (irregular network) of small veins as found in wings of Palæodictyoptera - a primitive fossil insect order of Carboniferous period - present day insects, with exception of mayflies and dragonflies show considerable simplification of wing venation by reduction in number of cross-veins.

Archelytridæ a family of fossil insects belonging to extinct order Protelytroptera of Permian period (Paleozoic era).

Archihymeidæ an old and obsolete name for the hymenopteran family Plumariidæ (Vespoidea: Apocrita).

Archiperlaria suborder of Plecoptera (Polyneoptera: Ptergota); primitive brightly coloured stoneflies mostly from Australasia and

South America; glossæ as long as paraglossæ; mandibles well developed, maxillary palpi filiform; both wings with abundant cross-veins. Nymphs with paired lateral gills on I, IV, V and VI abdominal segments - Diamphipnoidæ and Eustheniidæ.

Archostemata small suborder of Coleoptera (Coleopteroidea: Oligoneotera); showing several affinities in external features with those of Adephage; includes beetles having filiform or serrate antennæ; hypopharynx with characteristic darkened sclerome; notopleural sulcus present in prothorax; wings with Rs complete to base, oblongum distinct, apex of wing spirally rolled in repose. Larvæ with claw-bearing tarsus. Divided into two small families - Cupediadæ and Micromalthidæ.

Archtiidæ (Stephens 1829) Kirby 1892 family of stout, furry and hairy moths (Noctuoidea: Ditrysia); having well developed proboscis; wings banded or spotted with bright colours, forewings with M_2 basally approximated to M_3. Caterpillars large tuberculate and covered with dense tufted hairs; adults tiger moths - Bihar hairy caterpillar *Spilarctia obliqua* (Walker), Black hairy caterpillar *Pericalia ricini* (Fabricius), Red hairy caterpillar *Amsacta moorei* (Butler) et al.

Archulus arch-shaped cross-vein that connects the Radial and Median veins, in the wings of dragonflies (Anisoptera: Odonata).

Arctoperlaria suborder of stoneflies (Plecoptera) according to some recent classification, it comprises of Euholognatha (Nemouridæ and Scopuridæ) and Systellognatha (Perlidæ, Peltoperlidæ and Pteronarcidæ).

Arcuate arched or curved: bow-shaped.

Arculus arc-shaped vein forming a cross-connection between other veins in the wings of certain insects - radial and median veins in wings of dragonflies (Anisoptera: Odonata).

Argidæ Rohwer 1911 (Hylotomidæ) small family of Tenthredinoidea (Symphyta: Hymenoptera); 3-segmented antennæ, third segment long and often bifid; forewings without cross-vein 2r, mid and hind tibia often with periapical spurs. Larvæ with 6 to 8 abdominal legs; feed freely on woody agniosperms - *Dielocerus* species.

Argus common name of various small butterflies (Lycænidæ and Satyridæ) having upper surface blue or brown and under surface pale yellow or whitish-grey, with several eye-sports on the wings.

Argyresthiidæ a small family of Yponomeutoidea (Ditrysia: Lepidoptea) often included in family Yponomeutidæ; its tiny caterpillars are found in the shoots, leaf-buds, fruits etc. of many plants - *Argyresthia* species et al, found in Europe and North America.

Arista large bristle-like appendage on dorsal edge of apical antennal segment in true flies (Diptera) formed from numerous segements of flagellum.

Aristate antennæ of certain dipteran insects having bristle-like lateral outgrowth on flagellum.

Aristotle (384-322 B.C.) son of Nicomachus, Greek philospher - father of zoology - students of Plato (427-347 B.C) at Athens and tutor of young Alexander (Sikandar-e-Azam). wrote on zoology, physics, metaphysics, logic (which he invented) and rhetoric as also on ethics, politics and poetics.

Arixeniidæ Jordan 1909 very small family of earwigs (Arixeniia: Dermaptera) associated with bats, wingless insects having head, compound eyes not well developed, ocelli absent, cerci very short and somewhat arched - *Arixenia esau* Jorden, *A. jacobsoni* Burr et al.

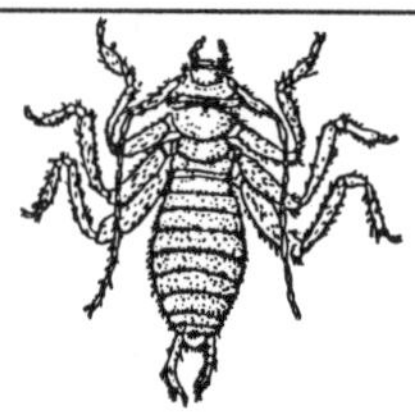

Arixenia esay Jordan

Arixenia jacobsoni Burr

Arixeniina Burr 1913 monofamilial suborder of Dermaptera (earwigs), elongate apterous insects, eyes vestigial; cerci not modified into forceps but short, arched and hairy; ectoparasitic, found in breast-pouches of certain bats (mammals); confined to Malaysia - Arixeniidæ.

Armyworm(s) noctuid caterpillars which often move in large number from plant to plant and field to field - *Pseudaletia separata* (Walker), *Spodoptera mauritia* Bosiduval et al.

Arolium median terminal lobe of pretarsus, arising between bases of claws.

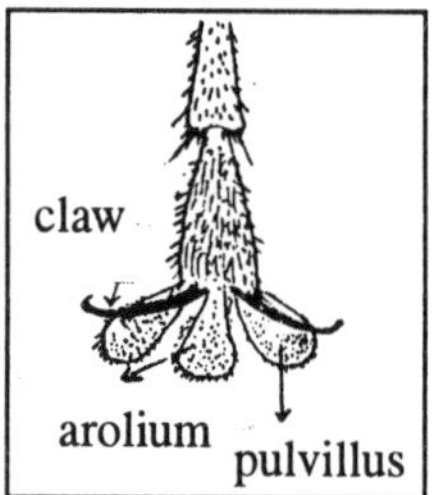

Aromia common and generic name of odorous species or musk beetles (Cerambycidæ: Chyrsomeloidea) having scent glands on hind coxæ taht secrete musk-like ester of salicylic acid - *Aromia moschata* (Linnæus) et al.

Arrestant(s) aggregation pheromones released mostly by bark beetles and *khapra* beetles; these cause insects to aggregate or congregate.

Arrhenophanidæ a very small family of small moths (Tinæoidea: Ditrysia) having wings entire and fringed.

Arrhenotocia parthenogenetic production of males alone; found in queen honeybee (Apoidea), *Trichogramma* and allied species (Chalcidoidea).

Arrhenotoky mode of reproduction in which offsprings of both sexes are produced by mated female; egg when fertilised produced viable diploid female but if unfertilised, it becomes viable haploid male: production of males from unfertilised eggs - drone bees (Hymenoptera).

Artery blood vessel that carries blood away from the heart to the tissue.

Arthropleona Borner 1901: suborder of Colembola; body with or without scales and segements distinctly defined; head horizontal, anternnæ in middle or distal half; furea present or absent; comprises to two superfamilies - Podureidea and Entomobryoidea.

Arthropoda largest phylum of invertebrate animals (Coelomata: Triploblastica) having annulate body and segmented appendages invested with chitinous exoskeleton 1,110,000 species), includes five subphyla Trilobita (extinct), Onychophora (most primitive), Cheliceratea (Merostomata, Pycnogonida and Arachnida), Crustacea and Uniramia (Myriapoda and Insecta); some entomologists consider thee as independent phyla. Besides, there are about 500 species of minute unsegmented, uncertain arthopods grouped into two classes - Pentastomida (tongue worms) and Tardigarda (water-bears).

Articulation movable point of contact between two sclerotic parts of body wall.

Artificial diet an unfamiliar feed that has been formulated, processed and developed in the laboratory for rearing the particular insect species in captivity, for research and experimental purposes.

Ascalaphidæ Schneider 1845 large cosmopolitan family of owl-flies (Nyremeleontoidae: Panipennia) active day fliers that fly in straight line; head covered with long fine hairs, antennæ very long and clubbed; wings broader but not so elongated, often partially pigmented, hind-wings shorter and

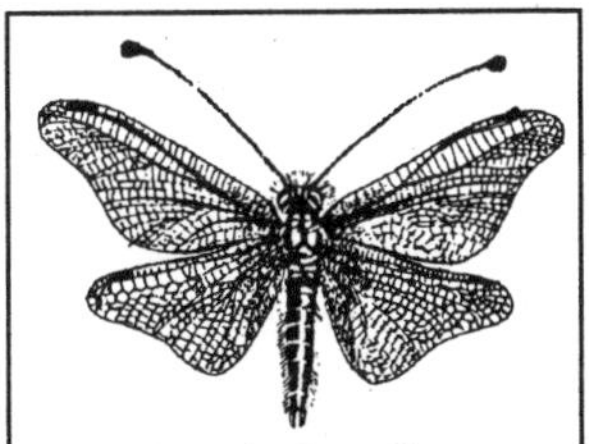

Ascalaphus fly

wider than forewings – *Ascalaphodes canifrons* (Westwood), *Helicomitus diax* (Walker), *Ogeogaster segmentator* Westwood, *Hybris* spp., *Suphalacsa* spp., et al.

Aschelminthes (Nemathelminthes) heterogenous phylum of Pseudocoelomata (Triplobastica: Anmalia) that includes small to microscopic worm-like animalcules which are mostly unsegmented and covered with cuticle. About 18000 species grouped into five classes – Rotifera (wheel animalcules), Gastrotricha (hairy stomach-worms), Kinorhyncha (jaw-moving worms), Nematoda (round worms) and Nematomorpha (horse-hair worms). Of late these classes have been elevated as independent phyla.

Aschiza Becker 1882 group (section) of dipteran insects belonging to suborder Cyclorrhapha; characterised by frontal suture lacking, lunule indistinct or absent; prilinum wanting; forewing with cell Cu elongated (except in Phoridæ) and extending more than half way to wing margin; three superfamilies – Lonchopteroidea, Phoroidea and Syrphoidea.

Asexual without sexual process.

Asexual reproduction reproductive process that does not involve the union of gametes.

Ashinagidæ very small family of Yponomeutoidea (Ditrysia: Lepidoptera); moths having wings covered with soft scales, anal regions small; venation well developed.

Asilidæ (Leach, 1815, 1819) very large family of robber-flies (Asiloidea: Brachycera); antennæ projecting forward, prominent tuft of hairs forming mouth-beard; eyes protruding, vertex depressed between the eyes; ocelli present; proboscis short, stout and horny adapted for piercing and sucking; legs prehensile adapted for catching the prey; mostly predaceous on other insects (500 species) – *Allocotosia aurata* Fabricius, *Asilus occidentalis* Hine, *Phellus glaucus, Philodicus femoralis* Ricardo, et al.

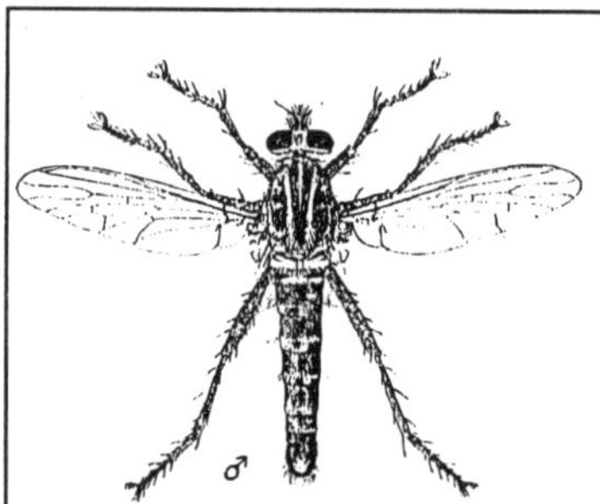

Philodicus femoralis Richardo

Asiloidea large superfamily of dipteran flies (Bachycera); antennæ 3-segmented and shorter than thorax, eyes and ocelli prominent;

maggots with vertically biting mandibles. Pupa obtect - Therevidæ, Acroceridæ, Scenopinidæ, Mydidæ, apioceridæ, Nemestyrinidæ, Asilidæ and Bombyliidæ.

Asparagus beetle herbivorous diurnal beetle, *Crioceris asparagi* (Linnæus) Chrysomelidæ: Polyphaga), about 6 mm long and conspicuously coloured yellow with one longitudinal and two vertical dark brown stripes, and outer border reddish; grubs gnaw shoots of asparagus, causing extensive damage.

Asparagus fly spotted-winged fly, *Platypara pæciloptea* (Schrank) [Tephritidæ (Trypeticæ): Dipteral; small 6 to 8 mm long, flies that are major pest of *Asparagus* sp., maggots mine and feed on young edible shoots.

Aspiceridæ a small family of Cynipoidea (Parasitica: Apocrita) often regarded as subfamily of Figitidæ - *Aspicera* species and its allies are parasites of Syrphidæ (Syrphoidea: Aschiza).

Aspidothoracidæ a family of Megasecopytera that include fossil insects occurring in the Upper Carboniferous period of France and England; characterised by an elongate cerci.

Assassin fly aerial predators (Asilidæ: Diptera); flies having powerful forelegs with which they capture their prey (bees, dragonflies etc.) on the wings. Also called robber-flies.

Asteiidæ family of small dipteran flies (Anthomyzoidea: Acalyptratæ); having characteristic long narrow wings with costa complete (not broken), Sc vestigial (not reaching C) and anal cells missing - *Asteia* species, *Liomyza* species, et al.

Asterolecaniidæ Berlese 1898 family of pit scales (Coccoidea: Homoptera); females small, elongate-oval and convex; antennæ reduced; legs vestigial or absent; often embedded in mass of yellow wax and lie in shallow pits on mid vein of leaves and leaflets - *Asterolecanium varisolosum, Cerococcus hibisci, C. quercus* and *Lecaniodiaspis azadirachtæ* are some common pests of forest tress (oak, olive etc.)

Asthenobiosis theory or doctrine of diapause or arrested growth; it propounds that there is accumulation of waste products in the tissues that prevents normal development of the individual; during the resting stage these waste products gradually disappear as excretion continues while normal metabolism slows down.

Asthenohymenidæ an important family of fossil (Paleozoic) insects belonging to order Megasecoptera that abound during the Upper Carboniferous period of France, England and North America.

Asymmetrical having no kind of symmetry - bilateral, radial nor spherical.

Asymmetry want of symmetry.

Asymptote stage in growth of population at which numerical stability is reached.

Atomiser devise for breaking up liquid stream into very fine droplets by high pressure of compressed air. Small hand sprayer that gives out liquid in a fine continuous spray; used for small laboratory experiments, greenhouse plants, other potted plants or even in kitchen gardens.

Atoxic non-poisonous.

Atrial orifice external opening of atrium.

Atrium spicular chamber formed by secondary invagination of body wall; opens into trachea by primitive tracheal orifice and to exterior by secondary artial orifice.

Atrophied reduced considerably in size; rudimentary: vestigial.

Atrophy wasting away through imperfect nourishment or lack of use.

Atropidæ Kolbe 1884 (Trogiidæ Enderlein 1911) family of book-lice (Trogiomorpha: Psocoptera); soft-bodied insects without scales; prothorax large and divided into three parts, meso- and metathorax separated by a suture, tarsi 3-segmented; wings usually absent but in some species vestigial, forewings represented by small scales without ventation - *Atroos pulsatorium* (Linæus), *Lepinotus inquilinus* Heyden, pests in libraries (books) and museums.

Attactidæ old and equivalent name for lepidopteran family Saturniidæ (Bombycoidea: Ditrysia); that includes giant silkworms and wild silkworms.

Attavidæ old and equivalent name for family Yponomeetidæ (Yponomeutoiedea: Ditrysia); that includes beautiful mining moths.

Attelabidæ family of Curculionoidea (Polyphaga: Coleopetra); having clubbed (not genticulate) antennæ, maxillary palpi 3-segmented; female cuts leaf-lamina, roll it up and lays an egg in the centre of the roll (300 species) - *Apoderus tranquebaricus* Fabricius, *Cænorhinus germanicus, Deporaus betualæ, Zamia* spp., et al.

Attractant substance that by some mechanism like gustatory, olfactory, tactile, vision, etc., lured certain insects for feeding or oviposition - Citral, linalol, linalyl acetate, terpinyl acelate etc.

Atychidæ very small family of Tortriiucoidæ (Ditrysia: Lepidoptera); small-sized moths with wide wings, crepascular in habit, confined to temperate regions.

Auchemorrhyncha Dumerial 1806 heterogenous group of paraneopteran insects (Homoptera), active forms capable of free locomotion; antennæ very short with terminal arista; rostrum arising distinctly from head; tarsi 3-segmented; four superfamilies - Fulgoroidea, Cercopoidea, Cicadoidea and Cicadelloidea.

Augmentation manipulation within an insect population with a view to increasing its efficiency through mass production technology and genetic engineering.

Aulacidæ Schuckard 1841 small family of Ichneumonid wasps (Hymenoptera), parasitic on grubs of cerambycid and other beetles - *Aulacus bituberculatus* Cameron.

Aulacogasteridæ family of small dipteran flies (Anthomyzoidea: Cyclorrhapha), very near to Drosophilidæ - *Aulacigaster leucopeza* (Meigen).

Austroperlidæ (=Penturoperlidæ) most archaic filipalpian family (Plecoptera); stoneflies having wings with many cross-veins except in anal lobe of hind-wings and more than 6 anal veins in each hind-wing; often included in family Gripopterygidæ; confined to Australasia and South America (12 species) - *Klapopteryx* species, *Tasmoanoperla* species et al.

Austroserphidæ very small family of hymenopteran insects (Proctotrupoidea: Aopocrita) resembles Heloridæ except that the antennal scape is produced dorsally into long point; confined to Australia.

Autecology study of relationship between individual species of organism and its environment *Cf.* synecology.

Auto- self.

Autochthon a fossil occurring where the oraganism (insect) once lived.

Autocide pest control by sterile-male technique.

Autocopulation self-copulation, occurs occasionally in certain hermaphrodite worms.

Autoecious completing entire life-cycle on one host.

Autogamy process of self-fertilisation, resulting in homozygosis, occurs in fungi and protozoans, besides in certain flowering plants.

Autogenesis process of self generation.

Autogeny spontaneous generation.

Autohemorrhage voluntary, exudation of poisonous blood (nauseous) by certain insects as a defence against predators.

Autolysis self dissolution that tissues undergo after death of their cells.

Autoparasitism type of parasitism wherein males of parasitic species develop as secondary parasites on female primary parasites of their own species – *Venturia* (*Nemeritis*) *canescens* (Ichneumonidæ: Hymenoptera), *Coccophagus* spp. (Eulohidæ: Hymenoptera) et al.

Autosome any chromosome other than sex chromosome.

Autotype (heutotype) specimen identified by the original author as illustration of his species.

Ausxiliæ small plates beneath bases of pretarsal claws, bearing pulvilli.

Aves class of **phylum** Chordata that includes all birds; biped, feathered, warm-blooded, oviparous vertebrate animals, having brain and eyes well developed; no teeth, instead a tough horny beak; no larynx but syrinx for producing sound; forelimbs modified as wings (10,000 species). Majority of birds (Neognathæ) fly; a few species (Palaeognathæ) are flightless, these include cassoway, emu, kiwi, moa, ostrich and rhea; while penguin (Impennæ) is sea-bird having its wings modified into peddles, which help in swimming.

Avian (avine) of birds.

Aviarist person who keeps an aviary.

Aviary very large cage(s) for keeping birds.

Avicolous living on birds.

Aviculture bird-rearing specially breeding of wild species in captivity.

Avifauna assemblage of birds found in certain region or environment.

Axenic rearing ot one or more individuals of a single known species in or on a non-living medium or synexic.

Axiidæ very small family of Geometroidea (Ditrysia: Lepidoptera), having tympanal organs in VII abdominal segment; only 10 species, confined to Mediterranean region.

Axillary cord thickened corrugated posterior edge of asticular membrane of wing base.

Axillary plate posterior sclerite of wing base supporting subcostal, radial, medial, cubital and vannel veins - Odonata.

Axillary region small triangular portion of wing base bearing axillary sclerites; base of triangle in hinge of wing with body of insect.

Axillary selerite(s) sclerites of axillary region in wing flexing insects.

Axon nerve fibre of neurone that conducts nerve impulse away from nerve cell body.

Axymyiidæ (Mamaev and Krivoscheina, 1966) very small family of Bibionoidea (Nematoceara: Diptera); includes only three species allied to bibnbionidæ and Anisopodidæ except that wings have no discal cell. Maggots have long, terminal tube-like extension and pupæ with a strong head shield and prothoracic horns.

AYYAR, Dr. T.V. Ramakrishna born 1880, graduated from Madras Christian College in 1898; after two years of post-graduation, took up teaching work. In 1904 appointed as an Assistant of H.M. Lefroy, the then Imperial Entomologist at Pusa (Bihar); transferred to Coimbatore in 1906, to work with T.B. Fletcher and became Senior Assistant Entomologist in 1920. During 1927' and 1928, studied at the Stanford University, California and obtained Ph.D. degree in entomology. On return, served as Lecturer in Zoology and Entomology at College of Agriculture, Coimbatore till 1930, when he became the Entomologist to Government of Madras. Retired in August 1935 but maintained interest in entomology till death (February 13, 1952). The Handbook of Economic Entomology for South India (1940/1963) is a lasting monument to his eminence in the field of entomology.

Azadirchtin biopesticide based on margosa or neem-seed (*Azadirachta indica*); effective against a large number of incest pests, as also some species of mites and nematodes; harmless to bees, birds and mammals.

Azinphos an organophosphatic general purpose insecticide that is effective against mites also; white solid, soluble in most organic solvents; acute oral and dermal LD_{50} for rats 8 to 9 and 280 mg/kg respectively. Available in two forms – azinphos-ethyl (Gusathion-A) and azinphos-methyl (Gusathion-M).

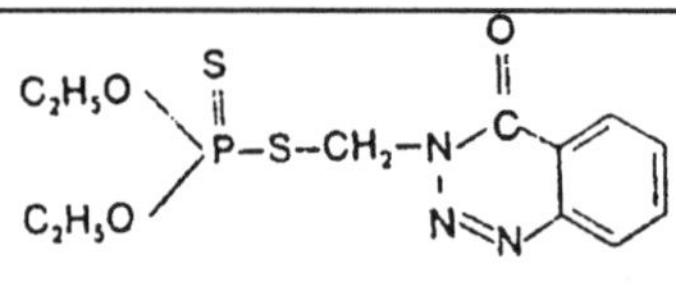

Azinphos-ethyl

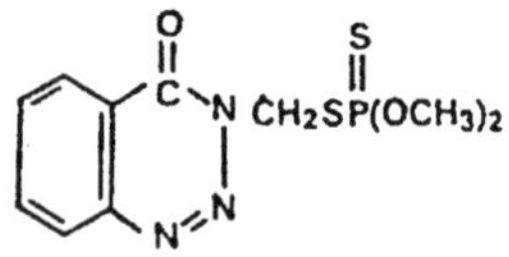

Azinphos-methyl

Aziridines derivatives of ethyleneimine (aziridine) used as chemosterilants - Tepa etc.

Azobenzene $C_6H_5N{:}NC_6H_5$; an organic dyestuff, mostly yellow, orange or reddish in colour; insoluble in water but soluble in most organic solvents - recommended as an ovicide.

Azoic without life; before the existence of animal life.

Azygote individual produced by haploid parthenogenesis.

Azoth old name for mercury (hydrargyrum, Hg); no more in usage.

Azygous unpaired: not yoked or joined with another.

B

Bacillidæ Brunner 1893 family of Phasmoidea (Phasmida: Polyneoptera); apterous insects with flattened or long and slender body; first abdominal segment distinct and shorter than metanotum - *Bacillus gallicus* Charpentier, *Setosa versicolor* Redtenbacher, *Acanthoclonia sævissima* (Westwood) et al.

Back dorsum: upper surface of inset body extending from head to anal end.

Backswimmer(s) (water boatmen) carnivorous water-bugs (Notonectidæ: Heteroptera), swim on their convex back by means of flattened hairy hind legs that move in unison like oars; also dive into water, taking some air under their wings and occasionally fly as well - *Notonecta glauca* Linnæus, *N.undulata* Say, *Enithares indica* Fabricius, et al.

Bacteriophage virus or such similar agent that can kill bacteria.

Bacterial (disease) caused by bacteria.

Bactericide toxin that can destroy or check growth of bacteria - Terramycin etc.

Bacteriidæ Brunner 1893 family of apterous or alate long slender stick like species (Bacterioidea: Phasmida); characterised by heaving first abdominal segment as long as or longer than metanotum - walking straw (*Palophus titan* Sjostedt) of Australia is more than 25 mm long.

Bacteriocyte modified fat cell found in certain insects that contains bacterium-shaped rods believed to be symbiotic bacteria.

Bacteriodidea Brues and Melander 1932 superfamily of Phasmida characterised by having carina of middle- and hind-tibia continued to apex, without sunken arolia; two families - Bacteriidæ and Bacunculidæ.

Bacteriologist person well-versed in bacteriology.

Bacteriology biological science dealing with study of bacteria.

Bacterium (pl. **bacteria**) microscopic unicellular or filamentous organism (kingdom Monera) having neither chlorophyll no nuclear membrane. Discovered in 1683 by Dutch naturalist and microscopist Antonvan Leeuwenhoek (1632–1723). According to their shape, bacteria are arranged into four groups - Bacilli (rod-shaped), Cocci (spherical), Commas (twisted) and Spirilli (spiral).

Bacunculidæ Brunner 1893 family of apterous, long, slender stick-like insects having long antennæ and forelegs; first abdominal segment shorter than metanotum - giant walking-stick, *Megophasma dentricus* (Stal), of Southern USA is about 150 mm long.

Bætidæ Klapalek 1909: family of Bætoidæ (Ephemeroptera: Palaeoptera); characterised by clear wings; forewing with few cross-veins and MA not forked, hind-wing greatly reduced or wanting; only two caudal filaments. Naiads elongated with long antennæ and legs; leaf-like gills and three caudal fringed filaments; inhabit cataracts and water-falls - *Cloeon dipterum* Linnæus, *C. pulchellum* Banks.

Bætiscidæ Ulmer 1920 very small family of Heptatenioidea (Ephemeroptera: Palaeoptera); having very small pronotum; forewings with Cu_1, Cu_2 and 1A almost equal in length and running parallel; hind-wings circular with many long intercalary veins. Naiad having huge 4-spined carapace formed by backward prolongation of thoracic notum - *Bætisca obesa* Walsh.

Bætoidea Ulmer 1920 superfamily of Ephemeroptera characterised by hind tarsi with four or less movable segments; fifth segments; fifth segment, if present, always fused with tibia; M_{1+2} and Cu_1 of forewing parallel at their bases; includes families - Bætidæ, Cænidæ, Ephemerellidæ, Leptophelediidæ and Oligoneuriellidæ.

Bagworm(s) psychid moths (Psychoidea: Ditrysia); caterpillars construct cases (cover of silken strands and plant fragments around themselves that are seen hanging on twigs and leaves of host plants; the cases differ in appearance and structure, according to species and sex; cases harbouring female are larger than those containing males; pupation takes place within these cases; in most cases in female moths are wingless and stay in the bags throughout their life, eventually laying the eggs therein, on hatching the caterpillars leave the case, to build their own cases - *Acanthopsyche minimora* Hampson, *Kophen caprea* Moore, *Eumeta crameri* (Westwood) et al.

Bait eatables used for enticing insects: food particle put on a hook to allure and catch fish.

Balancers (halteres) reduced or modified hind wings of all true flies (Diptera), shaped like short dumb-bells; during flight these vibrate and act as gyroscope stabilisers.

Bamboo shoot hole borer(s) *Dinoderus brevis* and *D. ocellaris* Lesne (Bostrychidæ: Bostrychoidea); small (3 to 4 mm long), cylindrical, brownish beetles with minute bristles and sculptures on elytra bore deep into felled bamboo and bamboo furniture; making the same useless and unserviceable.

Banchidæ small family of long tailed wasps (Ichneumonoidea: Hymenoptera).

Bar eye a mutation of *Drosophila* species (Drosophilidæ: Diptera) in which the number of optical units in the compound eye is greatly reduced.

Baroniidæ small family of Papilionoidea ('Ditrysia: Lepidoptera).

Basal pertaining to or located at the above.

Basal apodeme apodeme of phallobase.

Basal plate(s) sclerites of phallobase.

Basal metabolism sum total of anabolic and catabolic activities of an organism in the resting state, providing just enough energy to maintain vital functions.

Basement membranes inner non-cellular memebranous lining of any epithelial layer.

Basic manufacturer chemical industry which produces the basic or active ingredient used in a pesticide formulation - Bayer, Ciba, Hoechst, Sandoz, Voltas etc.

Basiconic receptor olfactory or other sense organ of insect comprising of a small projecting cone whose base is connected to a nerve fibre.

Basicosta proximal submarginal ridge of inner wall of a leg segment.

Basicostal suture external groove of a leg segment forming basicosta.

Basicpodite basal segment of telopodite.

Basisternum main part of sternum or ventral cuticular plate on each segment of an insect body.

Basitarsus proximal segment of tarsus.

Basketworm equivalent term for bagworm (Psychoidea: Ditrysia).

Basophilic staining strongly with basic dye.

Batesian mimicry resemblance of harmless insect species to a distasteful or poisonous species; it serves as protection against the predators, as they avoid eating all such species.

Bathosecopidæ family of Cicadoidea (Auchenorrhycha. Homoptera).

Bathyscinæ subfamily of Anisotomidæ (Staphylinoidea: Polyphaga), cave-d welling beetles having narrow head and thorax, globose elytra and long appendages; often blind with long tactile setæ.

BAWEJA, Rai Bahadur, Dr. Keshau Dass born July 14, 1896; graduated Punjab University 1920; joined Punjab Agricultural

College, Lyllapur; deputed to work in Locust scheme at Quetta. Studied at Rothamsted Experimental Station (England) 1935-37 and got Ph.D. Assistant Professor of Entomology at Lyallpur till 1940; joined IARI as Assistant Entomologist, promoted as Deputy Locust Entomologist; deputed as Entomologist to Government of Rajasthan in 1949, from where he retired in 1951. Honorary Professor of Entomology, Banaras Hindu University for next five years; Principal, S.K.N. Government Agricultural College, Jobner (Rajasthan) till 1961. Died in accident July 27, 1962.

Bedbug(s) common name for *Cimex* spp. (Cimicidæ: Cimicoidea); hemipterous insects having dark reddish-brown, broadly oval, flattened body; secondarily wingless, characterised by presence of stink glands that emit foul smell. Love humid climate, absorbs water vapour from air through their integument; shun light and hide during day in cracks and crevices in beds and other furntiture, feed intermittently on human blood at night (during darkness), causing sleeplessness and even rashes on skin of their victims; some are ectoparasites of bats and birds - *Cimex hemipterus* Fabricius, *C. lecturalius* Linnæus.

Bedelliidæ old and equivalent name for small lepidopterous family Lyonetieæ (Tinæoidea: Ditrysia) that includes minute moths.

Bee(s) small to large, smooth or hairy, dark or brightly coloured, social or solitary insects (Apoindea: Apocrita); social forms live in colonies (beehive) while solitary ones burrow into wood or stem and nest therein; almost all assist in pollinating various crops, while some yield honey and beeswax (20000 species; found all over the World) - carpenter bees, honeybee, rockbee etc.

Bee(s), Bumble (humble-bees) common name for several large social bees *Bombus* spp. (Bombinaæ/Bombidæ: Apoidea) that differ from honeybees in being larger and more hairy, besides their colony is annual and no honey is stored; most members of colony die off during Winter, leaving fertilised queen to hibernate and restart new colonies in the following year; excellent pollinators of blossoms - *Bombus hortorum* (Linnæus), *B. ruderatus* (Fabricius), *B. terrestris* (Linnæus), *B. tunicatus* Smith, et at.

Bee(s), Carder hoop-shaver: common name of some small hairless bees (Megachilidæ: Apoidea), the females (smaller than males) have a peculiar habit of stripping hair from moss and other plants, roll these hair into a small ball and carry it to their nest for lining the transparent cell-walls of their nests; valuable pollinators - *Anthidium emarginatum* (Say), *A. manicatum* et al.

Xylocopa varipuncta Patton

Bee(s), Carpenter nickname of various solitary bees (Xylocopidæ: Apodea) that burrow into wood to make their nest; resemble bumble bees but are larger, less hairy and more flattened; found in Tropical regions - Indian species include, *Xylocopa iridipennsis* Lepeletier, *X. æstuans* (Linnæus), *X. latipes* (Drury) and *X. varipuncta* Patton.

Bee(s), cuckoo parasitic bees that build no nest or hive of their own but like cuckoo (bird) lay eggs in other bees' nests; on hatching these grubs feed on food available in that hive and later on hosts' eggs and freshly hatched larvæ of the host - *Melecta* spp. (Melectidæ: apoidea), *Nomada* spp. (Nomadidæ: Apoidea).

Bee(s), Homeless cuckoo bees (Nomadidæ: Apoidea) resembling wasps with black and yellow stripped abdomen; legs without pollen baskets; do not make or store honey. These bees lay eggs in the underground nests of mining bees (solitary bee, Andrenidæ and Halictidæ) and on hatching the larvæ feed on the stored food in that nest and later even devour the eggs and larvæ of the host.

Megachile anthracina Smith

Bee(s), Leaf cutting solitary bees (Megachilidæ: Apodea); body stout and densely haired; nests in cavities of rotten wood and cut-stems, lining the same neatly with cut out circular pieces of green leaves - *Megachil anthracina* Smith, *M. disjuncta, M. lanata* etc.

Bee(s), Mining solitary bees (Andrenidæ: apoidea); similar in appearance to honeybees but smaller in size and net deep in the ground A female digs a tunnel 60 cm or more deep, constructs a cell at the end of burrow wherein she brings a ball of pollen and honey, lays her eggs on the ball and closes the cell; makes another cell higher up the tunnel and provision that in the same way; after completing 4 to 6 such cells, she comes out to the surface and die - *Andrema forbesi* Robertson, *A. vicina* Smith, *A. weedi* Viereck, *Dasypoda hirtipes* (Fabricius).

Bees(s), Potter (Flower) solitary bees (Anthophoridæ: Apoidea) of variable colours; nest in sandy or clayey ground or burrow in walls, construct cells of clay, fill these with honey and pollen, then lay an egg and close the cell or pot – *Anthophora abrupta* Say, *A. retusa* Linnæus, *A. urbana* Cresson, et al.

Bee(s), Stingless mosquito bees (Meliponinæ: Apidæ: apoidea), resemble honeybees – live in colonies, collect nectar and pollen, nesting usually in hollow trees; differ from honey-bees in being much smaller and having no sting – *Tringona minima* Gribodo, *T. pallida* (Latreille) et al.

Bee(s), Yellow-faced primitive apoids (Colletidæ: Apoidea), which nest in (sandy) soil or in cavities in the stems or wood.

Bee assassin(s) reduviid bugs, *Apoiomrus crassipes* (Fabricius) of North America and *Prishthesaucus papuensis* Stal of Australia and Papua, that feed upon bees which they capture in flowers.

Bee bread mixture of pollen and nectar stored by worker bees in beehive and used as food for the developing larvæ.

Bee chafer chafer beetle, *Trichius faciatus* (Scaraæidæ: Polyphaga); having shaggy yellowish hair on head and thorax; elytra dully yellow with three black bands; give bee-like appearance when in flight; adults frequent flowers during summer specially on sunny days.

Bee eater(s) group of spectacular birds found in warmer parts of Old World (Meropidæ: Coraciformes: Aves); having curved beak and bright plumage; nest in sandy banks and feed on bees, wasps and even locusts.

Bee flower flower pollinated by bee(s).

Bee fly small, robust dipteran flies (Bombyliidæ: Brachycera); resembling bees; maggots parasitic on eggs and nymphs of grasshoppers and beetles etc, adult flies feed on nectar of flowers, thus help in pollination – *Bombylis major* Linnæus, *Argyramoeba distigma* Wiedemann et al.

Bee louse (pl. **bee lice**) neither bee nor louse but a minute dipteran fly, *Braula coeca* Nitzch (Braulidæ: Pupipara) that infests beehives and honeybees; secondarily apterous; maggots burrow into honeycomb and feed on pollen and honey; number of adults may stick on body of a single bee, causing considerable irritation.

Braula coeca Nitzch

Bee milk old name for royal-jelly.

Bee moth small pyralid moth, *Aphomia sociaella* (Ditrysia: Lepidoptera), male differs from female in having brown and white markings on greenish forewings. Caterpillars infest nets of bumble-bees and wasps feeding first on disused cells and other waste products, finally devouring the honeycomb and the larvæ of bees and wasps; hibernate during Winter and pupate thereafter in the beehive itself.

Bee sting modified ovipositor, lies in a pocket in VII abdominal segment of female bee from where it an be protruded at will. Piercing part of stylet consists of number of valves formed from long drawn-out horny outgrowths of VIII and IX abdominal segments. Venom is produced by acid gland while alkaline gland secretes an oily non-toxic liquid to clean and lubricate the sting. The worker bee stings only in its defence.

Beeswax $C_{30}H_{61}O.CO.C_{15}H_{31}$; a mixture of esters and paraffin hydrocarbons, all of high molecule weight, grayish-brown with yellowish tinge; secreted by worker bees from special glands under their abdomen, masticated and mixed with cephalic gland fluids; used by the bees for constructing their cells (honeycomb); commercially used in cosmetics, pharmaceutics, polishes etc.

Beetle any coleopterous insect – hytophagous, carnivorous, scavenging and parasitic forms – head heavily sclerotised, mouthparts mandibulate, eyes conspicuous but no ocelli; prothorax freely movable; forewings modified into elytra and are useless in flight but cover and protect membranous hind-wings. In some species, called weevils, frons and vertex are produced in front to form rostrum or snout that bears antennæ and mouthparts – some (ptiliidæ) do not exceed 0.25 mm in length while some scarabæoids like, *Dynasters hercules, Goliathus regius* and *Megasoma elephas* attain length of over 150 mm.

Beetle, Areanut small dull coloured weevil, *Aracerus fasciculatus* de Geer (Anthribidæ: Curculionoidea); body covered with pubescence and snout is short. Larvæ white grubs, found feeding on Arecanut trees, dead wood etc.

Beetle, Atlas giant beetle of Tropical Asia, *Chalcosoma atlas* (Linnæus) Dynastidæ: Scarabæoidea); grotesque shining dark reddish-brown in colour, about 100 mm long including two long curved pronotal horns, also a long stout horn on head and another short median one on pronotum.

Beetle, Bacon (Hide) *Dermestes lardarinus* (Dermestidæ: Dermestoidea) a small beetle having elongated elytra with parallel sides; black with a yellowish-brown broad band; both grubs and adults feed on bacon, hides and other fatty substances.

Beetle, Black fungus phytophagous scavenger, *Alphitobius piceus* Olivier (Tenebrionidæ: Heteromera: Cucujoidea); nocturnal in habit, feeds on fungi and decaying vegetative matter.

Beetle, Blister (Oil) conspicuous in appearance with warning colouration, *Epicauta hirticornis* (Hagg-Rutenburg), *Lytta versicatoria* (Linnæus) *Mylabris cichorii* (Linnæus), *M. phalerata* (Pallas) and *M. sidæ* (Fabricius) are some species of Meloidæ (Cucujoidea); that excrete acrid oil from openings in apices of their femora; this oil contains cantharidine that has irritant properties and produces blisters on the skin of anyone touching these beetles.

Beetle, Bombardier small carabid beetle (Adephaga: Coleoptera) about 10 mm long; red with black elytra; when disturbed they discharge a few drops of acrid volatile liquid fron their abdomen with a minute explosion - *Anthis* spp. (Africa and India), *Brachinus* spp. (North America).

Beetle, Cardinal large, beautifully coloured beetles (Pyrochroidæ: Coleoptera) found in colder regions; grubs under bark of trees and adults also on twigs, leaves and flowers - *Pseudopyrochroa cardoni* Fairmaire, *P. subcostulata* Fairmaire, et al.

Beetle, Carpet wooly bears or skin beetles (Dermestidæ: Polyphaga) are constant menace to fabrics and articles containing hairs, fur, leather, skin, wool etc. as also dried insects and stuffed animal specimen in museums - black carpet beetles (*Attagenus gloriosæ* Fabricius and *A. piceus* Olivier); spotted carpet beetles (*Anthrenus coloratus* and *A. flavipes* Waterhouse) are the most common Indian species.

Beetle, Carrion burrying or sexton beetles (Silphidæ: Staphylinoidea); medium-sized, 12 to 30 mm long; always found in pairs (one male and one female) removing the soil from below carcasses of small animals, to burry it down and then feed and breed on the same - *Aclypea sculpturata* Grouvelle, *Necrophorus nepalensis* Hope, *Silpha rufithorax* Wiedamann, et al.

Beetle, Churchyard (Cellar) *Blaps* spp. (Tenebrionidæ: Heteromera: Cucujoidea); dull-black nocturnal slow-moving, flightless beetles, 20 to 30 mm long, charcterised by having wing cases fastened together; adults protect themselves by emitting strong smelling irritant fluid; inhabit caves and cellars - *B. indica* Bates and *B. orientalis* are Indian species while *B. mortisaga* Linnæus is European species.

Beetle, Cigarette (Tobacco) Old World species, *Lasioderma serricorne* (Fabricius) (Anobiidæ: Bostrychoidea), now cosmopolitan in distribution, is pest of cigarettes, cigars, cured tobacco, dried spices and meats, cured fish etc. Adult reddish-brown in colour, oval in shape, 2 to 3 mm long have minute hair on elytra. Grubs white, fleshy, crescent shaped are densely haired.

Beetle, Click skip-jacks; elongate flattened beetles (Elateridæ: Elateroidea) have hard body and wing cases; dark brown to black in colour, head small and bent down, antennæ toothed; legs short and thin. Gurbs resemble wireworms - *Cardiophorus stolatus* Erichson, *Heteroderus macroderes* Chaudoir, *Penia escholizi* Hope, et al.

Beetle, Colorado (potato) World famous pest, *Leptinotarsa decimlineata* (Say) Chrysomelidæ: Chrysomeloidea); native to USA, is herbivorous and diurnal. Adult 12 to 14 mm long, clay-yellow in colour with one black spot on head, several irregular black spots on thorax and ten longitudinal black stripes on elytra.

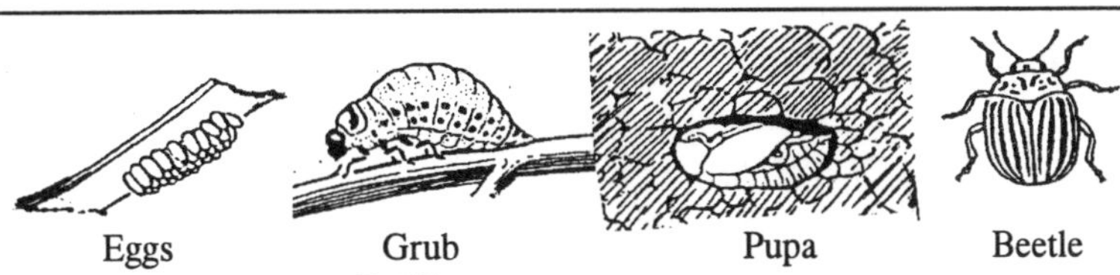

Colorado beetle

Beetle, Death-watch notorious, large (6 to 9 mm long), stout and sluggish woodborer, *Xestobium rufovilosum* de Geer (Anobiidæ: Bostrychoidea) that shuns light and is found under bark of old trees and boring woodwork in old houses. At mating time, sexes call each other by tapping their jaws on the wood; this sound heard in stillness of night has given the beetle its name.

Beetle, Drugstore small, 2.5 to 3.5 mm long, cylindrical, light brown beetle, *Stegobium paniceus* (Linnæus) Anobiidæ: Bostrychoidea) have striate elytra. Grubs white, fleshy, crescent-shaped and smooth (not hairy). Major pests of drugs and dried species; also damage books, papers and dried vegetable matter.

Beetle, Dung oval convex beetles with membranous mandibles (Coprinæ: Scarabæidæ), subterranean, live almost entirely on dung. Adults usually nocturnal and sttracted to light; during day they roll dung of mammals and carry these small dung balls to their underground nests to feed their future grubs - *Catharsius* spp., *Heliocopris bucephalus* Fabricius *H. gigas* Linnæus, *Sisyphus longipes* Olivier, et al.

Beetle(s), Fire luminous elaterids, *Pyrophorus luminosa* Illiger of West Indies, *P. noctilucus* (Linnæus), *P. phosphorescens* Castlenau and *P. pellucens* Eschscholtz of Tropical America emit brilliant greenish light from the pale yellow spot on either side of pronotum and reddish light from venter of abdomen. Buprestid, *Merimna atrata* Laporte and Gory of Australia attracted to brush fires and lights is also called fire beetle.

Beetle(s), Furniture *Heterobostrychus æqualis* Waterhouse, *Lyctus africanus* Lense, *Dinoderus minutus* Fabricius and *D. ocellaris* Stephens (Bostrychidæ: Bostrychoidea) are some of the common pests of wooden furniture in India.

Beetle, Giraffe robust, reddish-brown and black coloured big beetles, *Cladognathus giraffa* (Lucanidæ: Scarabæoidea); found in South India and Java, have very formidable looking toothed mandibles as long as their body; fly from dusk onwards and are attracted to light.

Beetle, Goldsmith *Cetonia aurata* (Linnæus) (Cetonidæ: Scarabæoidea); a brilliant golden yellow species, dorsally flattened and 14 to 20 mm long; occurs in Eastern Siberia.

Beetle, Harlequin very large longhorn beetle, *Acrocinus longimnus* (Linnæus) (Cerambycidæ: Chrysomelidæ), body length upto 90 mm with antennæ about twice the body length, thorax and wing-covers orange with red and velvety black markings. Grubs feed on fig leaves; found throughout tropical North and South America especially Brazil.

Beetle, Hercules one of the longest beetle, *Dynasters hercules* (Linnæus) (Dynastidæ: Scarabæoidea); found in Tropical North and South America; male about 155 mm long characterised by having an enormous horn projecting forwards from pronotum and curing downwards and another from head curving upwards to meet the upper one in the middle.

Beetle, Japanese black turtle bug or barley black bug *Scotinophara lurida* (Burmeister) (Podopidæ: Scutelleroidea), is pest of paddy as also other cereals (barley, maize, wheat etc.); found in Indian subcontinent, China, Japan and South-east Asia; brownish-black bug, about 10 mm long, with a prominent scutellum and pronotum having spines on either side; attacks the young tillers at milky stage of the crop, affecting adversely the maturity and yield of grain.

Beetle(s), Jewel *Stigmodera roei* Hope and *S. gratiosa* Chevrolet (Buprestidæ: Buprestoidea; Polyphaga); deeply pitted, brilliant green and glowing like jewels; often being set in mountings and used as jewellery.

Beetle(s) ***Khapra*** *Trogoderma granarium* Events (Dermestidæ: Dermestoidea); brownish in colour, oval in shape, 2 to 3 mm long, stout build beetles, lay eggs among the stored grains, newly hatched grubs begin feeding on soft germ point of the grain and continue feeding deeper into kernel, come out to pupate among the grains - serious pest of wheat, also infests other grains and dry fruits.

Beetle(s), Lady-bird attractive, conspicuous and popular beetles (Coccinellidæ: Clavicornia: Cucujoidea); head small, hidden by prothorax, fits smoothly into rounded elytra. Both larvæ and adults are predaceous (beneficial); their bright and bold colours warn the predators (birds) that these insects are distasteful - *Coccinela septempnctata* Linnæus and *Menochilus sexmaculatus* (Fabricius) prey on aphids; *Brumus suturalis* Fabricius on mites; *Rodolia cardinalis* (Mulsant) successfully introduced in India from Australia to control *Icerya purchasi* Maskell.

Beetle(s), Leaf-miner small 4 to 7 mm log, elongate beetles (Hispidæ: Chrysomeloidea); body covered with log stout upright spines. Grubs mine inside the leaves, feed by scraping leaf tissues and pupate therein; adults feed on tender leaves - *Asmangulia cuspidata* Maulik on sugarcane, *Dicladispa armigera* (Olivier) on paddy, *Onchocephala tuberculata* Olivier on sweet potato, etc.

Beetle, Long-headed flour *Latheticus oryzæ* Waterhouse (Tenebrionidæ: Heterocera: Cucujoidea); a common household pest destructive to cereals (flour and grains).

Beetle, Ox large reddish-brown beetle, *Stratægu antæus* (Dynastidæ: Scarabæoidea); male having three large horns on thorax while female has only one.

Beetles, Powder-post small, narrow, elongated, flattened, obscure insets (Lyctidæ: Bostrychoidea) attack every king of wooden articles that contain sapwood, such as packing cases, planks and plywood etc., wood attacked, is reduced to powder - *Lyctus africanus* Lasne and *L. malayanus* Lesne also attack mango trees.

Beetle(s), Red flour *Tribolium ferrugineum* (Fabricius) and allied species *T. confusum* (Duval) (Tenebrionidæ: Cucujoidea) are cosmopolitan household pests, infesting flour and stored cereals; these insects eat less but spoil more, due to their presence and excreta.

Beetle, Rhinoceros black palm beetle, *Oryctes rhinoceros* (Linnæus) (Scarabæidæ: Scarabæoidea); indigenous to India, found in entire

South-east Asia. Adults 40–50 mm long, robust, glossy-dark-brown to black in colour; pygidium in males long, smooth and rounded, in females short hairy and pointed. Grubs and pupæ found in decaying organic debris, manure pis etc., adults destructive to palm trees as also banana, pineapple etc.

Beetle(s), Rover medium-sized (8 to 24 mm long) slender, soft-bodied insects (Staphylinidæ: Staphylinoidea) having very short elytra; run rapidly often lifting the end of abdomen; feed on decaying animal and vegetable matter, thus acting as scavenger (beneficial) - *Coproporus minimus* Motschulsky, *Staphylinus semipurpureus* Kraatz, et al.

Beetle, Smoke *Melanophila consputa* Leconte (Buprestidæ: Bupresoidea); 10 to 15 mm long, black with 12 small yellowish elytral spots; founding Pacific states, attracted to forest fires and smelter plants; bite painful.

Beetle(s), Spider long-legged, round-boided, spider-like small (< 5 mm) beetles (Ptinidæ: Bostrychoidea), practically cosmopolitan, often abundant in grainaries, mills and warehouses feeding on dry animal and vegetable matter - *Ptinus fur* Linnæus, *Sphæricus gibboides* Boiled, *Niptus hololeucus* Faldermann, *Ptinus brunneus* Duftschmid, et al.

Beetle(s), Stag elongate-oval robust beetles (Lucanidæ: Coleoptera), dark brown to black in colour; male with conspicious large mandibles that project forwards; feed on dead and decaying wood - *Lucanus lunifer* (Hope), *L. cervus* (Linnæus) et al.

Beetle(s), Storehouse cosmopolitan household pest, *Gibbium pyslloides* (Czempinski) (Ptinidæ: Bostrychoidea); small 2.5 to 3.0 mm long, shining brown beetles that feed on opium cakes, dry spices as also vegetable and animal foodstuffs. Also called spider beetles.

Beetle(s), Tiger predaceous beetles (Cicindelidæ: Caraboidea); brightly coloured with white or creamy spots or bands, enormous powerful mandibles that overlap each other when not in use; diurnal in habit and found in all tropical and substropical countries; dig deep burrows (about 300 mm), await at the entrance, sieze the prey, drag it to bottom of the burrow to be devoured - *Cicindella sexpunctata* Fabricius, *Collyris bonelli* Guerin, *Prothyma paradoxa* Horn, et al.

Beetle(s), Whirling small shiny black beetles (Gyrindidæ: Caraboidea: Adephaga); predaceous, surface-swimming and air-breather; found gyrating in groups on surface of stagnant or slow moving water;

when disturbed they atonce dive down. The beetles have two pairs of eyes for seeing both above and below the water; legs well adapted for swimming – *Dineutus unidentatus* Aube, *Orctochilus gangeticus* Wiedmann, et al.

Behaviour relative responses/reactions of an organism to its environment.

Behningiidæ a family of Ephemeroidea (Ephemeroptera: Ephemeroidea); mayflies having wing venation highly specialised with main longitudinal veins arranged in parallel pairs and the legs reduced. Nymphs highly modified into burrowing type having fore- and mid-legs fossorial and hind-legs adapted to protect the gills – *Behninga* species et al.

Belidæ small coleopteran family of weevils (Curculionoidea: Polyphaga); antennæ filiform, gular sulci short but separate, tarsi 4-segmented; phytohagous; found in Australasia and South America (180 species).

Belitidæ family of small slender wasps (Serphoidea: Apocrita); characterised by having antennæ 14- to 15-segmented, labial palpi 3-jointed, trochanters 2-jointed and hind-wings with basal cell. Larvæ parasitic.

Belostomatidæ (Leach 1815) family of large, flat, brown aquatic bugs (Cryptocerata: heteroptera); head broad, antennæ short, 4-segmented and inconspicuous; eyes large, ocelli absent, rostrum short, 5-segmented and curved; wings leathery, hemelytra large with well defined veins; feed on insects, snails and even small fish; air is stored beneath the wings in reservoir formed by concave dorsum of abdomen (200 species) – giant water-bugs (*Lethocerus indicus* Stal, *L. grande* (Dimmock), *L. americanus* (Leidy) are 120 to 150 mm long.

Bembicidæ (Latereille 1702) Stephens 1829 family of large, robust, semi-social and predaceous sand wasps (Sphecoidea: Apocrita); body smooth or finely pubescent, head large and transverse; labrum large and well developed; eyes large and naked,, ocelli aborted and represented by scars; prothorax large; wings large, veins distinct; pterostigma much reduced; abdomen sessile: burrow in sandy places, sting and paralyse their prey and carry the victim to their nest for feeding their larvæ – *Bembix rostara* (Linnæus), *B. spinolæ* Lepeletier, *Steniolia* spp. et al.

Berlese organ copulatory pouch in females of some bugs (Cimicidæ: Cimicoidea); that receives sperm-cells from the male and secretes a substance that helps to activate the sperm-cells.

Berothidæ Handlirsch 1908 small family of beaded lacewings (Hemerobioidea: Planipennia); body small and slender with long filiform anternnæ and no ocelli; wings variable in shape, hairy and often with peculiar seed-like scales attached to important veins of females (40 species) – *Berotha* spp., *Lomamyia flavicornis* Walker, *L. texana* Banks, et al.

Berytidæ (= Neididæ) family of small delicate bugs (Gymnocerata: Heteroptera); body and appendages long and slender; head constricted in front of eyes, ocelli present; first antennal segment very long and bent just beyond its base; fomora elavate at apices; sluggish, found in meadows and undergrowth – stilt bugs (*Neides tipularis, gampsocoris delhiensis* Menon and Ghai, et al).

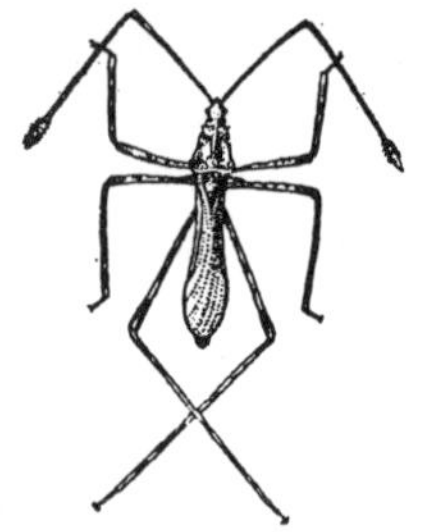
Gampsocoris delhiensis Menon and Ghai

Beta taxonomy level of taxonomic study dealing with the arrangement of species in lower and higher texa.

Bethylidæ Haliday 1840 family of Bethyloidea (Apocrita: Hymenoptera); metallic-bronze in colour; head elongated and depressed; eyes naked or hairy; mandibles large and strong; wings with reduced venation, stigma present. Larvæ mostly parasitic in or on caterpillars and grubs – *Pristocera orientalis* (Cameron), *Epyris californicus* (Ashmead), *Goniozus* antipodium (Ashmead) et al.

BHATTACHARYA, Dr. Arbinda – born September 2, 1917; M.Sc. Calcutta University 1942; served as inspector, Silk Reeling in Directorate of Industries Government of Bengal, 1943–45; Research Assistant Indian Veterinary Research Institute, Mukteshwar 1945–47; joined Indian Lac Research Institute, Ranchi 1947; deputed for higher training at Imperial College of Science and Technology, London, 1954–57 and obtained Ph.D. and DIC; became Lac Entomologist in 1964 and worked in that capacity till death (May 21, 1974).

BHC (HCH) $C_6H_6Cl_6$; benzene hexachloride was first synthesised in 1825 by Michael Faraday; its insecticidal properties ware discovered by A.P.W. Dupier (France in 1941 and F.D. Leicester (England) in 1942. It acts as stomach and contact poison as also fumigant to some extent.

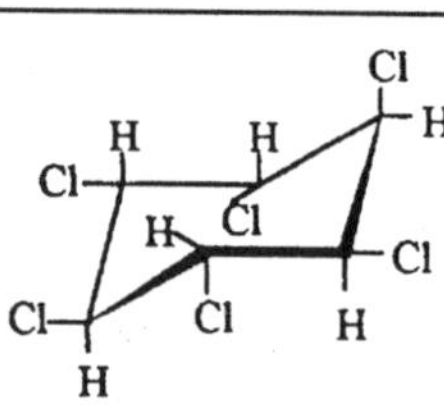

HCH

Effective against a wide insect pest spectrum, both aerial as well as soil-borne; mostly non-phytotoxic except to cucurbits; may cause tainting in potato, lettuce etc. making the same unpalatable - Gammexane.

Bibindandy the most important silkworm of Malagasy *Borocera madagascariensis* (Lasiocampidæ: Bombycoidea) that yields silk used for making lambalandy. Caterpillars are also pests of citrus (*Citrus* spp.), guava (*Psidium guajava*) and peach (*Prunus persica*) trees.

Bibionnidæ Kirby 1837: family of robust flies (Nematocera: Diptera); often covered with hair; antennæ shorter than thorax, antennal segments bead-like and closely apposed; ocelli present; pulvilli large; wings large with anterior veins strikingly prominent and no intermedian cell; found in grassy or decaying vegetation. Maggots terrestrial and peripuentic - *Bibio obscuripennis* de Meijerae, *B. marci* Linnæus, *B. imitator* Walker, et al.

Biforous spiracle spiracle of certain coleopteran grubs having two pouches of atrium,

Bihar hairy caterpillar *Spilarctia (Spilosoma) (Diacrisia) obliqua* (Walker) (Arctiidæ: Ditrysia); medium-sized (wing span 43 to 50 mm) moth, body crimson with three longitudinal black stripes on abdomen and black spots on the sides; wings pale buff coloured with black spots. Polyphagous; caterpillar feed voraciously on the leaves.

Bilateral symmetry arrangement of body parts so that right and left halves of the organism are mirror images of each other.

Binapacryl $C_{15}H_{18}N_2O_6$; white crystalline powder, insoluble in water but soluble in most organic solvents. Non-systemic, non-phytotoxic, contact poison; safe, against bees but harmful to fish and livestock; used as acaricide and fungicide, mainly against red spider mites and also powdery mildew on fruit and nut trees - Acricid, Endosan, Morocide.

Binomial nomenclature standard scientific method of naming an organism (animal or plant); evolved by Swedish botanist Carl von Line (Carolus . Linnæus) vide **Critica Botanica** (1773) and **Philosophica botanica** (1750). Accordingly, every organism described is given a generic name and an specific name, generic and specific names are underlined or printed in italics, whereas author's name is not italisized or underscored. If subsequently, the species is transferred to some other genus, the original auther's name is given in paraenthesis. This is known as scientific or

Botanical/Zoological name and is accepted all over the World – common housefly is *Musca domestica* Linnæus and human being is *Homo sapiens* Linnæus.

Binucleate egg(s) egg having two nuclei with different gene contents, which may be fertilised by two different spermatozoa; it occurs in many insects and results in gynandromorphs – same individual having both male and female features.

Bio- life

Bioassay (biological assay) quantitative estimation of biologically active substance; determination of relative strength of drug, hormone or pesticide by comparing its effect on a test organism.

Bioblast hypothetical unit of living matter: minute granule of protoplasm.

Biochemist person who has specialised in biochemistry.

Biochemistry physiological chemistry: chemistry of living beings including insects.

Biochrome any naturally occurring animal or plant pigment.

Biocide general poison or toxicant: pathogenic micro-organism (bacteria, fungi, protozoa, virus and even nematodes) or their bye-products capable of causing destruction in the pest population – Bathurin, Dipel, Vectoback etc.

Biocoenosis community of animals **(zoocoenosis)** or plants **(phytocoenosis):** association of organism ecologically interdependent.

Biodegration any process whereby the micro-organism utilise a substance as a source of energy and bring about its destruction; usually applied to the biological processes in soil, water or sewage.

Biogen hypothetical protein molecule that is unstable but has essential properties of protoplasm (living thing).

Biogenesis doctrine that living organism are derived only from living beings. *Cf.* abiogenesis.

Biogenous parasitic.

Biogeny process of organic evolution: development of an individual or race.

Biogeochemistry science of plants and animals in relation to chemicals in the soil.

Biogeography geographical distribution of animals and plants.

Biological pertaining to biology: produced by physiological means: effected by living organisms.

Biological control method of checking the population of undesirable organisms (plants or animals) by help of parasites, predators and pathogens - may be natural or man-made.

Biological pest suppression biological control of pests. Utilisation of parasites, predators and pathogens for regulating the host (pest) population and keeping the same below the economic-injury-level.

Biological warfare international use of pathogens (disease causing organism) for destroying crops and killing animals (including human beings); adapted as unfair means by enemies during wartime.

Biologist person having sound knowledge of biology.

Biology science of living beings: broadly divided into two groups - **botany** (plants) and **zoology** (animals). The term 'biology' was coined by Lamarck and Treviranus in 1802; Jean-Baptiste Pierre Antoine, d'Lamarck (1749-1829), a French naturalist and founder of invertebrate zoology, popularised the word 'biology'.

Biology, Cell branch· of biology concerning structure, function, reproduction and life-history of the cells.

Biology, Molecular aspect of biology that deals with bio-molecules and their interactions: area of biology dealing with the synthesis, nature and interactions of chemical substances found inside the living beings.

Bioluminescence ability of animals to produce light; emission of light is the result of a chemical reaction that does not involve production of heat; the phenomenon is highly developed in certain marine fishes and some species of insects in Collembola, Homoptera, Coleoptera and Diptera - blinkies, fireflies, glow-worms, lighting bugs etc. It is also found in a few crustanceans, snails and squids as also in bacteria and fungi. The light is emitted by oxidising luciferm in presence of enzyme luciferase and there are four types of light emissions giving different glow-patterns - continuous glow (eggs and pupæ of some lampyrids or fireflies); intermittent glow (larvæ of *Photuris* sp.); pulsation (*Luciola* spp., *Pycrocoelia* spp., *Pycrocelia* spp.) and flash (adults of *Lucila, Photinus* and *Photuris* species).

Biomass dry weight of living mater in population of particular organisms in a particular region.

Biome biotic community of animals and plants.

Biomechanic study of action of forces external and internal - on living beings including insects.

Biomechanics study of mechanics of living organisms.

Biometerology study of relationship between living organisms and the atmospheric phenomenon.

Biomorph representation of a living organism in decoration

Bion an independent living organism.

Bionomics ecological effect on growth and development of insects: study of relationship between an organism (insect) and its environments.

Bionomy science that deals with laws relating to vital functions of life.

Biophagous feeding on living organisms.

Biophore Weismann's hypothetical unit of living matter.

Biophysics study of phenomenon of living organism by physical methods.

Bioplasm protoplasm.

Bioplast minute portion of protoplasm.

Biopoiesis development of living matter from complex organic molecules that are themselves non-living but self-replicating.

Biopotency capacity of a chemical substance (hormone etc.) to function in a biological system.

Biosynthesis synthesis of chemical compound by living organism.

Biosystematics branch of systematics that includes experimental study of diversity, specially at species level.

Biota fauna and flora of particular region or a geographical era/epoch.

Biotic pertaining to life.

Biotic agent(s) living environmental factors that favours well-being or bring about premature mortality of animals or plant.

Biotic community the population of different species of plants and animals found in a particular locality, together constitute a biotic community.

Biotic environment an organism's environment that also has other living organisms which may be beneficial or harmful to that organism.

Biotic factor influence of one living organism on another.

Biotic potential estimate of maximum rate of increase of any species of insect (animal) if left to itself and isolated from its natural calamities or other inhibiting factors. Single pair of houseflies may produce 192000000000000000000 flies in less than five months (April to August).

Biotype species of identical genotype.

Bittacidæ Enderlein 1910 family of scorpion-flies (Eumecoptera: Mecoptera); body cylindrical; head elongated, antennæ slender, ocelli present; labium and mandibles long and slender; tarsi with single claw; both alate and apterous - *Bittacus latipennis* Gerstacker (alate), *Apterobittacus apterus* McLachlan (apterous).

Biturritiidæ very small family of Cicadoidea (Ciocadomorpha: Auchenorrhynch); leafhoppers allied to Membracidæ. Only six described species.

Blaberidæ family of Blaberoidea (Blattaria: Dictyoptera); femora not spinose; hind-wing folded fan-like when at rest; viviparous; the female retains eggs in its gential chamber or brood sac till these hatch out - *Blaberus craniifer* Brumeister and *B. giganteus* (Linnæus) are the largest cockroaches (about 50 mm log), other species include *Nauphoeta cinerea* (Olivier), *Pycnoscelis surinamensis* Linnæus, et al.

Blaberoidea superfamily of suborder Blattaria (Dictyoptera: Blattopteroidea) that includes a large and diverse group of insects, divided into three families - Blaberidæ, Epilampridæ and Polyphagidæ.

Blackfly common name often applied to bean aphid (*Aphis fabæ* Scpoli) as also to various dipteran flies of Simuliidæ family.

Bladderwort(s) group of carnivorous aquatic floating plants having no roots - *Utricularia flexuosa* - leaves and stems are covered with tiny bladders with trap-door that allows minute aquatic insect larvæ to enter the bladder but not to leave; these larvæ are then digested by the plant.

Blasticotomidæ Rohwer 1911 family of Tenthredinoidea (Symphyta: Hymenoptera); characterised by the larvæ having no abdominal legs and found boring in the stems of ferns - *Blasticotoma filiceti* Klug.

Blastobasidæ a small family of Gelechiodea (Ditrysia: Lepidoptera); often regarded as part of family Oecophoridæ; caterpillars feed on dry refuse seeds etc. or live as parasites on scale insects (Coccoidea) - *Blastobasis crassifica, B. transcripta, Exinotis catachlora, Holcocera pulvera, Prosintis florivora* et al.

Blastocoele fluid-filled cavity within blastula.

Blatoderm surface layer of cells of blastula before gastrulation.

Blastomere cell produced by division of fertilise ovum.

Blastopore opening of gastrula into archentron or gastrulation cavity.

Blastula early stage of embryonic development in which the cells are arranged into a hollow sphere - blastoder.

Blattæformia Handlirsch, 1908 old and obsolete name for one of the subclasses of class Insecta that included cockroaches, mantids, temites, zorabids etc., now regrouped under superorder Blattopteroidea (Polyneoptera: Pterygota).

Blattaria Burmeister 1929: suborder of Dictyoptera that includes cockroaches; body broad and flattened, head covered with large shield-like pronotum; antennæ long slender and many jointed; forewings leathery (tegmina), protect the membranous hind-wings; over 20 families – Blattidæ, Corydiidæ, Phyllodromidæ, Blaberidæ, Nyctiboridæ, Panesthiidæ, Panchloridæ, Epilampridæ, Euthyrrhaphidæ etc.

Blattellidæ a small family of Blaberoidea (Blataria: Dictyoptera); comprising of ovo-viviparous cockroaches; the females carry their eggs during the incubation period, in a sclerotised on membranous ootheca - *Blattella germanica* (German cockroach), *Supella supellectilium* et al.

Blattelytridæ an extint family of small insects belonging to fossil order Protelytroptera from Permian period (Palæzoic era). of Kansas (USA); closely related to Blattaria (Dictyoptera: Bloctoopteridea).

Blattidæ Stephens 1829 family of Blattaria (Dictyoptera: Blattopteroidea); characterised by hind-wins of alate forms having large anal area with many folds; females having valves on VII sternite - most common and cosmopolitan cockroaches *Blatta orientalis* Linnæus, *Periplaneta americana* (Linæus), *P., australsiœ* (Fabricius) et al.

Cockroach

Blattoidea Handirsch 1903 superfamily (old classification) of order Orthoptera that included families Blattidæ (cockroaches) and Mantidæ (mantids), both of which have now been elevated as suborders Blattaria and Mantodea and placed under order Dictyoptera.

Blattopteroidea Martynov 1938 superorder of polyneopteran insects that are exopterygote; wings folded back when at rest; metamorphosis incomplete (heterometabolous); includes four orders - Probtoblattoidea (fossil), Dictyoptera, Isoptera and Zoraptera.

Blepharoceratidæ (Blepharoceridæ) small dipteran family of net-winged midges (Culicoidea: Nematocera); elongate flies, eyes holoptic in both sexes, mouthparts of female adapted for lacerting, thorax with distinct transverse suture and wings with complex network of folds or creases; weak fliers; found in mountainous or hilly regions near water (torrential streams etc.). Larvæ aquatic, onisciform with ventral suckers - *Blepharocera indica* Brunetti et al.

Blood extracellular fluid in insect body (usually transparent) that consists of plasma (hæmolymph) and blood cells (hæmocytes). Red viscid fluid circulating in vascular system of higher animals (including man).

Blood sucker insect that sucks blood of other animals (mammals) - bedbug, louse, mosquito etc.

Blood gills tubular outgrowths from bodies of some aquatic larvæ of gnats (Chironomidæ: Culicoidea) and a few endoparasitic insects. These contain blood and are mainly concerned with absorption of water and salts (not with respiration) - species of *Chironomus, Halirytus, Tanytarsus* etc.

Blood-nosed beetle a large, black, flightless leaf-beetle, *Timarcha tenebriscosa* round thorax; wing-cases rounded and firmly fused together, the adult ejects a red fluid from its mouth when disturbed.

Blood-worm worm-like aquatic larva of non-biting midge (Chironomidæ: Nematocera). The larvæ look bright red due to presence of hæmoglobin, a rarity in insects.

Boarmiidæ Guenee 1844 (Hemitheidæ Braund 1845) a small Australian family of Geometroidea (Ditrysia: Lepiodotera); moths having cryptic colour-pattern; forewings with areole absent, hindwing with Sc approximated to cell upto about the middle, vein M_2 weak or wanting. Caterpillars loopers, often resembling twigs - *Boarmia* spp. et al.

Body entire frame or structure of an animal: main part of an animal as distinguished from limbs or appendages.

Body cavity definite cavity of body and appendages; arthropods (including insects) have continuous body cavity.

Body wall integument of body, formed of ectoderm, consisting of epidermis, cuticula and basement membrane.

Boganiidæ a very small family of Clavicornia (Cucijoidea: Polyphaga) with vaguely cucujoid affinities; confined to Australia. Grubs feed on fungi.

Bohartilloidea superfamily of Strepsiptera (Coleopteroidea: Oligoneoptera); known only from a male species recorded from

Honduras (Central America); having antennal segments 5-6 flabellate; mandibles short; base of maxilla swollen and hind-wing without vein MA_1.

Bolitophilidæ a family of fungus gnats (Mycetophiloidea: Nematocera); often included in family Mycetophilidæ; characterised by well developed short antennæ, with an oval pellucid spot below the antennæ - *Bolitophila* spp. et al.

Bollworm(s) caterpillars of certain gelechid and noctuid moths (Lepidoptera) that bore and feed inside the cotton bolls and are pernicious pests all over the World -pink bollworm (*Pectinophora gossypiella* Saunders), American bollworm [*Helicoverpa armigera* (Hübner)], spotted bollworms (*Earias insulana* Boisduval and *E. vittella* Stoll).

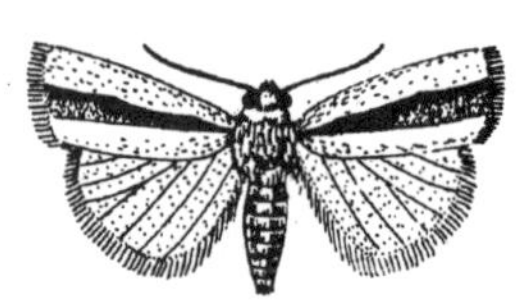

Earias vittella Stoll

Bombidæ (Lepeletier 1836) Kirb 1837 family of social bees (Apoidea: Apocrita); robust, densely hairy; head narrower than thorax; antennæ geniculate and longer in males than in females; eyes elongate, ocelli in straight line on vertex; tongue long and pubescent, mandibles large; legs slender in males and stout in females; first tarsal segment with large corbicula or pollen basket; wings long with subequal submarginal cells - bumble or humble-bees (*Bombus lapidarius, B. pratorum* et al.).

Bombycidæ Leach 1819 family of moderate-sized, stout-boided densely hairy or scaly moths (Bombycoidea: Ditrysia); antennæ bipectinate in both sexes; proboscis wanting, frenulum vestigial or absent. Caterpillars smooth with anal horn, pupae in thick silken cocoons- Chinese silkworm (*Bombyx mori* Linnaeus) is reared artificially on mulberry leaves for silk production.

Bombycoidea Dyar 1902 superfamily of Ditrysia (Lepidoptera: Mecopteroidea); that includes silk moths characterised by wings having soft scales and small anal regions, venation well developed, Cu_2 absent from all wings and frenulum rudimentary or absent; abdomen without tympanum - Bombycidæ, Brahmæidæ, Lasiocampidæ, Saturniidæ, Thaumetipoidæ, Lemoniidæ, Oxytenidæ and Zanolidæ.

Bombykol sex attractant pheromone of female silk-worm moth (*Bombyx* spp., Bombycidæ).

Bombyliidæ Van der Wulp (1877) a dipteran family of Asiloidea (Orthorrhapha: Brachycera) comprising of beeflies; medium-sized, robust bee-like oftern variegated; ocelli present; proboscis very long and often projected forward; thorax humped; wings usually infuscated with reduced venation; squamæ present. Adults feed on nectar of flowers but larvæ are parasitic or predaceous on eggs of beetles and grass-hoppers, larvæ of bees and wasps and pupæ of dipteran flies (2000 species) - *Argyramoeba trifasciata* Meigen and *Bombylius major* Linnæus are found parasitising the solitary bees while in Africa *Vilia lloydi* Austin is parasitic on tsetse fly *Glossina morsitans* Westwood (Glossinidæ: Mscoidea) and in USA *Anthrazx edilitia* Say is parasitic on crysalis of cutworm.

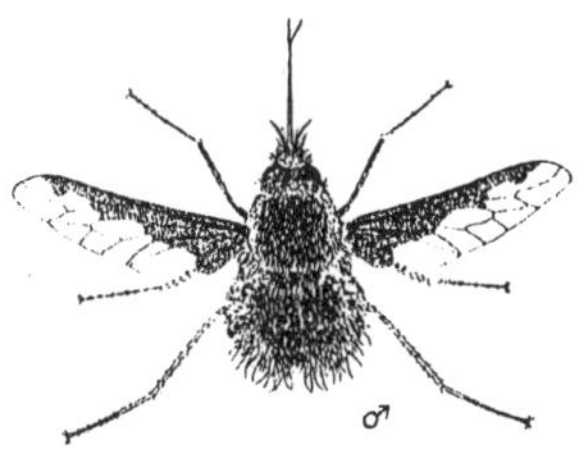

Bombylius major Linnæus

Boom spraying lance with several nozzles.

Boopiidæ Mjoberg 1910 small family of biting kangaroo-lice (Amblycera: Mallophaga); body clothed with stiff slender spines; antennæ capitate and 5-segemented; legs long and slender; tarsi with two claws; 13 species (4 genera), all confined to Australia, infesting kangaroos and wallabies, occasionally dogs as well - species of *Boopi* Piaget (8), *Latumcephalum* Le Souef (2) and *Heterodoxus* Le Souef and Bullen (2).

Borboridæ old and equivalent name for Sphæroceridæ (Cypselidæ) an Acalyptratæn family of Schizophora (Diptera) often abundant near decomposing organic matter.

Boreidæ Stephen 1829 family of carnivorous, snow scorpion-flies (Mecoptera) 2.5 to 3 mm long, having elongate beak-like head, no ocelli, rudimentary bristle-like wings in male and scale-like in female; legs extremely long and ovipositor conspicuous. Larva tiny, grub-like crescent shaped, hairless, white or creamish coloured - *Boreus hyemalis* (Linnæus), *B. unicoor* Hine, *B. wewtwoodi* Hagen, et al.

Bostrychiæ (Leach 1815, 1817) family of false powder-post beetles (Bostrychoidea: Polyphaga); elongate cylindrical body, head deflexed and covered with hood-shaped pronotum, antennæ 11-segmented with 3-jointed club; elytra smooth or sculptured, abdomen with five visible sterna of equal length; makes cylindrical burrows - *Dinoderus distinctus* Lesne is major pest of bamboo (*Bambusa* spp.); *Heterobostrychus equalis* Waterhouse and *Sinoxylon sudanicum* Lense are furniture beetles.

Bostrychoidea superfamily of Polyphaga (Coeloptera), terrestrial polyphagous, wood-boring beetles; integument hard, pronotum largely developed and hood-like; elytra covering most abdomen. Gurbs eruciform and short-legged. Pests of forest trees, bamboo and furniture; consist of for families - Anobiidæ, Bostrychidæ, ptinidæ and Lyctidæ.

Botanicals insecticides derived from plant material - nicotine ($C_{10}H_{14}N_2$) sulphate (highly toxic), rotenone, $C_{23}H_{22}O_6$ (toxic), pyrethrum (less toxic) etc.

Botfly common name for large hairy dipteran flies (Calyptratæ: Schizophora). Larva parasitic in mammals, living either in flesh or in internal cavities like head sinuses, throat or intesting - horse-botfly (*Gasterophius intestinalis*), nose-botfly (*G. haemorrhoidalis*) throat-botfly (*G. nasalis*) belong to family Gasterophilidæ while sheep-botfly (*Oestrus ovis*) and camel-botfly (*Cephalopsis titillator*) come under oestridæ (Techinoidea).

Bouten flabellum: small club-shaped enlargement or lobe at the tip of bee's proboscis (tongue).

Brachiopoda phylum of solitary, marine, bivalved coelomate (Trioploblasrica) animals, commonly known as lampshells; 350 species approtioned between two classes and five orders - Articulata (Rhynchonelida, Terebraulida and Thecideidida) and Inacrticulata (Lingulida and Acrotretida).

Brachy- short.

Brachycera Brauer 1880 suborder of Diptera; antennæ 3-segmented, shorter than thorax; arista if present, terminal; palpi one or 2-segmented projecting forward; forewing with discal cell, cubital cell closed or contracted before wing margin. Larva with head not well developed, retractile and with vertically biting mandibles. Three superfamilies - Tabanoidea, Asilodea and Epidoidea.

Brachycercidæ Lestage 1917 old and equivalent name of family Cænidæ (Bætoidea: Ephemerotera).

Brachycerous having short antennæ.

Brachyplatidæ old and equivalent name for Heteropteran family Plataspididæ (Pentatomoidea: Pentatomomorpha).

Brachpterous (insects) having short wings, that do not cover the abdomen fully.

Braconidæ Kirby 1837 family of tiny to small (2 to 10 mm long) stout-bodied wasps (Ichneumonoidea: Apocrita) that are attracted to flowers and light; antennæ not elbowed; eyes naked, ocelli three;

abdomen sessile, subsessile or petiolate; ovipositor as long as body; parasitic more commonly on lepidopterous insects - *Bracon brevicornis* Wasmann is effective parasite of coconut black-headed caterpillar (*Nephantis serinopa* Meyrick), *Chelonus narayani* Subha Rao and *Bassus* Species parasites caterpillars of *Helicoverpa armigera* (Hubner) while *Stenobracon deesæ* Cameron and *Apanteles flvipes* Cameron have been reported parasitising sugarcane borers.

Apanteles flavipes Cameron

Brahmæidæ very small family of tropical moths (Bombycoidea: Ditrysia); large sombre coloured with wing expense of about 150 mm; antenæ bipactinated, proboscis present, labial palpi large and upturned; wing pattern complex - *Brahmæa hearseyi* White, *B. wallichii* Gray, et al.

Brain insect's brain is a group of supra-esophageal ganglia situated on dorsal side of the collar of nerve fibres that surrounds oesophagus. It is formed by fusion of three pairs of segmental ganglia - protocerebrum, deutocerebrum and tritocerebrum - nerves from brain go to the principal sense organs of head while other connect the brain with rest of the **Central Nervous System** - a chain of ganglia running along the ventral part of insect's body.

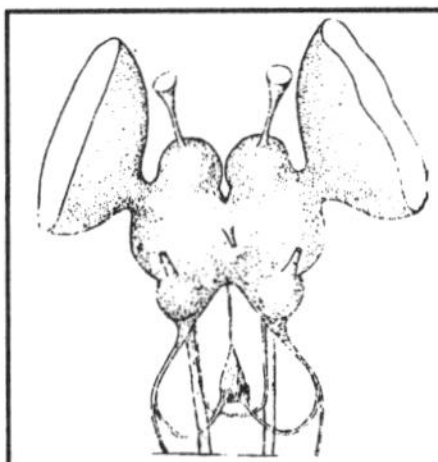
Brain (locust)

Brain hormone parathyrotropic hormone produced by pars intercerebralis of brain; the hormone stimulates thoracic glands to produce the moulting hormone or ecdysone.

Brassolidæ (Caligonidæ) a small family of Neotropical big butterflies (Papilionoidea: Ditrysia); having forewing with A_3 absent, discal cell of hind-wing closed and the under surface marked with eye-spots and intricate lines - *Caligo* spp. et al.

Brassolinæ subfamily of Nymphalidæ, often regarded as an independent family Brassolidæ (Papilionoidea: Ditrysia) that includes deeply and richly coloured big butterflies.

Brathinidæ a small family of grass-root beetles (Staphylinoidea: Polyphaga).

Braulidæ Egger 1853 monogeneric and monospecific family of bee-lice (Cyclorrhapha: Diptera); minute, compact hairy or spiny apterous insects having large head, short 2-segmented antennæ, short hypognathous mouthparts, vestigial eyes and no ocelli; legs large and hairy, tarsi 5-segmented and pulvilli pyriform. Adults ride from hive to hive on honeybees and are found in large number in weak and neglected hives; pass entire life (egg to adult) in beehives - *Braula coeca* Nitzch.

Breakdown of resistance inability of a plant variety or an animal species to maintain resistance when attacked by a newly selected insect biotype.

Brethid (=Brentidæ) family of Curculionoidea (Heteromera; Coleoptera); elongate liner body, usually bare, shining dull brown or ferruginous; antennæ 9 to 11-segmented, rarely clavate but never geniculate; head usually prolonged into horizontal rostrum, broad and small in males but long and slender in females; tarsi pilose below; found in decaying wood material (130 species) - *Ecotocemus cinnamonii, Eutrachelus* spp. et al.

Brindled beauty common name of a geometrid moth, *Lycia hirtaria* (Ditrysia: Lepidoptera) found in South England and Europe, having mottled brown colour well blended with trunk of the trees on which the moths rest during day. Caterpillars, stick-like, feed from May to July on apple, elm lime, pear, plum, willow and other fruit trees. Dark chestnut-brown pupæ overwinter in underground cells.

Brindled green nickname of a noctuid moth, *Dryobotodes eremita* (Dirtysia: Lepidoptera) found in oak woods of Britain and Europe; on wing at night specially during September October. Olive-green forewings of moths are variegated with black, brown and white; overwinters in egg stage; caterpillars feed from March to June on oak (*Quercus* spp.: buds and leaves and pupate in soil.

Brinjal leaf roller *Antoba* (*Eublemma*) *olivacea* (Walker) (Noctuidæ: Lepidoptera), medium-sized moth (wing span 28 to 5 mm) having forewings olive-green and hind wings whitish. Caterpillar purple-brown ornamented with yellow spots and hairs, folds the leaves from tip downwards and feed on the green matter within the roll.

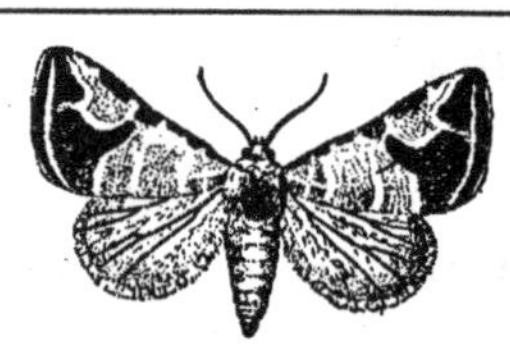

Antoba olivacea (Walker)

Brinjal leaf wing bug(s) *Urentius echinus* (Distant) and *U. sentis* (Distanct) (Tihgidæ: Cimicomorpha: Heteroptera); adult female oval and male more elongated in shape, pale ochraceous above and dark brown below; elytra reticulate consisting of irregular thick lines forming a frame work of cells; costal areas hyaline with strong spines on outer margin, hind-wings whitish and transparent. Specific pest of brinjal (*Solamum melongena* Linnæus).

Brinjal shoot and fruit borer *Leucinodes orbonalis* Guenee (Pyraustidæ: Ditrysia); medium-sized moth (22 to 26 mm wing span) having white forewings with conspicuous black and brown patches and dots, hind-wings opalescent with black dots along the margin. Caterpillar bores into fruit buds and fruits and feed within; the infested buds drop down while the fruits loose their market value.

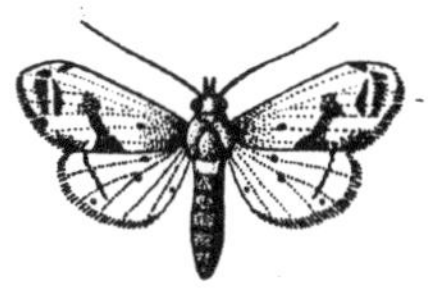

Leucinodes orbonalis Guenee

Brinjal stem borer *Euzophora perticella* Ragonot (Phycitidæ: Ditrysia); medium-sized moth (wing span 26 to 32 mm) having pale rufous forewing with a distinct dentate vertical black line beyond middle of the wing; hind-wing whitish. Caterpillar enters the stem and bores downwards; infested plants wither, growth is stunted and fruit yield adversely affected.

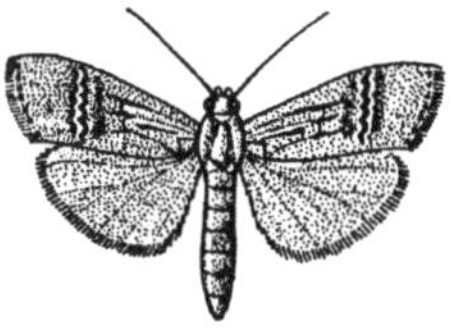

Euzophora perticella Ragonot

Bristle short stiff hair.

Bristletail(s) common name for many small, most primitive insets having spindle-shaped body; biting type mouthparts; long and filiform antennæ and multisegmented long cerci; in some a long median caudal filament is also present – two-pronged (Diplura) and three-pronged (Thysanura).

Broad-spectrum pesticide any pesticide that is effective against a number of unrelated pest species – DDT, phosphamidon, Bordeaux mixture, zineb etc.

Broadcast application application of a (soil) pesticide or fertiliser over an entire field rather than only along the rows or individual plants/ trees.

Brown fish a three-pronged bristletail, *Petrobius maritimus* Machiloidea: Thysanura); resembling silverfish but metallic brown in colour, living on rocky sea-shores above high tide mark; camouflaged with sea-weeds.

Brown planthopper *Nilaparvarta lugens* (Stal.) (Delphacidæ: Fulgoroidea: Homoptera); uniformaly brown in colour, male darker but slightly smaller than female. Both nymphs and adults feed at the base of rice plants sucking cell-sap, affected plants start drying and the symptoms are known as hopper-burn.

Bruchidæ Leach 1819 (Liariidæ Bedal 1891) family of small (5 to 7 mm) stout-bodied pulse-beetles (Chrysomelidæ: Polyphaga); dull grayish or brownish in colour; head hypognathous and produced into broad short snout anteriorly; elytra truncate, not covering the pygidium; legs short, femora thickened; abdomen thick-set. Gurbs feed exclusively on seeds of legumes and undergo hypermetamorphosis - *Callosbruchus chinensis* (Linnæus), *C. maculatus* (Fabricius), *Acanthoscelides obtectus* (Say), *Caryedon gonagra* Fabricius, et al.

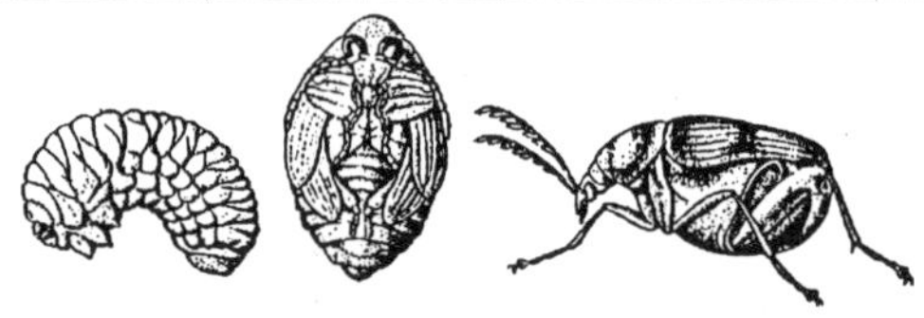

Grub Pupa Beetle

Bryozoa (Ectoprocta) phylum of moss animals (Coelomata: Triploblastica); sessile aquatic invertebrates with hardened exoskeleton, living in colonies About 4000 species, classified into three classes and four orders - Pylactolæmata (Plumatellida), Stenolæmata (Stanostomataæ or Cyclostomata) and Gymnolænata (Cheilostomata and Ctenostomata).

Buccal mouth or oral cavity of an animal including insect.

Bucculatigidæ old and equivalent name for lepidopterous family Lyohetiidæ (Tinæoidæ: Ditrysia).

Bucket sprayer a simple stirrup pump; fitted with a delivery pipe, lanse and nozzle. The pump is immersed in the pesticide kept in a bucket or similar container and while one person pumps the sprayer, the othe.: person holding the lanse does the spraying cheap and economical fora limited job; used in nerseries, kitchen gardens and small holdings, but not in orchards and big fields.

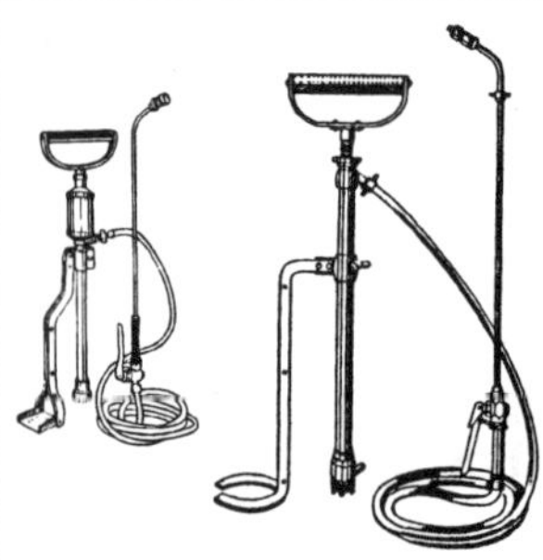

Bucket sprayers

Buckeye beautiful butterfly of North America, *Junonia cœnia* Hubner (Nymphalidæ: Ditrysia); having big eyespots on the wings.

Bug(s) any heteropteran insect characterised by very long proboscis adapted for piercing and sucking: name applied loosely to large number of small insects specially homoterans.

Bug, Alligator nickname given to American lanternfly *Fulgora lanternaria* (Fulgoridæ: Auchenorrhynch: Homoptera); on account of peculiar prolongation of its head; although believed to emit light, it is non-luminous.

Bug Ambush medium-sized, robust, slow-moving predaceous bugs (Phymatidæ: Gymnocertata) that inhabit flowers and foliage, lie in wait for other insects, including honeybees which they grasp with their strong forelegs and est alive - *Phymata monstrosa* Fabricius and *P. crassipes* Fabricius are confined to Europe while *P. erosa* (Linnæus) is cosmopolitan.

Bug, Aphid *Deræocoris ruber* (Linnæus) (Miridæ: Cimicoidea) of Europe, destroys noxious insects including aphids and is considered beneficial.

Bug, Arecanut *Carvalhoia arecæ* Miller (Mireidæ: Cimicoidea); pest of arecanut (*Areca catheca*) in South India. Both adults and nymphs suck sap from tender parts of arecanut palms; affected parts soon dry up leaving holes on the leaves and resulting in stunted growth and poor yield.

Bug, Assassin predators, (Reduviidæ: Gymnocerata); that suck blood of other species, animal and even man; posses short powerful proboscis with which they pierce their victim; some also transmit disease organism; make shrill sound by rubbing their proboscis against the presternum of thorax (3000 species) - *Acanthaspis rama* Distant, *Ectrychotes distpar* Reuter, *Pirates sanctus* Fabricius; species of *Rhodnius* and *Triatoma* transmit *Trypanosoma cruzi* that causes trypanosomiasis in human beings. Heath assasin bug *Coranus subapterus* is common in Britain feeding on insects and soldiers.

Bug, Bat (bat-fly) small, elongate, flattened insect (Polyctenidæ: Cimicoidea); ectoparasitic on bats; resemble bedbugs but lack eyes - *Polyctenes molossuc* Gigliolo occurs in India; *Hespeoctenes hermsi* Ferris and Usinger in North America, *H. impressues* Horvath in South America.

Bug, Beet common name for various heteropterous polyphagous plant-bugs (Piesamtidæ: Lygæoidea); small elongate-oval insects that damage and spread virus diseases of beetroot (*Beta vulgaris* Linnæus) - *Piesma quadratum* et al.

Bug, Bioculate predaceous bug, *Perillus bioculatus* (Fabricius) (Pentatomidæ: scutellerodea).

Bug, Black (Sugarcane) *Macropes excavatus* Distant, *Calocoris angustatus* Lethierrys (Lygæidæ: Lygæoidea: Pentatomomrpha); are sporadic pests of sugarcane, (*Saccharum officinarum* Linnæus).

Bug, Cashew(nut) *Paradasymus rostratus* Distant (Coreidæ: Coreoidea); the leaf footed, greenish-brown bug commonly found in Thiruvananthapuram in Kerala, causes shrivelling and drying up of tender cashewnuts (*Anacardium occidentale* Linnæus).

Bug, Chinch *Blissus gibbus* Fabricius (Lygæidæ: Pentatomomorpha); is common in India on grasses and sugarcane leaves, but not harmful. *B. leucopterus* (Say) is serious pest of cereals in North America and *Ischnodemus sabuleti* is found on grasses and weeds in England.

Bug, Copra (Ham) small, 3 to 5 mm long, *Necrobia rufipes* (de Geer) (Cleridæ: Polyphaga: Coleoptera); iridescent greenish-blue beetle with bases of antennæ and legs reddish, is pest of dried and stored products.

Bug, Damsel medium-sized, slender, predaceous bugs *Nabis ferus* Linnæus (Nabidæ: Gymnocerata) is dull-grey, about 8 mm long insect, that occurs throughout Northern hemisphere, while brachypterous, *N. apterus* Fabricius is palæarctic.

Bug, dusky cotton *Oxycrenus lactus* Kirby (Lygæidæ: Lygæoidea); attack open cotton bolls and sucks sap from the seeds; also stains the lint by getting crushed while ginning raw-cotton.

Bug, Earhead *Calocoris angustatus* Lethierry (Miridæ: Cimicoidea); small (about 8 mm long_), slender greenish-yellow insect is pest of sorghums and millets; feeds on green earheads when the grains are in milky stage, affecting adversely the yield of grain.

Bug, flower *Anthocoris nemorum* Linnæus and allied species (Anthocoridæ: Cimicoidea) are widely distributed from subarctic into palæarctic region; abundant in flowers and beneficial as these feed on aphids, leafhoppers, pysllids, scale insects, whiteflies etc., found in flowers. *A. kinigi* Brumpt sucks even human blood.

Bug, Gundhi paddy stink bug: *Polychrysa acuta* (Thunberg) (= *Leptocorisa varicornis* Fabricius) (Coreidæ: Coreoidea); notorious pest of paddy, appears at grain (milky) stage of the crop and sucks juice from soft grains; also emits peculiar foul smell – hence the name.

Polychrysa acuta (Thunberg)

Bug, Kissing small robust assassin bug, *Reduvius personatus* (Reduviidæ: Gymnocerata); predaceous mostly on bedbugs, named after newspaper account in 1899 mentioning that one of these bugs had pierced the lips of a lady (in England).

Bug, Lace small plant bugs (Tingidæ: Tingoidea); with beautiful, lacy pattern on body and hemelytra - *Monosteria minutula* Montandon, *M. globulifera* Walker, *Stephanitis typicus* (Distant) et al.

Bug, Litchi *Chrysocoris stolli* (Wolffenstein) (Scutelleridæ: Scutelleroidea) is pest of litchi (*Litchi chinensis*) in North Bihar, Myanmar, South China and Taiwan. Adult bug is conspicuous, being green in colour with six spots on thorax and seven on scutellum that covers the abdomen completely. Active almost round the year except during severe Winter.

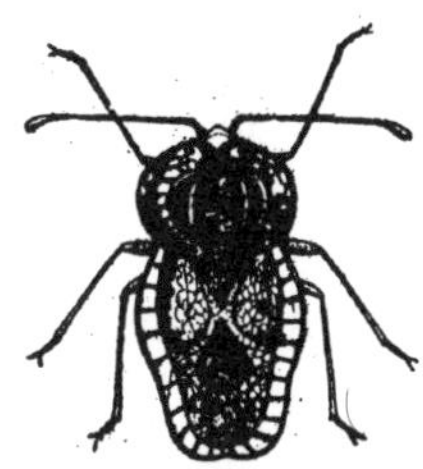

Monanthia globulifera Walker

Bug, Marine minute (< 33 mm long) *Æpophilus bonnairei* Signoret (Æpophilidæ: Gymnocerata); lives in crevices and under rocks and debris in and below intertidal areas.

Bug, Negro small shiny black species (Thyreocoridæ: Scutelleoidea); looking like small beetles, often found in large numbers in Spring flowers - *Thyreocoris* spp.

Bug, Painted small, about 6 mm long, pentatomid bugs (Gymnocrata: Heteroptera); characterised by predominant colour of adults and nymphs having conspicuous red and yellow markings all over the body - *Bagrada cruciferarum* Kirkaldy and *B. hilaris* (Burmeister) are pests of cole crops.

Bug, Peanut beautifully coloured magnificent non-luminous lanternflies *Laternaria* spp. (Fulgoridæ: Auchenorrhycha) of South America, having peculiarly or even grotesquely formed head that looks like peanut - *L. servillei* Spinola, *L. phosphorea* Linnæus, et al., the latter having wing span of about 155 mm.

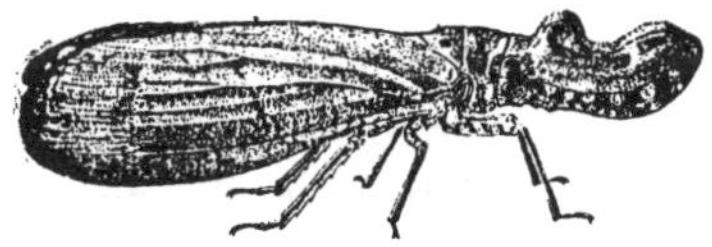

Latermaria servillei Spinola

Bug, Poultry *Hæmatosiphon inodorus* (Duges) (Cimicidæ: Cimicoidea), serious pest of poultry and a source of nusiance in Central America.

Bug, Red cotton cotton stainers, *Dysdercus koenigii* (Fabricius), *D. cingulatus* Fabricius and allied species (Pyrrhocoridæ: Pyrrhocoroidea); injurious to cotton, okra and other malveceous plants; nymphs and adults suck sap mostly from bolls (fruits) and leaves - *D. albidiventris* Stal is common is USA and *D. sidæ* Montrouzier in Australia.

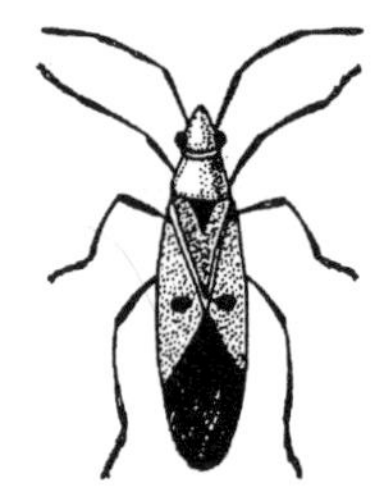

Dysdercus koenigi (Fabricius)

Bug, shield medium- to large-sized, broadly convex and oval bugs (Scutelleridæ: Pentatomorpha), scutellum very prominent looking like a shield - *Callipara imperialis* Fabricius, *Homæmus parvulus* (Germar), *Eurygaster alternatus* (Say) and *Scutellara nobilis* Fabricius.

Bug, (Jowar) shoot *Dicranotropis maidis* (Fulgoridæ: Homoptera), a pest of sorghum is widely distributed in Indian subcontinent; congregation and feeding of nymphs in top whorls of young plants, coupled with secretion of honeydew, results in rotting of plant parts.

But, Shore small, delicate, oval insects (Saldidæ: Aradoidea: Pentatomomorpha), predaceous upon aquatic and semi-aquatic insects and mites specially gnats and mosquitoes; living in bogs, marshes or along shores of streams, lakes or ponds - *Saldula saltatoria* (Linnæus), *S. orthochila* (Fieber) and allied species.

Bug, Squash *Anasa tristis* de Geer (Coreidæ: Pentatomomorpha); of America and *Gorizus hyoscyami* and allied species of Europe are pests of squashes and other gourd plants.

Bug, Stilt elongated, long-legged, sluggish plant bugs (Berytidæ: Lygæoidea); resembling cranefly - *Gampsocoris delhiensis* Menon and Ghai lives on growing tips of bottle gourd plant (*Lagenaria siceraria* (Molina) Standley).

Bug, Sunnhemp *Ragmus importunitas* Distant Miridæ: Cimicopidea: Geocorisæ) is pest of green-manuring crops specially sunnhemp. Both nymphs and adults suck sap from tender foliage resulting in stunted and malformed growth of leaves, tender buds and inflorescences.

Bug, Tea mosquito *Helopeltis theivora* Waterhouse (Miridæ: Geocarisæ); an active small, elongated bug having pale olive-greenish to black body, thorax bearing dorsally a characteristic knobbed process. Both nymphs and adults suck sap from young leaves, buds and tender shoots and inject toxic saliva therein; affected parts dry up, curl and become malformed.

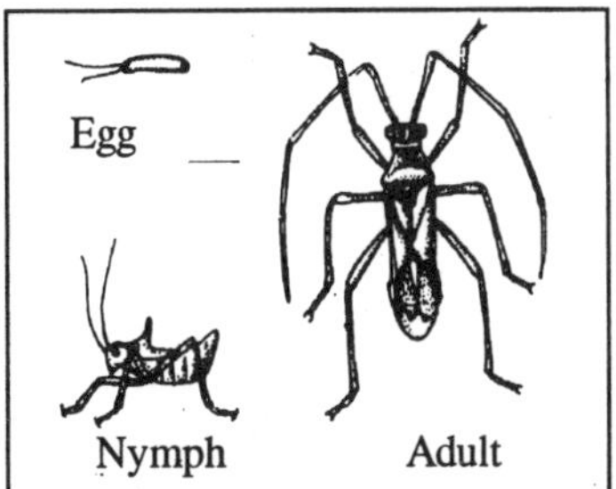

Helopeltis theivora

Bug, Toad small, rugose bugs (Gelastocoridæ: Cryptocerata); resembling miniature toads in appearance and habits - *Gelastocoris oculatus* Fabricius, *G. barberi* Torre-Bueno, *G. variegatus* Guerin and *Nerthra stygica* Say are some American species.

Bug, (Giant) Water largest heteropteran insects (Belostomatidæ: Cryptocerata), upto 150 mm in length; hind legs adapted for swimming; feed on young fish, tadpoles etc., and are attracted to light - *Lethocerus* spp., *Sphæropdema* spp., are common in India. Female of *S. molestum* Dufour forcibly attaches her eggs on dorsum of the male.

Bug, Water-cricket common Neotropical and Oriental regions (Veliidæ: Amphibicorisæ); inspite of its name, this bug, found, often gregariously, in streams and ponds, is silent; preys both by sight and by reacting to ripples *Velia caprai, Microvelia capitata et al.*

Bug, Water-scorpion small, conspicuous bugs (Nepidæ: Cryptocerata); with long tail, which is not a sting bug breathing tube; sting (bite) with beak and it is painful; found submerged in water of weedy stagnant ponds; adults poor swimmers; feed on insects, tadpoles and small fish - *Nepa cinerea* Linnæus (European), *Ranatra quadride* and *R. americana* Montandon (North America) et al.

Buprestidæ Leach 1815, 1819 family of injurious beetles (Buprestoidea: Polyphaga); conspicuous metallic insects having 11-segmented short serrate antennæ; thorax abdomen firmly united and five visible sterna, first two connate, rest stem-borer (*Sphenoptera gossypii* Cotes), Mango stem borer (*Belionota prasina* Thunberg), okra leaf-miner (*Trachys herilla* Obenbeger) et al.

Buprestoidea monofamilial superfamily of Polyphaga (Coleoptera: Coleopteroidea); common in Tropics; varying from 5 mm long delicate insects to 25 mm long robust beetles; integument hard and strong head partly sunk in thorax; wings large and functional in flight; main distinguishing character is presence of a prosternal process extending back into a mesosternal cavity – Buprestidæ.

Bursa copulatrix copulatory pouch of female.

Bush killer common name for various herbicides or weedicides.

Butalidæ old and equivalent name for lepidoteran family Scythrididæ (Yponomeutoidea; Ditrysia).

Butane C_4H_{10}, hydrocarbon of methane series; colourless gas, used as portable supply of fuel.

Butterflies in one's stomach to feel nervous: nervous tremors in one's stomach.

Butterfly lepidopterous insect (Papilionoidea) mostly diurnal, characterised by slender body, broad colourful wings, body and wings covered with scales; antennæ clubbed and wings held vertically above the body when in repose – *Catopsialia crocale* Fabricius and *Junonia almana* (Linnæus) are minor pests of paddy; *Eurema blanda silhetana* Wallace defoliates shade trees while *Euthalia garuda* feeds on mango leaves.

Butterfly, Alfalfa *Colias aurytheme* (Biosduval) (Pieridæ: Paplionoidea); yellow polymorphic species of various shades of pale yellow to bright orange-yellow, with four median orange spots and dusky outer wing margins; wing expanse 40 to 60 mm; common and abundant throughout North America.

Butterfly, Apollo group of alpine (mountain welling) butterflies (Parnassidæ: Paplionoidea), related to swallotails; found in Asia, Europe and North America - *Parnassius apollo* (Linnæus), *P. mnemosyne* (Linnæus) and *P. phoebus* (Fabricius); Alaskan species *P. eversmanni* Menetries is sexually dichromatic (♂ yellowish, ♀ whitish).

Butterfly, Ber *Taraucus theophrastus* (Fabricius) (Lycænidæ: Papilionidæ), dark brown butterfly having beautifully patterned wings (wings expanse 20 to 30 mm). Caterpillar fleshy and flattened with pale ochreous head and green body feeds on leaves of *Zizyphus* spp. and *Citrus* spp.

Butterfly, Bird-wing most magnificent, brilliantly coloured butterflies (Papilionidæ: Ditrysia) of South-east Asia; forewings narrow and very large in comparison to small hind-wings – *Troidea amphrysus* (Cramer), *T. brookiana* (Wallengren), *T. primas* (Linnæus), *T. alexandræ* of New Guinea is the biggest butterfly in the world with wing expanse of 300 mm.

Butterfly, Blue small swift-flying butterflies (Lycænidæ: Paplionoide), characterised by upper surface of wings metallic blue specially in males, females often differently coloured. Caterpillars resemble woodlice, secrete sweet liquid much liked by ants - small blue (*Cupido minimus*), large blue (*Maculina arion*), spring azure (*Lycæna pseudargolus* Boisduval and Leconte).

Butterfly, Brush-footed brightly coloured large butterflies (Nymphalidæ: Papilionoidea) characterised by short tibia clothed with long hair - *Danais plexippus* Linnæus (Danainæ), *Mycalesis perseis* Fabricius (Satyrinæ), *Telchinia violæ* (Fabricius) (Morphinæ) et al.

Butterfly, Camberwell beauty *Nymphalis antiopa* (Linnæus) (Nymphalidæ: Papilionoidea) brownish-black butterfly with yellow win borders; familiar in Holarctic region. Caterpillars feed on elm (*Ulmus wallichiana* and allied species) and willow (*Salix* spp., Salicaceæ); hibernate in adult stage.

Butterfly, Dog *Colias cæsonia* (Pieridæ: Paplionoidea) found in Southern states of North America; having black and yellow wings; colour pattern on forewings resemble side view of dog's head.

Butterfly, Leaf Oriental butterflies *Kallima* spp. (Nymphalidæ: Papilionoidea), mimicking old dry leaves; upper surface of wings is as usual beautifully coloured but the undersurface (wings folded above he body), resemble dead and dry leaf - *K. inachis* is Indian leaf butterfly and *K. phiarchus* is from Sri Lanka.

Kallima species

Butterfly Lemon (Citrus) large conspicuously coloured swallowtails, *Papilio* spp. (Papilionidæ Ditrysia), wings ornamented with yellow and black markings; wing expanse 80 to 100 mm. Eggs laid on dorsal surface of leaves of *Citrus* trees; young caterpillars look like bird's excreta; full grown ones are green with brownish side stripes; feed voraceously on leaves, leaving behind only midribs; fond throughout Indian subcontinent. *P. demoleus* Linnæus is more common.

Butterfly, Marbled White *Melanargia galathea* (Satyridæ: Paplionoidea); rare butterfly, quite different from others of its family; found occasionally in England and Wales, confined to meadows with tall grasses. Once a completely black and a completely white speciment were caught and auctioned for £110/ - in 1943.

Butterfly, Migratory a large number of butterflies (Danaidæ: Paplionoidea) are known to migrate from mountainous regions to plains and vice-versa or from temperate regions to tropics and back – Indian migrants include, *Catopsilia crocale* Cramer, *C. pyranthus* Linnæus, *Colias fieldi* Menetries, *Danaus genutia* Cramer, *D. septerntrionis* Butler, *Euploea core* Cramer, *Lampides boeticus* Linnæus, *Venessa cardni* Cramer et al.

Butterfly, Milkweed cosmopolitan, large, beautiful, orange butterfly with black and white markings, *Danaus menippe* (Hubner) (= *archipus* Linnæus, *plexippus* Fabricius); wing expanse 90 to 10 mm, fond throughout the year; lays eggs on leaves of milkweed (*Asclepias* spp.).

Butterfly, Monarch migratory butterfly of North America, *Danaus manippe* (Hubner) (= *D. plexippus* Fabricius) (Daninæ: Nymphalidæ); orange-coloured with black and white markings, having wing expanse of 90 to 100 mm; flies in large numbers each Autumn from Canada to Florida and Mexico (about 2400 km) and back home in Spring.

Butterfly Painted lady large and beautifully coloured migratory butterflies (Nymphalidæ: Lepidoptera) – *Vanessa cardui* (Linnæus) is cosmopolitan, whereas *V. indica* Herbet is Indian species.

Butterfly, Peacock *Vanessa (Nymphalis) io* (Linnæus) (Nymphalidæ: Lepidoptera) of Europe, having beautifully patterned wings with four conspicuous eye spots (one on each wing); Lepidoptera appears in July and remains on wings, until they hibernate in Winter.

Butterfly, Pomegranate *Virachola isocrates* (Fabricius) (Lycænidæ: Lepidoptera), glossy bluish-violet (♂) to brownish-violet (♀) in colour; female having conspicuous orange patch on forewings; hind-wings in both sexes with tail-like prolongation; polyphagous pest active throughout the year. Caterpillars bore into various fruits specially pomegranate (*Punica granatum* Linnæus).

Butterfly, Purple Emperor *Apatura iris* (Linnæus) Nymphalidæ: Lepidoptera); one of the largest and most magnificent butterflies of Britain having lovely iridescent, purple sheen in male's upper wings; powerful fliers, confined to oak-woods. Overwinter i n caterpillar stage (September to March) and pupate in June.

Butterfly, Rice dusky butterflies having peculiar habit of suddenly settling with closed wings and tilting slightly; Asian species include, *Mycelesis blasius* Fabricius, *M. leda* (Linnæus) and *M. perseus* Fabricius (Satyridæ: Papilionoidea) as pests of paddy.

Butterfly, Swallowtail conspicuously coloured butterflies (Papilionidæ: Lepidoptera), powerful fliers, most species having hind-wings with tail-like prolongation - *Papilio* spp., *Tros aristolochiæ* Fabricius, *Zetides agamemnon* Linnæus, and many more.

Butterfly, White Admiral misnomer for 'admirable' - a beautiful butterfly with blackish-brown and white upper wings and orange-brown underwings, spotted and banded with white; *Limenitis camilla* (Nymphalidæ: Lepidoptera), confined to woods of southern England, flies gracefully around bramble blossoms during July-August. Caterpillars overwinter in honeysuckle leaves.

Butterfly, White mountain *Oeneis melissa* Fabricius (Satyridæ: lepidoptera), marble dark brown and pale gray in colour, wing-span 40 to 45 mm; inhabits high mountains of Northen Europe.

Butterfly, Yellow *Colias phiodice eurytheme* Boisduval (Pieridæ: Lepidoptera), of USA and Southern Canada. Adults has orange-yellow wings with black outer borders, a black spot in centre of forewings and two reddish-orange spots on each hind-wing, in females the black border has yellow spots; wing span about 50 mm. Caterpillars fed on alfalfa (*Medicago sativa* Linnæus), hence also called alfalfa butterfly.

Butterfly-archis: not an insect but any orchid (*Orchis* species, Orchidceæ), with flowers resembling a butterfly.

Butterfly-bow a bow whose loops and ends are spread like butterfly's wings.

Butterfly-fish: a slimy fish *Blennius* species (Chætodontidæ: Acanthopterygia) having an eye-spot on dorsal fin.

Butterfly-flower any flower adapted for pollination by butterflies.

Butterfly-nut (-screw) a nut or screw turned by winged finger-grips.

Butterfly- (breast-) stroke a swimming-stroke, performed on the breast, the arms working simultaneously with an overarm action.

Butterfly-weed pleurisy-root (*Asclepias* species; Asclepiadaceæ); used as diaphoretic and expectorant.

Butterwrot small insect-eating herb, *Pinguicula alpina* (Lentibulariaceæ); having cup-shaped violet flowers and rosette of fleshy sticky leaves covered with numerous glands; when an insect alights on any leaf for feeding or relaxing, the leaf curls and rolls in wards trapping the insect it unrolls agains after the prey has been digested and assimilated.

Pinguicula alpina

Byke nest of wasps or wild bees.

Byrrhidæ Erichson 1846 family of Byrrhoidea (Polyphaga: Coleoptera); very small, black, hemispherical beetles having head deflexed with no visible clypeus; antennæ gradually thickened to apex; legs stout, front coxæ transverse, subcylindrical and open behind; tibia dilated tibial spurs distinct, last tarsal joint small and quadrate. Occurs in moss and under bark, clods etc. (500 species) - *Amphicyrta dentipes* Erichson of USA, *Byrrhus gigas* of Europe and North America, *Syncalypta curimoides* Champion of India.

Byrrhoidea superfamily of Polyphaga (Coleoptera: Coleopteroidea); that comprises of pill-beetles, having head with no distinct frontoclypeal sulcus and mid coxæ widely separated and hind coxæ close together with broad femoral plates - Byrrhidæ and Georyssidæ.

Bythoscopidæ small family of Cicadoidea (Homoptera); robust insects characterised by having entirely membranous wings; found in forests of deciduous trees and shurbs.

Byturidæ Thomson 1859 small family of Clavicornia (Cucujoidea; Plyphaga); that includes raspberry fruitworms; oval pubescent beetles related to Dermestidæ and differs from Byrrnidæ in having front coxæ globular or conical with its cavities closed behind; tarsi 5-5-5 with toothed claws - *Byturus æstivus* (Linnæus) and *B. tomentosa* (Fabricius) of Europe and *B. unicolor* Say of North America are major pests of raspberry (*Rubus idæus* Linnæus) and allied fruits. Adults feed on blossoms and grubs in fruits.

C

Cabbage a winter vegetable, *Brassica oleracea capitata* Linnæus (Brassicaceæ: Magnoliopsida), grown all over the World; used as salad as also as cooked vegetable.

Cabbage borer *Hellula undalis* Fabricius (Pyraustidæ: Pyralidoidea) is cosmopolitan in distribution. Freshly hatched caterpillar mines the leaf along the side of veins and makes it a white papery structure full of excreta caterpillar of 3rd instar onwards feed on the leaves, then bore inside the stem, leaf petiole or cabbage and cauliflower head. Moth has greyish-brown forewings with wavy grey markings and pale dusky hind-wings; wing span 60 to 70 mm. Life-cycle 2-3 weeks.

Cabbage butterfly (large whites) group of oligophagous pests, *Pieris* spp. (Pieridæ: Papilionoidea) having snow-white wings with black distal margins and apical spots; found in Temperate regions as a serious pest of cruciferous crops. In India the pest passes Winter in the plains and migrate to hilly regions during Summer – *P. brassicæ* (Linnæus, *P. canidia* (Sparrman), *P. napi* (Linnæus), *P. rapæ* (Linnaus) et al.

Cabbage butterflies

Cabbage flea beetle *Phylloterta cruciferæ* Goeze and allied species (Alticidæ: Chrysomeloidea: Polyphaga); small (< 2 mm long), elongated, metallic bluish-green beetle; very active, often seen hopping from plant to plant; a major pest of seedlings; grubs feed on tender roots while adults cut holes in tender leaves.

Cabbage gall-weevil a small weevil *Ceuthorrhynchus pleurostigma* (Curulioniadæ: Curculioniodea) that lays its eggs on roots of cole crops. On hatching the larva feeds on the roots causing the formation of large round galls (not found in India).

Cabbage leaf weber *Crocidolomia binotalis* Zeller (Pyraustidæ: Pyralidoidea); moths pale ochreous; forewings suffused with black lines, specks and dark edge white spots; hind-wings whitish. Caterpillars about 20 mm long and greenish in colour; feed gregariously for couple of days on chloropyll of leaves, then segregate, spin webs on ventral surface of leaves and fed therein on leaf-tissues; also feed on flower buds and bore into fruits (cabbage head, cauliflower or mustard pods).

Cabbage maggot (cabbage root fly) larva of small fly, *Hylemya brassice* (Anthomyiidæ: Muscoidea), native to Europe and found all over the World Eggs laid on cruciferous plants close to the ground; on hatching, the maggots feed on stem and main root just below the ground level and pupate in the soil.

Cabbage moth a noctuid moth, *Mamestra brassicæ* (Noctuoidea: Ditrysia); grayish brown with whitish or cream coloured markings. Caterpillar feeds on leaves of cabbage and other cole crops (not intercepted from India).

Cabbage semi-looper caterpillar of a common noctuid moth, *Trichoplusia ni* (Noctuoidea: Ditrysia); found all over India as pest of cabbage, cauliflower, safflower, opium poppy, tomato etc. Adult moth stout, brown coloured with gray forewings having light wavy white lines. Life-cycle about a month.

Cabbage whitefly (cabbage snowfly) *Aleurodes brassicæ* (Aleyrodidæ: Aleyrodoidea); body and wings covered wit white powdery wax. Eggs laid on cabbage leaves; on hatching, nymphs suck the plant sap and exude hoeydew (not reported from India).

Cabbage worm caterpillar of cabbage butterfly, *Pieris* spp. (Pieridæ: Papilionoidea); oligophagous pest that feeds on leaves of cabbage, cauliflower and allied species (cole crops) causing enormous damage. Full grown caterpillar 38 to 44 mm. long, velvety bluish-green in colour with black dots and yellow dorsal and lateral stripes covered with white hairs.

Caddis-worm aquatic larva of caddisfly (Trichoptera: Mecopteroidea).

Cæciilidæ Kolbe 1884 family of foliage-haunting psocids (Psocomorpha: Psocoptera); cosmopolitan in distribution; distinguished by tarsi 2-segmented and claws without teeth - *Caecilius aurantiaxcus* Hagen, *Lachesilla pedicularia* (Linnæus) et al.

Cælifera suborder of Orthoptera characterised by short antennae having less than 30 segments; tympanal organs, when present situated at the base of abdomen; comprises of two superfamilies - Acridoidea and Tridactyloidea.

Cænidæ Klapalek 1909 family of mayflies (Cænoidea: Ephemeroptera); having forewings milky with few cross-venis and fringed hind margins; no hind-wings; three caudal filaments; male claspers horn-shaped and one-segmented - *Brachycerus* spp., *Cænis* sp. et al.

Calamoceratidæ McLachlan 1877: a small family of caddisflies (Trichoptera: Mecopteroidea); characterised by lack of ocelli and the principal fork of the median vain in forewing with a closed cell. Larvæ inhabit still or flowing streams and construct cylindrical or subcylindrical cases lined with sand, pebbles or debris - *Ganonema americana* Walker, et al.

Caligonidæ old and equivalent term for lepidopterous family Brassolidæ (Papilionoidea: Ditrysia) that includes butterflies.

Callicertatidæ (Ceraphronidæ) small family of very of very small, slender wasps (Serphoidea: Apocrita); characterised by having short pointed ovipositor at the anal end of abdomen; parasitic on other insects.

Calliduliadæ family of day-flying moths (Calliduloidea: Ditrysia); having simple antenæ; long palpi; tympanum not developed and hind-wings fringed; found in Indo-Malaysian region.

Calliduloidea small superfamily of diurnal moths (Ditrysia: Lepidopteera); found in the Oriental and African regions; antennæ not pectinated; forewings with M_2 basaly nearer M_3; comprises of only two families - Callidulidæ and Pterothysanidæ.

Callimorphidæ old name of Hypsidæ (Noctuidea: Lepidoptera).

Callipharixenidæ Pierce 1918 small family of Corioxenoidea (Strepsiptera: Coloeopteroidea); male having 5-segmented antennæ, 3rd and 4th segments being laterally flabellate; female having cephalothorax greatly enlarged and two pairs of spiracles; endoparasitic on pentatomid and scutellerid bugs; reported from Africa and Thailand - *Callipharixenos* spp., and *Corioxenos* spp.

Calliphoridæ Brauer 1889 small family of blue-bottles, green-bottles, etc (Tachinoidea: Calyptratæ); look like houseflies; body usually pollinose, proboscis fleshy; hypopleural bristles present; arista strongly plumose. Larvæ saprophagous, flesh feeders or parasitic - *Calliphora vicina* Robineau-Desvoidy and *C. erythrocephala* (Meigen) are common blue-bottles; *Lucilia sericata* (Meigne) and *L. cuprina* are known as green-bottle flies.

Callipterina group or subtribe of Aphidinæ (Aphididæ: Homoptera) in which fundatrigeniæ (form of aphids) are absent as in species of *Callipterus, Drepanaphis, Myzocallis* et al.

Callirhipidæ very small family of beetles (Artematopoidea: Polyphaga) essentially Tropical in distribution; antennæ pectinate or flabellate; tarsi without membranous lobes below and elytra without a hook-like process beneath the apex (100 species) - *Callorhipis* spp. are common in India.

Calistopteriadæ small family of book-lice (Psocomorpha: Psocoptera) characterised by tarsi 2-segmented and hind wings with forked media – only one specie found in New Guinea.

Callow (teneral) freshly emerged adult from pupa specially of worker ant (Formicoidea; Hymenmoptera), soft-bodied and not fully coloured.

Calobatidæ small family of flies (Acalyptratæ: Schizophora); characterised by having arista dorsal, propleura bare in the middle, oral vibrissae absent and legs long and slender.

Caloneurodea an order of fossil insects of Upper Carboniferous to Permian periods; antennæ long and slender without pronotal lobes; wings membranous, homonomous with numerous cross-veins; hind legs longer than fore- and mid-legs; cerci short. The important families included are – Caloneuridæ, Synomaoptilidæ, Euthygrammidæ, Paleuthygrammidæ, Permobiellidæ, Pleisiongrammidæ and Anomalogrammidæ.

Calopterygidæ old and equivallent name for family Agriidæ (Zygoptera: Odonata) that comprises of damselflies.

Calopteryx common name of some damselflies, *Agrioj* spp. (Agriidæ: Zygoptera).

Calotermitidæ enderlein 1909 old and equivalent name for termite family Kalotermitidæ (Isoptera).

Calypter(es) squamæ: membranous lobe-like structures at posterior angle of wings, covering the halteres of certain dipteran flies; well developed in houseflies.

Calyptomeridæ old and obsolete term for a small family that was placed under superfamily Sphærioidea (Myxophaga: Coleoptera).

Calyptratæ Macquart 1883 large group of dipteran insects (Schizophorea: Cyclorrhapha); characterised by having second antennal segment with distinct external groove above, theca developed at the base of proboscis; sub-costal vein complete and squamæ well developed; comprises of three super families – Muscoidea, Protachinoidea and Tachnoidea.

Camera lucida an optical drawing device with a mirror that works as a prism. The accessory when attached near the eye-piece of a microscope, reflects the image of the slide under study and projects the same on the paper kept adjoining the microscope; thereby facilitating in tracing the picture of the object on the slide.

Camillidæ very small dipteran family (Acalyptratæ: Schizoptera) of no economic importance.

Camouflage cryptic colouration, one of the most effective method of defence specially for insects which are sought after as food – close resemblance to natural surroundings is the most common form of camouflage.

Campodeidæ Westwood 1873 family of dipluran insects (Campodeoidea: Rhabdura); having delicate, slender flattened body with long conspicuous antennæ; first abdominal segment without styli and with lobe-like appendages; cerci many jointed and imperforate apically; found in damp decaying vegetation – *Campodea staphlinus* Westwood, *C. montgomeryi* Silvestri, *C. philpotti* Tillyard, *Lepidocampa weberi* et al.

Campodeoidea superfamily of Rhabdura (Diplura: Diplurata); slender primitive insects, antennæ with trichobothria from 3rd segment onwards, styles on II to VII abdominal segments and no abdominal spiracles; cerei long slender and closed apically, divided into two families – Campodeidæ and Procampodeidæ.

Caneaciedæ family of dipteran seashore flies (Acalytratae: Schizophora); characterised by having ocellar triangle large and shining, postocellar bristles divergent or absent and anal cell present.

Cannibal one who eats flesh of its own species.

Cannibalism practice of eating one's won species – caterpillars of gram pod-borer, *Helicoverpa armigera* (Hübner) (Noctuidæ: Ditrysia).

Cantharidæ Heyden, Rreitter and Wiese 1883 large family of leather-winged or slodier beetles (Cantharoidea; Polyphaga); head fully exposed, epipleura narrow at the base; elytra soft and not reticulate; no phosphorescent organs; marked sexual differences, female wingless (1400 species) – *Dasytes* spp., *Podabrs* spp., *Thagoncha* spp. *Silis* spp. et al.

Cantharidiformia old name for superfamily Cantharoidea (Polyphaga: Coleoptera).

Cantharoidea Ritter 1906 (Cantharidifomia; Malacodermata) superfamily of Polyphaga (Coleoptera: Coleopteroidea); beetles having soft integument with various parts loosely articulated; grubs predaceous, suck body juice of the victim.

Capniidæ Klapalek 1905 family of Filipalpian stoneflies (Plecoptera) found in Northern hemisphere; wing-venation and anal lobe of hind-wing reduced, distal cross-veins lacking; cerci long and multi-articulate (200 species) – *Allocapnia* spp. *Capnia* spp. et al. *Notonemura* spp. and its allies found in sounthern hemisphere have been grouped under separate family, Notonemuridæ.

Carabidæ Leach 1815, 1817 large, widely distributed family of hard and heavily armoured elongate and flattered, swift running, ground beetles (Caraboidea: Adephaga); head free, prognathous and narrower than thorax, antennæ 11-segmented, mouthparts well developed; clypeus not extending beyond base of antenæ; mandibles and maxillae large and strong; legs long, slender and cursorial (2500 spicies) - *Anthia sexguttata* and *Selina westermanni* are common in India. *Amara brucci* is found at 5300 metres height on the Himalayas and *Bembidion vivicola* occurs at 500 metres amidst ice and snow on Mt. Everest (Nepal). *Illaphanus stephensi* Macleay of Australlia and *Tachus lævis* Say of North America are about one mm long wheels *Carabus gigas* Creutz of Europe, *Catadromus elseyi* White and *Hyperion schroetteri* Schreiblers of Australia and *Tefflus dispar* Sternberg of West Africa are 45 to 55 mm long.

Caraboidea Leng 1920 superfamily of Adephaga (Coleoptera: Coleopteroidea); beetles having antennæ long and filiform; eyes entire; tarsi 5-segmented and abdomen with 6 or more visible sternites. Mostly predaceous, a few have become secondarily vegararian; found under stones, clods, bark, rotten wood etc., as also snow capped mountains. Important families include, Carabidæ, Cicindelidæ, Amphizoidæ, Halipidæ, Dytiscidæ and Omophronidæ.

Carbaryl $C_{12}H_{11}NO_3$; contact carbamate groups of insecticide with slight systemic action, colourless crystalline solid, slightly soluble in water but readily soluble in most organic solvents, compatible with all but highly alkaline pesticides, like lime sulphur and Bordeaux mixture. Oral and dermal LD_{50} values for rats are 400 and >500 mg/kg respectively - Sevin® etc.

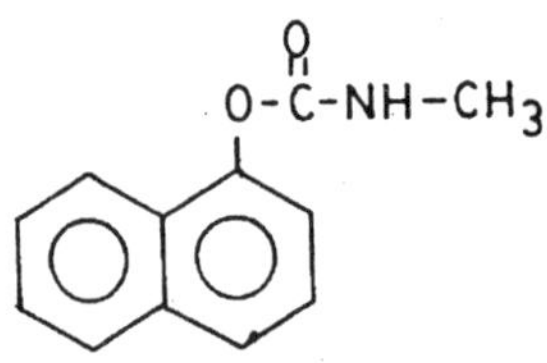

Carbaryl

Carbofuran $C_{12}H_{15}NO_3$; colourless crystalline solid; effective against soil-inhabiting pests as also thrips, shoot-flies, mites etc. Acute oral LD_{50} to rats 8 to 14 and dermal LD_{50} to rabbits 10200 mg/kg - Furadan® etc.

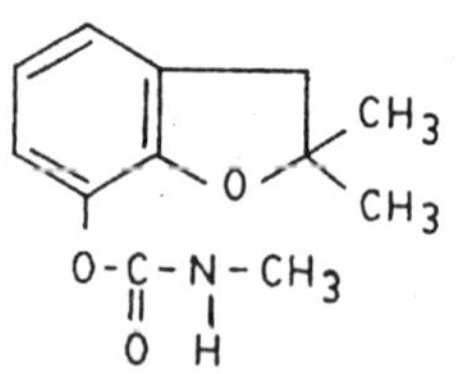

Carbofuran

Carbon bisuplphide (dislphide) CS_2; colourless liquid with faint disagreeable smell of hydrogen sulphide; bareley soluble in water and readily soluble in various organic solvents; used as fumigant for control of stored-grain pests in godowns.

Cardia anterior part of ventriculus.

Cardiac pertaining to heart.

Cardiac sinus channel of haemocoele, dorsal to alimentary canal.

Cardioblast cells of upper edges of embryonic mesoderm that form dorsal blood vessel.

Cardo Proximal subdivision of maxilliary appendage which articulates with the heads.

Carnivore (carnivorous) organism (plant or animal) that feeds on living animals.,

Carnosauria group of giant flesh-eating dinosaurs (Theropisda: Saurischia: Archosauria); lived during Cretaceous period (Mexozoic era), grew upto 15 metres and weighed about 4 quintals; head measured about one mtere having wide gaping jaws with sharp stabbing teeth; forelimbs much reduced, the reptile walked on its hind limbs, each bearing three forward toes and one backward; preyed upon herbivorous dinosaurs (Sauropoda) – *Tyranosarus* species.

Caropodite fifth segment, generalised appendae of arthopoda: tibia in insects.

Carposinidæ a microlepidoteran family (Copromorphoidea: Ditrysia); characterised by forewing having raised scale tufts and hind tibia with long hair-like scales. Found in Australia and Hawaii – *Meridarchus reprobaa, M. scyrodes* et al.

Carrier material serving as diluent and vehicle for active ingredients, usually in dust formulations talc, China-clay etc.

Cassididæ a small family of Chrysomeloidea (Polyphaga: Coleoptera); small oval shaped beetles characterised by body and elytra broad enlarged; head concealed under prothorax; antennæ short. Both grubs and adults phytophagous, scrapping and biting holes in leaves of various plans.

Cassidinæ subfamily of Chrysomelidæ, often regarded as an independent family of Chrysomeloidea, includes tortoise beetles having flattened shield-like body with extremely brilliant colouration; many are pests of economic crops – *Aspidomorpha miliaris* Fabricius, *Chirida sexnotata* Fabricius and *Metriona circumdata* Herbest infest sweet potato (*Ipomoea batatas* (L) Lamk), *Cassidea exilis* Baly attack amaranthus leaves and *Oocassida obscurea* (Fabricius) feeds on ber (*Zizyphus* spp.) leaves.

Castniidæ a small family of Castnioidea (Ditrysia: Leidoptera); that includes about 180 species of brightly coloured day-flying moths, confined to Tropical America, Australian and Indo-Malayan regions – *Castnia licus* is pest of sugarcane in America, *Synemon* species bore and feed inside the roots of sedge (*Carex* spp., cyperaceae) in Australasia.

Castnioidea Handlirch 1925 a superfamily of Ditrysia (Lepidoptera: Mecopteriodea); diurnal moths having antennæ clavate; maxillary palpi present, tympanal organs wanting; tibial spurs present; forewing with Cu_2 but hind-wing with or without Cu_2 – Castniidæ and Tascinidæ.

Castor capsule borer *Dichocrocis punctiferalis* (Guenee) (Pyraustidæ: Ditrysia); medium-sized, brownish-yellow moth with numerous black dots on the wings; wing-span 40-50 mm. Caterpillar bores into fruits, bud or shoot and feeds within.

Catalyst substance that accelerates chemical reaction without becoming part end product.

Category rank or level in hierarchic classification.

Caterpillar lepidopterous larva – polypod or urceiform – having well developed head; three thoracic and 10 evident abdominal segments, nine pairs of spiracles borne on prothoracic and first eight abdominal segments. Thorax bears three pairs of jointed (true) legs, one on each segment and abdomen usually has five pairs of (false) prolegs situated on III, IV, V, VI, and X segment. Caterpillars of family Geometridæ have only two pairs of prolegs (loopers) on VI and X abdominal segment while those of Plusiinæ (Noctuidæ) are semiloopers having first or first and second pairs of prolegs aborted. The caterpillars of *Phyllocnistris* spp. (Gracillariidæ) are apodous. Larvæ of sawflies (Tenthredinoidea: Hymenoptera) are also called caterpillars but these are not true caterpillars. All caterpillars are herbivorous, majority begin voracious leaf-defoliators.

Caterpillar, Bihar hairy *Spilarctia* (*Spilosoma*) *obliqua* (Walker) (Arctiidæ: Noctuoiea); full grown caterpillar stout, 30 to 40 mm long, have seven orange coloured transverse bands with tufts of yellow hair that are black at both ends. Moths have crimson coloured body with black dots; antennæ black pectinate in males and filform in females; wings pinkish buff-coloured with numerous black spots, wing span 40 to 50 mm. Notorious pest of jute and large number of other crops.

Caterpillar, Black hairy *Estigmene lactnea* (Cramer) (Arctiidæ: Noctuoidea); black in colour and densely hairy caterpillars are voracious feeders and polyphagous pest of various economic crops. Adult large white moth with crimson margins on head body and wings.

Caterpillar, Brown hairy *Cricula trifenestrata* Helfer (Saturniidæ: Ditrysia); stout, smooth, dark brown, wild-silk caterpillars about 50 mm long when full grown, appear sporadically, defoliating ber (*Ziziphus* spp.), Cashew (*Anacardium accidentale* Linnæus), mango (*Mangifeera indica* Linnæus), pepper (*Piper* spp.) and tea (*Camellia sinensis* Linnæus); pupate in golden-yellow silken cocoons on the stems and branches. Moths reddish-brown, females having three transparent eye-spots near centre of forewing and one on hind-wing; wingspan 70 to 92 mm.

♀

Cricula trifenestrata Helfer

Caterpillar, Crab (lobster) *Neostauropus alternus* (Walker) (Notodontidæ: Lepidoptera); stout pale grayish-brown, medium-sized moth, suffused with darker shades and marginal row of reddish-brown and pale spots on both wings; wing expanse 35 to 45 mm (♂) and 50 to 60 mm (♀). Curious looking caterpillar, having long and slender thoracic legs and last abdominal segment swollen ears a pair of slender curved appendages instead of anal claspers; Polyphagous leaf defoliator.

Catterpillar, Earhead common name for various caterpillars that feed on earheads of cereal crops, maize, millets etc; most common and destructive pests include, *Anatruchyntis simplex* Walsingham (Cosmopterygidæ: Gelechiodea), *Celama internella* Walker, *Eublemma silicula* Swinhoe and *E. hemirrhoda* (all Noctuidæ: Noctuodea).

Catterpillar, Red hairy *Amsacta albistriga* (Walker) and *A. moorei* (Butler) (Arctiidæ: Noctuoidea); full grown caterpillars of both species look alike, 50 to 70 mm long, densely hairy with pinkish-orange hairs. Phytophagons but more destructive to groundnuts. Moth of *A. moorei* have white head with crimson line behind it; thorax white; wings white, forewing having a scarlet fascia along the costa and a black speck at each angle of cell; wing expanse 45 to 55 mm. *A. albsitriga* differs from *A. moorei* in being smaller in size; with yellow line on white head; thorax tinged with ochreous and costal stripe on forewing being yellow.

Caterpillar, Red-humped *Schizura concinna* (Abbot and Smith) (Notodontidæ: Noctuodea); full grown caterpillar 30 to 35 mm long, red or yellow coloured with six or eight black tubercles on each segment, entire body lined with white, brown, red and black; head bright red; IV segment also red in colour with a prominent dorsal hump and two conspicuous black tubercles. Feed gregariously on deciduous trees and shrubs, a pest of walnut (*Juglans regia* Linnæus) and other fruits trees grown in temperate regions. Moth reddish-brown and gray in colour with wing expause of 30 to 35 mm.

Caterpillar, Rice-hairy *Psalis pennatula* (Hübner) (Lymantriidæ: Noctuoidea); yellowish-brown coloured, thickly haired caterpillars (35 to 40 mm long) are voracious defoliator s but minor pest of paddy. (*Oryza sativa* Linnæus). Adults are stout, straw-coloured moths.

Caterpillar, Tobacco *Spodoptera (Prodenia) littoralis* (Fabricius_) (Noctuidæ: Ditrysia, stout, medium-sized moth (wing-span 30 to 35 mm), forewings dark reddish-brown bearing wavy white markings and hind-wings white. Caterpillar stout, cylindrical, pale greenish-brown with dark markings and stripes, feed gregariously in early stage, then disperse and feed voraciously on foliage during night and remain hidden under soil during day.

Caterpillar, Yellow-necked *Datana ministra* (Drury) (Notodontidæ: Noctuoidea), black caterpillar with yellow longitudinal stripes and long white soft hairs; head and prothoracic shield yellow with reddish tinge. Minor pest of fruit trees of temperate regions especially almond and apricot (*Prunnus* spp.; Rosaceæ) and walnut (*Juglans regia,* Linnæus) trees.

Caterpillar hunter nickname of various carabid beetles (Coleoptera) that feed on caterpillars and are regarded as beneficial insects.

Catocalla generic and common name for a group of beautiful underwings (Plsiidæ: Noctuoidea: Lepidoptera); forest inhabiting species, having forewings cryptic and hind-wing brightly coloured. Caterpillars green in colour, matching the foliage on which they feed – *Catocala* spp. (caterpillars bluish-green while feeding on green foliage, become bluish-gray in dark-coloured environment.

Caudal anal end of body.

Caudal gills respiratory organs in some aquatic insects such as naiads of damselflies (Zygoptera: Odonata); these are thin-wall outgrowths from anal or hind end of abdomen.

Caudate having caudal appendage: tailed.

Cavenicolous inhabiting caverns: (insects) found living in caves.

Cebrionidæ family of Elateroidea (Polyphaga: Coloeptera); large beetles having head with a transverse ridge above the antennal sockets; labrum free, mandibles prominent and bent; female often apterous does not leave the subterranean burrow where she is sought out by the active male. Grub (wireworm) with large gula and eversible ventral cervical membrane - 200 species.

Cecidosidæ a microplepidopteran family (Tineodea: Ditrysia); having mouthparts undeveloped, fore- and hind-wings each with single anal vein and short fringe. Caterpillars gall-makers. Confined to South-America.

Cecidomiidæ Macquart 1938 family of gall midges (Myceltophiloidea: Nematocera); minute delicate flies with comparatively long antennæ and legs; antennæ spatula or breast-bone present midventrally on thorax; win-venation greatly reduced, a few longitudinal veins and no cross-veins; suck plant juices and often cause small galls on host plants - *Allassomyia tenuispatha* Keiffer, *Erosomyia indica* Grover, *Procontarina mangiferæ* (Grover) and *Udumbarie mainiensis* Grover infest mango (*Mangifera indica* Linnæus) trees in India, while *Lasiopera falcata* Felt causes galls on stems of bitter-gourd (*Momordica charantia* Linnæus) plants.

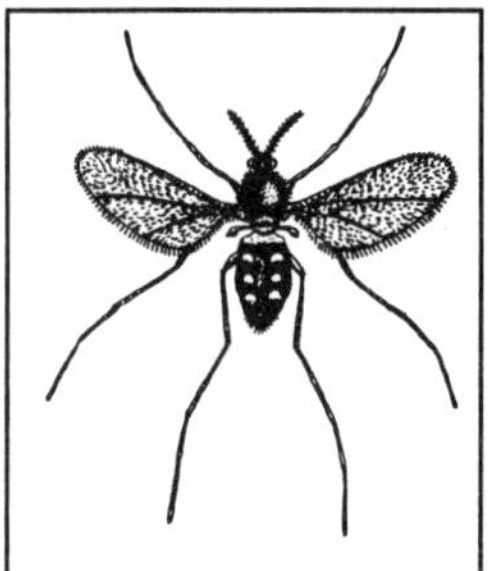

Lasioptera falcata Felt

Cell are of wing membrane between veins and cross-veins. Microscopic unit of living matter enclosing a membrane. A device for generating electric current by Chemical action.

Cell, Close area of wing membrane entirely surrounded by veins/cross-veins.

Cell fusion formation of cell with nuclei and cytoplasm of two different cells.

Cellular any structure composed of cells.

Celphos a solid fumigant; tablets and pallets made of aluminium phosphide, ammonium salts and some biding agent like paraffin wax etc. On exposure to air, within one to four hours the tablets decompose and release phosphine (PH_3) gas, which is highly volatile and has excellent insecticidal efficacy; used widely for controlling stored-grain pests; to various burrowing-pests like crabs, rodents, snakes and even mammals.

Cephaloidæ a very small family of Heteromea (Cucujoidea: Polyphaga); beetles akin to Meloidæ but the head id less deflexed and the neck broader. No hypermetamorphosis. About 12 species found in East Asia and North America.

Cephalothorax head and thorax fused together, as in mites and spiders (Arachnida) and scorpions (Crustacea), posteriorly it is attached to abdomen.

Cephidæ Haliday 1840 small family of stem-sawflies (stem borers) (Cephoidea: Symphyta); slender, narrow-bodied, dark coloured insects, ocelli vestigial; prothorax exceptionally large and movable articulated with the following segment; abdomen terminates in small retractile point or spine which arises from a fleshy protuberance on last segment above the anus. Larvæ apodous; bore into stems of grasses, wheat etc. - *Cephus cinctus* and *C. pygmæus* Linnæus are pest of wheat (*Triticum aestivum* Lamark) in USA, *Janus interger* is pest of currants (*Ribes* spp., Grossulariaceæ).

Cephoidea monofamilial superfamily of symphyta (= Chalastogastra) (Hymenoptera) characterised by abdomen constricted between first and second segments and cencheri absent - Cephidæ.

Cerambycidæ (Leach 1815, 1817) large family of phytophagous, longicorn (long-horned) beetles (Chrysomeloidea: Phytophaga); having elongate body, often several centimetres long and attractive colouration, antennæ very long and inserted on frontal promineneses. Grub soft, fleshy, off-white to yellowish in colour, elongate cylindrical in shape, apodous and often pubescent; bore into trunk, stems and roots, destructive to shade, forest and fruit trees. About 1300 species divided into several subfamilies including Lamiinæ, Prionæ and Cerambycinæ.

Cerambycinæ large subfamily of Cerambycidæ (Chrysomeloidea; Hateromera); adults large, flattened longicorn beetles having antennæ atleast two-thirds as long as body. Grubs feed on all parts of plant/ tree. - *Demonax balyi* (Pascoe), *Plocaedenus obesus* Gahan, *Xylotrechus semei* Laporte and Gory, et al.

Cerapachyinæ subfamily (carnivorous) of Formicidæ (Formicoidea: Apocrita); that includes predaceous ants having distinct constriction between first and second segment of gaster; conical dorsal spine or pedicel present, sting well developed.

Ceraphronidæ only family of Ceraphoronoidæ (Prototurpoidea: Apocrita); small black wasps having antennæ 10- or11-segmented arising from junction of clypeus and face, generally apterous, if alate, venation greatly reduced; mostly secondary parasites of aphids and coccids through various braconid and chalcid primaries - *Lygocerus cameroni* Kieffer, *L. testacei* Kieffer *L. manus* Kieffer parasites larvæ and pupæ of *Aphidus* spp. (Braconidæ).

Ceraphornoidea monofamilial superfamily of protorupoid group (Hymenoptera); antennæ 9- to 11-segmented inserted very low down, at clypeal margin; fore tibia with two spurs; sides of gaster rounded and without speiracles - Ceraphronidæ.

Ceratacombidæ (Dipsocoridæ) small family of ground bugs or jumping bugs (Dipsocoroidea: Heteroptera); minute delicate insects; nocturnal, living among mosses and dead leaves; antennæ long and filiform; hemelytra with embolium; 50 species.

Ceratinidæ (Latreille 1802) Ashmead 1899 small family of Apoidea (Apocrita: Hymenoptera); that includes small, metallic bees, almost bare except for-legs; nest in pith of the plants - *Ceratina* (*Clavicera*) *cyanea* and allied species.

Ceratocampidæ old name for a small family Citheroniidæ (Bombycoidea: Lepidoptera); that includes Royal moth (*Citheronia regalis* and allied species) of north America.

Ceratophyllidæ Dampf 1910 (Dolichopsyllidæ Baker 1905, Vermipsylidæ Wagner 1889) the largest family of Ceratophylloidea (Siphonaptera: Siphnapteroidea); fleas having nogenial comb, frontal margin evenly rounded; metanotum with marginal spinelets and abdominal tergites each with at least two rows of seatæ - ground squirrel flea *Diamanus montanus,* (Baker), European rat flea *Nosopsyllus faciatus* (Bosc), European chiken flea *Ceratophyllus gallinæ* (Schrank), Western chicken flea *C. niger* Fox, *Canadian chicken flea C. gibsoni* Fox, et al.

Ceratophylloidea superfamily of oligeneopteran order Siphonaptera: wingless insects characterised by presence of outer internal ridge of mid-coxa; pointed external apical tooth in hind tibia and sensilum with 14, more often 16 or more pits on each side. Include bulk of fleas parasitic on birds, bats, mice etc. - bat fleas (Ischnopsyllidæ), mole fleas (Hystrichopsyllidæ), rat and chicken fleas (Ceratopsyllidæ). *Cf.* Pulicidea.

Ceratopogonidæ Malloch 1817 family of biting midges Nematocera: Diptera); minute flies, one to 5 mm long, having antennæ 12- to 15-segmented, plumose in males and pilose in females; ocelli atrophied or wanting; mouthparts biting and piercing; wings narrow, naked or

Culicoides dovei Hall

hairy with few veins; mostly predaceous, females blood suckers burrow into heavy coat of cattle; maggots of most species semiaquatic, breed in water - *Culicoides dovei* Hall (*C. macrostoa, C. pattoni* are of medical and veterinary importance); *Forcipomyric tipulivora* Edwards parasitic on craneflies, *Pterobosca latipes* Edwards and *P. fidens* Edwards attach themselves to wings of dragonflies.

Ceratotheca part of an insect's pupal skin that encloses the antenna.

Caratothripidæ Bagnall 1912 family of palearctic thrips (Thripoidea: Teebrantia); characterside by 6th and 7th antennal segment very small and style-like and much smaller than segment V; usually with one- or two-segmented apical style.

Cercopiæ Leach 1815 small family of Cicadoidea (Auchenorrhyncha: Homoptera); small (< 13 mm) jumping insects, commonly called frog-hoppers or spittle bug; having two ocelli, one or two stout spines on hind tibia, hind coxae short and conical. Nymphs remain in a frothy spittle mass - *Clovia lineaticollis* (Motschulsky), *Cosmoscrata relata* Distant, *Phymatostetha deschampis* Lethierry. Sugarcane froghopper, *Tomaspis saccharina* Distant causes froghopper blight of sugarcane in Trinidad.

Cercopod alternate name of cercus, a filamentous appendages on hind part of abdomen: foot-like appendage on last abdominal segment.

Cercus (pl. **cerci**) usually a pair: many-jointed antennæ-like appendages found at anal end of abdomen in various insects.

Cerocophanidæ a very small family of Bombycoidea (Ditrysia: Lepidoptera); moths having a short proboscis and no frenulum; only a few species confined to South America.

Cerocophanidæ very small family of Elateroidea (Polyphaga: Coleoptera); about 20 species having labrum concealed and hind coxae without femoral plates; found in hollow trees.

Cervix (pl. **cervices**) neck; comprising of posterior non-sclerotised part of labial somite and anterior part of prothorax.

Cerylonidæ small family of Clavicornia (Cucujoidea: Polyphava); earlier placed under family Colydiidæ; having palpi with cylindrical segments; tarsi 4-4-4, mid coxal cavities broad and closed externally by sterna (200 species) - *Cerylon* spp., *Murmidius* spp. et al.

Cetoniidæ Mac Leay 1819 family of flower beetles (Scarabæoidea: Polyphaga); that are pollen feeders; flattened with large scutellum; brilliantly coloured, dorsally flattened with large scutellum; madibles not expanded and invisible from above, labrum concealed (2600 species) - *Chiloloba acuta* (Wiedemann) on cole crops, *Oxycetonia* *versicolor* (Fabricius) on falsa (*Grewia subinaequalis* de Candolle) while *Protactia aurichalcea* feeds on pollen-stores of *Apis indica* Fabricius hives.

Chaerocampinæ subfamily of Sphingidæ (Siphingoidea: ditrysia); having large proboscis, apex of forewing but slightly produced; male with small lateral expansions to abdomen. Caterpillar with anterior somites retractile; third one swollen; horn variable. Pupa without external sheath for proboscis – *Hippotion alector* (Linnæus), *Daphnis nerii* Linnæus, *Theretra actea* Cramer, et al.

Chæta (pl. **chaetæ**) bristle or chitinous setæ.

Chætognatha phylum of free-living, bilaterally symmetrical, unsegmented marine, planktonic, arrow-worms (Coelomata: Triplobastica); body torpedo-shaped, comprises of head, trunk and postnasal tail; hermaphrodite, self- or cross- fertlisation by reciprocal copulation. About 70 species classified into two orders – Phragmorphora and Aphragmorphora.

Chætopod any worm that moves with the help of setæ or bristles.

Chaætosema Jordan's organ: cluster of sensory bristles on head of some adult lepidopterous insects; may be replacement of ocelli.

Chætotaxy arrangement and nomenclature of macrochætæ setæ or differentiated bristles, on insect exoskeleton, often valuable for classificatory purposes.

Chafers common name of large herbivorous beetles (Melolonthidæ: Scarabæoidea), also called cockchafers, beetle-grubs or May (June) beetles. Grubs very destructive to a wide range of crops, feeding of roots.

Chalastogastra Konow 1897 old and equivalent name for the suborder Sessiliventres Haliday 1840 or Symphyta Gehrstecker 1867 (Hymenoptera); Symphyta is the preferred name.

Chalcid any small wasp belonging to superfamily Chalcidoidea (Hymenoptera: Hymenopteroidea).

Chalcididæ (Leach 1830) Westwood 1840 family of small chalcid wasps (Chalcidoidea: Apocrita); head small, antennæ simple and short; eyes large, ocelli in a row on vertex; hind cosæ long and cylindrical, hind femora short and straight; wings ample, not folded longitudinally, veins reduced to the subcosta. Mostly primary or secondary parasites on other insects, some hypeparasites – *Dirhinus pachycerus, Spilochalcis fletcheri, Stomatoceras ayyari* et al.

Chalcidoidea Ashmead 1897 largest superfamily of Apocrita (Hymenoptera) antennæ elbowed having less than 14 segments;

pronotum ventrally rounded, in alate forms forewings with pterostigma and proximal part of costa missing, venation much reduced and hind-wing with no anal lobe. Mostly parasites and hyperparasites, a few phytophagous.

Chalcosiinæ a subfamily of Zygænidæ (Zygænoidea: Ditrysia) that is essentially Tropical; may species having slender bodies and broad large wings look like butterflies; in some species hid-wings are tailed - *Eleysma* spp., *Histia* spp. et al.

Chalcosiinæ largest subfamily of Zygænidæ (Zygnoidea: Ditrysia); characterised by antennæ bipectinated to the tips in both sexes, more heavily in males than in females; tip of antenna flabellate; palpi minute and porrect; mid-and hind tibia with minute terminal pairs of spurs. Caterpillars short and thick with tubercles, from which arise short hairs - *Agalope hylina* Kollar, *Soritia leptalina* Kollar, *Chalcosia thallo* Linnæus, *Campylotes histrionicus* Westwood et al.

Chalybion iridescent, steel-blue wasp of North America; builds its clay-nest under eaves of houses.

Chamæmyiidæ Hendel 1910 (Ochthiphilidæ Zetterstedt 1890) family of minute gray flies (Lauxanioidea: Schizophora); having wings with Sc complete, vibrissæ absent and females with four spermathecæ, predaceous on aphids and coccids - *Leucopis griseola, L. lutericornis* et al.

Chameleon fly *Stratiomys chamaeleon* (Stratiomyidæ: Brachycera); abdomen flat, broader than thorax and with black and yellow bands like a wasp. Maggot mosquito-like, aquatic and breath through tail-siphon. Flies feed on umbelliferous plants.

Chaoboridæ Edwards 1920 family of Culicoidea (Nematocera: Diptera); often regarded as subfamily of Culicidæ; these mosquitoes have mouthparts short, concelaed and not adapted for piercing (cannot suck blood). Larvæ transparent and apneustic are predatory and feed on other mosquito larvæ - *Chaoborus crystallius, C. astictopus* Dyar and Shannon and allied species.

Charipidæ (Allotriidæ) family of tiny wasps (Cynipoidea: Apocrita); often place as subfamily (Charipinæ) under Cynipidæ; body smooth and impunctate, antennæ not elbowed; forewings without stigma and having reduced venation - *Charips* and allied species hyperparasites of aphids mostly through *Aphidius* spp.

CHATTERJEE, Dr. N.C. - born December 1890; B.Sc. Allahabad University (1911) and D.Sc. Banaras Hindu University (1938). A.D. Imms appointed him as his assistant at Forest Research Institute,

Dehradun in 1911; became Assistant Forest Research Entomologist in March 1920 and retired in that capacity in December 1945; reappointed as Forest Entomologist in March 1948 until November 1949. He carried out research on lac insects under Imms and insect pests of sandal (*Santalum album* Linnæus) under Beeson. Died February 20, 1950.

Cheese skipper larvæ of cheese-fly, *Piophila casei* (Linnæus) Piophilidæ: Diptera); pest of stored cheese and other proteinous food-stuff. Full grown maggot can jump considerable distance by flexing its body, grasping its tail with mouth hooks and suddenly releasing the same.

Chelicera (pl. **cheliceræ**) first pair of appendages in Arachnida, used for biting, seizing, crushing or piercing.

Chelicerata subphylum of Arthropoda (now considered as phylum) having chitinous exoskeleton and segmented appendages: chelicerae modified as pincers. 63000 species in three classes - Merostomata (horseshoe crabs), Pycnogonida (sea-spiders) and Arachnida (mites, spiders, scorpions etc.).

Chelisochidæ Burr 1907 family of earwigs (Forficulodea: Forficulina); ædeagus unpaired; X tergum well developed, XI tergum and telson vestigial or absent - *Chelisoches morio* preys upon aphids, leaf-hoppers and leaf-mining hispid larvæ.

Chelonariiæ very small family of Dryopoidea (Polyphaga: Coleoptera); that includes subaquatic beetles having head with an occipital keel, anternnæ and legs retractile into grooves and pronotal hind margin crenulate; found in Tropical Asia and America.

Chemistry science that deals with properties of matter, its structure, composition and susceptibility to changes.

Chemosterilant(s) chemical compounds that when administered to the insects, reduce or entirely eliminate the capacity to reproduce viable progeny: produce partial or complete sterility without killing the insect.

Chemotaxis movement or organisms in different direction in response to chemical stimulus.

Chemotropism orientation by differential growth in response to chemical stimulus.

Chemurgy applied chemistry dealing with use of agricultural products.

Chermidæ (Fallen 1814) a homopteran family synonym of Psylidæ Latreille 1807.

Chermides a homopteran family erected by Fallen in 1814, name later corrected to Chermidæ (Fallen 1814) - genus *Chermes* Linnæus 1758 is synonymous to *Psylla* Geoffroy 1762.

Chermoidea superfamily of Sternorrhyncha often considered as synonym of Psylloidea (Homoptera) erected to include the single family Chermidæ (= Psyllidæ).

Chiasma crossing of nerve tracts within a nerve centre.

Chimney sweep small diurnal geometrid moth, *Odezia atrata* (Ditrysla: Lepidoptera); having body and wings completely black.

Chiroomidæ Rondani 1841 family of midges (Culicoidea: Nematocera); small delicate gnat-like flies; had small often concealed by thorax, mouthparts poorly developed, antennæ conspicuously plumose in males and pilose in females; wings not scaled, anterior wing-veins more strongly marked than the posterior ones – *Chironomus cubicularum, C. dosalis, Clunio takahashi, Corynoneura celeripes, Tanytarsus dissimilis* et al.

Chirothripoididæ Bagnall 1912 a small family of Phloethripoidea (Tubuylifera: Thysanoptera) comprising of thrips characterised by VIII abdominal segment with long peg-like projections along the posterior margin and the last abdominal segment very short and stout – *Chirothrips hamatus, C. manicatus* et al.

Chitin $(C_8H_{13}O_5N)n$; organic chemical compound, insoluble in alcohol, ether, water and other solvents; resistant to acids and alkalies; forms exoskeleton and endocuticle of insects.

Chlidanotidæ a small microlepidopteran family (Tortricoidea: Ditrysia); moths have pointed palpi; forewing veins R_3 and R_4 stalked or coincident and hind-wing with M_2 and M_3 parallel. Confined to Indo-Australian region – species of *Archimaga, Childanota, Eletracma, Matrenis* and *Trymalitis.*,

Chlordane $C_{10}H_6Cl_8$; chlorinated hydrocarbon; contact and stomach poison with slight fumigant action; insoluble in water, slightly soluble in organic solvents and unstable in presence of alkalies; non-phytotoxic but harmful to bees. Its insecticidal properties were discovered by C.W. Kearns and his associates in 1945, since then it was used against household pest and some soil-borne insects. Oral and dermal LD_{50} values for rates are 283 and > 1600 mg/kg respectively.

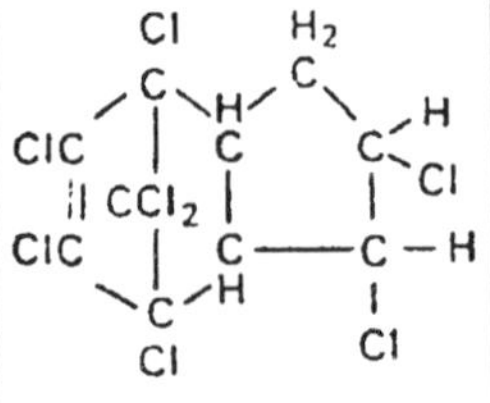

Chlordane

Chlorfenvinphos $C_{12}H_{14}O_4CI_3P$; derivative of phosphoric acid, amber coloured liquid with mild chemical odour; slightly soluble in water, miscible in acetone, alcohol, kerosene, xylene etc., contact insecticide with slight systemic action; non-phytotoxic; acute oral and dermal LD_{50} values for rat are 10 to 155 and 108 mg/kg respectively – Birlane®, Supona® etc.

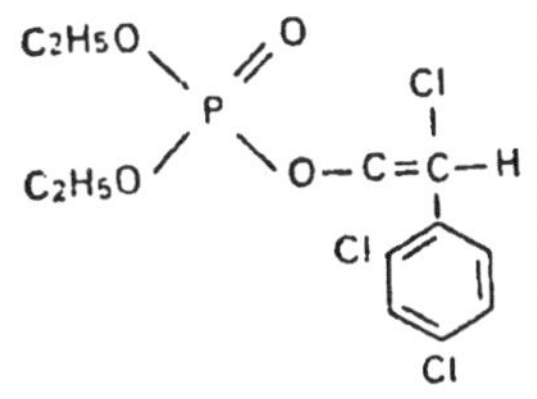
Chlorfenvinphos

Chlorinated terpens insecticides produced by chlorination of naturally occurring terpenes – Strobane ($C_{10}H_{11}Cl_7$), Toxaphene ($C_{10}H_{10}Cl_8$) etc.

Chlorobenzilate $C_{16}H_{14}Cl_2O_3$; non-systemic acaricide, effective against all stages of phytophagous mites; insoluble in water but soluble in benzene, methyl-alcohol etc. and gets hydrolysed in strong acids and alkalies. Oral and dermal LD_{50} values for rats are 700 to 3200 and > 5000 mg/kg respectively – Akar®, Folbex® etc.

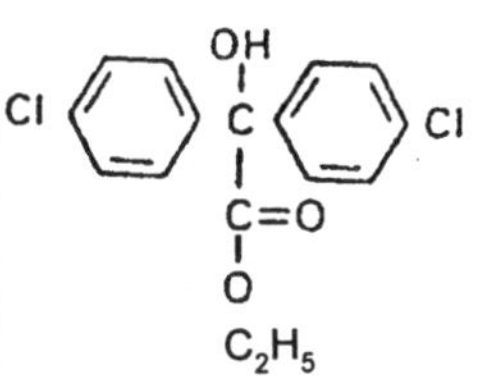
Chlorobenzilate

Chloroperlidæ family of Perloidea (Setipalpia: Plecoptea); small specialised stoneflies having reduced wing-venation and anal lobes; found in North America and Eurasia (about 120 species).

Chlorophyll green pigment in plants that is essential for photosynthesis.

Chloropidæ Ronadani 1856 a family of Acalyprtatæ (Schizophora: Diptera); comprising of minute, active flies having head somewhat angular, antennæ prominent and porrect; wings with subcosta reduced, second discal and basal cell united, anal cell wanting. Maggots bore into stems or mine the leaves while adults of some species irritaste both man and the domestic animals by hovering around their eyes – *Chloropisca circumdata* (Meigen) and *C. oryzae* Matsumura are pests of rice in Japan; *Oscinosoma frit (Linnæus)* a widely distributed holarctic species is an economic pest of cereals and grasses while *Siphunculina funicola* de Meijere is the common, eye-fly of India.

Chlorpyriphos $C_9H_{11}Cl_{13}NO_3PS$; white crystalline substance, with mild mercaptan odour; contact and stomach poison with some fumigant action; relatively sable to hydrolysis and oxidation; acute oral LD_{50}

to rats 135 to 163 mg/kg and dermal for rabbit 1000 to 2000 mg/kg – Dursban® etc.

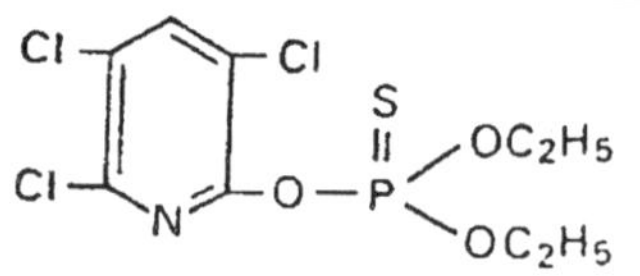

Chlorpyriphos

Chordata Largest phylum of higher animals characterised by notochord, nerve cord and gill slits or pouches in some stages of development. About 100000 species classified into two main division, Acraniata and Craniata (Vertebata); the former embraces three subphyla – Urochordata (Larvacea, Ascidiacea Thallacea), Cephalochordata (Leptocardii) and Hemichordata (now considered as phylum) while Craniata includes two subphyla – Agnatha (Cyclostomata) and Gnathostomta that includes two superclasses, Pisces (Chondrichthyes, Osteichthyes) and Tetrapoda (Amphibia, Reptilia, Aves and Mamalia).

Chorion non-chitinous shell of insect egg; secreted by follicle cells.

Chorionin seleroprotein that forms the main part of the shell (chorion) of insect's egg.

Choristidæ Tillyard 1926 a very small family of Eumecoptera (Mecoptera: Mecopteroidea); scorpion-flies having head and rostrum short, antennæ as long as or longer than the forewings; labrum broad, mandibles pointed and male with bulbous genitalia. Only two genera, each with two species, confined to Australia – *Chorista australis* Klug, *Tæniochorista pallida* Peterson et al.

Chromatogram preparation made on paper with the help of chromatograph for detection of amino-acids etc. present in any sample.

Chromatography physico-chemical technique for separating inorganic and organic constituents of a substance; first applied successfully by a Polish botanist Tswett in 1906, it is now universally used for separating antibiotics, proteins, amino acids carbohydrates, steroids, vitamins and several other natural and synthetic substances.

Chromomere individual chromatic granule(s) in a chromosome.

Chromosome hereditary bearing gene; tiny (microscopic) thread-like deeply staining structures in the nucleus of a cell; occurs always in pairs, their number varying from species to species.

Chromosome translocation phenomenon in which two non-homologous chromosomes in the same cell undergo breakage (due to irradiation) followed by rejoining of fragments with wrong partners.

Chrysalis

Chrysalis pupa or pupal case of butterfly (Lepidoptera).

Chrysididæ (Latreille 1802) Dahlbom 1854 family of cuckoo-wasps or ruby-tail wasps (Bethyloidea: Apocrita); having bright shining green or red body; head orthognathus, antannæ 125- or 13-segmented; pronotum short but wide. Lay eggs in burrows of other solitary wasps, bees or sawflies and their larvæ feed on those solitary bees, wasps and sawflies or on the caterpillars stored as food by those wasps (1400 species) - *Chrysis ignita* and allied species, *C. shanghaiensis* lays its eggs on mature caterpillars of *Monema* spp. (Limacodidæ).

Chrysis ignita

Chrysomelidæ (Leach 1815, 1819) large family of Chrysomeloidea (Polyphaga: Coleoptera); oval, convex, brightly coloured beetles, 4 to 12 mm log; antnnæ widely separated at the base; eyes not prominent; prothorax laterally margined; pygidium not exposed beyond elytra; tarsal segment not bilobed and forecoæ transverse: 29000 species spread in various subfamilies - Hispinæ, Cassidinæ, Halticinæ, Sagrinæ, Galerucinæ, Doinaciinæ, Chrysomelinæ, Oricerinæ etc,., most of these are being recognised as independent families.

Chrysomeloidea superfamily of Polyphaga (Coleoptera: Coleopteroidea); beetles having head not rostrate, gular sulci distinct and separate; antennæ without 3-segmented club - vast collage of phytophagous and xylophagous beetles grouped under Cerambycidæ, Bruchidæ and Chrysomelidæ families.

Chrysopidæ Hagen 1866 family of aphis-lions and lacewings (Planipennia: Neuroptera); green or yellowish-brown body, less than 25 mm long, large iridescent compound

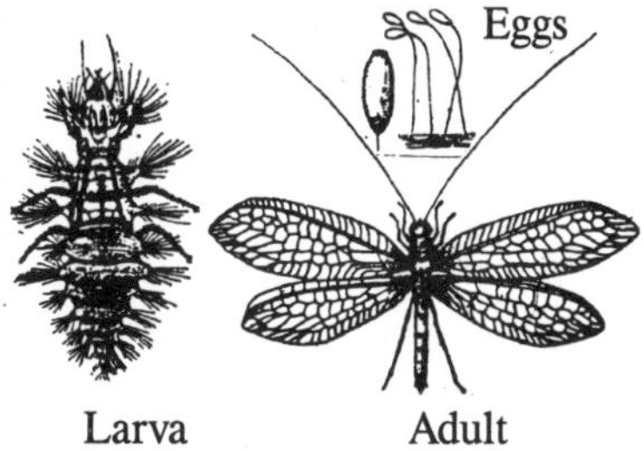

Chrysoperla scelestis Banks

eyes; antennæ filiform, longer than forewings; trichosors absent. Larval body with long hooked hair, empodium trumpet-shaped; feed voraceously on homopteran insects especially aphids - *Chrysoperla cymbele* Banks, *C. modesters, C. scelestis* Banks, et al.

Chrysopolomidæ (Ectropidæ) very small family of Zygænoidea (Ditrysia); palpi present; hind-wings with vein Sc + R_1 anastomoses with the cell - only 25 species placed under two genera.

Cibarium special preoral food chamber between base of hypoharynx and under surface of clypeus.

Cicada homopteran plant-bugs (Cicadidæ: Cicadoidea) of Tropical regions having long piercing mouthparts and four membranous wings. Males short-lived, make a very loud rattling noise to attract the females; these mating calls can be easily head at the distance of half a kilometre. Eggs are inserted into the stems of shrubs and small trees. Young nymphs fall to be ground and live subterranean life, feeding for a few weeks on roots of various plants and the remain inactive for 5 to 17 years. Fully mature nymphs emerge from the soil, become adults, feed, sing, mate, oviposit and perish before the onset of Winter - *Magicicada septendecim* (Linnæus) (Periodical cicada) remains underground for 13 to 17 years.

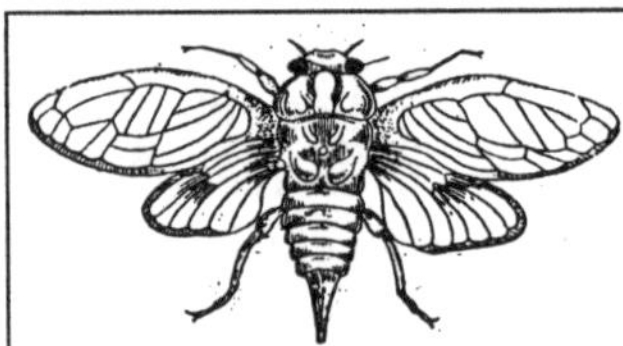

Cicada

Cicada, Chinese blistering *Huechys sanguinea* (de Geer) of China, Japan and parts of South-east Asia is also known as Red medicinal cicada because of its brilliant red and black colours and its uses as a vesicant.

Cicada, Dog-day *Tibicen linnei* Smith and Grossbeck (Cicadidæ: Cicadoidea); a large robust green and black species, about 32 mm long with wing expanse of 115 mm; makes shrill noisy songs, late in Summer.

Cicada, Great Borneo *Pamponia imperatoria* (Westwood) (Cicadidæ: Cicadoidea) is about 75 mm long with wing span of over 200 mm. It sings in the evening and is relished as food in the Malay Peninsula.

Cicada, Putnam's *Platypedia minor* (Uhler) (Cicadidæ: Cicadoidea), a common Western species is 16 to 20 mm long and makes clicking cocophony with the wings.

Cicada, The oil *Graptopsaltria coorata* (Stal) (Cicadidæ: Cicadoidea) of Japan makes sounds like hot frying fat, hence the common name.

Cicada, Vineyard commonest European species *Tibicen hæmatodes* (Scopoli) (Cicadid: Cicadoidea) that starts to sign at the beginning of grape harvest, early in August and stops by tradition on Holi Friday in the middle of October.

Cicadellidæ (Latreile 1802) (Jassidæ Fieber 1866): large family of small leafhoppers and sharpshooters (Cicadoidea: Auchenorrhyncha) having minute bristle-like antennaæ inserted in front of and between the eyes, two ocelli; hind tibæ with a double row of spines. Both nymphs and adults have characteristic habit of running sidewise; suck plant juices, causing during and wilting of tissues, some species also transmit plant diseases – *Cicadula fascifrons* Stal, *Emposca notata* Melicher, *Kolla mimica* Distant, *Nephotettix virescens* (Distant) are pets of paddy.

Pycna repanda (Linnæys)

Cicadidæ (Latreille 1802) (Cicadariæ Latreille 1802) family of cicadas or harvest-flies (Cicadoidea: Auchenorrhyncha); robust insets having three ocelli in a triangle between the eyes, antennæ small bristle-like 5- or 6-segmented; anterior femora thickened and spined beneath; claws without empodia. Male usually with stridulatory organs at the base of abdomen; produces shrill sound that can be heard half a kilometre away (1500 species) – *Pycna repanda* (Linnæus) is pest of apple trees and *Pltypleura octoguttata, Pomonia fusca* and *Huechys thoracica* are other Indian species. *Pomponia adusta* (Walker) of Java is about 80 mm long.

Cicadina suborder of Homoptera in certain system of classification, equivalent to suborder Auchenorrhyncha of others. It includes Cicadidæ (cicadas), Fulgoridæ (planthoppers), Cercopidæ (frog-hoppers), Jassidæ (leafhoppers) and Membracidæ (treehoppers).

Cicadoidea Ashmead 1904 superfamily of Auchenorryncha (Homoptera: Hemipteroidea); characterised by antennal pedicel without sensill and flagellum, with sensory grooves; tegulae absent; wings with ambient vein 1A and 2A not forming Y-vein; middle coxæ short, hind coxæ mobile – several families.

Cicindelidæ Leach 1815, 1817 family of tiger beetles (Caraboidea: Adephaga); being voracious and fierce of all insects; head vertical, wider than thorax; eyes large, mandibles long, dentate and curved, clypeus extending on each side beyond the base of antennæ; lacinia terminating in a movable segment or hook; legs long, slender and tarsi 5-segmented; elytra beautifully coloured, usually covering the body completely. Very alert an active; swift fliers; occur in damp places and along waterways. (20000 species) – *Amblycheila cylindrifomis* Say (40 mm long), *Manticora congoensis* Peringuey (45 mm) *Megacephala carolina* (Linnæus) (18-20 mm) and *Prothyma leptalis* (5-6 mm).

Ciidæ (Cisidæ, Cioidæ) family of Heteromeran beetles (Cucujoidea: Polyphaga); head deflexed more or less covered with pronotum; fore-coxæ transverse, their cavities open behind and internally. Grub with paired urogomphi. Feed gregariously on fungus-impregnated wood (380 species) – *Cis fuscipes* Mellie, et al.

Cimbicidæ (Leach 1817) Kirby 1837 small family of large, robust sawflies (Tenthredinoidea: Symphyta) found in forested Temperate regions of the World; having antennæ clavate; wings with redial cell divided by a cross-vein; fore-tibia and hind tarsi with spatulate hairs; abdomen with distinct pleural sclerites; bearing spiracles – Species of *Abia, Amasis, Cimbex, Trichiosoma etc.*

Cimicida Leach 1815 and old and obsolete name of Heteropteran family Cimicidæ (Cimicoidea: Geocorisæ) that includes bedbugs.

Cimicidæ (Latreille 1804) (Acanthidæ) small family of Cimicoidea (Cimicomorpha: Geocorisæ); small, flat, oval, blood-sucking bus with rostrum lying in ventral groove, ocelli absent; tarsi 3-sengmented; hemelytra, if present, very short. Ectoparasites of bids and mammals – bedbugs (*Cimex lectularius* Linnæus. *C. rotundatus* Signoret) on human beings; *Oeciacus hirundinis* Jenyns on bats and birds.

Cimicides Latreille 1804 old and obsolete term for Heteropteran family Cimicidæ (Cimicoidea: Geocorisæ).

Cimicoidea superfamily of Cimicomorpha (Geocorisæ: Gymnocerata: Heteroptera); bugs characterised by having antennæ not geniculate, 2nd segment largest; pronotum tapezoidal; if alate, hemelytra with costal fracture and cuncus – Nabidæ, Cimicidæ, Anthoicoridæ, Polyctenidæ, Velocipedid, Microphysidæ, Isometopidæ and Miridæ.

Cimicomorpha section of Geocorisæ (Gymnocerata: Heteroptera); land-dwelling bugs, having 4-segmented antennæ; hemelytra often with costal fracture and cuncus; hind-wing with veins R and M joining distally; abdomen ventor; without trichobothria – Tingoidea, Reduvioidea, Cimicoidea and Dipsocoroidea.

Circulate go round a circuitous course: (of blood) flow from one part of body (heart) to other parts of body and back.

Circulation movement of blood from and to heart.

Circulatory system arrangement or process of blood circulation – closed and open.

Circulatory system, Closed system in which blood flows from the heart to various parts of body through the arteries and capillaries and back through the veins – common in most animals.

Circulatory system, Open system in which blood flows through body cavity, irrigating the various tissues and organs, as in insects and some other lower animals.

Cistelidæ old name for Coleopteran family Alleculidæ (Heteromera: Cucujoidea): no more in usage.

Citrus mealybug *Plaococus citri* (Risso) (Pseudococcidæ: Coccoidea); oval flat body, covered with white waxy coating; anennæ present, legs well developed. Polyphagous pest that prefer limes, lemons and organges; generally attacks leaves and tender shoots, sometimes even the roots.

Citrus psylla *Diaphorina citri* Kuwayana (Psyllidæ: Cicadellidæ); very small, 2 to 3 mm long, brownish plant -louse with transparent wings that sit with head almost touching the leaf surface and est of the body raised up. Both nymph and adults suck plant sap and inject certain toxins in the plants, besides excreting honeydew; also act as virus-vector causing citrus decline diseases.

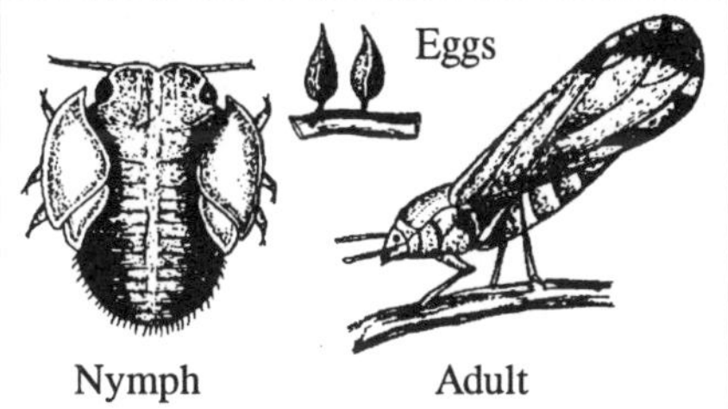

Diaphorina citri Kuwayana

Citrus red scale *Aonidiella aurantii* (Maskell) (Diaspididæ: Coccoidea); polyphagous pest; native to India but by now cosmopolitan; is distinguished by its conspicuous, reddish circular scale. *A. taxus* Leonardi of China, Japan and East Asia is known as Chinese red scale, while *Crysomphalus ficus* Ashmead id Florida red scale.

Citrus yellow scale *Aonidiella citrina* (Coquillett) (Diaspididæ: Coccoidea); polyphagous, but prefers *Citrus* spp., found in India, Japan and USA; the females have yellowish, round scale covering.

Cixiidæ Spinola 1839 widely distributed small family of primitive bugs (Cicadoidea: Auchenorrhyncha); elongated body with no prolongation of head in front, rostrum short; antennæ arise below the eyes, 2 or 3 ocelli present; sides of clypeus not keeled; wings broad, transparent, not reticulated apically, frequently along the veins ornamented with sports. Plant feeders; nymphs of most species subterranean, sucking sap from roots of grasses – *Oliarus hodgarii* and *O. walkeri* Stal occur on grasses in India, *O. felish* in Australia and *Cixius* spp. in Britain.

Oliarus walkeri Stal

Class taxonomic rank between phylum or subphylum and order or superorder.

Classification systematic arrangement of objects in series or groups showing there agreement in structure and/or other characters. Modern classification of living organisms was first suggested by Swedish botanist Carl von Line (Carolus A. Linnæus, 1707-1778) in 1737. He described about 3000 insect species under class Insecta and listed these in seven classis (orders) - Aptera, (Apterygota) Angiosptera (Neuroptera), Piezata (Hymenoptea), Eleutherata (Coleoptera), Glossata (Lepidoptera), Ryngota (Hemiptera) and Antliata (Diptera, Anoplura and Acarina); based on form and structure of wings only (alary system).

Clavae (claviform) club-shaped.

Clavicornia a large heterogeneous group (section) of beetles (Cucujoidea: Polyphaga) comprising of as many as 22 families; all having club-shaped antennæ and are phytophagous.

Clavigeridæ small family of Staphylinoidea (Polyphaga: Coleoptera) having short elytra that do not cover the abdomen - *Claviger* species found in nests of ants and termites.

Cleavage division of egg or its nucleus: segmentation of egg (cleave = cut in two).

Cleavage cells cells formed during cleavage.

Cleg common name for blood sucking flies – gadfly or horsefly, *Haematopota pluvialis* and allied species (Tabanidæ: Diptera).

Cleonymidæ a small family of Chalcidoidea (Parasitica: Apocrita), very small, metallic-coloured wasps characterised by pronotum shorter than mesonotum and never quadrate; mesepisternum large; fore femora usually swollen and dentate – of late reduced to subfamily Cleonyminæ (Pteromalidæ: Chalcidoidea).

Cleptidæ (Leach 1815) Dahlborn 1854 small family of Bethyloidea (Apocrita: Hymenoptera); often regarded ass subfamily (Cleptinæ) of Chrysideidæ; primitive chalcid wasps with more gastral segments exposed and integument weakly sculptured. Larvæ of *Cleptes* spp. develop on mature sawfly larvæ in their cocoons; *Mesitiopterus* species has been bred from phasmid eggs.

Cleptobiosis form of symbiosis, wherein one species of ant live in or near the colony of another species and feeds on the stored food in that colony. Also known as cleptoparasitism.

Cleptoparasitism robbing: multiple parasitism: parasitism wherein the adult parasitoid preferentially appropriates for its own progeny, the previously paralysed host of another parasitoid. Slave driving ants (Formicidæ: Apocrita), regularly capture the pupæ from nests of other species of ants, carry these to their own nest and compel the workers emerging from these pupæ to bereed their own brood. Beelouce *Braula coeca* Nitzsch (Braulidæ: Lauxaniodea) invade the nests of bees and devours the stored food in the beehive.

Cleridæ (Billberg 1820) Kirby 1837 family of checkered beetles (Cleroidea: Polyphaga); soft-bodied; head partly sunken into prothorax; eyes conspicuous; antennæ 11-segmented with three terminal segments greatly enlarged, last segment dilated; elytra leathery. Both grubs and adults are predaceous and often considered as beneficial. *Dasyceroelerus erinceus* and *Ommadius nilgiriensis* are predaceous on scolytid beetles; *Enoclerus sphegeus* (Fabricius) on barkbeetles; *Trichodes alvearius* (Fabricius) and *T. apiarius* (Linnæus) invade beehives to eat the bee larvæ.

Climatic migration locomotory activity bringing the organism to a preferred climatic zone.

Clistogastra konow 1905 old and equivalent name of suborder Euembioptera (Embioptera: Polyneoptera); small uncommon insects having soft, elongated flattened body; like most embids, the males have two pairs of wings and well developed venation while the females are wingless – *Clothoda* spp. et al.

Club enlarged apical segment of an insect antenna, all butterflies, skippers and many other insects have their antennæ clubbed.

Clypeus facial area of cranium just above labrum, separated from frons by epistomal suture; often subdivided into anteclypeus and post-clypeus.

Coccinellidæ Latreille 1807 a widely-distributed family of Clavicornia (Cucujoidea: Polyphaga); adult beetles mall, convex, hemispherical, brightly coloured and variously spotted (2 to 28); mostly predaceous on aphids, coccids, aleyrodids etc. (lady bird beetles); a few phytophgous and destructive to various economic crops specially vegetable crops (*hadda* beetles).

Cocc(o)idologist an entomologist who has specialised in the study of coccids (Coccoidea). V. Prabhakar Rao (1910-1983) and Sukhdew Raj Wadhi (1932-1986) were the famous Indian Coccidologists. Both had also specialised in Plant Quarantine.

Coccina suborder of Homoptera in some system of classification; equivalent to superfamily coccoidea of suborder Stenorrhyncha of others.

Coccoidea Handliersch 1903 superfamily of Sternorrhyncha (Homoptera: Hemipteroide); minute, often inconspicuous and highly specialised insects that are scale-like, gall-like or covered with waxy exudate; sexually dimorphic, females apterous and degenerate having sucking mouthparts; males often rare, have one pair of wings and mouthparts vestigial. Honeydew secreted by these insects often render them attractive to ants - mealy-bugs and scale insects.

Cochlidiidæ (= Cochliopodidæ) old and obsolete name for lepidopteran family Limacodidæ (Zygænoidea:); Ditrysia); includes slug-caterpillars distinguished by their slug-like movement as all their legs are vestigial or rudimentary.

Cochliopodidæ an equivalent name for lepidopteran family Cochlidiidæ (Zygænooidea: Ditrysia) which is now known as Limacodidæ (= Eucleidæ, Heterogenidæ).

Cockroach oldest living insect (Blattideæ: Dictyoptera); that has thrived for last over 300 million years (Carboniferous period till today); most species are nocturnal about 50 mm long; brown to black in colour, body broad and flattened, head bent down and concealed from above by pronotum; cerci well developed. Common

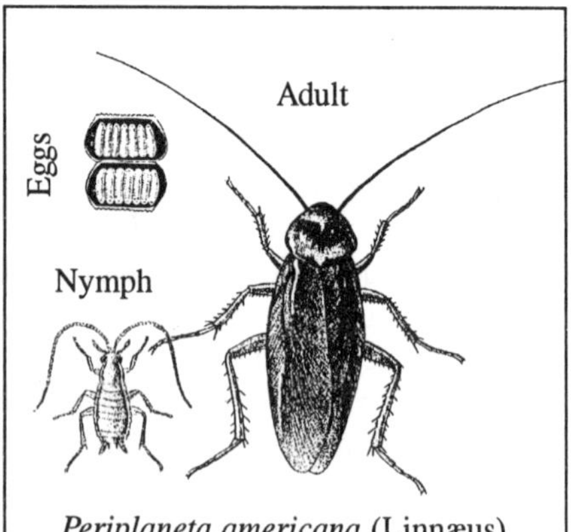

Periplaneta americana (Linnæus)

household pest – some African and Oriental species (Nocticolinæ) are cavernicolous – *Blatta orienalis* (Linnæus) and *Periplaneta americana* (Linnæus) are most common and cosmopolitan, while *P. australasiæ* (Fabricius) is confined to Australasia.

Cockroach, Dusky *Ectohius lapponicus* (Blattidæ: Blattaria) small woodland cockroach of Europe 10 to 14 mm long and brown in colour, male has long functional wings (tegmina) covering the abdomen completely but female has much shorter forewings (non-functional) and vestigial hind-wings.

Cockroach, Wood-boring *Cryptocerus punctulatus* Scudder (Cryptoceridæ: Blattaria: Dictyoptera); wingless cockroach of North America, found in rotting wood and under the bark of dead trees; digestion of wood carried out by means of symbiotic protozoa in the instestine, as these insects do not produce any enzymes.

Coconut scale *Aspidiotus destructor* Signoret (Diaspididæ: Coccoidea) armoured scale; yellowish-brown, translucent nymph (crawler); ovoid and 2 mm long, get covered with waxy material and become immobile; adult female circular, semi-transparent and pale brown. Both suck plant sap and devitalise the plants; polyphagous but prefer various palm trees.

Cocoon protective case formed by larvæ of various insect speices within which they pass their pupal stage.

Coelenterata (Cnidaria) phylum of Radiate (Enterozoa: Metazo); two-layered, many-celled animals, having sac-like body cavity and tentacles with stinging cells. About 10000 species placed under three classes – Hydrozoa (freshwater jellyfish), Scyphozoa (marine jellyfish) and Anthozoa (soft-coral, sea-fan, sea-pen etc.).

Coral colony Red coral Sea–fan Sea–Pen

Coelentrata

Coeloconic receptors olfactory or other sense organs of insect that are mor advanced than basiconic receptors and are distinguished in having the terminal process of sense-cell sunk in small pit.

Coelome(e) body cavity.

Coenagriidæ Tillyard 1926 (= Coenagrionidæ Karsch 894) a large family of worldwide distribution (Coenagrioidea: Zygoptera); large broad-winged damselflies with base of wings without stalk and discoid cell very short, anterior since being much shorter than the posterior side – species of *Pyrrhosoma* are fond in Europe; those of *Erythromma* occur in North America while *Coenagrion, Ischnura* and *Enallagama* are cosmopolitan.

Coenagrioidea Tillard 1926 (Coenagrionidea) a superfamily of Zygoptera (Odonata: Odonatoptera); includes damselflies characterised by narrow petiolate wings having 2 to 4 antinodal cross-veins and post-nodals in a line with the cross-veins behind them; boundary between meso- and metathorax extending only upto spiracle – Platystictidæ, Protoneuridæ, Platyenemidæ, Coenagriidæ, Psuedostigmatidæ and Megapodagriidæ.

Coleophoridæ (Hubner 1816) Staintion 1864 (= Euphistidæ, Haplopitilidæ) a family of Gelechioidea (Ditrysia: Lepidoptera); very small pale-coloured moths having head smooth, antennæ long, rough scaled and porrect in repose; maxillary palpi rudimentary or absent; wings slender, often pointed apically with rear marginal fring; wing span 7 to 15 mm. About 1200 species found in Southern Europe and USA – *Celeophora fletcherilla* Fernald, *C. malivorella* Riley, et al.

Coleoptera Linnæus 1758 largest insects order comprising of bettles and weevils (Coleopteroidea: Oligoneouptera); first appeared during Triassic period (Mesozoic era) and believed to be descendent of Protocoleoptera that flourished during Palaeozoic era. Hard bodied, having biting and chewing mouthparts, feebly movable prothorax with large pronotum, small mesothorax and large metathorax. Forewings modified into elytra, that though unfit for flight support the body during flight and protect the membranous hind-wings during repose Metamorphosis indirect or complete (holemetabolus). Larvæ campodeiform or eruciform, seldom apodus. Pupæ adecticous and exarate rarely obtect. Over 350000 species spread into four suborders – Archostemata, Myxophaga, Adephaga and Polyphaga. Smallest beetle *Nanosella fungi* (Ptiliidæ) is 0.25 mm long whereas male of hercules beetle is 155 mm.

Coleopterist an entomologist who has specialised in the study of Coleoptera (beetles and weevils); Tahilram Gianchand Vazirani (1925-84) was the famous Coleopterist who worked at Zoological

Survey of India, Kolkata and later at the British Museum (Natural History), London.

Coleopteroidea Handlirsen 1908 superorder of Oligoneoptera; includes holmetabolous insects that are almost equally at home in water, air or on land; divided into two orders – Coleoptera and Strepsiptera.

Coleorrhyncha monofamilial suborder of paraneopteran insects (Homoptera: Hemipteraoidea); primitive forms having short 3-segmented antennæ without terminal arista and concealed beneath the head; straight beak, formed at anteroventral extremity or the face, propleura forming shield for the base of the beak; tarsi trimerous; found in dam moss – Peloridiidæ.

Collembola Lubbock 1870 order of very small (< 5mm) apterygote insects (Oilgoentomata), antennæ 4- to 6-segmented; compound eyes; when present, comprise of eight ommatida on each side; entognathus chewing type mouthparts; tarsi and tibia usually fused; abdomen with only six segments; collophore beneath first abdominal segment, tenaculum below III segment, majority with springing-organ situated ventrally near anal end of abdomen; metamorphosis primitive or lacking: flourishes well upto 6000 metres height as also submerged in sea; inhabits surface of stagnant water in lakes and ponds, snow and ice as well as decaying mater and nests of ants and terminates; about 2200 species listed under two suborders – Arthropleona and Symphypleona. Earliest known fossil insect, *Rhyniella praecursor* that live 370 million years age (Devonian period) a collembolan.

Colleterial glands accessory glands of female insect, secreting adhesive substance to fasten her eggs together.

Colletidæ Bingham 1897 family of Apoidea (Aculeata: Apocrita); to 16 mm long, hairy, solitary bees resembling honeybees but primitive and having antennæ simple, short 12- or 13-segmented, ocelli in straight line across the vertex; legs short and hairy with pollen baskets; forewings with one radial and 3 submarginal cells; abdomen broadly triangular and segment VI exserted – plumed bees; yellow-faced bees, etc.

Colloidal formulation solution in which particles stay dispersed indefinitely and the size of particles is less than 6μ.

Collophore characteristic appendicular organ of first abdominal segment of collembolan insects; large thick abdominal tube of body wall projecting ventrally and functioning as an adhesive organ or glue-bearer.

Colobathristidæ small family of Heteropteran bugs often regarded as subfamily of Bertidæ (Lygaeioidea: Pentatomomorpha) but differs in having long legs and basally constricted abdomen; found mainly in Neoptropical regions. Over 70 species – *Phenacantha saccharicida* is pest of sugarcane in Java.

Colon large intestine: posterior part of anterior intestine between ileum and rectum.

Colonial inhabitant of a colony: pertaining to nature of colony.

Colonici insects that live in colonies, specially the first generation of wingless female aphids that are fond in large numbers on various host plants.

Colonise to form or establism a colony.

Colonisation controlled release of parasites/predators in a favourable environment with a view to establishing the same.

Colony localised population or individuals of same species – ants, aphids, bees, termites etc.

Color (Colour) appearance: hue: sensation produced by stimulation of optic nerve by particular light-vibrations. Some of the shades are:–

— **Albus** – white
— **Argenteus** – silvery
— **Aureus** – golden-yellow
— **Azure** – faint blue, sea-blue
— **Caesious** – bluish-gray or grayish-green
— **Candidus** – shiny white
— **Castaneus** – chestnut: light reddish-brown
— **Cerise** – light or clear red; cherry-coloured
— **Cerulean (cærulcus)** – sky-blue: dark bluish
— **Cineous** – ash gray
— **Coccineous** – scarlet: bright red
— **Cruentous** – blood red
— **Cyan** – greenish-blue
— **Ferruginous** – rust-red
— **Flavescent (flavus)** – yellowish or yellowish-brown
— **Fulvous** – brownish-yellow: tawny
— **Fusco-piceous** – pitch black with brownish tint
— **Fusco-rufous** – reddish-brown: liver coloured
— **Fusco-testaceous** – dull reddish-brown
— **Fuscous** – dark brown approaching black
— **Glaucous** – sea-green: grayish-blue with greenish tinge
— **Griseus** – pearl-grey
— **Hepatic** – liver-coloured
— **Incanus** – gray

— **Incarnatus (carmeus)** - flesh-coloured
— **Luridus** - pale yellow
— **Luteus** - deep yellow
— **Niger** - black
— **Nivalis (niveus)** - snow white
— **Ochreous** - yellowish with slight tinge of brownish
— **Ochroleucous** - yellowish-white
— **Piceous** - pitch black
— **Prasinous** - bright green
— **Pyrrhour** - reddish-flame-coloured
— **Roseus** - pink
— **Ruba (ruber)** - red
— **Rubens** - reddish
— **Rufous** - brownish-red
— **Sorrel** - pale reddish-brown or light chestnut
— **Tarquoise** - greenish-sea-blue: peacock-coloured
— **Testaceous** - dull yellowish-brown: bile coloured
— **Varicolor** - variously coloured
— **Virens (viridus)** - green

Colydiidæ tropical family of Heteromera (Cucujoidea: Polyogaga); small elongate or cylindrical, bark or ground beetles; antennæ 10- or 11-segmented including one or two-segmented club: tarsi 4-segmented first 3-4 abdominal sternites fused (1500 species) - species of *Bothrideres, Monoedus, Sarothias* et al.

Commensalism association between organisms of different species wherein one is beneficial and other is neither benefited nor harmed; the association may be internal or external.

Community complex group of living organisms that inhibit a particular area and interact with each other.

Compatibility ability of pesticides to get mixed up without any physical or chemical interactions that would lead to reduction in biological efficacy or produce any other adverse effect.

Competition interference occurring among individuals utilising a common resource which is insufficient to satisfy the needs of all.

Compound eye composite optic organ made up of numerous ommatidial found only in crustaceans and insects.

Compressed flattened from side to side.

COMSTOCK, Professor John Henery - born February 24, 1849 at Janesville, Wisconsin. Graduated in 1874 from Cornell University; higher education 1874-75 at Yale University and 1888-89 at

University of Leipzig; Assistant Professor of Entomology 1875-78 at Cornell University; U.S.A Entomologist (1879-81) at Washington, again at Cornell as Professor of Entomology till 1914 and Professor Emeritus thereafter. Described bulk of North American species of Coccidæ and authored An Introduction to Entomology (1920/1949). Died in 1931.

Concentrate spraying direct application of pesticide concentrate, without dilution.

Concentration proportion of active ingredient (a.i.) in pesticide formulation – before or after dilution.

Concolor uniform in colour.: uniformly conloured.

Coniopterygidæ Enderlein 1905 family of dusty-wings (Coniopterygoidea: Planipennia); very small fragile insects covered with fine whitish powdery wax; antennæ long, having 16 to 43 bead-like segments; eyes large, ocelli, absent and wing-venation simplified by reduction with very few cross-veins. Larvæ pyriform; (50 species) – *Coniopteryx pusana* Withycombe, *Nimboa basipunctata* Withycombe, *Parasemidalis flaviceps* Banks, *Piloconis guttata* Enderlein, *Semidalis aleyrodiformis* (Stephens) et al.

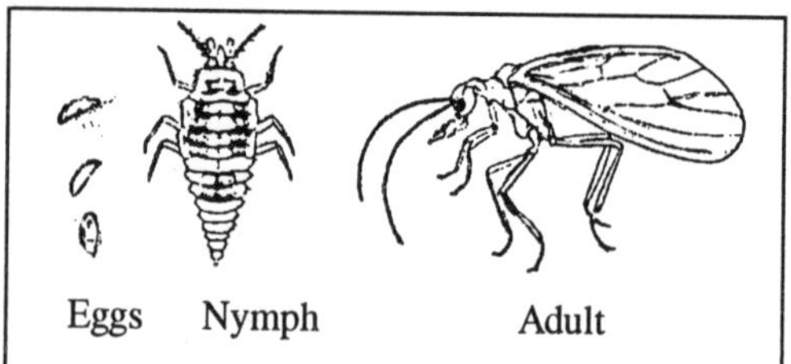

Parasemidalis flaviceps Banks

Coniopterygoidea monofamilial superfamily of Planipennia (Neuroptera: Neuropteroidea); very small insects covered by whitish powdery exudation; wing expanse 4 to 10 mm; forewings with Rs having only two branches, veins not bifurcating near the margin – Coniopterygidæ.

Conjugation temporary union of two sexes of same species of animals for exchange of sex cells.

Connective longitudinal cord of nerve fibre connecting successive ganglia.

Conopidæ (Leach 1815) Stephens 1829 family of thick-headed wasp-flies long and conspicuously jointed; palpi reduced; wings hyaline or ducky, anal cell closed. Maggots endoparasitic on bees and wasps (600 species) – *Conops erythrocephala* Fabricius, *Physocephala* *bicolor* Brunnetti are found in India, *C. flavipes* Linnæus in Europe and *P. affinis* (Williston) in USA.

Conservation environmental modification or manipulation of the habitat and multiplication with a view to encouraging the survival of a articular organism (insect) without disturbing the natural balance of ecosystem.

Conspecific individuals of same species.

Consumer nurtitional grouping in food chain of an ecosystem, composed of heterotrophic organism or population that eats dead or living organic matter.

Contact poison chemical that kills by contact action; usually through penetration from body cuticle – insecticides such as nicotine sulphate, HCH, DDT etc.

Contamination pollution: presence of unwanted organisms may be in air, food or even in an insect population.

Control, Insect checking the insect population from going beyond certain limit, it may be natural or applied.

Control, Applied checking the insect population by adapting various measures - mechanical, cultural, physical, chemical, biological, legislative, integrated etc.

Control, Biological control of pests by means of parasites, predators and disease producing organisms.

Control, Cultural regular farm operations with a view to destroying pest population or preventing the attack thereof- clean cultivation including removal of dead branches, decaying fruits, vegetable trash and weeds; adjusting field operations including sowing and harvesting dates; shifting the location of crop; crop rotation and growing resistant (tolerant) varieties.

Control, Genetic method of pest control in which selected strains of target species, that have genetic abnormalities are released into target population wherein their mating with normal individuals produce sterile or less viable progeny.

Control, Legislative various laws and regulations enacted by the Government to prevent the introduction of foreign pests and retard spread and increase of certain pests within the country.

Control, Natural checking of insect population by nature – influence of climatic conditions, topography; presence of parasites, predators and pathogens on the distribution and welfare of insects.

Control, Physical extremes of temperature and humidity – high or low - and effective barriers in natural spread and control of insects;

these as also light and electricity have been usefully applied by man - light traps; heating and cooling orchards; using smoke, radiation etc.

Copeognatha Enderlein 1903 old and obsolete name for insect order Corrodentia (Psocoptera Shipley 1904) that included psocids or book-lice.

Copper(s) group of widely distributed small, swift-flying butterflies (Lycænidæ: Papilionoidea); upper surface of wings iridescent copper spotted and margined with black or dark brown and undersurface dull and cryptic. Caterpillars flattened and slug-like - American copper *Lycæna hypophlæas* (Boisduval), bronze copper *L. thæ* (Guerin), large copper *L. dispar,* small copper *L. phlæas,* et al.

Copridæ equivalent name for Scarabæidæ (Polyphaga: Coleoptera) that includes chafer-beetles.

Copromorphidæ a small family of Compromorphoidea (Ditrysia: Lepidoptera); moths having head smooth-scaled; forewing having raised scale-tufts; hind-wing with all three branches of M present. Found only in Australia and South America. Caterpillars of certain species have been reported attaching shoots and stems of fig trees (*Ficus* species) in Australia.

Copromorphoidea (Alucitiodea) a superfamily of Ditrysia (Lepidoptera); characterised by haustellum not scaled and wings often deeply cleft or divided into plume-like divisions. The group is sometimes placed outside the tineoids. The families included are - Copromorphidæ, Orneodidæ, and Carposinidæ.

Coprophagy feeding on dung or excreta.

Coprophilous living in dung.

Coptosomatidæ old and equivalent name for heteropteran family Plataspididæ (Pentatomoidea; Pentatomomorpha).

Cordate heart-shaped.

Cordulegasteridæ Banks 1829 small family of large yellow and black dragonflies (Æschnoidea: Anisoptera); hind-wing with clearly marked anal loop; pterostigma moderately long and narrow; female having long ovipositor with lateral valves vestigial and remaining hypertrophied (30 species) - *Cordulegaster dorsalis* Hagen (Pacific coast of USA) is 75 to 80 mm long with wing span of 95 to 100 mm; *Anotogater* spp. are Palæarctic and Indo-Malaysian species.

Corduliidæ Banks 1892 large anisopteran family (Libelluloidea; Odonata); strong fliers having tibiæ usually coppery to metallic bluish green and keeled in male; triangle of forewing with upperside more than half as long as basal side and anal angle of the wing rounded (200 species) - *Somatochlora* spp. *Cordulia* spp. *Macromia* spp. et al.

Cordyluridæ Macquart 1835 family of leaf-minning, stem-boring and root-feeding flies (Schizophora: Diptera); closely related to Muscidæ and often placed as subfamily (Cordylurinæ); under Muscidæ. Adults predaceous and grubs mostly scatophagous or phytophagous; includes species of *Cordylura, Hydromyza Leptopa, Norellia* et al.

Coreidæ leach 1815 (= Coreidæ Samonelle 1819, Stephens 1829, Westwood 1840, Stal 1872, Reuter 1875, 1910) family of leaf-footed or squash bugs (Coreoidea: Gymnocerata); antennæ 4-segmented; rostrum 4-segmented and not curved; eyes well developed ocelli present, scutellum small, tarsi 3-segmented and having pulvilli, hemelytra with corium, clavus and membrane; polyphagous and emit foul smell (2000 species) - rice gundhi bug (*Leptocorisia varicornis* Fabricius) is serious pest of paddy in Indian subcontinent; squash bug (*Anasa tristis* de Geer) is major pest of cucurbits in North America.

Coreidea Leach 1815 old and obsolete spelling of heteropteran family Coreidæ (Coreoidea: Pentatomomorpha).

Coreoidea superfamily of Pentatomomorpha (Geocorisæ; Gymnocerata); elongated or elongated-oval body; antennæ 4-segmented; clavus extremely beyond scutellum; membrane with more than 5 longitudinal veins that are often branched; legs strong claws with pulvilli; scent-gland openings conspicuous (emit bad odour); splitted into give families - Coreidæ, Alydidæ, Rhopalidæ, Stenocephalidæ and Hyocephalidæ.

Coriaceous having leather-like surface.

Corioxenoidea a superfamily of Strepsiptera; include species that are connecting link between the primitive forms and the higher families; endoparasitic on bugs (Heteroptera) - Callipharixenidæ and Stichotrematidæ.

Corixidæ (Leach 1815) dominent family of aquatic heteropteran bugs (Corixoidea: Hydrocorise), commonly called water boatman; having same characters as superfamily; remain at the bottom of water, most species microphagous - *Arctocorixa abdominalis* (Say), *A. mercenaria* (Say) et al.

Corixoidea monofamilial superfamily of Hydrocorisæ (Pentatomonorpha: Geocoriæ); aquatic bugs having depressed body; antennæ 3 or 4 segmented, rostrum 1 or 2 segmented; ocelli present or absent; fore tarsi spatulate and armed with a row of pegs in males, hind tarsi also clawless - Corixidæ.

Corizidæ (Mayr 1866) old and equivalent name for heteropteran family Rhopalidæ (Coreodea: Pentatomomorpha).

Cornea outer transparent portion of each ommatidium of a compound eyes, admits light rays but prevents ultra-violet rays.

Corneagen cells layer of transparent epidermal cells beneath the cuticular lenses (cornea) in a compound eye of an insect or a crustacean. As the lenses are shed at each moult, the same are renewed by means of these corneagen cells.

Cornicle(s) pair of tubules present on V abdominal segment of most aphids (Aphidid - wanting in Phylloxerinæ), secretory channels excreting waxy fluid not honeydew) for self protection; often erroneously called honey-tubes.

Corpora allatum (pl. **corpora allata**) a pair of small ductless glands (endocrine structure) behind the brain of insect larvæ or other immature arthropod that secretes juvenile hormone (neotenin) which maintains the insect in its larval state at each moult.

Corrodentia (Burmeister 1839) Comstock and Comstock 1895 alternate and equivalent name for insect order Psocoptera Shipley 1904 that includes book-lice.

Corydalidæ Burmeister 1839 (Corydalida Leach 1815) family of dobsonflies (fishflies) (Megalopteran: Neuroptera); aquatic insects having prognathous head with three ocelli. Long filiform antennæ, biting mouthparts; 4th tarsal segment cylindrical; wing-span 50 to 130 mm. Larvæ with eight pairs of unsegmented lateral abdominal gill-filaments and a pair of hooked apical anal prolegs but no terminal anal filament; predaceous on naiads of dragonflies, mayflies, stoneflies and other small aquatic creatures (100 species) - *Corydalus cornutus* (Linnæus) of North America is 120 to 130 mm long; besides species of *Chlorina, Neuronus, Nigronia, Parachauliodes, Platyneuromus* et al.

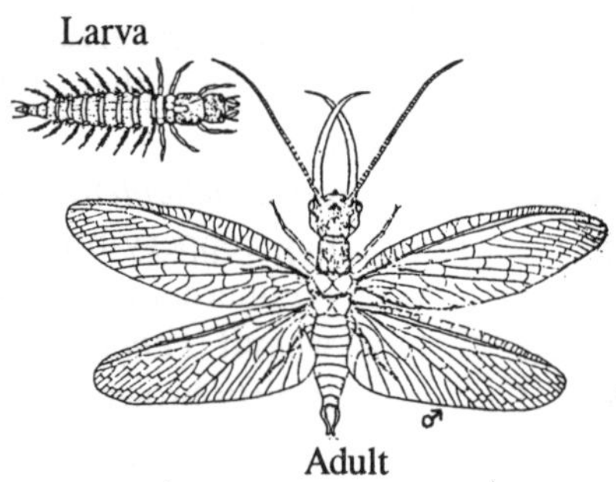

Corydalus cornutus (Linnæus)

Corydiiæ Bunner 1865 family of Blattaria (Dictyoptera: Blattopteroidea): comprising of heterogenous group of brachypterous and apterous, large and small winged species; hind-wings of alate forms with small anal area having one or two folds – Pacific roach (*Euthyrhpha pacifica* Coquerel)), erratic roach [*Arenivage erratica* (Rehn)] et al.

Cosmopolitan having world-wide distribution: species occurring in almost al major regions of the World.

Cossidæ Leach (1815) Walker 1855 a family of Cossoidea (Ditrysia: Lepidoptera); hairy and scaly moths having antennæ smooth or plumose; proboscis and maxillary palpi absent, labial palpi short or vestigial; wings with primitive type of venation, frenelum present and wing span 40 to 180 mm – carpenter moth, goat moth, leopard moth et al.

Cossoidea Tillyard (1926) a superfamily of Ditrysia (Lepidoptera: Mecopteroidea); moth with probosus absent; in both pairs of wings Cu_2 present and vein M furcate within the cell; areole present in orewings, caterpillars wood borers—cossidae.

Costa (C) costal vein: fist vein of insect's wing, usually running along the anterior margin and connected basally with humeral plate.

Costal margin anterior margin of insect's wing.

Cotton leaf roller *Sylepta derogata* Fabricius (Pyraustidæ: Lepidoptera); medium-sized moth with yellowish wings having brown wavy markings. Green caterpillar rolls the leaf and feeds on leaf tissues from within.

Sylepta derogata Fabricius

Cotton shoot roller *Phycita infusella* Meyrick (Phycitidæ: Lepidoptera); small, delicate moths with elongate wings, hind-wings broad bear pectin of hair on dorsal side. Caterpillars attack top shots of cotton, the attacked leaves wither, dry up and drop of, thereby the plant growth is arrested.

Cotton stem borer *Sphenoptera gossypii* Cotes (Buprestidæ: Coleoptera); adult a shining coppery-brown beetle, 8 mm long; grub bores into stem feeds on internal tissues and pupates within the larval gallery.

Cotton stem weevil *Pempherulus affinins* Faust (Curculionidae Coleoptera); dirty brown weevil, adult (3 mm long) feeds on bark of plants while grub bores its way inward reaching in between the bark and the stem where it makes tunnels and feeds on the soft tissues.

Coumarin (Cumarin) $C_9H_6O_2$; substance that deters feeding by certain insect pests.

Cow-killer large bright red coloured wasp with black and white strip, *Dasymutilla occidentalis* (Linnæus) (Mutillidæ: Scolloidea), 20 to 30 mm long; female wingless, covered with dense velvety hair and having a very strong sting; also called valvet-ant.

Coxa (pl. **coxæ**) first or basal segment of insect leg that articulates with the body; designated as coxapodite in other arthropods.

Coxal pertaining to coxa.

Coxal corium articular membrane surrounding base of coxa (insects leg).

Coxal gland excretory gland opening at the base of leg in some insects.

Coxal plate plate-like extension of hind-legs of certain water beetles (Coleoptera).

Coxite (coxopodite) primary basal segment of generalised arthropod appendage (leg), implanted in membranous pleural wall of body segment.

Coxopleurite sclerite of thoracic pleuron adjacent to dorsal margin of coxa and bearing dorsal coxal articulation: eutrochantin.

Crabronidæ (Leach 1815) Dahlbom 1845 (Thyreopidæ) family of square-headed wasps (Sphecoidea: Apocrita), < one to 17 mm long, head quadrate or subquadrate; clypeus transverse and usually hirdute; mandibles acute, simple, bidentate or tridentate; wings hyaline or fuscous – *Crabro latipes, C. 4-cinctus* et al.

Crambidæ Duoncel 1844 family of Pyralidoidea (Ditrysia: Lepidoptera), earlier a subfamily (Crambinæ) of Pyralidæ; small, slender, finely and closely scaled grass-moths, crepuscular or nocturnal or nocturnal in habit, proboscis well developed and variable in length; ocelli prominent; labial palpi long an straight; maxillary palpi strongly triangular and dilate with scales at extremity; tympanal organs present. Caterpillar serious pests of various graminaceous crops – *Chilo simplex* Butler (rice-borer), *C. partellus* (Swinhoe) (sorghum-borer), *Chilotræa saccharalis* (Fabricius) (American sugarcane-borer). *D. zeacolella* Dyar (American corn-stalk-borer) et al.

Cranefly daddy-long-legs; slender-bodied, extremely long-legged flies (Tipulidæ: Nematocera); having conspicuous mouthparts and prominent halteres (balancers), body 4 to 40 mm long with wing-expanse upto 100 mm (*Tipula brobdignagia* Westwood); maggots often nick-named as leathery-jackets.

Cranefly, Banded European species, *Pachyrhina maculata* Meign; adult fly having yellow abdomen conspicuously marked with a row of black bands; injurious to bulb and root crops.

Cranefly, Gaint *Holorusia rubiginosia* Loew of North America; 35 mm long with a wing span of 55 mm; lives among roots and stems or in mud of aquatic plants in ponds or slow running streams of Pacific coast.

Cranefly, Leaf-eating *Cylindrotoma splendens* Doane of Alaska (USA), 9 to 11 mm long; oviposits in leaf tissues on which the magots feed.

Cranefly, Snow snow-gnat, *Chionea nivicola* Doane is the smallest cranefly, about 4 mm long, apterous; found in snow-clad mountainous regions of Canada.

Cranium sclerotic skull-like part of head.

Crawler active first instar larva of most scale insects and whiteflies.

Cremaster hooked or spine-like appendage at posterior end of lepidopterous pupa; used for attachment.

Cremastochilini group (tribe) of sombre-coloured nocturnal beetles (Cetoniidæ: Scarabæoidea), grubs and adults living in nest of ants (myrmecophilous) and termites (termitophilous).

Crepuscular active during twilight or dusk (evening and dawn (early morning) - dragonflies.

Crickest(s) common name of various orthopteran insects belonging to families Gryllidæ (crickets), Gryllotalpidæ (mole crickets), Stenopelmatidæ (sand crickets) and Tridctylidæ (pigmy mole crickets); closely related to long-horned grasshoppers but differ in having brown colour and long unsegmented anal cerci. Chirping (singing) sound is produced by males by rubbing together the bases of their forewings.

Cricket, Australian King *Anostostoma australasiæ* Gray (Stenopelmatidæ); brown apterous species, about 75 mm long; male having enormous head and huge projecting labium and mandibles; found mainly in eastern Australia.

Cricket, Black tree *Homoeogryllus japonicus* Haan (Gryllidæ: Ensifera) is native to Japan, found also in China.

Cricket, Burrowing *Brachytrypes portentosus* (Lichtenstein) (Grylliadæ: Ensifera) found in Indo-gangetic plains is nocturnal and forages on casuarina (*Casuarina equisetifolia*) seedlings.

Cricket, Bush erroneously called long-horned grasshoppers, these species of Tettigonidæ (Tttigoniodea: Ensifera) differ from true grasshoppers (Acrididæ) in having very long thread -like antennæ, and from other crickets (Gryllidæ) and mole crickets (Gryllotalpidæ) in having tarsi 4-segmented; hearing organs situated on foretibia; males produce peculiar sound by rubbing toothed rib of left forewings against the hind margin of right forewing – *Leptophyes punctatissima* (speckled), *Conocephlus dorsalis* (short-winged), *Meconema thalassinum* (Oak), *Metrioptera brachyptera* (Bog), *Pholidoptera griseoaptera* (Dark), *Tettigonia viridissima* (Great green) et al.

Cricket, Chinese species of Gryllidæ (Grylloidea: Ensifera) found commonly in China, include *Gryllus conspersus* Schaum (Besprinkled cricket), *G. mitratus* Burmeister (Mitred cricket) and *Loxoblemmus taisun* Saussure (Broad-faced cricket).

Cricket, Desert *Gryllus desertus* Pallas (Gyralidæ: Gryloidea) has been reported from South Europe, North Africa and West Asia.

Cricket, Field *Gryllus campestis* Linnæus (Gryllinæ: Gryllidæ) is the most common species found in North Africa, Asia (including India) and Europe.

Cricket, House *Acheta (Gryllus) domesticus* (Linnæous) (Gryllinæ; Gryllidæ) is cosmopolitan in distribution, having long antennæ; 3-segmented tarsi; auditory organs on foretibia; ovipositor long, slender and needle-like; cerci extremely long and unsegmented; great nuisance in houses and also bit boles in clothers. *Gryllodes melanocephalus* (Seville) is another species reported from India.

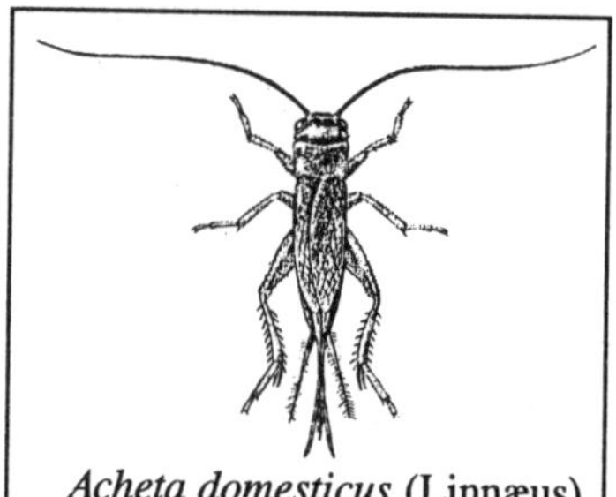

Acheta domesticus (Linnæus)

Cricket, mole *Gryllotalpa africana* Palisot de Beauvois (Gryllotalpidæ: Grylloidea); sort antennæ; eyes reduced; 2 or 3-segmented tarsi,

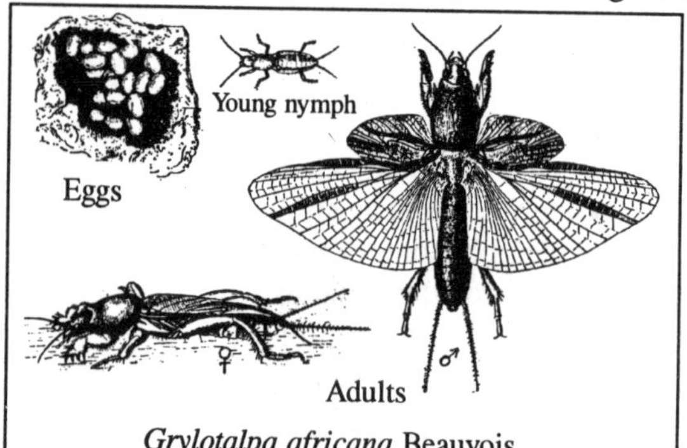

Grylotalpa africana Beauvois

broad spade-like toothed forelegs adapted for diggings; forewings short; stridulatory organs wanting; ovipositor vestigial; subterranean; cause serious damage to potato (*Solanum tuberosum* Linnæus) in field and storage. *G. gryllotalpa* (Linnæus) and *G. vineæ* are rare species, subterranean and found in damp meadows; while *G. borealis* is common in North America, *G. vulgaris* Latreile in North Africa and *G. ornata* Walker in India.

Cricket, Oceanic *Gryllus oceanicus* Le Guillou (Gryllinæ: Gryllidæ) reported from Japan, Pacific Island and Hawaii.

Cricket, Shell *Gryllus testaceous* Walker (Gryllinæ: Gryllidæ) is found in the Indian subcontinent, Japan and South-east Asia.

Cricket, Shrill house *Brachtrypes megacephalus* Lefebure (Gyrallidæ: Grylloidea) has been recorded from Sicily and North Africa, makes a noise so penetrating that it can be heard even at the distance of two kilometres.

Cricket, Spanish *Gryllus hispanicus* Rambur (Gryllinæ; Gryllid) confined to Spanish peninsula.

Cricket, Surface burrowing *Platygryllus melanocephalus* (Serville) (Grylidæ: Grylloidea) reported from the Indian subcontinent nibbling leaves of mango trees; does not make deep burrow in the soil.

Cricket, Two spotted *Acheta bimaculata* (de Geer) (Gryllidæ: Grylloidea); omnivorous, frequents hot dry places in North Africa, South Europe and West Asia.

Cricket, Water actually not crickets but aquatic bugs (Velidæ; Cettoidea), found all over the World. *Velia*. spp. occur in slow moving streams, *Rhagovelia* spp. in swift steams while *Halovelia* spp. are marine; found in Indian and Pacific oceans.

Cricket, Wood *Nemobius sylvestris* (Gryllidæ: Orthoptera) smallest species of European crickets, about 10 mm long dark brown in colour and having short flap-like forewings and no hind-wings; active during day. These run and jump with a curious jerk; live among dead leaves and rotting wood; commonly found in New forests of England.

Crista acustica organ of Siebold: chordotonal organ situated on the outer face of anterior trachea: an elongated ridge composed of a large number of scolopophores of subintegumental type that gradually decrease in size towards the distal extremity of the tibia.

Crochet(s) hooked spines at tips or prolegs of caterpillars, arranged in characteristic (diagnostic) manner.

Crop enlarged and modified part of oesophagial region where food is held in anticipation of its admission to gizzard or·stomach. It may be a simple dilation as in cockroaches etc., or a lateral diverticulum connected to oesophagous by a narrow tube as in butterflies, mothes etc.

Cross-vein(s) short veins located at definite position between longitudinal veins in the wings of insects.

Crumera minute pouch or pocket lined by thin membrane found in Coccoidea (Homoptera), situated between central nervous system and ventral body wall; connected with channel of labium.

Crura cerebri para-oesophageal connectives; two cords one on each side of oesophagus that connect the brain and suboesophaeal ganglion.

Crustacea class of phylum Arthropoda (now considered as independent phylum); mainly aquatic mostly marine animals; exoskeleton often massive, head bears two pairs of antennæ, one pair of compound eyes, one pair of lateral mandibles, two pairs of maxillæ and five pairs of walking legs: carapace often present, breathe mostey by gills, rarely by body surface. About 39000 species placed under nine subclasses - Cephalocardiam, Remipedia, Brachiopoda, Ostracoda, Mystacocardia, Copepoda, Branchiura, Cirripedia and Malacostraca; the last one includes, crabs, crayfishes, lobsters, prawns, shrimps etc.

Cryolite Na_3AIF_4; sodium fluoaluminate; colourless crystalline stomach poison, soluble in alkalies and dilute acids: first used as insecticide by Morcovitch and Stanly in 1929; effective against biting and chewing insects, it was cheaper and less toxic to higher animals; oral LD_{50} value for rates being 13.5 only.

Crytocerata Fiebre 1851 equivalent name for Hydrocorisæ, a subborder of Heteroptera comprising of aquatic or semi-aquatic bugs found all over India but not of any economic importance - backswimmers, giant water-bugs, water-boatmen, water-scorpions, toad-bugs etc.

Cryptostemmatid (Ceratocombid) old and equivalent name for heteropteran family Dipsocoridæ (Cimicomorpha: Gymnocerata); very small (<2 mm); delicate bugs; head porrect; eyes small, ocelli present; antennæ 4-segmented, basal two very small last two long and pilose, third swollen at the base; metathorax without scent glands and tarsi 3-segmented - jumping ground bugs.

Ctenophora phylum of comb-jellies (Radiata: Metazoa); small marine organisms having eight rows of comb-like plates, used for locomotion. About 100 species distributed in two classes and even orders - Tentaculata (Cydippida, Ganeshida, Lobata, Thalassocalycida, Cestida, Platyctidd) and Nuda (Beroida).

Ctenuchidæ (Amatidæ, Syntomidæ) a family of Sphingoidea (Ditrysia: Lepidoptera); brilliantly coloured, tropical, medium-sized moths, largely diurnal and inactive, with well developed proboscis characterised by forewing much longer than hind-wing and having Sc + R_1 fused with Rs. Many species wasp-like in appearance. Caterpillars small in size with verrucæ bearing numerous setæ; freed on grasses or lichens and pupate in cocoons of felted hairs and silk. About 2000 species - *Amata passalis, Ceryx godarti, Psychotoe duvauceli* et al.

Cubitus (Cu) fifth vein of insect wing primarily two branched (Cu_1 and Cu_2), which break up into number of secondary branches.

Cucubano local name in West Indies for fire-beetle, *Pyrophorus luminosa* Illiger (Elateridae: Polyphaga); over 25 mm long, emits a brilliant green light from the two pale yellow prothoracic spots and reddish light from venter of abdomen.

Cucujidæ (Leach 1815) Stephens 1829 family of flat bark beetles (Clavicornia: Cucujoidea); red, brown or yellow coloured; antennæ 11-segmented, often filiform, sometimes distally enlarged; elytra covering abdomen; first tarsal segment much shorter than second; found under tree bark in decaying wood; majority predaceous on coccids and mites, a few infest stored-grains - *Cryptolestes pusillus* Schonherr feed on biscuits; *Oryzæphilus surinamensis* (Linnæus), *Silvans unidentauts* Fabricius and *Cathartus quadricolis* Guerin are destructive to dried fruits and cereals (in storage), while *Laemophooleus* spp. are both predaceous and grain-eating.

Cucujo common name for fire-beetle, *Pyrophorus noctilucus* (Linnæus), a luminous elaterid (Polyphaga: Coleoptera) of Tropical North and South America, metallic green in colour with two pale yellow prothoracic spots that emit brilliant green light while reddish light appears from venter of its abdomen; more or less similar to Cucubano of West Indies.

Cucujoidea sharp 1912 superfamily of Polyphaga (Coleoptera: Colepteraoidea); having antennæ filiform or clubbed rarely serrate and ocelli absent; abdomen normally with five visible serenities, if with VI or VII, tarsi always heteromerous; includes large number of families grouped under two sections - Clavicornia and Heteromera.

Culicidæ Stephens 1829 family of mosquitoes (Nematocera: Diptera); small, slender, delicate flies having no ocelli and thin long legs covered with scales; wings also fringed with scales along the posterior margin and the veins. Sexual dimorphism very prominent - female having slender piercing proboscis and antennæ sparsely hairy (pilose) while the male has much reduced mouthparts and antennæ densely plumose: females suck blood and also transmit various diseases while males feed on nectar - *Anopheles culicifacies* and *A. stephensi* caused malaria, *Culex fatingans* transits filariasis and *Aedes ægypti* (Linnæus) is responsible for yellow fever and dengue.

Culicide pesticide (toxin) that can kill mosquitoes.

Cuneus wedge-shaped area that is part of corium - the thickened basal portion of hemelytra.

Cupedidæ (Cupesidæ: Cupidæ) small family of reticulated beetles (Archostemata: Coleoptera); antennæ 11-semented; metasternum with a deep ante-costal suture and abdomen with five free ventral segments - not reported so far, from India; Mesozoic fossils are common in Britain rocks.

Cupidinidæ old name for lepidopteran family Lycænidæ (Papilionoidea: Ditrysia).

Curculionidæ Leach 1817: family of snout-beetles or weevils (Curculionoidea: Polyphaga); had prolonged into snout that is curved downwards and bears mouthparts at its tip, labrum rarely present, palpi small, short and rigid concealed within the mouth; eyes well developed; antennæ elbowed (geniculate), trochanters rarely long; body usually covered with scales or hair. Two subdivisions - Adelognatha and Phanerognatha - *Myllocerus maculosus* Desbrochers, *Tanymecus circumdatus* (Wiedemann) et al.

Curculionoidea Hopkins 1911 largest superfamily of Polyphaga (Coleoptera: Oligoneoptera), head characteristically elongated into rostrum (except Scolytidæ); antennæ usually clubbed; phytophagous and includes a large number of notorious pest of various economic crops (over 60000 species).

Cursoria Westwood 1839 an insect order in old classification that included running orthopteran insects comprising of wingless, short-winged and long-winged forms, grouped into three families - Blattidæ (cockroaches), Mantidæ (mantids) and Phasmidæ (walking-sticks) now placed under order Dictyoptera (Blattaria and Mantodea) and Phasmida.

Curtonotidæ small, unimportant dipteran family of Acalyptratæ (Schizophora: Cyclorrhapha).

Cutellus small knife-like proboscis of some blood sucking fleas.

Cuterebridæ Brauer 1889: myodarian cyclorrhaphous dipteran family of large hairy, robust-flies (Calyperatæ: Schizophora); had with receptacle for receiving the proboscis; proboscis reduced; oral opening large; palpi very small; scutellum elongate; wings and squamæ large, vein M sharply bent. Maggots living under the skin of living mammals. About 5 species, confined to New World – human botfly [*Dermatobia hominis* (Linnæus)] et al. human botfly *Dermatobia hominis* (Linnæus); *Cuterebra* spp. infest various domestic animals including rabbits and rodents.

Cuticle complex protective non-cellular layer secreted by hypodermis; its major components are chitin 25 to 60 per cent of dry weight and protein it forms the outermost investment of body and its appendages as also provides the ling of tracheal system, some glands and parts of alimentary canal and reproductive tract; it consists of two main layers thin outer epicuticle (exocuticle) and thin inner procuticle (endocuticle), when newly formed it is flexible and elastic.

Cuticula outer non-cellular layer of body wall.

Cuticulin inner most non- chitinous layer of tracheal cuticle.

Cutworm(s) common name for some nocturnal noctuid caterpillars specially *Agrotis* spp., due to their habit of cutting the tender stems and shoots of young plants and seedlings at ground level, then feeding on he same. Polyphagous and cosmopolitan in distribution. Most common species *A. ipsilon* (Hufnagel) moth have grayish-brown wavy lines and spots on forewings and creamy-white hid-wings with a dark terminal fringe; wing expanse 42 to 56 mm; also known as greasy cutworm. Other cutworms include, spotted cutworm *Amathes c-nigram* (Linnæus); dingy cutworm *Fellia subgothica* (Haworth) and variegated cutworm *Peridroma margaritosa* (Haworth).

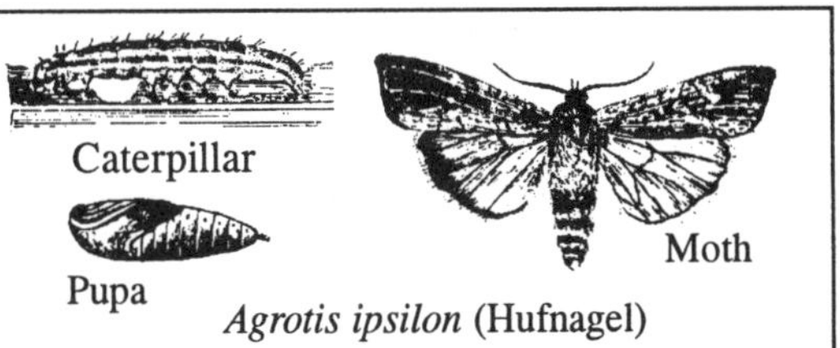

Agrotis ipsilon (Hufnagel)

Cyanogas a fumigant, available as fine dark-gray powder that contains about 20% calcium cyanide, $Ca(CN)_2$; used as a rodenticide, it is to be pumped into rat-burrows in the field or orchard; the chemical on coming into contact with soil moisture, liberates draconic acid (HCN), which is a deadly poisonous gas and kills all the rodents therin instantaneously.

$$Ca(CN)_2 + 2H_2O \rightarrow 2\ HCN + Ca(OH)_2.$$

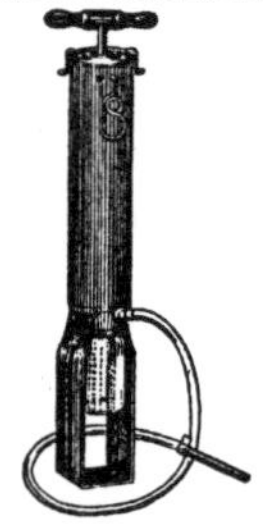
Cyanogas foot pump

Cyanogas foot pump a hand operated pump, to be held in position by the foot of the operator; used for fumigating the live rat-burrows in the field or orchard. Cyanogas powder is filled in the small container, the nozzle at the end of rubber tubing is pushed into the rat-burrow and sealed temporarily with earth; a few strokes of the pump push a small quantity of the powder into the burrow to kill all the occupants inside.

Cyclorrhapha Brauer 1880 suborder of Diptera; head with a frontal lunule and usually a ptilimum,; antennae 3-segmented; forewing with discal cell always presents, cubital cell contracted or closed. Maggots having vestigial had and pupation inside he puparium. Divided into three section – Aschiza, Schizophora and Puripara.

Cyclotornidæ very small family of Tinæoidea (Ditrysia: Lepidoptera); includes parasities; confined to Australasia – freshly hatched caterpillars of *Cyclotorna* species parasitise the bugs (Homoptera) white in their second instar, these live in ants' nests.

Cycnodidæ old and equivalent name for lepidopterous family Elachistidæ (Trinæoidea: Ditrysia).

Cydnidæ (Billberg 1820) large family of ground and burrowing bugs (Gymnocetrata: Heteroptera); dark coloured insects antennæ 5-segmented, eyes and ocelli present; scutellum large; hemelytra normal with coriaceous clavus; tibiæ spinose, tarsi 3-segmented; abdomen with six connexinia visible dorsally; found under dead wood and stones – *Cydnus* spp., *Pangæus* spp, *Sehirus bicolour* et al.

Cylindrachetidæ small family of orthopteran insects (Tridactyloidea: Cælifera); elongate-cylindrical body, short antennæ, reduced ocelli form eyes; forelegs fossorial and hind legs not saltatorial; confined to Australsia and adapted to burrowing subterranean life – *Cylindracheta kochi* Saussure.

Cylindrocccidæ very small family of Coccoidea (Homoptera); body elongate-cylindrical, pygidium absent; hind legs 2 to 3 times as long as forelegs; spiracles lacking. Confined to Southern hemisphere.

Cymatophoridæ Harrich-Schæffer 1847 (Polyplocidæ, Noctuobombycidæ) small family resembling Noctuidæ, also includes Thyatiridæ Grote 1890, (Geometroidea: Ditrysia), characterised by forewing with M_2 not nearer to M_3 than to M_1 at the base and tympanal organs situated at the base of abdomen. Confined to Northern hemisphere - Buff arches (*Habrosyne derasa*), peach blossom (*Thyatira batis*) gooseberry spanworm (*Cymatophora riberaria* Fitch) et al.

Cymbidæ small group of moths (Chloeopnorinæ) earlier placed under Arctiidæ (Noctuoudea: Ditrysia), now regarded as a separate family or as subfamily of Noctuidæ; usually green in colour; found among herbage of trees and shrubs; caterpillars stout-bodied and never prominently hairy, pupa boat-shaped - *Earias insulana* Boisduval a destructive pest of cotton (*Gossypium* spp.), widely distributed in the Tropics.

Cynipidæ (Halida 1840) Westwood 1840 family of gall-wasps (Cynipoidea: Apocrita); narrow-waisted hymenopterous insects, upto 6 mm in length; antennæ 11- to 16-segmented; thorax rounded and succinctly sculptured; wings usually well developed , sometimes vestigial or absent specially in females; forewings without stigma and hind-wings with 2 or 3 nervures; reproduction bisexual or parthenogenetic, majority produce galls and live within the same - *Callirhytis semicarpifoliæ* Cameron, *Onychia striolata* Cameron, *Neuroterus saltatorius* (Edwards) et al.

Cynipoidea Ashmead 1899 superfamily of minute to small insects (Apocrita: Hymenoptera); characterised by simple antennæ, trochanters undivided; wign-venation greatly reduced with not more than 5 closed cells and without stigma in forewings and anal lobe in hind-wings; ovipositor retractile and coiled; mostly gall-producing but some small families parasitic on other insects (1700 species).

Cypermathrin $C_{22}H_{19}NO_3Cl_2$ synthetic pyrethroid, used as contact insecticide against various sucking pests - Cymbush®, Polytrin®, Ripcord® etc.

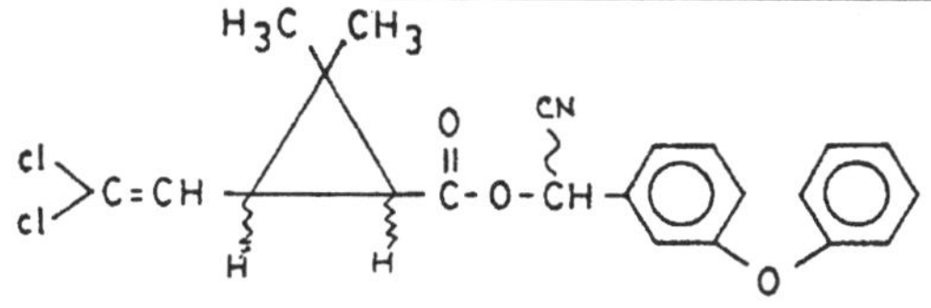

Cypermethrin

Cypselidæ old and equivalent name for dipteran family Borboridæ (Schizophora: Cylorrahapha).

Cyrtidæ old and equivalent name for Acroceratidæ (Brachycera: Diptera) that includes bristleless flies.

D

Dacnonypha one of the four suborders of Lepidoptera (Mecopteroidea: Oligoneoptera); forewing with fibula, hind-wing have some costal spines; females have sternite VIII and IX with long apodemes while males have sternite IX nearly square with two short anterolateral apodemes. Caterpillars opodous; pupæ exarate with hypertrophied mandibles. Only one superfamily - Eriocraniodea.

Dacnostomata Westwood 1839 equivalant and obsolete term for insect order Mantodea Burmeister 1838, often relegated as suborder of Dictyoptera Leach 1818.

Dactylopiidæ Moulton 1931 family of Æolothripoidea (Terebrantia: Thysanoptera); thrips having ovipositor well developed, saw-like and curved upwards.

Dactylopiidæ Maskell 1887 a very small family of Cochineal insets (Coccoidea: Sternorrhyncha); female elongate-oval, 2-3 mm long, convex, with small 5 to 7-segmented antennæ; legs small hidden beneath the body. Male having three pairs of ocellanæ two long caudal filaments and a minute stylus. Only 9 species (2 genera) confined to Mexico; feed on various cacti (Cactaceæ). Dyestuff cochineal extracted from the bodies of females is used for colouring cosmetics, drinks and food-stuffs - *Dactylopius coccus* Costa, *Epicoccus* spp., et al.

Dactylopodite terminal segment of generalised arthropod appendage, typically claw-like; pretarsus in insects.

Daddy-long-legs common name for several long-legged craneflies (Tipulidæ; Nematocera); having thin body, narrow wings and six long fragile legs; larvæ thick-skinned, live in soil, feed on plant roots and are called leather-jackets. The name is also applied to some long-legged spiders.

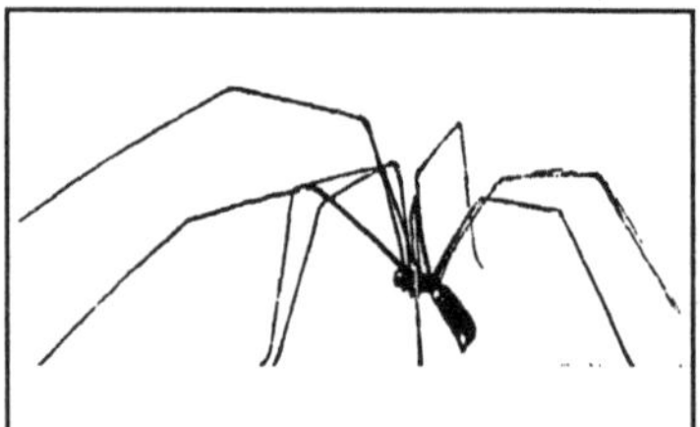

Daddy-long-legs (spider)

Damselfly group of brilliantly coloured, medium-sized, predaceous insects (Zygoptera: Odonata); akin to dragonflies (Anisoptera) but smaller, more delicate and weak fliers; having two subequal pairs of membranous wings usually held together over the body while resting – *Coenagrion puella* and *Ischnura elegans* are broad-winged damselflies; *I. verticollis* Say, found in the rocky mountains is 25-30 mm long, while *Lestes vigilax* Hagen attains length of 53 mm and wing expanse of 56 mm.

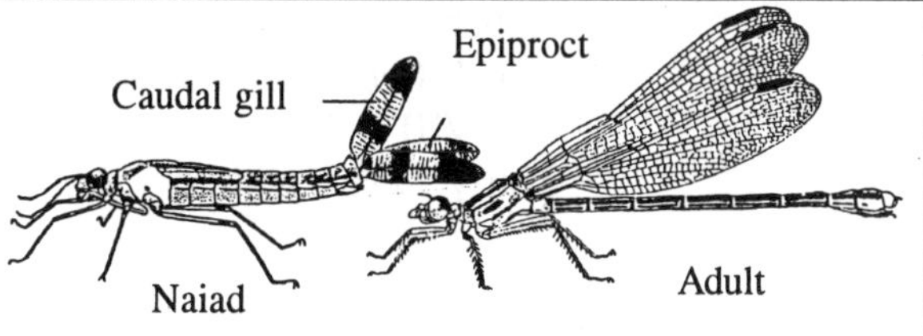

Damselfly (*Lesters* species)

Damselfly, Banded European species, *Agrion splenderns* (Agriidæ: Agrioidea); having iridescent coloured patches on males's wings give the insect banded appearance; flutter like butterflies in sunshine.

Damselfly, Blue *Enallagma cyanthingerum* (Coenagriidæ: Coenagrioidea) have VIII and IX abdominal segments of male prominently blue coloured; though cosmopolitan, more common in Britain and Europe, often swarming around lakes, rivers and small waters. Nymphs slim, found hiding among the aquatic plants.

Damselfly, Demoiselle comparatively large damselfly *Agrion virgo* (Agriidæ: Agrioidea) is European species: males having irridescent signs which flash deep-blue, purple· or green in sunshine.

Damselfly, Green small metallic bluish-green or green-bodied damselfly, *Lestes sponsa* (Lestidæ: Lestinoidea) fond all over the World, while American species *l. vigilax* Hagen is about 50 mm long with wign-expanse of 55 mm.

Damselfly, Red European species, *Ceriagrion tenellum* having red instead of black legs and males with red eyes – Large red damselfly, *Pyrrhosoma nymphula* (Coenagriidæ; Coenagriodea) is another species found all over Britain and Europe.

Danaidae Hampson 1918 (Acraeidæ, Duloeidæ, Limnadidæ, Maniolidæ) a family of Papilionoidea (Ditrysia: Lepidoptera) often regarded as a subfamily of Nymphalidæ (Papilionoidea: Ditrysia); large brightly coloured butterflies; antennæ scaleless with club small or vestigial; forewing with subcostal vein forked at the base. Dimorphism, seasonal a well as sexual is common. Caterpillars of Indian crow *Euploea core* Cramer, feed on leaves of cluster fig or *gular (Ficus glomerata* Roxburgh), *peepal* (*F. religiosa* Linnæus), oleander (*Nerium indicum* Miller) etc., while those of plain tiger *Danaus chrysippus* (Linnæus) feed on rocket-salad (*Eruca vesicaria sativa* Miller); monarch butterfly *D. plexipus* Fabricius is found in North America.

Darner (daring needle) large dragonflies (Anisooptera: Odonata), body length about 100 mm and wign span 120 mm; feed on smaller dragonflies and other insects - *Anax junius* Durry, *A. walsinghami* McLachlan (Æshnidæ: Æshnoidea) are common green darner and giant darner of USA, being 100-110 mm long with wing span of 125 mm.

Darter(s) large flat bodied, broad-winged dragonflies (Libellulidæ; Libelluloidea) that dart out on brief flight with great speed - *Libellula depressa, L. quadrimaculata, Sympetrum danæ, S. sanguineum, S. striolatum* et al, are Holarctic species. Snakebird *Anhinga melanogaster* (Pelicaniformes: Aves) that swims under water and hunts for fish.

DARWIN, Charles Robert (1809-82) British naturalist (biologist); conducted survey and research on board the HMS Beagle and published, Voyage of the Beagle (1839). The theory of evolution proposed by him in On the origin of Species by means of Natural Selection (1859) still holds. He also published Descent of Man (1871) wherein he has described the evolution of man from ape.

Charles Robert Darwin

Darwin theory hypothesis that different species arise by process of natural selection.

Darwinism doctrine of 'Origin of Species' propounded by British biologist, Charles R. Darwin (1809-1882).

Dascillidæ family of Dascilloidea (Polyphaga Coleoptera); medium-sized beetles found feeding on roots of various grasses including paddy. Grubs aquatic, feed on minute algae, protozoa and organic debris.

Day-fly alternate name for mayfly (Ephemeroptera), as most of the adult live for a few hours *i.e.* longevity is said to be one day.

DDT $C_{14}H_{19}Cl_5$; dichloro-diphenyl-trichloro-ethane; organochlorine which is contact and stomach poison, non-systemic, highly persistant and non-phytotoxic except to curcurbits. Colourless waxy crystals, insoluble in water, moderately soluble in petroleum oils and readily soluble in most aromatic and chlorinated solvents. Toxic to bees, fish and livestock. Oral LD_{50} value for male rats 113 mg/kg. First synthesidsed in 1874 by German chemist Othnar Zeiolla, its insecticidal properties were discovered in 1939 by Paul Muller. DDT is disapproved by environmentalists as its is not bio-degradable and as such enters the foodchain of birds, animals and human beings.

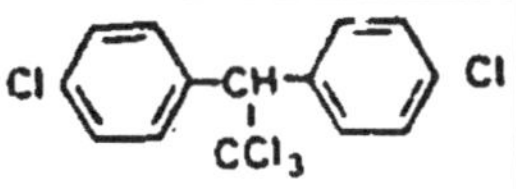

DDT

Dead without life.

Dealation loss of wings, as in case of reproductive castes of ants, termites etc., the mating takes place during night, up in the air, after which the insects rub off or shed their wings and the fertilised female (queen) enters the ground for laying eggs and starting a new colony.

Death cessation of life.

Deccan wingless grasshopper *Colemania spehenarioides* (Bolivar) (Acrididæ: Acridoidea: Cælifeara); green and red adult measures 35 to 40 mm in length; hoppers and adults feed on leaves, flowers and ripening ears of millets and maize crops; also on pulses.

Colemania sphenarioides (Bolivar)

Decticous Pupa exarate pupa with robust mandibles that help the insect to come out from its cocoon.

Defensive secretion(s) caustic irritant or foul-smelling fluids discharged as means of defence – irritant secretins from hair of certain caterpillars, formic acid from some ants, blistering chemicals from cantharidin beetles; butyric acid from carabid beetles; salicylalde-hyde from some leaf beetles; poisonous vapours from bombardier beetles and so on.

Definite (definative) host an insect in which a parasite attains sexual maturity.

Deflocenalting agent material added to spray suspension to delay sedimentaion.

Defoliant substance that induces premature leaf-fall.

Defoliate to deprive of leaves.

Defoliator insect – beetle, caterpillar, grasshopper etc. – that feeds voraciously on leaves and completely strips the plant or tree.

Delphacidæ Spinola 1839 (Aræopidæ) largest family of Fulgoroidea (Auchenorrhyncha: Homoptera) characterised by apical segment of labium being distinctly longer than its width and hind tibia each with a mobile, apical spur; the costal cell wanting and clavus non-granulate (1200 species) – sugarcane leafhopper (*Perkinsiella saccharicida* Kirkaldy), brown backed planthopper [(*Nilaparvata lugens* (Stal)] et al.

Deltamethrin (Decamethrin) $C_{22}H_{19}Br_2NO_3$; synthetic pyrethroid, a third generation insecticide; having quick knock-down effect low mammalian toxicity, high photo-stability and minimum atmospheric pollution; acute oral LD_{50} value for rat 70 to 140 mg/kg – Decis®, etc.

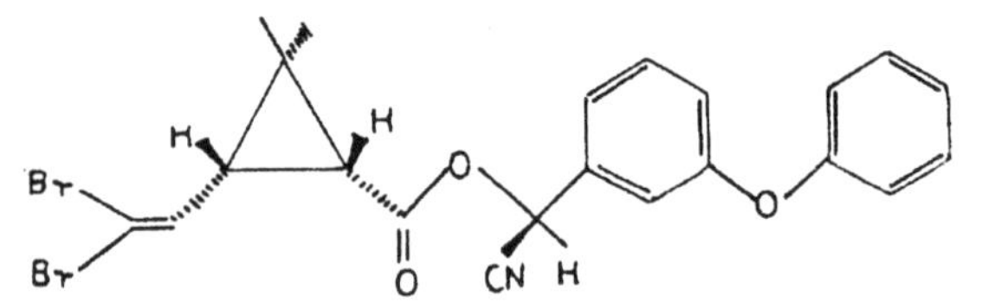

Deltamethrin

Denticulate bearing a row of very small teeth.

Deposit amount and pattern of spray or dust deposited per unit area of plant surface.

Deposition velocity velocity at which the spray impinges on the target.

Depressed flattened dorso-ventrally.

Deratoptera Clairville 1798 old and obsolete term for order Mantodea Burmeister 1838, which is often considered as suborder of Dictyoptera Leach 1818.

Derbidæ family of very delicate, Tropical, long-winged bugs (Fulgroidea: Homoptera); apical segment of labium as long as wide and anal area of hind-wing not reticulate (750 species) *Proutista moesta* (Westwood), *Melenia* spp., et al.

Dermaptera Leach 1815 (de Geer 1773) order of Dermapteroidea (Polyneuroptera; Pterygota); long slender insects with chewing mouthparts; wings absent or present, forewings leathery, devoid of veins and shorter than hind-wings; hind-wings membranous and semi-circular; pair of forcep-like cerci at anal end; metamorphosis incomplete. About 1500 species of earwigs; grouped into three suborders – Forficulina, Heminmerina and Arixaniina.

Dermapteroidea Martynov, 1925 superorder of Polynuropteran insects; primary winged and hemimetabdous; comprises of two orders – an extinct Protelytroptera and a living Dermaptera (earwigs).

Dermesidæ (Gylenhal 1808; Leach 1815) family of carpet beetles or skin beetles (Dermestoidea; Polyphaga); medium-sized oval or hemispherical beetles clothed with fine hairs or scales having short, clavate, 11-segmented antennæ and only one ocellus. larvæ caraboid, have tufts of long hair on their body. Both adults and larvæ feed on dried animal and vegetable matter and are destructive to bacon, cheese, furs, hides, horns, museum specimen, wool etc., some are stored-grain pests while a few are predaceous on mantids – *Anthrenus scrophulariæ* (Linnæus), *A. verbasci* (Linnæus), *Dermestes vulpinus* Fabricius, *Trogoderma granarium* et al are all cosmopolitan.

Dermodermaptera Verhoeff 1902 old insect order erected to include hemimerids that are now placed under Hemimerina, a suborder of Dermapera.

Dermoptera mammalian order (Tetrapoda: Chordata); nocturanals having fore- and hind-limbs connected by a fold of skin (patagium) that helps in flying – flying lemurs (*Cynocephalus* species).

Derodontidæ small family of Dermestoidea (Polyphaga: Coleoptera); characterised by having two ocelli; forecoxal cavities closed behind; hind coxæ extending laterally beyond the outer edge of metepisternum (15 species) – Ten Holarctic species (*Laricobius*), whose grubs live in slime-fungus are predatory on adelgids (aphids).

Derris $C_{23}H_{22}O_6$; rotenone; vegetable preparation that was once used as insecticide.

Desicent chemical that can kill or destroy vegetation by inducing water loss.

Desmergate(s) individual of an ant colony that is intermediate between the soldier and worker caste.

Detrivorous organisms that feed on dead animals or partially decomposed organic matter.

Deuterophlebiidæ a small monogeneric dipteran family of mountain midges (Orthorrhapha: Nematocera) having antennæ very elongate, filiform and 6-segmented; ocelli and mouthparts

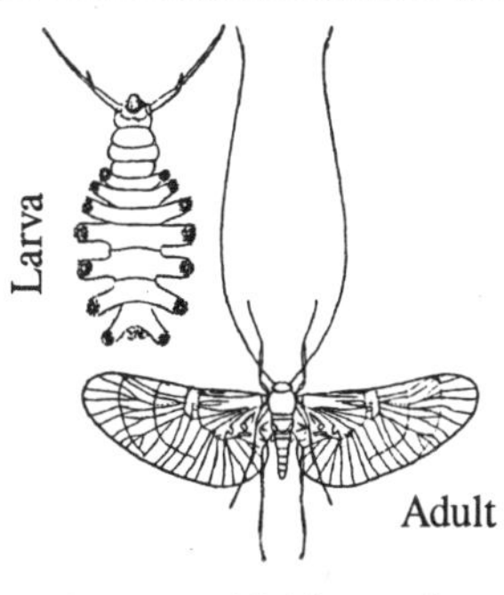

Deuterophlebia species

lacking; wings with a network of crease clothed with fine pubescence and almost all the viens obliterated - *Deuterophlebia* species found in North America and North-west Himalayas.

Deuterotaxy (ampherotokous) parasitoids exhibiting thelotoky, but also producing a few females.

Deuterotoky parthenogenetic reproduction in which females produce progeny comprising of both sexes.

Deutocerebrum midbrain, second or middle portion of arthropod brain containing first antennal nerve centers.

Deutonymph third instar nymph of a mite (Acariformes: Arachnida).

Devil's Coach-horse large European beetle, *Ocypus olens* (Staphylinidæ: Polyphga), over 30 mm in length; body dull black, pointed at hind or posterior end; large hind-wings, folded beneath short elytra; when disturbed, it gives out strong foul-smelling liquid from the glands located at its hind end.

Dexiidae Macdqart 1835 family of Calyptratæ (Schizophora); antennæ arising half way between the eyes and moth; hairy or feathered arista; strongly convex metanotum and ventral abdominal joints concealed below the tergites. Maggots endoparasitic on caterpillars and grubs - *Dexia* spp. on grubs of Melotlonthidæ and *Zophomyia* spp. are predaceous on microlepidoptera.

Diamond back moth *Plutella xylostella* (Linnæus) (Pluterllidæ: Ditrysia), small grayish-brown moth, wing expanxe about 15 mm, characterised by three pale whitish triangular markings on hind margin of each forewing, when at rest, by joining both forewings, three diamond-shaped spot a pear clearly on the dorsum; hind-wings have a fringe of long fine hairs. Caterpillars, pale yellowish-green in colour, are major pest of almost all cruciferous crops.

Diamphipnoidæ very small family of Archiperlarian stoneflies (Plecoptera); differs from Eustheniidæ in having hind margin of hind-wings indented at end of Cu_2; Rs in forewing branched several times. Nymphs with first four abdominal segments bearing tufted paired gills. Confined to Chile (South America), five species in two genera - *Diamphipnoa* and *Diamphipnosis*.

Diapause period of dormancy: slowing down of metabolism, often due to seasonal changes in atmospheric conditions; sometimes due to an inborn rhythm wherein the insect fails to produce normal hormones necessary for continued growth or development.

Diaphragm one of the horizontal partitions of body cavity - dorsal and ventral diaphragms.

Diapriidæ a small family of Proctotrupoidea (Proctotrupoid: Apocrita); minute wasps distinguished by having 11- to 15-segmented antennæ inserted near the middle of the face; mandibles dentate; pterostigma absent and tip of abdomen not stylate; adults winged but the venation greatly reduces. Primary parasites of dipteran flies - *Diapria conica* attacks *Eristalis tenax* Linnæus (Syrphidæ) while *Cinetus fulvus* parasitises *Arachnocampa liminosa* Skuse (Mycetophilidæ).

Diaspididæ Maskell 1878 largest family of Coccoidea (Sternorrhyncha: Homoptera); adult female small soft-bodied, flattened, disc-like apodous insect that remains freely under the scale covering formed of wax secreted by its body; male winged and short-lived. Reproduction bisexual or parthenogenetical being oviparous or ovo-viviparous - a large number of economically important pests including San Jose scale *Quadraspidious perniciosus* (Comstock) an abnoxious pest of fruit trees all over the World.

Diastatidæ a very small family of Drosophiloidea (Acalyptratæ: Schizophora), stalk-eyed flies that are often included in the family Ephydridæ (Acalyptratæ).

Diazinon $C_{12}H_{21}N_2O_3PS$; non-systemic organophosphatic, contact insecticide with some acaricidal action; colourless oil, insoluble in water but soluble in petroleum oils and miscible with acetone, ethanol and xylene; compatible with most pesticides except copper compounds (fungicides); phytotoxic at higher doses, harmful to bees, fish, poultry and livestock. Oral LD_{50} value for male rats 108 mg/kg - Basudin®.

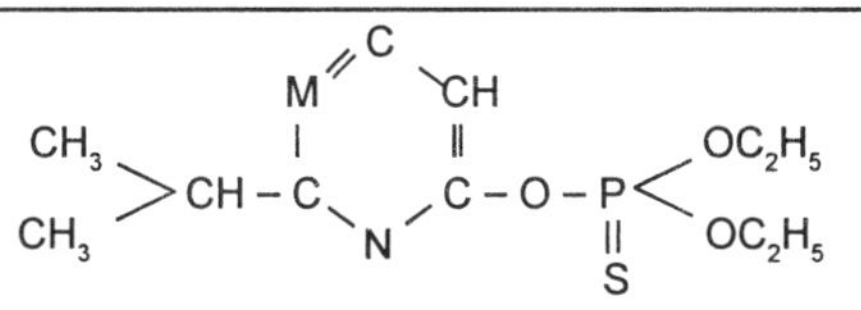

Diazinon

Dichlofenthion $C_{10}H_{13}Cl_2O_3PS$; non-systemic, organophosphoric nematocide, discovered by M.A. Manzelli in 1955; effective against nematodes. Acute oral LD_{50} value for rats 250-270 mg/kg; dermal for rabbit 6000 mg/kg - Nemacide.

Dichlorvos (DDVP) $C_4H_7O_4Cl_2PS$; organophosphate, contact and stomach poison with fumigant action; brings about quick knock down of insects. Soon after spraying on foliage it gets hydrolysed to dimehyl phosphoric acid and dichloroacetaldehyde which decomposes and evaporates leaving behind no residue. Highly toxic to bees. Oral and dermal LD_{50} values for rats 80 and 75 to 900 respectively - Nogos®, Nuvan®, Vapona® etc.

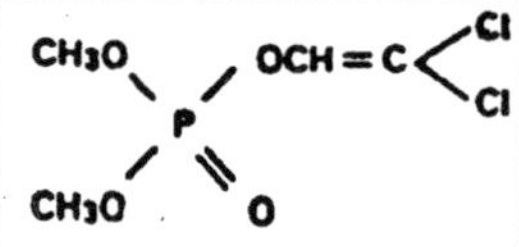

Dichlorvos

Dichoptic (flies) dipteran flies having their two compound eyes markedly separated from each other.

Dicofol $C_{14}H_9Cl_5O$; organochlorine, non-systemic acaricide; white solid, insoluble in water, soluble in most aliphatic and aromatic solvents; hydrolysed by alkali, compatible with all except highly alkaline pesticides. Oral and dermal LD_{50} values for rats 575 to 830 and 1000 to 1230 mg/kg respectively - Acarin, Kelthane®, Tik-tok, etc.

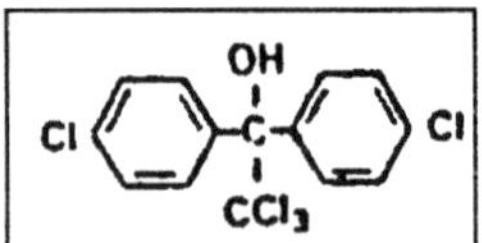

Dicofol

Dicondylic joint joint with two points or articulation between adjacent leg segments.

Dicrogeniidæ a small hymenopteran family of Procgotrupoidea (Proctotrupoid: Akpocrita); very small wasps characterised by 15-segmented antennæ inserted above clypeus on frontal shelf. Parasites of dipteran flies.

Dicrotophos $C_8H_{16}O_5PN$; organophosphatic systemic insecticide, effective as contact and stomach poison; bright yellow to brown coloured liquid with weak ester smell; miscible with acetone, alcohol and water; hydrolyses rapidly in alkali medium. Toxic to bees but safe against insect parasites and predators. Acute oral LD_{50} value for rat 22-45 mg/kg - Bidrin®, Carbicron® etc.

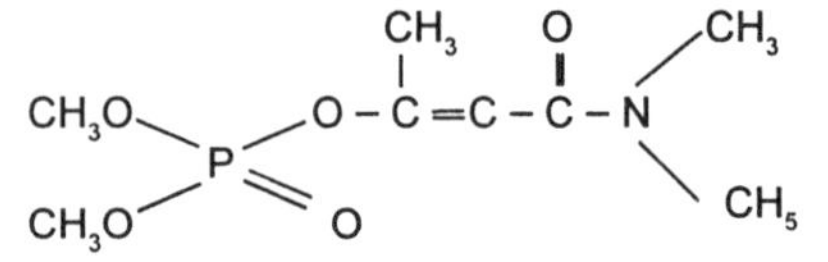

Dicrotophos

Dictyophoridæ Spinola 1839 family of Fulgorioidea (Homoptera), medium-sized bugs characterised by strongly modified head bearing a distinct process and apical segment of labium longer that its width (500 species) - *Rotiala viridis, Dictyophora sauropsis, Thanatodictya lineata* et al.

Dictyoptera Leach 1818 order of Polylneuropteran insects (Boattopteroidea: Pterygota); antennæ filiform and many jointed; mouthparts mandibulate and strongly chitinised; head hypognathous;

forewings modified into thickened tagmina; male genitalia asymmetrical and complex; IX abdominal segment bears a pair of styles; cerci many jointed; ovipositor reduced; metamorphosis incomplete; eggs contained in ootheca. About 8000 species, placed in two suborders-Blattaria (cockroaches) and Mantodea (mantids).

Dieldrin $C_{12}H_8Cl_6O$; chlorinated hydrocarbon named after German chemist Otta Deils; an epoxide of aldrin, soluble in water, moderately soluble in organic solvents and stable in alkali. It has long residual action and is used mainly for checking soil-borne insect pests. Acute oral and dermal LD_{50} values for rats are 40 and 100 mg/kg respectively - Alvit®, Dieldrex® Endosil® etc.

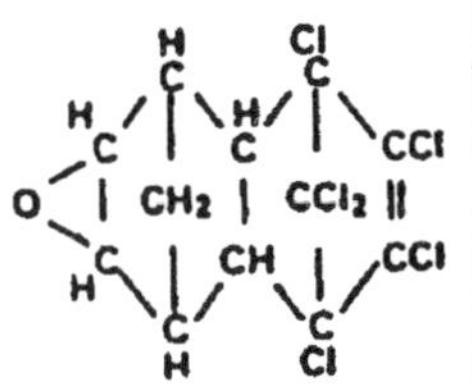

Dieldrin

Dihydroxyphenylalanine (dopa) $C_9H_{11}NO_4$ an amino acid that can be formed by oxidation of tyrosine.

Dilaridæ Handlirsch 1906 family of Mantispoidea (Planipennia: Neuroptera) includes small Pleasing lacewings characterised by three tuberculated and prominent ocelli; vein Sc not connected to R_1 and Rs and having only 3 main branches; male antennæ pectinate; female with long exserted ovipositor. About 35 species - *Dilar americana* McKachlan, *D. japonica* McLachlan, *Lidar* spp., *Nepal* spp. et al.

Diluent component of dust or spray which reduces the concentration of active ingredient (a.i.); it may help in mechanical application but does not directly affect toxicity - talc added to BHC or DDT.

Dilution gene any modifier gene that acts to reduce the effect of another gene.

Dilution method a technique wherein series of cultures is tested with varied concentrations of an antibiotic with a view to determining the minimum inhabiting concentration of the antibiotic.

Dimerous composed of two parts.

Dimthoate $C_5H_{12}NO_3PS_2$; organophosphate, systemic and contact insecticide and acaricide; white solid, slightly soluble in water, readily soluble in most organic solvents, stable in aqueous solution and to sunlight but readily hydrolysed by aqueous alkali, incompatible with alkaline pesticides. Harmful to bees, fish and livestock. Oral and dermal LD_{50} values for rats 215-250 and 700-1150 mg/kg respectively -Cygon®, Nugor®, Rogor® etc.

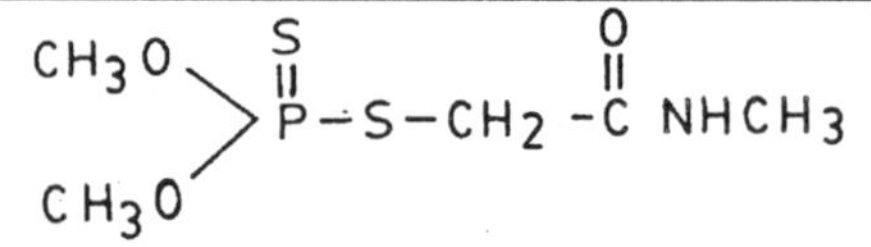

Dimethoate

Dimorphidæ (Astatidæ) a small hymenopteran family of Sphecoidea (Aculeate: Apocrita); small vari-coloured wasps characterised by middle tibia with two apical spurs; radial cell truncate, three cubital cells present; abdomen depressed with dorsal surface flattened. Predaceous on other (harmful) insects.

Dimorphism difference in size, structure, form, colour etc., between two types of individuals of same species.

Dinergate(s) soldier-ant; particular caste in an ant colony, the members having head and jaws greatly enlarged; used for breaking seeds and defending the colony.

Diopsidæ dipteran family of stalk-eyed flies, Diopsoidea (Schizophora: Cyclorrhapha); head bristles educed; eyes borne on long stalks; vibrissae absent, anterior femora thickened with spines beneath, aedeagus short, saprophaghous or phytophagous (150 species) - *Diopsis thoracica* and *D. tenuipes* pests of paddy in Africa and the Orient.

Diopsoidea superfamily of Schizophora (Cyclorrhapha: Diptera); characterised by having no inferior orbital bristles; postverticals reduced or lacking; costa not broken, Sc and anal vein continued to margin; cross-vein *im* present - Diuopsidæ, Syringogastridæ.

Dioptidæ a small family of Noctuoidea (Ditrysia: Lepidoptera); slender moths closely allied to those of Notodontidae, having brightly coloured wings with transparent windows; forewing with M_3 and Cu_1 stalked together and hind-wing with first anal vein distinct - *Phryganidis* spp. occasinally defoliate the oak trees in USA.

Diplatyidæ family of Pygidicranioidea (Forficulina: Dermaptera) characterised by having paired ædeagus with four gonopores. Immature stages have long multi-articulated cerci while the adults have true forceps - *Deplatys* spp., et al.

Diploblastic lower invertebrates having two germ layers - ectoderm and endoderm. *Cf.* Triploblastic.

Diploblastica major division of Animalia kingdom that includes non-chordate lower animals having only two germinal layers during their development; includes three phyla - Porifera, Coelenterata and Ctenophora.

Diploglossata Saussure 1879 monofamilial, smallest known insect order of hemimerids, in old classification; now regarded as suborder of Dermaptera (Hamimerina). Small, 8 to 10 mm long, depressed, apterous insects covered with short hairs, biting mouthparts, atrophied eyes and long unsegmented cerci. Ectoparasites on different species of rats; confined to South Africa - Hemimeridæ.

Diplura Borner 1904 order of Diplurata (Apterygota); includes two-pronged bristletails, large sized (50 mm) wingless insects hanging prognathous head with long many-jointed, moniliform anntennæ; entognathous biting type mouthparts, mandibles and maxillae retracted into a pouch just above labium; eyes wanting; XI abdominal segment with a pair of conspicious cerci that are either segmented or forcep-like. Metamorphosis slight (700 species), embracing two suborders - Rhabdura (*Campodea folsomi* Silvestri) and Dicellurata [*Japyx (lapyx) diversiungius* Silvestri]

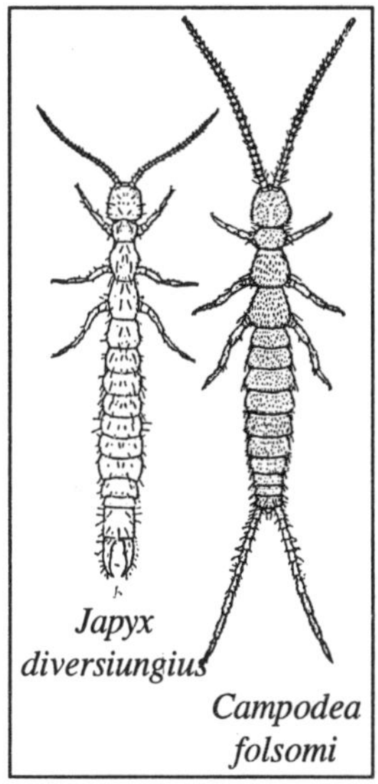

Diplurata superorder of Entognatha (Apterygota); includes single order Diplura.

Diprionidæ Rowher 1911 (Perreyidæ, Pterygophoridæ, Loboceratidæ, Lophryidæ) family of Tenthredinoidea (Strophandria: Symphyta); includes conifer sawflies having antennæ more than 9-segmented, serrate in females and pectinate or bipectinate in male; forewing with cross-vein 2*r* absent, hid-wing with cross-vein *m-cu* enclosing a median cells, anal cell closed. Conspicuous spotted larvæ live gregariously and defoliate the coniferous trees - *Diprin pini* et al.

Dipsocoridæ (= Cryptostemmatidæ, Ceratocombidæ) only family of Dipsocoroidea (Cimicomorpha: Geocerisæ); bugs with head porrect, first two antennal segments short and thick, last two long and pilose; mainly nocturnal, live in damp concealed habitats (50 species) - *Schizoptea* and related genera.

Dipsocoroidea monofamilial superfamily of Cimicomorpha (Geocorisæ: Heteropetera); very small bugs; antennæ with two thick basal segments, rostrum 3-segmented; forewing having costal fracture and cuncus fused with membrane - Dipsocordiæ.

Diptera Linnæus, 1758 order of true flies and mosquitoes (Mecopteroidea: Oligoneuroptera); mouthparts piercing and sucking; forewings membranous, hind-wings modified into balancers or halteres; metamorphosis complete. 98000 species grouped into three (four) suborders – Nematocera, Brachycera and Cyclorrhapha (Aschiza, Schizophora). The flies came into being some 150000000 years ago (Cretaceos period of Mesozoic era); families Protoryhyphidæ, Eoptychoptera and Architipulidae are by now wholly extinct.

Discolor not uniformly coloured.

Disinfect to kill or inactivate the pests or pathogens present on surface of plant parts or their immediate vicinity like air, soil etc. To make the plants or plant parts free form infection, by destroying the pests or pathogens established in or on those parts.

Disperasal spread of species to new areas in various directions; scattering of individuals out of a population.

Dissogeny having two sexually mature stages - larva and adult - in life of an individual.

Dissosteria Caroline locust; large long-winged grasshopper (Acrididæ: Cælifera) found in Southern USA; not a true locust.

Disulfoton (Thiodemeton) $C_8H_{19}O_2PS_3$; phosphorothiolothionate compound systemic acaricide and insecticide; colourlesss, oily liquid with characteristic odour; soluble in water and most organic solvents (acetone, alcohol, ether et.); boiling point 62°C. Acute oral and dermal LD_{50} to rats 4 and 50 mg/ kg respectively - Disyston®, Solvirex® etc.

Ditomyiidæ a small family of Mycetopiloidea (Nematocera: Diptera); small flies characterised by having nine pairs of spiracles, otherwise resembling fungus-gnats of the family Mycetophilidæ - *Ditomyia* species found living in the fungi.

Ditrysia largest suborder of Lepidoptera; mandibles vestigial or non-functional.; maxilla without lacinia; forewings without jugum or fibula, hind-wings with frenulum or the coupling is amplexiform; venation of fore- and hind-wings different. Caterpillars with not more than 5 pairs of crochet-bearing abdominal lcgs. Pupa adecticous, obtect, without functional mandibles - moths and butterflies.

Diurnal active during daytime. *Cf.* nocturnal.

Dixidæ (Van Wulp)(Brauer 1880 family of Culicoidea (Nematocera: Diptera) characterised by being devoid of hair or scales; antennæ elongate, 16-segmented and filiform apically; resembles mosquitoes but have smaller proboscis, not adapted for piercing and no scales on wings. Larvæ aquatic and metapneustic inhabit shady pools and often mistaken for mosquito (*Anopheles* spp.) larvæ - *Dixa* spp., et al.

Dobsonfly *Corydalus cornutus* (Linnæus); large predatory neuropteran insect (Cordalidæ: Megaloptera); having soft-body, long antennæ, biting mouthparts, large horn-like mandibles and two pairs of primitively veined membranous wings; wing-span upto 150 mm. Common in North America.

Dog, Orange nickname of giant lemon butterfly *Papilio thoas* Linnæus (Papiliionidæ; Lepidoptera) found in USA, defoliating leaves of citrus trees.

Dog-day harvest-fly colloquial name for various large cicadas (Cicadidæ: Homoptera) of USA that appear as adults during late Summer and Autumn having started their life in the ground about two years earlier - *Tibicen linnei* Smith and Grossbeck, *T. auletes* Germar) are 30-35 mm long, wing span 110-120 mm.

Dolichoderinæ (Forel 1887) Dalla Torre 1893 subfamily of Formnicidæ (Scolioidea: Apocrita), ants having very small single-segmented pedicel and constriction between first and second segments of gaster wanting; emit foul-smelling liquid from anus; habit pliable integument that allows great distension when their crops are gorged with food - *Prenolepis imparis, Azteca chartifex* Forel, *Tapinoma sessile* (Say), *Iridomyrmex detectus* Smith, et al.

Dolichopodidæ family of small bristly metallic green or bluish-green flies (Empiodoidea: Brachycera); having dorsal or terminal arista; short, fleshy proboscis; cell M and 1st M_2 confluent and cell Cu very short; two pulvilli herbage (> 2000 species) - *Aphrosyus* spp., *Dolichpus popularis, Melanderia* spp. et al.

Dolichopsyllidæ Baker 1905 old and equivalent name for the siphonapteran family Ceratophyllidæ Dampf 1910 (Ceratophylloidea) that includes fleas parasitic on birds and mammals.

Donaciidæ family of long-horned, leaf-beetles (Chrysomelidea: Polyphaga); elongate slender body, metallic black or greenish in colour, with head produced and having long antennæ, widely separated at the base; eyes prominent; prothorax laterally rounded and pygidium not exposed beyond elytra. Aquatic in pre-imaginal stages - larvæ of *Donacia aeraria, D. lenzi* and *D. provosti* feed on roots of paddy (*Oryza sativa* Linnæus) in Japan and Korea.

Donaciinæ subfamily of Chrysomelidæ (Chrysomelioidea: Polyphaga), now regarded as independent family of Chrysomelioidea.

Doritis medium-sized swallowtail butterfly (Kpapilionidæ: Lepidoptera) found in mountains of Balkans and Turkery; pale yellow with mauve body and mauve-bordered wings.

Dormant alive but not growing: quiescent or resting stage.

Dorsal back or upper side of organ or organism.

Dorsal diaphragm membranous and muscular sheets of tissue extending from dorsal blood vessel to latrodorsal parts of body wall, separating dorsal sinus from perivesceral sinus.

Dorsal sinus space of body cavity above dorsal diaphragm.

Dorsal vessel dorsal blood vessel comprising of pulsatile heat and non-pulsating aorta.

Dorsum entire back of an animal: in insects, the upper portion of the abdomen.

Doryoinæ subfamily of formicidae (Formicoidea: Apocrita); includes driver and legionary ants of the tropics; characterised by lack of ocelli, eyes also absent or vestigial; queens apterous; build no nests; mostly carnivorous; various specie march in long lines, foraging on either side; feeding on cockroaches, spiders and other small animals - *Dorylus labiatus* Swinhoe predaceous on other ants (*Pheidole* spp.) as also potato tubers in field while *D. orientalis* Westwood damage, cauliflower, potato etc.

Dose quantity of pesticide applied per individual or per unit area, volume or weight.

Drepanidæ (Drepanulidæ) a small family of hook-tips moths (Geometroidea: Ditrysia); found in the Indo-Malayan region of the Orient; tips of forewings are turned back to form a blunt hook. Caterpillars hump-backed with warts and swellings on the skin and no claspers; rest with head and tail directed upwards - *Drepana* spp., *Cilix* spp., et al.

Drepanoidea Forbes, 1932, a superfamily in old classification of Lepidoptera, that comprised of such families as Callidilidæ, Drepanidæ and Thyatindæ. All these families are now placed under superfamily Geometroidea of Ditrysia.

Drift dust or spray carried away by air or water currents beyond the target area.

Drilidæ a family of Cantharoidea (Polyphaga: Coleptera) that includes false firefly beetles; grubs predaceous, suck the body juices of their victims.

Drop spectrum distribution, by number or volume of drops of spray into different droplet sizes.

Drosophilidæ Loew 1862 small family of pomace-flies or vinegar-flies (Drosophiloidea: Schizophora); pale yellowish or tawny coloured with large red eyes; vibrissæ present; feed on fruit juices, vinegar, beer, wines etc., some are predaceous - *Acletoxnus indicus* (larvæ predaceous on aleyrodids), *Gitona* spp. infest tender fruits, also feed on decaying fruits and fungi while *Drosophila melanogaster* Meigen is used in genetic studies.

Dryinidæ Haliday 1833, 1840 family of parasitic wasps (Bethyloidea : Apocrita); head transverse, antennæ 10-segmented pronotum extending back to tegulae, anterior tarsi in females chelate; parasitises mostly nymphs of Homoptera (500 species) - *Aphelopus* spp., *Echthrodelphax fairchildi* Perkins, *Gonatopus contortulus* Patten, *Lestodrynius pyrillæ* Keiffer, et al.

Lestodrynius pyrillæ Keiffer

Ductus ejaculatorius median ectodermal outlet tube of male genital system.

Dun sub-imago or first winged stage of mayfly (Ephemeroptera) resembles adult in size and shape but is dull brown coloured; it casts off its skin almost immediately to emerge as a brightly coloured shining adult (spinner).

Dusky cotton bug *Oxycarenus laetus* Kirby (Lygæidae: Pentatomorpha); very small, uniformly dusky-brown coloured bugs. Both nymphs and adults feed gregariously on juice of immature seeds in partly open or damaged boll; affected seeds do not mature; the insects go to ginnaries and get crushed along with seed-cotton, thereby staining the lint. Also known as Egyptian cotton-stainer.

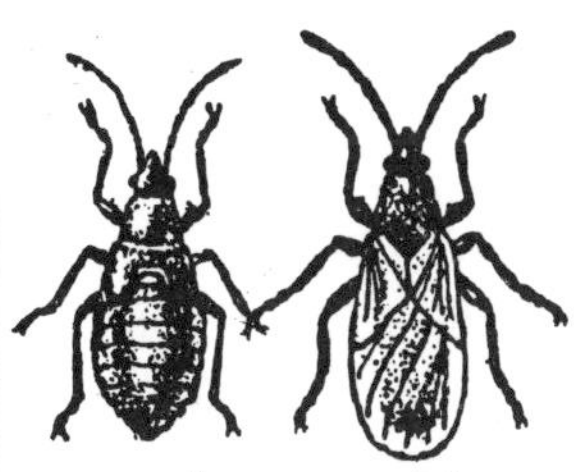

nymph adult

Oxycarenus laetus Kirby

Duster equipment for applying dust formulation of pesticides - manual, power driven and aerial.

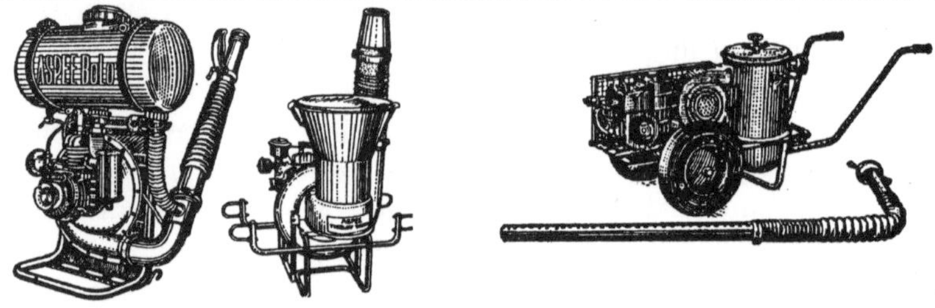

Power dusters

Dusting gun simple plunger type air-pump made of tin or any other light metal; body cylindrical, about 600 mm long (may be 800 or 900 mm) with a diameter of about 100 mm; discharges dust cloud with each stroke of the pump, flow of dust is intermittent and not continuous; useful for dusting against household pests as also in kitchen gardens and nurseries.

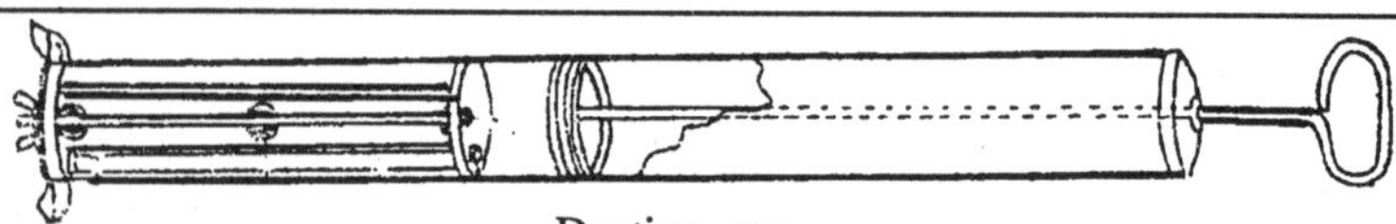

Dusting gun

Dwarf atypically small form or variety of a plant or animal.

Dyar's law an empirical rule based on the observation that the head capsule of caterpillars grow in geometric progression, increasing in width by a constant ratio at each moult.

Dynastidæ MacLeay 1819 family of rhinoceros beetles (Scarabæoidea: Polyphaga); large smooth shiny black insects, horns and tubercles present on head and pronotum respectively; exhibit great sexual dimorphism - *Oryctes rhinocerus* (Linnaes is pest of palm trees while *Heteronychus poropygus* Bates and *Xylotrupes gideon* (Linnaeus) are pests of paddy (*Oryza sativa* Linnæus).

Oryctes rhinocerus (Linnæus)

Dyspolusia abnormal cell growth.

Dystrophic environment which does not supply sufficient nutrition - a lake with over-acidic water.

Dytiscidæ Leach 1815, 1817 family of aquatic carnivorous beetles

(Caravoidea: Adephaga); large oval body, with powerful hind legs, flattened, fringed and adapted for swimming; males distinguished by broad front tarsi which form cup-like suckers for holding female while mating (1400 species) - *Dytiscus marginalis* Linnæus, the giant diving beetle is 35 mm long; *Cybiser tripunctatus* (Olivier) from China and *Eretes sticticus* Linnæus from India are edible; others include *Copelatus indicus, Hydraticus vittatus* et al.

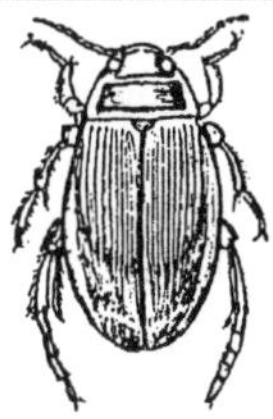

Cybister tripunctatus (Olivier)

D. dimidiatus Bergstrasser

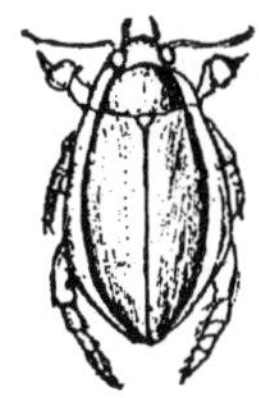

Dytiscus marginalis Linnæus

E

Eardrum tympanium: a thin membrane exposed on one side to external air and associated internally with sensory nerves; responds to sound waves; found in most orthoptera, Homoptera and Lepidoptera.

Early brown nickname of common European stonefly *Protonemura meyeri* (Nemouridæ: Plecoptea) found during early Spring (February-March) near water.

Early leafhopper largest British lafhopper, *Ledra aurita* (Cicadelidæ: Cicadoidea); about 15 mm long, grayish-green in colour, body flattened and two ear-like projections from thorax; found in lichen-covered oak trees.

Early thorn European geometrid moth, *Selenia bilunaria* having yellow buff-coloured wings ragged or serrated at the edges. Caterpillar stick-like, brown in colour, feed on blackthorn and other shrubs.

Earwig Dermapteran insect (Insecta: Arthopoda), alate or apterous, distinguished by having a pair of forcep-like horny cerci at anal end of abdomen that also exhibits sexual dimorphism. Live concealed in the soil or under clods, stones, and bark; abundant in the tropics, mostly scavengers, a few are ectoparasites – *Euborellia annulipes* (Lucas) is pest of groundnut (*Arachis hypogaea* Linnæus) in India: *Forficula auricularia* Linnæus is European earwig.

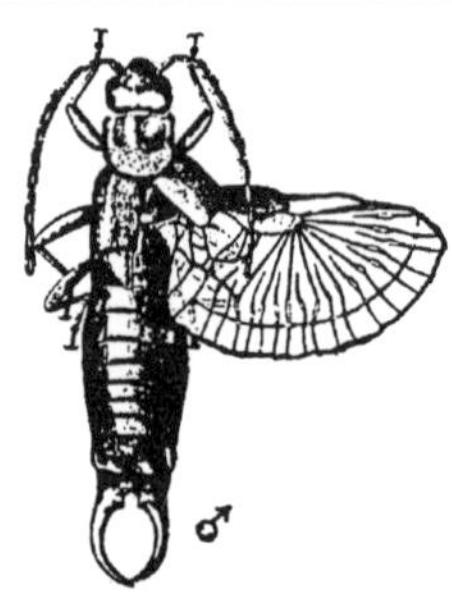

Forficula auricularia Linnæus

Earwig, Black *Chelisochus morio* (Fabricius) (Chelisochidæ: Forficulina); native to the Oriental Asia, now cosmopoliton; about 25 mm long, and shiny black in colour. In Hawaii (USA) predaceous on sugarcane leafhopper and its parasites.

Earwig, Seaside; apterous dark-brown species, *Anisolabis maritima* (Gene) (Labiduridæ: Forficulina) maritime earwig, 18 to 25 mm long, with pale legs and telson greatly reduced; inhabits ocean beaches and forages along the water's edge; commonly fond on both the Atlantic and Pacific seashores of USA having been introduced from the Old World.

Earwig, Lesser small black species, *Labia minor* (Linnæus) (Labiidæ: Forficulina); cosmopolitan in distribution; adults, active fliers; having tarson degenerated and pygidium small but well developed; found in rough herbage; often attracted to light.

Ecad population of a species having same genetic stock but distinct morphological characters imposed upon by particular habitat; these characters are only somatic variations and are reversible with change of habitat of environment.

Ecadysis (pl. ecdyses) act of moulting or casting off an integument or cuticula - common in arthropods, lizards and snakes.

Ecdysone steroid hormone secreted from prothoracic glands initiating ecdysis.

Ecdyuridæ Klapalek 1909 an old and equivalent name for family Heptageniidæ Bengtsson 1913 (Heptagenioidea: Ephermeroptera); often misspelled as Eudyonuridæ or Ecdyonurus.

Ecdyuroidea Klapalek 1909 an old and equivalent name of superfamily Heptagenioidea Ulmer 1920 (Ephemeroptra: Ephmeroidea).

Echinodermata phylum of slow-moving, free-living, spiny skinned animals (Coelo-mata: Triploblastica); larvæ bilaterally symmetrical and adults radially symmetrical (majority pentamerous); exclusively marine, mostly oviparous, a few viviparous. Over 600 species grouped into two subphyla comprising of six classes - Eleutherozoa (Asteroidea, Ophiuroidea, Echinoidea, Holothuroidea, Ophiocistioidea) and Pelmatozoa (Crinoidea) that include brittle stars, sea--cucumber, sea-lily, sea-urchin, star-fish etc.

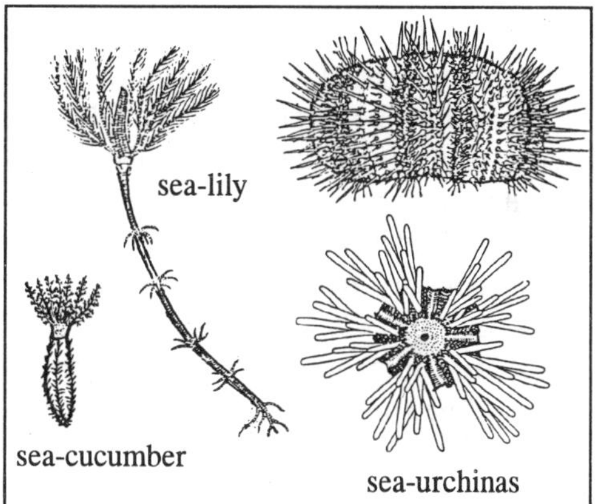

Echinoderms

Echiur(id)a phylum of small to large bilaterally symmetrical, unsegmented benthic marine worms (Coelomata: Triplohlastica); body sausage-shaped or sac-like, highly extensible but not aversible, sexes separate. Only 150 species classified into three orders – Heteromyota, Echiuroida and Xenopneusta.

Ecological pertaining to ecology.

Ecological niche status of an organism in a community vis-a-vis its response and behavioural pattern; several overlapping and interacting ecological niches make up an ecosystem.

Ecological succession history of gradual changes in an ecological community.

Ecologist person having sound knowledge of ecology.

Ecology science of inter-relationship of organism in and to its environments – both biotic and abiotic: study of living organism in relation to its environment.

Ecomorphosis effect of environment, resulting in appreciate structural differences as exhibited by some collembolan species.

Economic damage amount of damage or loss that justifies the cost of adopting the control measures.

Economic-injury-level lowest pest population density that will cause economic damage: degree of crop damage at which economic losses become significant.

Economic–thresh-hold density of pest population, at which the control measures need be adopted to prevent the pest population from reaching the economic-injury-level.

Ecosystem interacting system of living organisms in an area and their abiotic (physical environment.

Ectadenia accessory glands that arise as ectodermal invagination of ejaculatory duct – Coleoptera.

Ecto- outside. *Cf.* ento- (endo-).

Ectobiinæ subfamily of Epilampridæ (Blattaria: Dictyoptera); mostly confined to temperate regions of the Old World – *Ectobius* spp., et al.

Ectoderm outer cell layer of embryo that gives rise to epidermis: epiblast.

Ectogenous capable of living independently; outside the body of the host.

Ectohormone substance secreted externally by an organism to influence behaviour of other organism – pheromone.

Ectoparasite parasitoid that develops externally on body of its host.

Ectophallus outer phallic wall: walls of phallobose and ædeagus together.

Ectophyte vegetable ectoparasite: a plant that lives externally on another organism.

Ectoplasm outer layer of cytoplasm of a cell.

Ectozoon (pl. **ectozoa**) animal that lives externally on another organism.

Ectrephidæ very small family of Bostrychoidea (Polyphaga: Coleoptera); earlier placed under Ptinidæ - only five species (myrmecophilous) all confined to Australia.

Effector organ gland or muscle of insect body that is activated by nerve stimuli.

Efferent carry or conducting away.

Efferent nerve nerve that conducts from nerve centre towards periphery; axon of motor neuron.

Egest to pass out/discharge undigested food/residue from alimentary canal: defæcate.

Egg non-motile female gamete: insect egg consists of outer shell (Chorion) containing a delicate membrane surrounding a thin cortical layer of protoplam and a dense central yolky mass; presence of one to many openings (microphyles) on one end of the egg, facilitate the entrance of the sperm. The eggs vary greatly in shape, form and size; these may be spherical, oval, elongate, barrel-shaped or pear-shaped; smooth, ridged or sculptured; may be laid singly or in clusters, enclosed in gelatinous mass or in the ground as egg-pods or retained in oothecæ.

Egg, Centrolecithal egg, in which yolk lies in central position; common in eggs of insects and other arthropods.

Egg chamber follicle of ovarial egg tube containing an oocyte.

Egg guide median process of subgential plate behind gential opening, present in most acridids.

Egg tube middle section of ovariole that contains germ cells.

Eginiidæ newly erected family of small flies (Protochinoidea: Muscoidea); earlier regarded as subfamily Eginiinæ (Anthomyiidæ) and later placed under Muscidæ (Muscoidea: Schizophora).

Eglandular without glands.

Ejaculatory duct ecodermal outlet duct of male genital system.

Elacatidæ (= Othniidæ) family of Heteromera (Cucujoidea: Polyphaga); characterised by fore-coxæ being small and rounded with prosternal process between them usually broad -*Eurystethus* spp. (earlier under Eurystethidæ), *Othnius* spp. et al.

Elachistidæ a family of very small moths (Elachistooidea: Ditrysia); having forewings with 1A usually present and anal area ample; hind-wings narrow. Caterpillars mine the leaves specially of grasses. Over 200 described species - *Elachista* spp., et al.

Elachistoidea Forster 1856 a lepidopteran superfamily of suborder Frenatæ in the old classification. It comprised of Elachistidæ, doglasiidæ and Heliozelidæ; now place under superfamily Gelechiodea of suborder Ditrysia (Lepidoptera).

Elasmidæ a family of Chalcidoidea (Apocrita: Hymenoptera); small, black or brown coloured wasps having enlarged compressed hind legs. Primary parasites of lepidopterous insects or hyperparasites on braconids and ichenumonids that are parasitic on caterpillars - *Elasmus brevicornis, E. indicus, E. zehnineri* are larval parasites while *E. albomaculatus* is hyperparasite of *Apanteles malveolus.*

Elateridæ Leach 815, 1817 family of click beetles (Elateroidea: Polyphaga); head small, partly sunken into thorax, labium visible,; posternum lobed in front; prothorax loosely joined to mesothorax and freely movable; visit flowers and feed on buds of deciduous frit trees - *Quasims minitissimus* Germar of Europe (2 mm), *Alausgibboni* of Australia (50 mm), *Tetralobs flabellicornis* (Linnæus) of Africa (50 to 70 mm). In India *Drasterius* species damages potato tubers (*Solanum tuberosum* Linnæus) in fields.

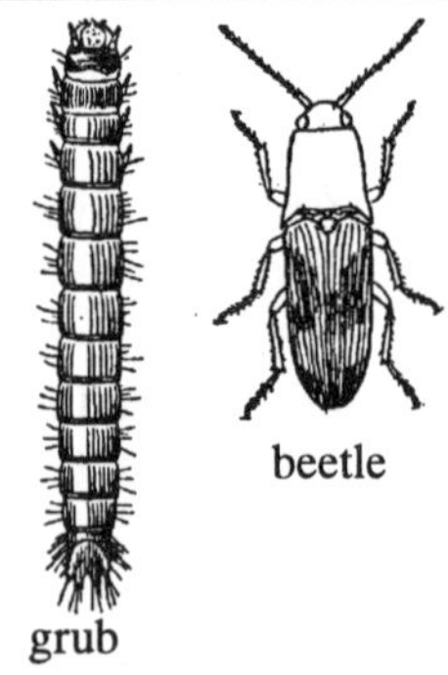

Drasterius species

Elenchidæ a family of very small twisted-sing parasites (Styloptoidea: Strepsiptera); male possesses 4-segmented antennæ and 2-segmented tarsi; in female the brood-passage opening is broad and semi-circular. Parasitic on Fulgoridæ and Delphacidæ.

Eleutherata insect order proponed by Linnæus (1735: Systema Naturæ). It included all beetles and weevils (Coleoptera Linnæus 1758) as also earwigs (Dermaptera Leach 1815). The term is no more is usage.

Ellipura Borner 1910 an order errected for certain apteran insects which were later placed under order Protura Silvestri (Myrientomata: Entotropha); no more in usage.

Elmidæ family of small subaquatic beetles (Dryopoidea: Polyphaga); 2 to 4 mm long found crawling about on pond weeds (not swimming); body covered with numerous bent hydrofuge hairs that enable it to take the air down for breathing; feed on aquatic plant debris. Larva onisciform, clings to weeds About 300 species.

Elmoidæ a family of fossil order Megasecoptera from Upper Carboniferous period of France, England and North America.

Elythroptera Latreille 1809 old and obsolete term suggested to include mantids, now placed under order Mantodea Burmeister 1838, which is often relegated as suborder of Dictyoptera Leach 1818.

Elytron (pl. **elytra)** thickened forewing of Coleopteran insects (beetles and weevils) covering and protecting membranous hind-wings.

Embidina Hagen 1861 (= Embidoptera Lameere 1900, Embiidina Enderlein 1903) old and equivalent terms for Embioptera Shipley 1904 that include embiids and web-spinners; no more in usage.

Embiid(s) web-spinners; members of insect family Embiidæ (Embioptera: Embioideaa) - *Embia tyrhenica, Embolyntha batesi, Metembia ferox, Pseudoembia immsi* et al.

Embiidæ Burmeiser largest family of Euembioptera (Embioptera: Embiodea) that includes web-spinners having wing-venation more complex with R_{4-5} forked and M and Cu distinct and branched; X tergite in males asymmetrical (hemitergites) and entire in females - *Embia termitophila* are fond in termitaria; other species include *Embia major, Pseudombia flava* et al.

Embioptera Shipley 1904 order of little known fragile insects (Polyneoptera) mostly with soft thin cuticle and weak power of flight, living in silken tunnels; mouthparts mandibulate, ligula 4-lobed; tarsi 3-segmented; cerci 2-segmented; males winged with gradual metamorphosis, females apterous and having less pronouced metamorphosis; on starvation both sexes become cannibalistic. About 400 species placed under tow suborders - Protembioptera, Euembioptera.

Embolemidæ a small hymenopteran family of Bethyloidea (Aculeata: Apocrita); minute, black parasitic wasps, characterised by 16-segmented antennæ inserted on a prominence high above clypeus; thorax large and elongated; abdomen with a short petiole. Female apterous, male winged with complete venation - North American *Ampulicomorpha* species has been bred from *Epiptea* species (Achilidæ: Fulgoroidea).

Embolium small costal region of the hemelytron or horny basal half of the forewing of certain heteropteran insects belonging to families Anthocoridæ and Ceratocombidæ.

Embonychidæ monogeneric and monospecific family of Euembioptera (Embioptera: Insecta) – *Embonycha interreupta* is the only species found in North Vietnam.

Embryo organism in early stage of development.

Embryogenesis formation and development of an embryo from an egg.

Embryogeny origin and development of an embryo.

Embryologist specialist in embryology.

Embryology biological science that deals with study of growth and development within the egg.

Emigration movement of individuals out of a population.

Empididæ (Latreille 1804, 1809) Fallen 1817 family of Empidoidea (Brachycera: Diptera); small-sized, carnivorous, gray, yellowish or dark coloured flies having horny proboscis adapted for piercing; cell M and 1st M_2 separate; empodium linear, membranous or setiform. (3000 species) – species of *Empis, Hilara* and *Rhamphomyia* may often be seen dancing or swarming in the air.

Empidoidea superfamily of Brachycera (Diptera: Oligoneoptera); small carnivorous empid flies, some of which have interesting mating habits, the male kills a small insect, wraps it up in silk thread and presents to the female; while she is unravelling and eating the same, mating takes place. Two families – Empidudæ and Dolichopodidæ.

Empodium (pl. **Empodia)** small appendage between the apical pulvilli and associated with the claws in true flies (Diptera).

Emulsifiable concentrate (EC) liquid formulation of pesticide(s) in aliphatic hydrocarbon solvents with or without cosolvents, containing surfactant or emulsifier to permit dilution with water; when added to water will spontaneously disperse as fine droplets to form emulsion.

Emulsifier spray additive that permits formation of stable suspension of oil droplets in aqueous solution.

Emulsion stable dispersion of oil droplets in aqueous solutio.

Enchirinæ subfamily of Enchiridæ (Scarabæoidea: Polyphaga); males have forelegs very long and thick with a number of projecting spies which make the movement extremely clumsy. Found in South-east Asia, feeding on fermenting juices of palm trees. – *Propomacrus jansoni* of China is about 60 mm long.

Encyrtidæ Walker 1837 extensive family of chalcids (Chalcidoidea: Hymenoptera) that live as parasites of ova, larvæ or pupæ specially of Homoptera and Lepidoptera, mid-tibia spur greatly enlarged, mesosternum also much enlarged. Species of *Ageniaspis, Litomestrix* and *Copidosoma* parasities eggs of Lepidoptera and exihibit polyembryony. *Cheiloneurus pyrillæ* (Mani) is egg parasite of *Pyrilla* spp.

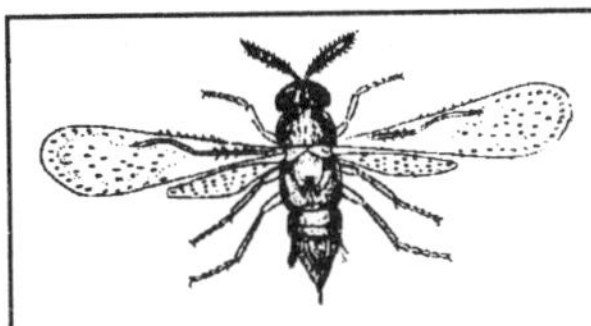

Cheiloneurus pyrillæ (Mani)

End chamber germarium of gonadial tube.

Endemic prevalent or regularly found in a particular area.

Endo- (ento-) inside. *Cf.* ecto- (exo-).

Endocapsulation immune response of certain host insects to eggs and young larvæ of non-adopted parasites; gradually these eggs and larvæ become encysted by blood cells of host insect and die away.

Endocrinology scientific study of normal and pathological physiology of glands of internal secretion.

Endocuticula secondary cuticula: inner softer layer of cuticula.

Endoderm (entoderm) inner most cell layer of embryo.

Endomychidæ (Leach 1815) Samouelle 1819 a family of small to medium-sized convex bodied beetles (Clavicornia: Cucujoidea); distinguished by 4-4-4 tarsi, 3rd segment concealed in deeply bilobed 2nd segment. Larva without branched setæ on the body and mandibles not falcate. Over 600 species found mainly among fungi on timber in Tropical forests as also in dung and vegetable refuse - *Endomychus coccinens, Mycetæa hirta* et al., the former also known as ladybird-mimic, differs from ladybird beetle (Coccinellidæ) in having long thin unclubbed antennæ.

Endoparasite parasitoid which develops within the body of its host.

Endophallus inner chamber of phallus invaginated at end of ædeagus, into which ejaculatory duct opens.

Endoskeleton internal cuticular ridges and processes of body wall.

Endosulfan $C_9H_6Cl_6O_2S$; organochlorine, its insecticidal properties (contact and stomach poison) were first described by W. Finkenbrink in 1956. Brown crystalline solid, insoluble in water but soluble in xylene and other organic solvents; its acute oral and dermal LD_{50} values for rat are 110 and 74 to 130 mg/kg respectively - Thiodan®, Thiokill etc.

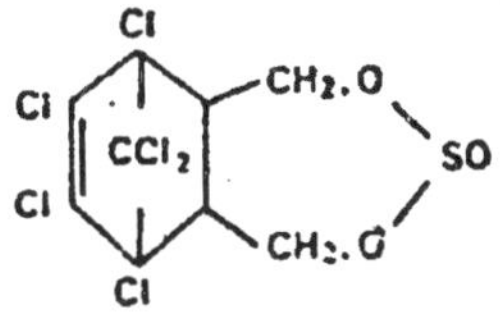

Endosulfan

Endotheca inner wall of phallotheca.

Endromididæ (Endromidæ) a mono-generic and mono-specific family of Bombycoidea (Ditrysia: Lepidoptera); large forest-dwelling, day-flying moth having no proboscis and hind-wing with Sc + R_1 connected to cell by a cross-vein. Male rust-brown with black and white pattern on forewings, fly long distances in large numbers seeking the females by their scent. Female dull in colour, spends most of its time immobile on tree truck. Found in forests of North and Central Europe. Caterpillars feed on beech (*Fagua sylvtica*), birch (*Betula* spp.,), hazel (*Corylus colurna* and allied species) and lime trees - *Endromis versicolor.*

Enicocephalidæ family of gnat-bugs (Geocorisæ: Gymnocerata); small to medium-sized predaceous insects; head two-lobed marginated behind he eyes; antennæ 4-segmented; eyes large, ocelli distinct; prothorax divided into three distinct lobes; legs round and flat, fore-legs raptorial tibia dilated apically, tarsi one-segmented; arolia absent; hemelytra wholly membranous with few longitudinal and cross-veins; some emit musty odour - *Enicocephalus formicina* (Uhler), *Systelloderes biceps* (Say), *S. iowensis* Darke and Harris, et al.

Ennominæ large superfamily of Geometridæ (Geometroidea: Ditrysia); slender-bodied moths having cryptic colour pattern; in some species (*Erannis, Theria* etc.) females are completely apterous. Caterpillars (defoliators) resemble sticks or twigs and often go unnoticed.

Enomychidæ (Leach 1815) Samonelle 1819 family of moderate-sized clavicornia fungus-beetles (Cucujoidea: Polyphaga), body convex; last segment of maxillary palpi not enlarged; tarsi 4-4-4-. Grub without branched setæ on its body and mandibles not falcate - *Endomychus coccinens, Mycetina hirta* et al.

Ensifera suborder of Orthoptera (Orthopteroidea: Pterygota); includes long-horned grasshoppers and crickets; antennæ many segmented and as long as body; tympanal organs, when present, on fore-tibia; stridulatory organs, if present, usually tegminal. Three superfamilies - Gryllacridoidea, Tettigonioidea and Gryloidea.

Entognathous having mouthparts in sockets from which these can be protruded, if and when necessary. *Cf.* ectognathous.

Entomic pertaining to or denoting insects.

Entomobryidæ Tomosvary 1883 a large cosmopolitan family of collembolan spring-tails (Entomobryoidea: Arthropleona); found throughout the Himalayan range, above the timber-line to over 6000 metres; body nearly cylindrical; scales often present, antennæ 4- to 6-segmented; pronotum concealed from above; furcula near anal end and often very long – *Entomobrya* spp. feed on fungi. the green glaceir springtail feed on rotting vegetation and is often found on the snow.

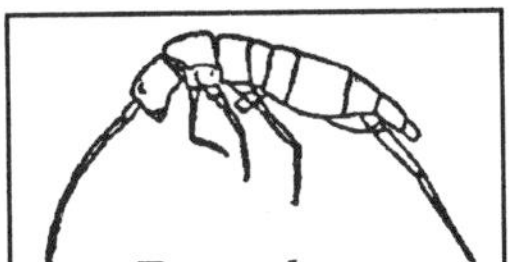
Entomobrya species

Entomologist person well versed in science of insects.

Entomology science of insects: branch of zoology that deals with origin of insects, their development, distribution, structure and bodily functions, habits, and habitat, behaviour, relation to one another as also environment, classification and economic importance.

Entomophagous insectivorous: organism – plant (pitcher plant, sundew etc.) or animal (lizard etc.) – that feeds on insects.

Entomophagous parasites insects that are parasitic on other insects. also called insect parasites.

Entomophagy consumption of insects by other organisms (plants or animals).

Entomophily pollination by insects.

Entoproct(s) small (< 5 mm long) sessile, solitary or colonial animals (Entoprocta: Metazoa), generally attached to solid substrates or epizoic on other animals; can be easily distinguished from hydroids by ciliation of the tentacles (arranged as a crown).

Entoprocta small phylum of Bilateria (Metazoa); mostly sessile invertebrates, formerly included in phylum Bryozoa but now separate as these have pseudocoelomate visceral cavity and anus opening inside circlets of tentacles. Except *Urnatella* spp. (freshwatr forms) rest are all marine, attached by a stalk to rocks, shells or pilings (150 species) – *Loxosoma* spp., *Pedicellina* spp. et al.

Environment sum-total of physical, chemical and biological factors to which an organism is exposed.

Enzyme(s) literally means 'in yeast' but the term is used as a general name for biocatalysis. Enzymes were discovered by German chemist Edward Buchner (1860-1917) in 1897 just by accident while preparing yeast extract for medicinal purpose.

Eosentomidæ Berlese 1909 small monogeneric family of Eosentomoidea (Protura: Myrintomata); tiny (< 2 mm long), primitive wingless insects having tracheæ and spiracles; all three pairs of abdominal appendages 2-segmented, VIII abdominal segment without dorsolateral combs (100 species) - *Eosentomon hyatti* Conde, *E. indicum* (Schepotieff), *E. transitorium* Berlese et al.

Eosentomoidea superfamily of apterygote insects (Protura) that have well developed tracheæ and spiracles; all legs have simple segmented tarsi ending in single claw. About 120 described species arranged in two families - Eosentomidæ and Sinentomidæ.

Epallagidæ family of medium-sized damselflies (Agrioidea: Zygoptera); having wings indistinctly stalked basally, antenodal cross-veins not continuous beyond the arculus; pterostigma long. Nymph robust with sacooid or triquetral terminal gills and also lateral abdominal gills.

Eprmeniidæ a small microlepidopterous family (Yponomeutoidea: Ditrysia); characterised by hind tibia and hind tarsus having stiff bristles. Caterpillar mines the leaves and later lives under a web. About 100 species widely distributed - *Epimarptis philocoma, Idioglossa triacma* et al.

Ephemerellidæ Klapalek 1909 widely distributed family of mayflies (Leptophlebioidea; Euphemeroptea); distinguished by forewing with Cu_2 bent almost at right angle to reach the hind margin. Naiads robust with dorsal rows of spines for tubercles - *Ephemrella* spp. et al.

Ephemerida Leach 1817 old and obsolete term for insect order Ephemeroptera Hæckel 1896 (Plectoptera Packard 1886).

Ephemeridæ Klapalek 1909 a cosmopolitan family of mayflies (Ephemeroidea: Ephemeroptera) having forewing with 1A unfrocked but connected to hind margin by many intercalaries. Naiads with gills held dorsally and manibular process pointed; found in lakes and ponds - *Hexagenia* spp., et al.

Ephemeroidea Ulmer 1920 superfamily of Ephemeroptera that includes mayflies having wings with copious venation and hind legs with four freely movable tarsal segments. Nymphs usually burrowing, have long mandibular processes and plumose gills - Behningiidæ, Potamanthidæ, Euthyplocidæ, Enphemeridæ, Polymitarcidæ and Palingeniidæ.

Ephemeroptera Hæckel 1896 (ephemerida Leach 1817, Plectoptera packard 1886) order of palæopteran insects having delicate soft-body; chewing mouthparts that are often rudimentary. Alimentary canal modified for storage of air (stomach serves as air-sac) to aid-in-flight. Wings with many cross-veins, forewings strongly triangular and much larger than hind-wings; abdomen having 2 to

3 long, many segmented caudal filaments; metamophosis incomplete. Nymphs (naiads) aquatic and herbivorous, feed on algal slime and live for several years. Adults longevity is normally less than a day. Over 1500 species of mayflies spread in five superfamilies – Ephemeroidea, Heptagenioidea, Leptophlebioidea, Cænoidea and Prosopistonmatioidea.

Epherida Leach 1817 (Ephemerida Leach 1817) old and equivalent term for insect order Ephemeroptera Leach 1817.

Ephydridæ family of myodarian cyclorrhaphan shoreflies (Acalyptratæ: Schizophora); small, black, long-legged flies found in damp and marshy places. Maggots mostly aquatic found in decaying vegetable matter, saltish or brackish ponds etc. *Hydrellia* spp. have been reported as pests or paddy (*Oryza sativa* Linnæus) from various countries; *H. philippina* and *H. sasakii* from India.

Epicopeiidæ an old and equivalent name for the lepidopteran family Uraniidæ (Geometroidea: Eitrysia).

Epicranial suture dorsal Y-shaped suture of cranium.

Epicranium upper part of cranium exclusive of frons.

Epicuticula nonchitinous, external thin covering of procuticula.

Epidemic widespread outbreaks of pest or disease.

Epidemiology systematic study of epidemics.

Epidermis outermost layer of cells (epithelium) or body wall: hypodermis.

Epididymis convoluted part of vas deferens.

Epimere dorsal process of phallobase.

Epimeron area of pleuron posterior to pleural suture.

Epimetabola primitive metamorphosis (Palæometabola) wherein no distinct external changes are evident except gradual increase in size – Apterygota (Thysanura, Diplura, Protura and Collembola).

Epineural sinus channel of embryonic hæmocoele beneath yolk or alimentary canal.

Epiophlebiidæ Muttkowski 1910 only family of Anisozygoptera (Odonata: Palæoptera); adults have zygopteran wing-venation and general body form of Anisoptera; only one genus and two species – *Epiophlebia superstes* Selys from Japan and *E. laidlawi* Tillyard from Himalayas (India and Nepal).

Epiphallus long, complex sclerite lying dorsally at the base of phallic organs in some orthopteran insects.

Epipharynx small median lobe often present on ventral surface of clypeus and labrum.

Epiplankton plankton occurring in the sea from the surface to a depth of about 100 fathoms (200 metres).

Epiplemidæ a small family of small inconspicuous moths (Geometroidea: Ditrysia); having tympanal organs ventral and frenulum present. Moths rest during day with their forewings rolled up and hind pair applied to the sides of the body, in this attitude these resemble spiders.

Epiproct small tergal plate above anus of insect.

Epipyropidæ a small family of Zygænoidea (Ditrysia: Lepidoptera); includes small moths whose caterpillars are predatory on various Auchenorrhyncha (Homoptera), living attached to the victim - *Epipyrops melanoleuca* Fletcher on coccids and sugarcane pyrilla (nymphs and adults), *E. fulginosa* on mango hoppers (*Idioscopus* spp.) et al.

Epipyrops melanoleuca Fletcher on *pyrilla* nymph

Episternum area of pleuron before pleural suture and above trochantin.

Epistone sclerite between labrum and epicranium.

Epithelium tissue or layer of cells that covers body surface.

Epizootic widespread occurrence of a disease in a large proportion of an animal population.

Ergate(s) worker ant (Formicoidea: Apocrita), apterous forms, normally the smallest members in the colony; most species are dimorphic and a few are polymorphic. Usually large individuals are called macregates and dwarfs micrergates.

Eriocraniidæ a small family of Erocranioidea (Dacnonypha: Lepidoptera); comprising of small primitive diurnal moths having untoothed mandibles but no maxillary palpi; proboscis not well developed and deep fringes on edges of wings. Occur in Eurasia and North America - *Eriocrania spermanolla* have golden-yellow wings with blue veins and yellowish-gray fringes, its apodous caterpillars are leaf miners on brich (*Betula alnoides* Buch-Ham.), hazelnut (*Corylus colurna* Linnæus), chestnut (*Castanopsis indica* Roxberg), oak (*Quercus* spp.) etc.

Erotylidæ (Leach 1815) Samonelle 1819 family of Clavicornia (Cucujoidea; Polyphaga), bright coloured (red and black) hemispherical beetles, upto about 25 mm in length, distinguished by club-shaped antennæ; fore-coxal cavitics closed from behind; tarsi 5-segmented (4th small), elytra epipleura complete. Found in warmer regions of America and called Pleasing-fungus-beetles, as most of these feed on fungi in rotting wood, a few are pests of vegetable crops (1600 species).

Erycinidæ old name of Riodinidæ (= Nemeobiidæ) a family of butterflies (Papilionoidea: Ditrysia) found in Central and South America.

Escape susceptible plant that shows no infestation or damage because the insect pest did not attack or feed on it: Pseudoresistance showing lack of insect infestation.

Escutcheon alternate name of scutellum especially of beetles; shield-covering over thorax.

Estivation æstivation: to pass the Summer.

Ethiology biological science dealing with comparative study of animal behaviour.

Ethmiidæ a small family of moths (Gelechioidea: Ditrysia); having wings often boldly marked; hind-wings not very narrow and vein Cu_2 clearly present. Caterpillar feeds on leaves living under a slight web.

Etiology science concerning study of causation: assignment of cause of a disease.

Euchari(ti)dæ family of chalcid wasps (Chalacidoidea: Apocrita) have sickle-shaped mandibles with one or two teeth; thorax very convex; scutellum produced backwards; hind tibia with two spurs, gaster compressed; usually ectoparasites of ants, found in Tropics - *Orasema* species, *Schizaspidia tenuicornis* et al.

Euchemidæ family of Elateroidea (Polyphaga: Coleoptera); small beetles having stout prominent mandibles, labrum concealed, all five abdominal sternites connate; found in warm climate attached to rotten wood (1000 species) - *Melasis bupestoideas* et al.

Euchiridæ a small family of beetles (Scarabæoidea: Polyphaga).

Euchromiidæ (=Syntomidæ) old and obsolete term for a noctuid family that is now merged with family Ctenuchidæ (Sphingoidea: Ditrysia) - hairy caterpillar, *Echromia polymena* Linnæus is a common pest of sweet potato (*Ipomoea batatas* Linnæus), while *E. gigantica* (Linnæus) of Brazil is one of the largest species (60 mm long and 25 mm broad).

Eucinetidæ very small family of Eucineloidea (Polyphaga: Coleoptera); long, narrow and boat-shaped beetles; postgena not keeled, hind-coxæ very large all tarsi 5-segmented (25 species) *Eucinetes* spp. et al.

Eucinetoidea superfamily of Polyphaga (Colepotera: Oligoneptera); includes beetles characterised by metasternal sulcus not extending to the margins; hind coxæ having posterior face vertical and deeply excavated; wings with not more that 4 anal veins - Clambidæ, eucinetidæ and Heloididæ.

Eucleidæ Comstock 1895 old name for lepidopteran family Limacodidæ (=Cochildiidæ, Heterogenidæ) (Zygænoidea: Ditrysia) that includes slug caterpillars.

Euclidimera generic and common name of Mother-shipton, *E.mi.* (Noctuidæ), small, day-flying owlet-moth having markings on forewings that resemble profile of an old woman; found in Britain, feeding on flowers.

Eucopelomata (Coelomata) subdivision of Triploblastica; comprising of all animals having true coelom or body cavity; includes all phyla above Aschelminthes.

Eucomsmidæ Durrant 1918 equivalent name for lepidopertran family Olethreutidæ (Torticoidea: Ditrysia).

Eugereon big-nosed dragonfly (Anisoptera: Odonota), fossil insect of Permain period; one of the oldest dragonflies with wing spread of 160 mm, different from present day species in having suctorial mouthparts.

Euglossinæ a subfamily of Apidæ (Apoidea: Apocrita); that includes solitary (some subsocial) bees, either metallic and bare or dark and densely hairy; nest is built either of mud or resin. Often regarded as an independent family Euglossidæ (= Chrysanthedidæ).

Euloeidæ an old and obsolete term synonym of Danaidæ (papilionoidea: Ditrysia); some consider it as subfamily of Euploeinæ (= Danainæ) under family Nymphalidæ (Papilionoidea; Ditrysia).

Eulophiæ Walker 1846 family of Calcidoidea (Parasitica: Apocrita); very small, parasitic or hyperparasitic plumed wasps, 1 to 3 mm long, maxillæ extend in front of anterior margin of sutellum; tarsi 4- or 5-segmented and abdomen sessile or with a distinct petiole - *Aphelinus mali* Haliday and *Trichospilus pupivora* Ferrier are commonly used for biological control of coconut black-headed caterpillars, *Nephantis serinopa* Meyrick and wooly aphid *Eriosoma lanigera* (Hausmann) respectively; *Tetrastichus pyrillæ* Crawford is egg-parasite of *Pyrilla* spp.

Tetrastichus pyrillæ Crawford

Eumastacidæ a family of small relatively primitive acridoids (Cælifera: Orthoptera); having head flattened anteriorly; wings often reduced or absent; tympanal and stridulatory organs absent. Over 300 species mainly Tropical and subtropical, found living in bushes.

Eumenidæ (Leach 1815) Westwood 1840 family of solitary wasps, commonly called mason-mud-or potter-wasps (Vesopidea: Apocrita); having dentate mandibles; wings distinctly plicated often fuscous with violet reflections; forewing with one radial and four cubital cells; middle tibia with one apical spine and claws dentate; only two forms (females and males) - *Eumenes coarctata* (Linnæus), *Odynerus geminus* Cresson, *Menobia quadridens* (Linnæus) et al.

Eumenes coarctata (Linnæus)

Euplæodictyoptera Jeannal 1945 Palæzonic fossil order of Palæodictyoptera (Palæoptera), large, pterygote insects with biting and chewing mouthparts; wings homonous with complete set of venation; phytophagous (all extinct).

Euparagidæ small family of Vespoidea (Aculeata: Apocrita) that includes medium-sized wasps which est above or below the ground surface.

Eupathithripidæ Bagnall 1915 family of thrips (Phloeothhripodea: Tubulifera); having sensory cones on antennæ usually long and pointed, each with long slender bristle or accessory cone; cone-bearing segments greatly swollen; eyes conspicuously large and contiguous.

Eupelmidæ a small group of parasitic wasps segregated recently from family Encyrtidæ (Chalcidoidea: Apocrita) having been bred from eggs of various insects including some Cecidomyidæ, Cocooidea, Coleoptera and Lepidoptera as also some *Arachnophaga* (Aracnida).

Euplantulæ pad-like structures on ventral surface of tarsal subsegments.

Eupleuron dorsal supracoxal sclerotisation of thoracic pleuron.

Euplexoptera Westwood 1839: old and equivalent term for insect order Dermaptera Leach 1815; no more in usage.

Euploeidæ old and obsolete term, synonym of Danaidæ (Papilionoidea: Ditrysia); some taxonomists consider it as subfamily Euploeinæ (= Danainæ) under family Nymphalidæ (Papilionoidea: Ditrysia).

Eurybathic living at the bottom of a body of water (sea, lake etc).

Eurybrachidæ a small family of Fulgoroidea (Auchenorrhyncha: Homoptera); allied to family Flatidæ with about 180 species confined to the Oriental and Ethiopean regions - *Eurybrachys tomentosa* is pest of sandal tree (*Santalum album* L.) in South India.

Eurygamous mating in flight - bees, various species of butterflies etc.

Eurygamy inability to mate in a confined space (captivity); phenomenon characteristic of many gnats (Mycetophitidæ and Trichoceridæ) and mosquitoes (Culicidæ: Culicoidea).

Euryhygric insect or any other organism that can tolerate a wide range of atmospheric humidity.

Eurymelidæ a family of leafhoppers (Cicadelloidea; Auchenorrhyncha); characterised by having ocelli on ventral side of head and tegmen with M_{1+2} extending upto the apex. About 400 species, confined to Australasia, feed mainly on eucalypts (*Eucalyptus* spp.), oaks (*Quercus* spp.) and mistletoe (*Viscum album* Linnæus), *Lorathus* etc., some (*Pogonoscopus* spp. live in ants' nests.

Eurymtopidæ Mjoberg 1910 family of Mallophaga in old system of classification, now forms a part of family Philopteridæ (Ischnocera: Mallophaga).

Eurypterida Paleozoic fossil order of Arthropoda, euroypterids lived from Camberian to Carboniferous periods and were more abundant during Siluràn and Devonian periods; acquatic, mostly freshwater or mud-inhabiting invertebrates - ;largest about two metres long - *Eurypterus* spp.

Eurytherm any organism (including insect) that is tolerant to a wide range of temperatures.

Eurytomidæ Walker 1833 family of minute gall-making chalcid wasps (Chalcidoidea; Apicrita); hind coxæ not enlarged and hind femur not toothed, pronotum wide and quadrate. Larvæ known as straw-worms - *Tetramesa grandis* is pest of wheat, *Eurytoma felis* is pest of citrus in Australia while *Bruchophagus gibbus* is major pest of legumes in Australasia and North America.

Eurytopic organism having a wide range of distribution.

Euschemoninæ a subfamily of Hesperiidæ (Hespterioidea: Ditrysia); includes skippers; once regarded as an independent family (Eusochemonidæ) of moths as the males posses a frenulum, but the larvæ and pupæ exhibit define hesperiid characteristics - *Euschemon* spp.

Eusocial insects insects living in co-operation with regard to laying of eggs, care of young ones, division of labour and presence of atleast two generations in the colony. These insects are commonly called social insects - ants, bees, wasps and termites.

Eusternum intrasegmental ventral plate of thoracic segment.

Eustheniidæ primitive family of archiperlarian stoneflies (Plecoptera), hind margin of hind-wings uniformly rounded (entire), anal lobe with arcedictyon and vein Rs branched only once. Nymphs predaceous with tubular gills on first 5 or 6 abdominal segments. Nymphal duration is of several years during which it undergoes more than 30 moults (instars). Found in Australasia and South America (17 species) - *Stenopoerla prasina* (New Zealand), *Eusthenia* species (South America) et al.

Euteliinæ small subfamily of Noctuidæ (Noctuoidea: Ditrysia); characterised by palpi with 2nd joint straight; abdomen longer than hind-wing and with a long tubula or paired anal tufts; forewing with vein 8 and 9 anastomosing to form the areole; hind-wing with vein 3, 4, 5 from lower angle of cell, 6 and 7 from upper angle. Caterpillars with four pairs of prolegs - *Angura lunulata* Moore, *Bombotelia jocosatrix* (Guenee), *Ingura subapicalis* Walker, et al.

Euthyplociidæ family of Ephemeroidea (Ephemeroptera, mayflies having forewing with many intercalaries between Cu_1 and Cu_2; male clasper single-segmented. Nymph with long maxillary palpi and setose mandibular processes.

Euthyrrhaphidæ a family of Blattaria (Dictyoptera), small insects having thorax pubescent, hind-wing with a pronounced thickening surrounding the ends of Sc, and anal area or fully winged forms, slender and folded once or twice only.

Eutrichosomatidæ a small family of wasps (Chalcidoidea: Apocrita) that are parasitic on coleopterous insects specially weevils.

Eutrochantin sclerite of thoracic pleuron adjacent to dorsal margin or coxa and bearing dorsal coxal articulation.

Evaniidæ (Leach 1812) Westwood 1840 a family of small black ensign wasps or evania-flies (Evanioidea: Apocrita); robust and short with antennæ inserted well above the clypeus, an anal lobe on hind-wing, short propleuron, long and abrupt petiole; compressed and short gaster and short unexserted ovipositor. Parasitic on oothecæ of cockroaches - *Evania albitarsis, E. appendigaster, E. curvicarinata, Brachygaster minuta* et al.

Evanioidea superfamily of Apocrita (Hymenoptera) that includes wasps distinguished by a wider costal cell and forewing and the gaster being attached by a slender petiole to the top of propodeum above the base of hind coxæ. The important families include, Aulacidæ. and Gasteruptiidæ.

Evolution a fundamental aspect of biology: process by which the organisms have changed through time, structurally and functionally. Doctrine or theory according to which higher forms of life have arisen gradually out of the lower forms.

Excretion removal of metabolic waste materials; it is mostly in form of uric acid.

Exeiridæ small family of Sphecoidea (Aculeata: Apocrita), medium-sized wasps considered beneficial as they always provision their nest with various insects, spiders etc. which they later consume.

Exocuticula chitinous outer layer of cuticula: primary cuticula.

Exoskeleton external supporting structure composed of sclerites; secreted by ectoderm.

Exotic outlandish: organism (mainly parasites and predators) introduce from foreign country.

Extant still existing or living.

Extinct no longer in existence: dead.

Exuvial glands glands of epidermis that secrete exuvial or moulting fluid.

Exuviate to shed: cast off the skin, shells or any other covering for a new covering or condition.

Exuviation act or process of exuviating.

Exuvium (pl. exuviæ) cast off skin after moulting. Fossil remains of an organism.

Eye photoreceptor or light perceptive organ: organ of sight or vision. Insects usually have simple eyes (ocelli) placed dorsally and compound eyes placed laterally on the head.

Eyepiece image forming lens(es): magnifying lens or system of lenses in optical instrument (microscope, telescope etc.); used for viewing the image formed by objective.

Eyespot simple eye or ocellus; organ of vision in many insects and other invertebrates.

F

FABRICIUS, Johann Christian (Fab.) father of entomology - born January 7, 1745 at Tonder (Denmark). Studied under Carolus A. Linnæus at Upsala (1762-64); worked as Lecturer political economic (1969-75) and Professor of natural history 1775 onwards. Founded system of insect classification based on mouthparts of the insects. His famous works include, **Systema Entomologiæ** (1775), **Genera Insectorum** (1776), **Philosophica Entomologia** (1778), **Species Insectorum** (1781), **Mantissa Insectorum** (1787), **Entomologia Systematica** volume 1 (1792), 2 (1793), 3 and 4 (1794) and Supplement (1798) and **Systema Piezatorum** (1804). Added insect order Odonata (1793) to already established seven orders of Linnæus and described another 10000 insect species. Died March 3, 1808 at Kiel (Germany).

Face front portion of head: in insects, region of head enclosed by frontal suture.

Dacet(s) tiny cuticular lenses covering the crystalline cones in the compound eye of an insect or crustacean. In insects their number (per compound eye) varies from less than dozen (worker ant) to over 20000 (dragonflies); some parasitic insects and those found in caves, may even have only one facet.

Facial pertaining to face.

Facillia facial ridges or vibrissal ridges that demarcate the face region enclosed by frontal suture in true flies (Diptera).

Factitious unreal: artificial.

Factitious host easily available organism used as laboratory host for mass culture of parasites or predators, which otherwise is not a host of that species under natural or field conditions.

Fæcal pertaining to fæces.

Fæcaloid resembling fæces freshly hatched and young caterpillars of lemon butterfly (*Papilo* spp.)

Falcate (Falciform) sickle-shaped.

False pseudo-

False katydid *Microcentrum retinerve* (Tettigoniidæ: Ensifera); angular-winged, green grasshoppers with long antennæ, found in North America.

False legs pseudolegs or prolegs; fleshy pad-like outgrowths from abdomen of various larvæ – Lepidopterous caterpillars usually have 5 pairs and those of sawflies (Hymenoptera) seven pairs. Maggots of many species of true flies (Diptera) also have false legs on their abdomen, either on ventral side (for walking) or in rings round the body (for pushing ahead).

Family principle taxonomic subdivision of an order of animals/plants, comprising of one to several similar genera. Family names of insects (usually Latin) terminate in -idæ, while those of plants end with -aceæ. Parents and children; all descendants of common ancestor.

Fannidæ family of Muscoidea (Schizophora: Cyclorrhpha); female orbits are wide and lack proclinate superior orbital bristles; vein Sc is bent and atonce well separated from R_1. Larva flat with paired segmental outgrowths; live in dung and decaying organic matter.

Fat body mass of fatty tissue filling most of the body cavity of an insect. Food stored herein includes fats, proteins and glycogen; which is made use of during hibernation or pupation.

Fatigue decreased ability of organism or one of its organs or tissue to function because of prolonged use; it arises when stimuli follows in too rapid succession to allow complete recovery.

Fauna assemblage of animal life in certain environment or place.

Feathered thorn medium-sized woodland moth (Lepidoptera) of Europe, *Himera pennaria,* having tawny-orange wings; forewings being crossed by two dark transverse lines and a round sport at the tip of each forewing.

Fecifork anal-comb or dung-fork: flattened fork-like appendage attached to anal end of abdomen of the larvæ of certain beetles (Coleoptera) and moths (Lepidoptera); used for shoveling up dung that is deposited on its back as protection against sun as well as parasites and predators.

Fecund fertile.

Fecundate impregnate: fertilise.

Fecundity number of eggs or offsprings that a female of a species can lay during her life-time.

Feed give food to: take food.

Feeding lesion damaged area of any part of plant resulting from feeding of insect(s).

Feeler(s) antenna(e).

Female (♀) an organism of feminine gender. sex (recipient) that has megasporangium (plants) or lays eggs/young ones (animals).

Femur third segment of insect leg: largest and strongest part of the leg.

Fenitrothion $C_9H_{12}O_5NPS$; phosphorothionate insecticide derived from thiophosphoric acid and acts as contact and stomach poison. Discovered in 1939 by Sumitomo Chemical Co., Osaka (Japan); it is yellowish-brown oily liquid, insoluble in water, but soluble in alcohol, ether and low boiling aromatics; phytotoxic to *Brassica* spp. and certain varieties of apple and cotton. Acute oral and dermal LD_{50} value for rat are 250 to 670 and 1500 to 3000 mg/kg respectively - Accothion®, Folithion®, Sumithion® etc.

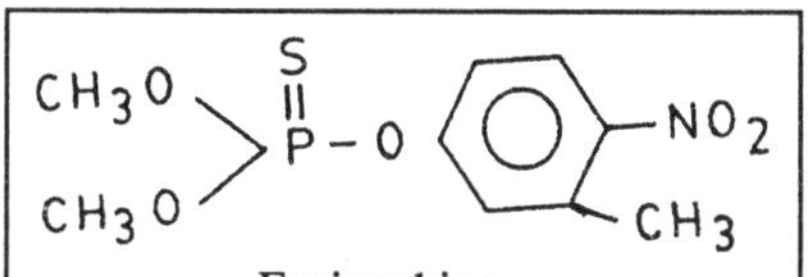

Fenitrothion

Fensuslfothion $C_{11}H_{17}O_4PS_2$; phosphorothionate - derivative of thiophosphoric acid - yellowish-brown liquid having contact and systemic action; used as insecticide and nematicide; slightly soluble in water but readily soluble in most organic solvents. Acute oral and dermal LD_{50} values for rat are 2 to 11 and 3 to 30 mg/kg respectively - Dasanit®, Terrcur®, etc.

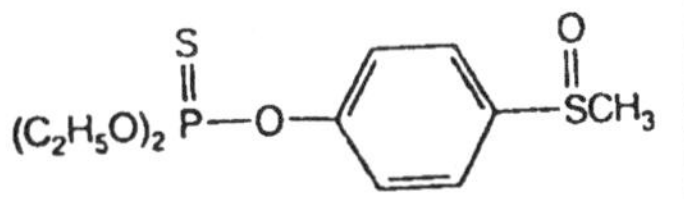

Fensulfothion

Fenthion $C_{10}H_{15}O_3PS_2$; phophorothionate insecticide, a derivative of thiophosphoric acid, having systemic action. Colourless oily liquid, insoluble in water, slightly sounble in petroleum ether and readily soluble in most organic solvent. Oral and dermal LD_{50} values fro rats being 200 to 350 and 1300 mg/kg respectively - Baytex®, Entex®, Lebaycid®, etc.

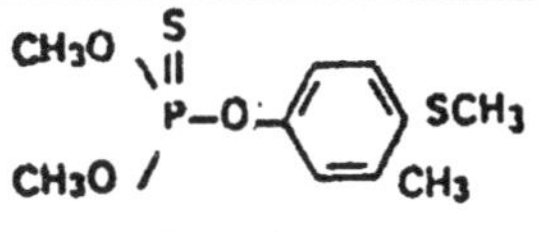

Fenthion

Fergusoninidæ very small family of Agromyzoidea (Schizophora: Cyclorrhapha); akin to Agromyzidæ; confined to Australia, where these flies make galls on *Eucalyptus* spp.

Fertile able to bear or reproduce.

Fertilisation coming together of an egg and a sperm and subsequent fusion of their nuclei.

Fertilise impregnate.

Fertility ability to reproduce; state of being fertile.

Fideliidæ family of Apoidea (Apocrita: Hymenoptera); similar to Magachilidæ except that forewings have three submarginal cells and dorsal mesepisternal groove present. Nest in the soil; cells without any linning - only two genera and a few species found in South America and South Africa.

Figitidæ family of Cynipoidea (Apocrita; Hymenoptera); includes small and dark (blackish) coloured wasps, distinguished by antennæ not elbowed, scutellum sculptured but without cup-like raised depression; second gastral tergite which, though large, is not half the length of gaster. Primary larva is eucoiliform followed by polypodeiform instar having 10 pairs of trunk limbs - *Aspicera* spp. parasities Syrphidæ; *Anacharis* spp. are found feeding on Neuroptera and *Figites* spp. on Diptera.

Filipalpia suborder of Plecoptera (Orthopteroidea: Poluneoptera); stoneflies having amphinotic distribution (Australasia to South America); glossæ as long as praglossæ; mandibles well developed, maxillary palpi filiform: differs from Archiperlania in having much reduced cross-veins. Nymphs without segmentally arranged gills. Two superfamilies - Leptoperloidea and Nemouroidea.

Filippi's glands accessary glands associated with silk glands in caterpillars of most species; paired organs, often voluminous, each opens by a separate duck into the silk duct of its side. In some species, these may be rudimentary, reduced to a group of follicles surrounding the silk duct.

First axillary sclerite that serves as anterior hinge plate of wing base associated with base of subcostal vein.

Flagellum part of antennæ distal to pedical, filamentous and multiarticulate.

Flareback pest resugence: rapid increase of target insect pest population subjected to an insecticidal treatment which also destroys (kills) the associated natural enemies of the pest.

Flat headed borer *Sphenoptera dadkhani* Obenberger (Buprestidæ: Buprestoidea); a serious pest of peach trees in North-western India; is sun-loving having distinct preference for trees that are more than five years old, causing upto 50% infestation; gurbs are active during warmer part of the year, hibernating during Winter; there are three generations in a year.

Flatidæ Spinola 1839 family of Fulgoroidea (Auchenorrhyncha: Homoptera), beautiful moth like species, clypeus usually without lateral carinæ or costal area of tegment without cross-veins. Nymphs and adults frequently rest gregariously; the former are largely covered with long curled waxy filaments; found in tropics and subtropics (100 species) - *Flata ocellata* Fabricius, (on jasmines), *Lawana conspersa* Walker (on tea) et al.

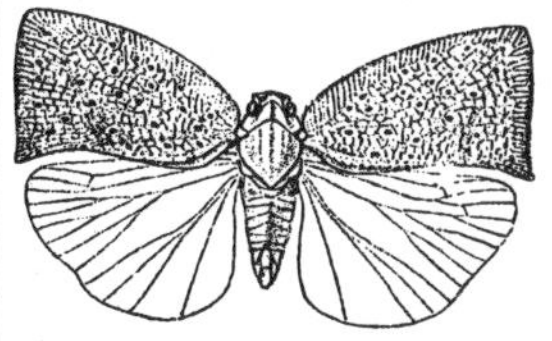

Flata ocellata Fabricius

Flavellum small spoon-shaped tip of the tongue (ligula) of bees (Apoidea: Apocrita).

Flea tiny, apterous, blood-sucking, ectoparasitic insect (Siphonaptera: Oligoneoptera) that transmits various diseases.

Flea, Cat *Ctenocephalides felis* (Bouche) found commonly on cats and dogs but occasionally on man and other domestic animals also.

Flea, Ground squirrel *Diamanus moutanus* (Baker) (Dolichopsyllidæ); confined to Western North America, infests ground squirrels and other rodents; also transmits bubonic plague from rodent to rodent.

Flea, Dog *Ctenocephalides canis* (Curtis) found feeding on blood or large number of mammals, dog being its preferred host.

Flea, Human *Pulex irritans* Linnæus, primarily parasite of human beings, also parasitises dogs, squirrels etc. and acts as carrier of plague.

Flea, Rat *Xenopsyla cheopis* (Rothschild), an ectoparasite of rats that attacks man also and spread plague.

FLETCHER, T. Bainbrigge - Assistant Entomologist, Government of Madras in 1904, Entomologist 1909 and Imperial Entomologist at IARI, Pusa (Bihar) 1913-34, devoted his attention to insect taxonomy and build up a very good library and valuable insect collection; organised five all India entomological meetings at Pusa (Bihar) and published their proceedings (1916-23); started publication of the **Catalogues of Indian Insects**, besides continuing with **Scientific Monographs and Memoirs**. His most outstanding contribution, **Some South Indian Insects** (1914) is still being consulted by all students and scholars of entomology and allied sciences.

Flexible plaint: that can bent easily.

Flight ability to fly: power of flying attained by strolling wings, working independently or in unison; most insects (alate) and majority of birds (Aves), besides flying fish (Pisces) and bats (mammal). Cockroach-like insects that lived in trees and ferns of Mesozoic forests (200 million years ago) were probably the first animals to use this mode of locomotion. Normal flying speed of locusts is about three metres per second while some other insects can fly upto 10 metres per second (short distances).

Flightless without power of flying – apterous insects and a few birds like emu, kiwi, ostrich, rhea, etc.

flightless insects most insects have two pairs of functional wings (true flies or dipteran insects have only one pair of wings), except primary wingless insects such as bristletails, springtails, silverfish, campodids and japigids that have never evolved upto winged stage. These come under the apterygote orders, Thysanura, Diplura, Protura and Collembola In insects like ants, (Hymenoptera), aphids (homoptera) and termites (Isoptera), both winged and wingless castes, but due to their mode of life subsequently host the wings, include bedbugs, fleas, lice, a few parasitic insects as also some beetles and scale insects.

Floral bud maggot cotton flower lint maggot, *Dasyneura gossypii* Felt (Cecidomyiidæ: Nematocera); a dipteran phytophagous, free-living larva that feeds in the buds, flowers and lint, it is an important pest of cotton in West Indies.

Floss rough silk covering cocoons of silkworms (Lepidoptera).

Fluidity ability or ease of a liquid to flow. *Cf.* viscosity.

Fluted scale alternate name for cotton cushion scale, *Icerya purchasi* Maskell, (Margarodidæ: Coccoidea); a small scale insect, native to Australia, found all over the World as a major pest of citrus and various other fruits trees.

Fly general term for many adult insects, but more strictly for true flies (Diptera) specially housefly *Musca domestica* Linnæus. Butterfly (Lepidoptera), damselfly and dragonfly (Odonata), mayfly (Ephemeroptera), stonefly (Plecotera) and whitefly (Homoptera) are not true flies.

Fly, Biting house- *Stomoxys calcitrans* (Lilnnæus) (Muscidæ: Muscoidea) breeds in horse-manure an stable refuse; loves absence of light and abundance of moisture. Maggots differ from those of house-fly in having hind spiracles smaller and rounded with three apertures on each plate only slightly curved instead of being sinuous.

Fly, Black common name loosely applied to various small insects specially bean-aphid, *Aphis fabæ* Scopoli (Aphididæ: Aphidoidea), as also to small stout hump-backed dark flies of family Simuliidæ (Oligoneura: Nematocera: Diptera).

Fly, Blow blue-bottle flies, metallic blue or green colour; *Calliphora erythrocephala* (Meigen) and *C. vomitoria* (Linnæus) (Caliphoridæ: Musccoidea) are common house pests, indoors during Winter and outside on warm days. These are characterised by weak development of macrochætæ; their maggots occur in carrion, flesh etc. *Lucilia* spp. (*L. caesar, L. serricata* etc.) are known as green-bottle flies as these are metallic green in colour.

Fly, Cluster *Pollenia rudis* (Fabricius) (Callihoridæ: Muscoidea); common parasite of earthworms; eggs are laid during Autumn end in the soil, maggot enters the earthworm, probably through the gential aperature and feeds inside for several months (including entire Winter) till all the contents of the host are consumed; then it pupates and the adult comes out after 5 to 6 weeks.

Fly, Fairy chalcid wasp (Mymaridæ: Chalcidoidea); World's smallest insects with wing spread of < 1 mm. Larvae are egg-parasites.

Fly, Fever *Dilophus febrilis* (Bibionidæ: Nematocera), day-flying dark coloured hairy flies, antennæ of males attached below the very large eyes; females have smaller eyes. Does not cause any fever. Maggots live in decaying plants and roots of verious crops and pupate in soil.

Fly, Fire beetles (Lampyridæ: Cantharoidea) that emit light at night; luminous organs are situated on VI and VII abdominal segments wherein luciferin in oxidised with help of enzymes luciferase and cold light of chemiluminescence is produced – *Lampyris tardus* (India), *L. noctiluca* (Linnæus) (Europe), *Luciola lusitanica* (Mediterranean).

Fly, Flesh grayish flies (Sarcophagidæ: Musocoidea); that are larviparous, parasites invertebrates like bees, grasshppers, spiders, snails etc., also breed on carcasses, decaying vegetation, farmyard manure etc. – *Sarcophaga* spp., *S. hæmatodes* however is coprophagus and *S. carnaria* (Linnæus) is the famous carrion fly.

Fly, Fruit small to medium-sized flies (Tephritidæ: Agromyzoidea); almost cosmopolitan in distribution; are easily recognised by conspicuous spots or bands on their wings. Eggs laid on fruits, maggots, bore inside and feed on fleshy fruits and pupate therein – *Bactrocra dorsalis* (Hendel), *B.*

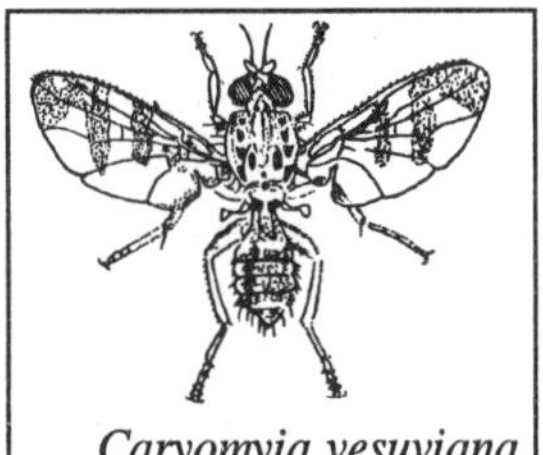

Caryomyia vesuviana Costa

cucurbitæ (Coquillett), *B. zonatus* (Saunders) and *Caryomyia vesuviana* Costa are some major fruit pests in India while Mediterranean fruitfly, *Ceratitis capitata* (Weidemann) and African mango fruitfly, *C. cosyra* (Walker) are confined to their respective regions and have not yet been intercepted from India.

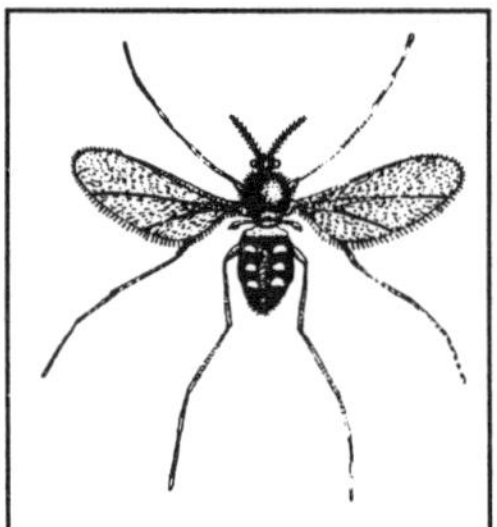

Lasioptera falcata Felt

Fly, Gall mite fragile cecidomyid fly (Mycetophiloidea: Nematocera); akin to gall midge having exceptionally long legs. Pests of cucurbits; maggots tunnel inside the shoots and feed within infested vines develop galls, which in turn cause stunted growth - *Lasioptera falicata* Felt. *L. cephalandræ* et al.

Fly, Hessain small fragile, gnat-like fly, *Mayetiola destructor* (Cecidomyidæ: Mycetophiloidea), cosmopolitan in distribution its sap-sucking maggots are serious pests of various economic crops especially cereals (wheat etc.).

Fly, House small, blackish flies, *Musca domestica* Linnæus (Muscinæ: Muscidæ); cosmopolitan household pest, abundant in hottest months; adult having reddish-brown big eyes, transparent wings and highly specialised moutherparts. Method of feeding is to regurgitate saliva or gastric juices over the food, this partly digest and soften the food which is then sucked up. Maggots headless and legless; pupæ coarctate. Serious menace to human health, being carrier of diseases like diarrhoea, typhoid etc.

Fly, Hover syrphid fly (Syrphoidea: Aschiza); black and yellow coloured, resembling wasp, or densely hairy like bumble-bee; nearly all species are attracted to flowers and are seen vibrating with extreme rapidity (hovering) over the flowering plants. Maggots may be phytophagous (*Mesogramma polita*), carnivorous (*Dasysyrphs tricinctus*) or saprophagous (*Volucella bombtlans* Linnæus).

Fly, Safflower fruit *Acanthiopilus helionthis* Rossi (Tephritidæ: Agromyzoidea); pale green flies; maggots damage buds and flowers and pupate therein, infested buds rot and give offensive smell.

Fly, Safflower stem *Melanagromyza obtusa* Malloch and *M. phasoeli* Coquillett; (Agromyzoidea: Schizophora); metallic-blue flies, major pest of pulse in India; maggots bore into stems resulting in small swellings; attacked stems wilt, droop and die away.

Fly, Snow snow-gnat, *Chionea nivicola* (Tipulidæ: Tipuloidea); a small (4 mm long), soft-bodied, long-legged fly of North America; extremely hardy living at high altitudes, appears on the snow in Winter.

Fly, Stable cosmopolitan blood-sucking pest of livestock, *Stomoxys calcitrans* Linnæus (Muscidæ: Schizoptera) resembling housefly but with stiff piercing proboscis; also transmits diseases like Anthrax, Tetanus etc., breeds in horse-dung impregnated with urine of farm-yard manure.

Fly, Warble small stoutly-built dipteran flies (Hypodermatidæ: Calyptratea) lay eggs on hairs of legs of various animals; maggots pierce the skin, feed inside and move upwards till they reach the dorsum of the animals; where they rest and cause small swelling (warble) in which they live - *Hypoderma bovis* (de Geer), *H. diana* (Brauer) and *H. lineatum* (de Villier) attack cattle and horses while *Elephantoloemus indicus* is serious menace in domesticated herds of Indian elephants (*Elephus* spp.) in India, Myanmar and Malaysia.

Focilinæ subfamily of Noctuidæ (Ditrysia: Lepidoptera); palpi sickle-shaped with 2nd joint curved and the 3rd usually long; tibia and tarsal joints short and stout; forewing with vein 6 from angle of cell 8 anastomosing with 9 to form the areole; hind-wing with vein 5 fully developed, 3, 4 and 6, 7 from angles of cell. Caterpillars semiloopers with first or first two pairs of prolegs rudimentary or absent - *Hyposemansis singha* Guenee, *Lopharthum comprimens* Walker, *Raparna digramma* Walker, *Avitta subsignans* Walker, et al.

Follicle egg-chamber.

Follicle cell(s) inner epithelial cells of ovarian egg tube.

Foramen magnum occipital foramen; opening from head into neck.

Forcep(s) claspers or pincers at posterior end of the abdomen of earwigs (Dermaptera); used for copulation or for defence.

Forcipate resembling forceps.

Forfex pincers of an earwig; forcipulæ.

Forficulate having forceps or pincers: like scissors.

Forficulidæ family of Forficuloidea (Forficulina: Dermaptera); distinguished by 2nd tarsal segment bilobed, parameres rounded and single median penis lobe - *Allodahlia* species are from Indo-

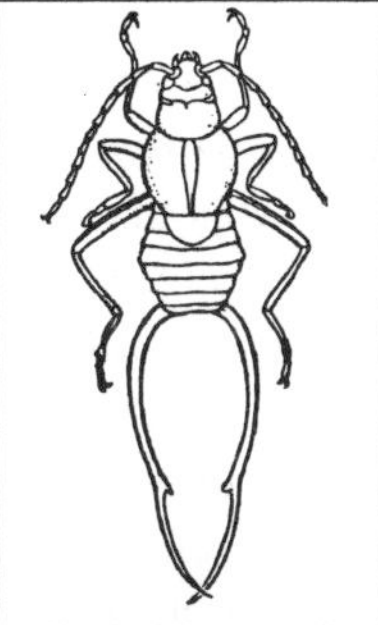

Allodahlia species

Malaysian region; *Forficula auricularia* Linnæus is predaceous on fly maggots in Mediterranean region but phytophagous in cooler regions of Northern Europe; *F. sjostedti* is found at elevations between 300 and 400 metres on Kilimanjaro (E. Africa) while *Anechura* spp. are apterous high altitude forms.

Forficulina Newman 1934 suborder of Dermaptera (Polyneoptera: Pterygota); alate or apterous, body without strong bristles; eyes well developed, cerci forcipate; 1st tarsal segment longer than 2nd; viviparous and free living - Pygidicranoidea, Labioidea and Forficuloidea.

Forficuloidea superfamily of Forficulina (Dermaptera: Pterygota), characterised by 2nd tarsal segment bilobed or extending beneath 3rd segment; ædeagus with long, simple distally rounded prameres; penis lobes paired or simple - Labiduridæ, Chelisochidæ and Forficulidæ.

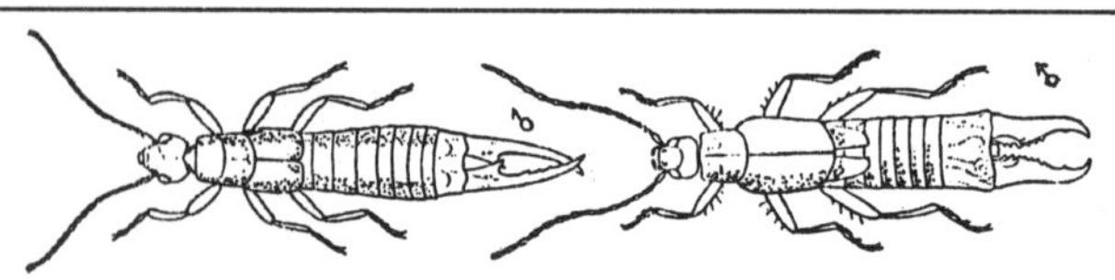

Nala figinii Burr (Labiduridæ) — *Adiathetus tenebrator* Kirby (Chelisochidæ)

Formicidæ (Latreille, 1802) Stephens 1829 large family of social insects (Scolioidea: Apocrita); antennæ elbowed comprising of 4 to 13 segments; compound eyes and three ocelli well developed in males but in females and workers these are usually reduced or vestigial. Males and futile female (queen) always winged, female wings are soon shed; workers wingless and with some posterior thoracic sutures (4600 species) - black ant [(*Oecophylla smaragdina* (Fabricius)], red ant (*Dorylus orientalis* Westwood), *Camponotus* spp. et al.

Formicoidea monofamilial superfamily of Apocrita (Hymenoptera: Oligoneoptera) in old classification, now included under superfamily Scolioidea. It includes all ants - Formicidæ.

Formothion $C_6H_{12}NO_4PS_2$; Organophorphate (Phosphorothiolothionate) acaricide and insecticide having systemic and contact action. Yellow viscous oil or crystalline mass, slightly soluble in water and miscible in alcohol, benzene, chloroform, ether and ketone. Stable in non-polar solvent but is hydrolysed by alkali, hence incompatible with alkaline pesticides. Acute oral and dermal LD_{50} values for rat are 375 to 353 and 400 to 1680 mg/kg respectively – Anthio®, etc.

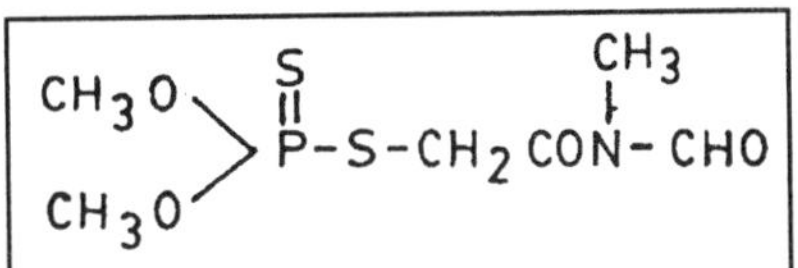

Formothion

Fortutous bio-control accidental movement of an exotic natural enemy to new area resulting in pest population suppression in that area.

Fossil recognisable remains of prehistorical animal or plant, once buried in earth and preserved mostly in segmentary rocks as impressions, casts or moulds. There are about 13000 fossil insect species recorded so far.

Fossil insects insects that lived during Palæzoic and Mesozoic eras. Over 13000 species have been described till today from various parts of the World; the oldest being fossil Collembola – *Phyniella præcursor* and *Rhyniognatha hirsti* – that existed during lower Devoniona period.

Fossore(s) solitary wasps (Hymenoptera) having their legs adapted for digging and running or for nest-building.

Fossorial adapted or modified for digging – legs of mole-crickets (Gryllotalpidæ: Orthoptera) or solitary wasps (Sphecoidea: Hymenoptera).

Foulbrood common name for various destructive bacterial diseases of honeybee larvæ.

Fourth axillary small sclerite of wing base, present only in some insects, intervening between third axillary and posterior wing process of tergum.

Fracticipita Oudemans 1909 one of the two suborders of Siphonaptera that covered all broken-headed fleas, but being of little systematic value, is no more in usage. *Cf.* Integricipita Oudemans.

Franklinothripidæ Bagnall 1926 family of Æolothripoidea (Terebrantia: Thysanoptera); thrips with saw-like ovipositor often placed under family Æolothripdæ; antennæ very long and slender; front ocellus vestigial or wanting; forewings narrow and without distinct cross-veins; mostly predaceous on other thrips–*Franklinothrips vespiformis* is predaceous on *Heliothrips rubrocinctus, F. myrmæciformis* on *Retithrips ægytiaca.*

Frass fæces and other refuse of boring insects including wood fragments.

Frentatæ Comstock 1892 (Hetroneura Tillyard 1918) suborder of Lepidoptera in old classification; body completely covered with hair or scales; antennæ variable, may or may not be hooked or clubbed, wings well developed, coupled or united by a frenulum or humeral area and characterised by venation that is dissimilar in fore- and hind-wigs; anamorphosis complex; includes both moths and butterflies and is divided into two divisions - Heteroneura (moths) and Rhopalocea (butterflies and skippers).

Frenulum long curved bristle or group of bristles arising near the base of hind-wing and extending forward and outward, passing under turf of scales on underside of forewing; present in most moths; holds the two wings together while in flight.

Fritillary common name for several butterflies (Nymphalinæ: Nymphalidæ), having beaurifully patterned wings resembling lily flowers (*Fritillaria* spp.), strong fliers, males more brightly coloured than females; more common in Britain and Europe.

Fritillary, Dark-green *Mesoacidalia aglaja* (Nymphalidæ: Papilionoidea); wing span about 65 mm, having patches of dark green scales on yellowish-brown and silver-spotted hind-underwings.

Fritillary, High brown *Fabriciana adippe* (Nymphalidæ: Papilionoidea); wing-span about 70 mm and buff-coloured underwings, is typical fritillary of woodlands, plentiful throughout England and Europe.

Fritillary, pearl-bordered *Clossiana auphrosyne* another Nymphalid butterfly (Papilionoidea: Ditrysia); lighter in coloure with far fewer silver spots on hind-underwings than the small pearl-bordered fritillary; widely distributed in Britain, appears early in May and is addicted to woodlands.

Fritillary, Silver-washed *Argynnis paphia* (Nymphalidæ: Papilionoidea); largest and most magnificant of British fritillaries, essentially woodland species, more common in new forests feeding from bramble and thistle flowers. Hind-wings are streaked or washed with silver on underside instead of being spotted. Female is larger and duller than male.

Fritillary, Small pearl-borderd *Clossiana selene* (Nymphalinæ: Nymphalidæ) easily confused with pearl-bordered fritillary; is darker in colour, with more silver spots and appears on wing a little later (June-July), found in marshy places throughout Britain.

Fritillary, Violet *Clossiana dia* is small nymphalid butterfly (Papilionoidea: Ditrysia) widely distributed in Eurasia (Central Europe through Western China). It has orange and black chaqueboard pattern characteristic of fritillaries but ventral surface of hind-wings is of violet shade instead of orange.

Froghopper plant-bug (Cercopoidæ: Auchenorrhyncha); having broad head, frog-like appearance and jumping ability. Some species because of their toxic saliva cause necrotic lesions on the leaves known as frog-hopper blight – *Philæhus spumarius* (Linnæus) (Britain), *Tomaspis* species (New World) et al.

Frons facial area of cranium between frontal and epistomal sutures.

Frontal ganglion median precerebral ganglion of stomodæal system.

Frontal lunule small crescentic sclerite, just above the bases of antennæ in Cyclorrhappha (Diptera).

Frontal triangle space between compound eyes and basal line of antennæ in holoptic flies (Diptera); the triangular region bearing ocelli and bounded by grooves or depression is **ocellar triangle.**

Frugivorous one who feeds on fruits and nuts.

Fulgoridæ Latreille 1807 tropical family of lantern-flies, brilliantly coloured plant-hoppers (Fulgoroidæ: Auchenorrhyncha), wing span upto 150 mm, hind-wings with reticulate system of supernumerary veins and cross-veins; ovipositor incomplete ædeagus with distinct theca surrounded by the penis; some species also secrete flocculent white wax (600 species) – *Eurybrachys tomenhtosa* Fabricius, *Trypetimorpha fenestrata* Costa, *Fulgora delesserti* et al.

Fulgoroidea Kirkaldy 1907 superfamily of Auchenorrhyncha (Homoptera: Hemipteroidea); head with postclypeus reduced; antennal pedicel with postclypeus reduced; antennal pedicel with numerous sensilla; tegulæ present; wings without ambient vein; middle coxæ elongate and placed wide apart; hind coxæ immoble. Over 9000 species arranged in 20 or more families.

Fulgoromorpha one of the two primary groups of Auchenorrhyncha (Homoptera) consisting of a single superfamily Fulgoroidea; the other group Cicadomorpha embraces Cercopoidea, Cicadoidea and Cicadelloidea.

Fulturæ sclerites in lateral walls of adoral surface of base of hypopharynx; suspensorium of hypopharynx.

Fumigant chemical exhibiting toxicity in vapour phase – carbon bisulphide (CS_2) etc.

Fundatride(s) apterous, viviparous, pathenogenetic female aphids (Aphididæ: Homoptera) that emerge in Spring from the overwintered eggs on primary host plant; antennæ, legs and sense organs not so well developed as in succeeding apterous generations.

Fundatrigeniæ progeny of fundatrices – apterous parthenogenetic viviparous female aphids (Aphididæ), that live on primary host.

Fungicide pesticide that can kill fungi – Bordeaux mixture, sulphur dust, thiram, zineb etc.

Fungiform mushroom-shaped.

Fungistat(ic) any substance capable of preventing or checking fungal growth without actually killing the fungi.

Fungivoridæ old and obsolete term for orthorrhalphous dipteran family Mycetophilidæ (Mycetophiloidea: Nematocera).

Fungivorous feeding in or on fungi.

Fungus (pl. **fungi)** a group of nucleated, usually filamentous, spore-bearing organisms (devoid of chlorophyll). These may be saprobic, symbiotic or parasitic . Earlier a subkingdom of Plantæ, it now forms and independent kingdom – Mycota – with about 100000 described species. A large number of species serve as causal agent for the various plant diseases that are often controlled by roughing or spraying suitable fungicides.

Fungus ants(s) ants that cultivated fungi on which they then feed – *Atta* spp. (Formicidæ: Scolioidea).

Fungus beetle(s) small hairy beetles (Mycotophagidæ: Heteromera); found feeding on fungi.

Fungus borer(s) a common name for number of unrelated insect species whose larvæ bore into fungi and feed therein. The insects may be coleopteran, dipteran, lepidopteran or even hymenopteran.

Fungus gall a small plant gall resulting from an attack of a parasitic fungus.

Fungus garden (cutlivation) workers of certain species of ants (leaf-cutting, parasol etc) and termite (Isoptera) carry pieces of leaves into their nests on which certain species of fungi grow an serve as food for young ones of that colony.

Fungus gnats small dipteran flies (Mycetophilidæ: Mycetophiloidea) whose maggots bore into mushrooms and allied fungi.

Funicle series of small similar segments between the ring joints and the club, in genicula antennæ of insects. Stalk of an ovule.

Furcasternum distinct part of sternum in some insects bearing furca.

Furcula springing organ of collembolan insects projecting from ventral surface of IV and VI abdominal segment, consisting of large median base (manubrium) and two long terminal prongs, contraction of adductor muscle of prongs, forcibly extends the organ throwing the insect upward and forward.

Furca (pl. **furcæ)** forked endosternal process of higher insects, formed on sternal apophyses.

G

Gadfly stout, bristlesless, blood-sucking fly (Tabanidæ: Brachycera); female attacks cattle, horses and other domesticated animals while male subsists on honeydew and juices of flowers; some species are carriers of Typanosomes and other parasites - *Tabanus bovinus, T. maculixcornis, T. striatus* and allied species.

Gadfly, Sheep sheep botfly, *Oestrus ovis* Linnæus (Oestridæ: Tachinoidea: Schizophora); large, 12 mm long, hairy dull-yellow, active fly, appears in the hot sunshine during late Spring and Summer. Females larviposit in the nostrils of sheep and goats; maggots inhabit the nostrils for about 10 months, causing extreme annoyance, pain and even death; when mature, the maggots drop to the ground, pupate in the soil from where the adults emerge.

Gæad young one of ametabolous insect (Apterygota), as these do not undergo any metamorphosis, the young ones are practically identical to adults, hence the name larva or nymph is inappropriate.

Galea hood-like outer endite lobe of amxillæ of certain insects, provided with muscle arising in stipes.

Galeate having a galea.

Galerucinæ subfamily of Chrysomelidæ (Chrysomeloidea: Heteromera); leaf-beetles having short-oval body and enlarged abdomen; grubs are highly injurious pests, mostly of woodlands trees, a few semi-aquatic feed on lily pods and other floating plants - *Galerucella xanthomelæna* feeds on elm leaves and *G. lineola* attacks willows. In India *Galerucida bicolor* (Hope) and *Raphidopalpa* spp. are destructive pests of vegetable crops.

Gall abnormal growth of tissues caused by stimulus: abnormal plant growth resulting from feeding activities of insect larvæ/nymphs or mites and eelworms as also fungi. Galls may be open or close, former are produced by aphids and coccids and these insects do not live within the galls whereas in case of close galls produced by beetles, flies, moths or wasps, the gall completely surrounds the larva which feeds and grows inside the gall and comes out as adult.

Gall, Bladder pouch-galls; globose, irregular, convoluted unilocular galls as caused by *Austrothrips cochinchinensis* Karny (Thripiæ:

Thysanoptera) in a climbing shrub, *Calycopteris florubunda.* Single independently occurring and closely clustered pouch-galls at the base of leaves of yellow myrobalan (*Terminalia chebula* Ritz.) caused by *Dixothrips onerosus* (Thysanoptera).

Gall, Blister leaflets attacked are converted into an aggregate of small irregular lobules so as to appear like an inflorescent; as caused by *Crotonothrips gallarum* (Thysanoptera) in *Secamone emetica.*

Gall, Finger horn-galls: tubular outgrowths from leaf surface as caused by *Liothrips ramakrishnae* (Tubulifera: Thysanoptera) in *Ficus* species.

Gall(s), Honey galls caused by *Disholcaspis eldoradensis* (Beutenmitter) (Cynipidæ: Cynipoidea: Apocrita) on oak twigs (*Quercus* spp.) in North America; remarkable, as during the development, the galls exude quantities of sweat nectar eagerly sought by honeybees.

Gall, Jamping tiny (1 to 2 mm in diameter), bead-like gall of *Neuroterus saltatorius* (Edwards) (Cynipidæ: Apocrita) formed in great numbers on ventral surface of leaves of oak trees (*Quercus* spp.); drop to the ground during late Summer and Autumn; the active larvæ within, throw themselves against the thin walls, causing the galls to jump several centrimetres. Common in Western USA.

Gall, Rosette galls in which the leaflets get enlarged and rolled irregularly to form a bunch of rosette-like structures; these are caused by the thrips, *Thilakothrips babuli* Ramakrishnan (Thrysanoptera) in white *kikar* trees (*Acacia leucophloea* Roxburgh).

Gall, Tubular galls formed all along the midrib region, with margins of leaf free as caused by the thrips, *Mesicothrips plicans* (Thysanoptera) in some evergreen trees (*Linociera* species).

Gall-midge minute delicate fly (Cecidomyiidæ: Mycetohiloidea: Nematocera); having long moniliform anternnæ with conspicuous whorls of hair, long legs and wing-venation without cross-veins – *Ameradiplosis echinogalliperda* Mani, *Dasyneura amaramanjarae* Grover, *Procontarinia matteiana* Kieffer and Cocconi, et al.

Gall-mite phytophagous mite (Europhyidæ: Acarina), microscopic arachnid that feeds on plant juices and causes galls – *Aceria mangiferae* Sayed responsible for mango malformation.

Gall-nut nut-like galls produced on oak trees (*Quercus* spp.) by gall-wasp; gallic and tannic acids are obtained from these gall-nuts which are used for tanning, dyeing and making ink – *Quercus infectoria* Olivier is coommonly called gall-oak or dyer's oak.

Gall-wasp minute black insect (Cynipidæ; Apocrita) usually alate but without pterostigma, produces galls on all plant parts (roots to flowers) of lives in the galls made by other insects or mites. Their main host is oak trees (*Quercus* spp.) though a few are also found on roses (*Rosa* spp.) Tannic acid and certain dyes are obtained from such galls - *Neuroterus lenticularis* et al.

Galleriidæ (Zeller 1848) Wallengren 1871 family of small fragile moths (Pyralidoidea: Ditrysia), proboscis well developed; maxillary palpi filiform and not traingularly scaled. Forewing with vein 1a separate from 1b and vein 7 present; hind-wing with median nervure pectinated on upperside, vein 7 anastomosing with 8 and frenulum multiple in females. Caterpillars feed on dried material, combs of beehives and nests of wasps—wax moth (*Galleria mellonella* Linnæus) feeds on wax in beehives, *Achroia grisella* Fabricius, et al.

Gallicolous (gallicolæ) insects that produce galls on leaves, stems etc. - gall-midges, gall-wasps, certain aphids and coccids as also some caterpillars.

Gallivorous insect larvæ that feed on tissues of galls.

Galvanotaxis movement of free-living organisms (including insects) in response to electrical stimulus.

Gamete reproductive cell - egg or sperm.

Gasteruptionidæ a small family of Ichneumonoidea (Parasitica: Apocrita); head free, compound eyes well developed, prothorax elongated greatly into a neck; forewing with one or more recurrent veins, radial cell long and narrow and cubital cell one or none; abdomen long, gradually clavate behind and inserted on thorax far above the base of hind coxæ. Parasitic on the immature stages of other insects.

Gastroceoele archenteron: cavity of endoderm.

Gastrotricha minor phylum of Pseudocoelomata (Triploblastica); very small, aquatic (freshwater and marine) animals, < 4 mm long, free-living, bilaterally symmetrical, elongated, flattened, vermiform (unsegmented) body, acoelomate, lacking circulatory and respiratory systems; mostly hermaphrodite, a few parthenogenetic. About 450 species placed in two orders - Chætonotida and Macrodasyida.

Gastrula embryo after gastrulation.

Gastrulation formation of endoderm by invagination of its wall or internal prolification of cells.

Gelechiidæ Stainton, 1854 a large family of Gelechiolidea (Ditrysia: Lepidoptera); small, cryptic and brightly coloured moths that are diurnal, crepuscular and nocturnal in habit; proboscis moderately long and scaled basally; antennæ rarely with basal pecten; maxillary palpi vestigial or absent, labial palpi slender, long, upturned and acute; hind tibia rough-haired; forewings trapezoideal, margins of hind-wings often concave and emarginate with pointed tips., About 5500 species, cosmopolitan in distribution - Angoumois grain moth *Sitotroga cerealella,* (Olivier), pink bollworm *Platyedra gossypiella,* (Saundrs), potato tuber moth *Phthorimæa opercullella* (Zeller) et al.

Phthorimæa operculella (Zeller)

Gelechioidea Mosher 1916 a superfamily of Ditrysia (Lepidoptera); moths having head usually smooth, proboscis moderately long, haustellum densely scaled at the base; maxillary palpi reduced or absent; labial palpi long and porrect; hind tibia usually rough-haired; wings with soft scaled and small anal regions; hind-wings broader than forewing. Caterpillars very small, phytophagous, often gregarious and many spin leaves and shoots.

Gena (pl. **genæ**) facial areas behind and beneath the eyes; lateral parts of parietals.

Gene fundamental physical unit of heredity that transmits information from one cell to another and thereby from one generation to another. The words 'gene', 'genotype' and 'phenotype' were coined by a Danish botanist Johannsen Wilhelm (1857-1927).

Generation period from any given stage in life-cycle (usually adult) to same stage in offspring.

Gengidæ a very small family of Fulgoroidea (Fulgoromorpha: Auchenorhyncha) allied to family Flatidæ; it includes only two species both from South Africa - *Gengis panoplites* and *Microeurybrachys vitrifrons.*

Genital chamber copulatory invagination cavity in female that receives median oviduct and duct of spermatheca into its anterior end — Bursa copulatrix.

Genitalia genital organs collectively, specially the external ones.

Genotype actual constitution of a zygote of genses. *Cf.* phenotype.

Genus (*pl.* **genera)** taxanomic rank, subdivision of a family, composed on one to several allied species.

Geocorisæ Latreille 1825 subdivision (series) of Gymnocerata (Heteroptera : Paraneoptera); includes terrestrial bugs having long conspicuous antennæ and en ejaculatory bulb in males. Comprises of two section — Cimicomorpha and Pentatomomorpha.

Geometer(s) earth-measurers or inchworms are common names of the looper (Geometridæ: Geometroidea) due to their peculiar method (looping action) of movement.

Geometridæ Stephens 1829 very large family of Geometroides (Ditrysia: Lepidoptera), having worldwide distribution and includes carpet, pugs, waves etc., distinguished by forewing having R_{2+3} connected with R_{4+5} hindwing with $Sc + R_1$ remote from Rs beyond the cell; tympanal organs situated ventrally, Caterpillars (looper) have only two pairs of prolegs on VI and X abdominal segments — *Comostola læsaria* (Walker), *hyposdira successaria* (Walker), *Thalassodes quadraria* Guenee.

Geometroidea Forbes 1923 superfamily of Ditrysia (Lepidopetra: Oligoneoptera); slender built moths with relatively large wings and cryptic colouration; maxillary palp vestigial or absent; vein Cu_2 missing from both wings; tympanal organs in abdomen — Drepanidæ, Geometridæ, Thyatiridæ, Uraniidæ, Axiidæ and Epiplemidæ.

Geophagous feeding on soil.

Geophilous living in or on soil.

Georyssidæ small family of small subaquatic beetle (Hydrophiloidea: Polyphaga); head deflexed, maxillary palpt short; antennæ 9-segmented, first three forming the club and tarsi 4-segmented instead of five. Found usually on sand or mud on banks of water bodies — *Georyssus* spp., et al.

Geotaxis tendency of crawl or move towards or into the earth (positive geotaxis) or away from the earth (negative geotaxis). Larvæ of many insect go upward (away from the earth) in search of food and come down, when full fed, to pupate on ground or in the soil.

Geotrupidæ family of dor beetles (Scarabæcoidea: Polyphaga); large convex insect of coprophilous habit; antennæ 11-segmented, last three forming a lamellateclub; scutellum distinct; hind tibia with two terminal spurs, elytra cover entire abdomen (300 species) — *Geotrupes* spp., *Lethrus* spp. (apterous) et al.

Gephyroneura Enderlien 1936 suborder of Diptera in an old system of classification, comprising of four family - Sciadocertidæ, Empididæ (Empidæ), Lonchopteridæ (Musidoridæ) and Dolichopodidæ (Dolichopids).

Germ band area of thickened cells on ventral side of biastoderm that become embryo.

Germ cell reproductive cell that is destined to become ova or spermatozoa.

Germarium apex of ovariole, below terminal filament; consist of mass of cells from which are differentiated primordial germ cell and nutritive or nursecell, all at first linked by cell-bridge.

Germicide substance (toxin) capable of destroying germs in general and bacteria in particular.

Gerridæ (Leach 1815) family of pond-skaters (Geocorisæ; Heteroptera); elongate, slender water bugs; rostrum 4-segmented, small median opening of scent gland on metathorax, hemelytra, when present, homogenous; mid and hind legs elongated; undersurface of body covered with fine hydrofuge hairs that trap cushion or air under the insect, preventing it from sinking; glide with fast jerky movement on surface on ponds etc., feed on any living or dead insect that falls into water (200 species) — *Gerris lacustris* Linnæus, *G. paludum* Fabricius, *G. najas* de Geer, *G. thoracicus* Schumacher, et a1.

Ghost larva phantom gnal: aquatic larva of *Chaoborua* spp. (Culicidæ: Nematocera); completely transparent and apneustic; possesses a pair of pigmented airsacs in thorax and second pair in VII abdominal segment, these help in floatation; feed often on mosquito larvæ.

Giant much above the normal size of its kind.

Giant chromosome exceptionally large (upto 0.2 mm in length), many threaded chromosome (s) found in salivary glands of *Drosophila* species (Drosophilidea: Schizophora) and some other dipteran flies.

Giant coccid ground -pearl: *Margarodes* spp.(Margarodidæ: coccoidea); covered with pearl-like waxy scales; female capable of undergoing prolonged quiescence (encystment): collected and strung into necklaces in South Africa and Bahamas and worn as ornament.

Giant lacewing *Osmylus fulvicephalus* (Osmylidæ: Planipennia) one of the largest Europtean neuropteran, slender dark gray or brown with wing-spread of about 50 mm, found near streams, Larvæ live among wet moss feeding on other insects.

Giant orange-tip large plerid butterflies, *Hebomoia* spp. (Papilionoidea: Ditrysia); *H, vossi* of Indonesia has bright yellow wing with orange tip and wing expanse of 110 to 120 mm; *H. glaucippe* white with orange-pink wing tips, bordered with black is slightly bigger in size, measuring 120 mm across the wings.

Giant skipper *Rhopalocampta benjamini* (Megathyminæ: Hesperiidæ); greenish-black wings with especially strengthened veins in males;

wings expanse 40 mm or more. Found all over India and South-east Asia.

Gill(s) respiratory organ(s) in water-breathing animals — fishes, aquatic insect etc.

Gill(s) Blood gills present in larvæ of some dipteran and trichopteran insect, tubular or digitiform (sometimes eversible) and contain blood; help in absorption of water and inorganic ions rather than respiration.

Gill(s), Spiracular extensions of spiracles or surrounding cuticle; forming hollow processes; found in pupæ of some beetle and dipteran flies; adapted for both aquatic and aerial respiration.

Gill(s), Tracheal organs of respiration, good of gaseous exchange take place through these; present in majority of aquatic larvæ; usually borne on abdomen, filiform or lamellate and well supplied with tracheæ and tracheoles.

Gizzard proventriculus or second stomach: part of alimentary canal, immediately behind the crop, where sifting of food takes place before digestion and absorption in the mid-gut; poorly developed in sucking insect, but in mandibulate insect it is usually enlarged and horny with powerful circular muscles and radial teeth for crushing hard food particles.

Glabrous smooth; hairless surface.

Glaphyrinæ subfamily of Scarabæidæ (Scarabæoidea: Polyphaga) often considered as independent family; characterised by very hairy body, antennæ 10-segmented and antennal club globrous; metepimera large.

Glossa (pl. glossæ) pair of median lingular lobes of labium each provided with muscle arising in prementum.

Glossata insect order suggested by Fabricius (1775) for Lepidoptera Linnæus 1758; no more usage.

Glosselytrodea order of fossil insect that appeared during Permian period (Palæozoic era) and disappeared during jurassic period (Mesozoic era).

Glossinidæ Malloch 1824 family of viviparous biting flies (Muscoidea; Schizophora), arista plumose with feathered hairs, proboscis needle-like, prosternum membranous; differs from houseflies in being slightly bigger in size and brownish in colour. Found in Central Africa, well known carrier of pathogens of certain virulent diseases of man and domesticated animals — tsetse flies (*Glossina* spp): *Glossina palpalis* (Robineau-Desvoidy) transmits *Trypanosoma gambiense* Dutton, that cause sleeping sickness in man; also transmits *T. vivax* causing surra disease in cattle and horses; *G. morsitans*

Westwood transmits *Trypanosoma* species causing sleeping sickness and nagana disease of cattle and horses.

Glow-Worm fire-fly, *Lampyris noctiluca* (Linnæus) (Lampyridæ: Polyphaga) that emits shining green light, at night, from ventral surface of its abdomen.

Glyphodate $C_3H_8NO_5P$; N-(phosphonomethyl) glycine; an off-white powder, broad spectrum non-selective, post-emergenc herbicide, highly effective on several grasses and broad-leaf woods.

Gnat common name for various dipteran flies (Nematocera) having biting mouthparts; males are often seen in dancing swarms — fungus gnat (Mycetophilidæ: Mycetophiloidea), phantom gnat (Chaoborinæ: Cullicidæ), Winter gnat (Trichoceridæ: Tripuloidea).

Gnat, Eye a tiny, active dipteran fly, *Hippelates pusio* Loew (Cholorpidæ: Acalyptraræ); an irritating or tormenting pest of the domestic animals (cattle etc.) and man, specially in Southeren USA; also transmits pink-eye disease of humans.

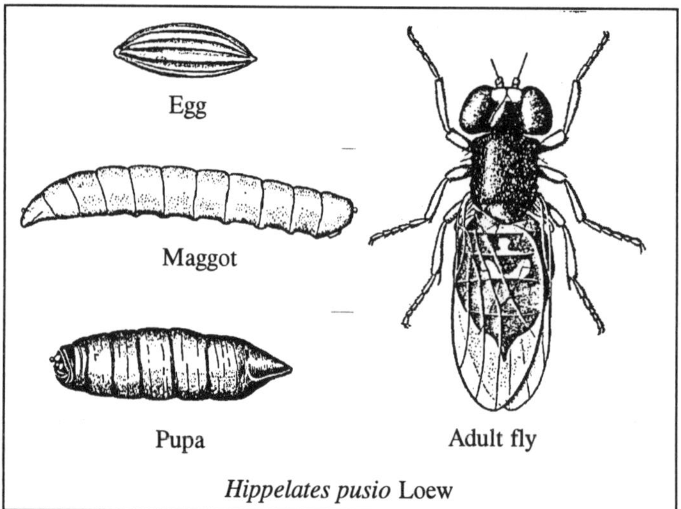

Hippelates pusio Loew

Gnathal segment somites of insect embryo, appendages of which are transformed into organs of feeding — mandibles and maxillæ.

Gnathocephalon part of insect head formed by gnathal segments, bearing mandibles, maxillæ and labium.

Gold-fringed borer *Chilo (Chilotræe) auricilia* (Dudgeon) (Crambidæ: Lepidoptera) often confused with *C. supressalis* Walker; medium-sized moths, light brown in colour, forewings brownish having five metallic spots with golden scales lying in a regular curve, hind-wings brownish- white, wing-expanse 18 to 24 mm (♂) and 22 to 28(♀). Caterpillar having five longitudinal violet stripes, the middle one being comparatively faint. Pest of paddy (*Oryza sativa* Linnæus) sorghums, sugarcane (*Scacharum* spp., Poaceæ) etc.

Gold-fringed moth *Hypsopygia costalis* (Fabricius) (Pyralididæ: Lepidoptera); in colour, forewings brownish having five metallic spots with golden scales lying in a regular curve, hind-wing brownish-white; wing; frequent hey stacks on which the caterpillar feeds; common in Europe and North America. Also called clover hayworm.

Gold-tail moth *Euprotis similis* (Lymantriidæ: Lepiopetera); a tussok moth having white wings and body with a bright yellow (golden) tip on the abdomen. Caterpillar red, black and white covered with irritant hairs; feeds on leave of hawthorn (*Crateæ*gus *oxyacantha* Linnæus) and pupates in silken cocoons.

Golden apple beetle *Chrysolina (Chrysomela* Auct.*) fastuosa* (Chrysomelidæ: Polyphaga); a small leaf-eating beetle about 6 mm long, having golden green convex elytra with dark blue stripe along the inner edge of each elytron.

Golden oak-scale scale insect, *Asterolecanium variolosum* (Asterolecaniidæ: Cocooidea), greenish-brown, convex, almost globular scale, 1.5 to 2.0 mm long; serious pest of oak trees (*Quercus* spp.) all over the world. Successfully controlled in New Zealand forests by introducing parasite wasp *Habrolepis dalmani* from North America.

Golden swift moth *Heplialus hecata* (Linnæus) (Hepialidæ: Monotrysia), golden-brown moth having short antennæ and no proboscis; male having three oblique white lines on forewings; common in Europe attacking roots of ferns.

Golden-eye alternate common name for green lacewing *Chrysopa oculata* Say (Chrysopidæ: Neuroptera); a North American species, having small slender pale green body, less than 25 mm long with long antennæ, prominent reddish-golden eye and large finely veined green wings, carried roof-wise over the body, when at rest. Larvæ feed on the injurious insects particularly aphids.

Gonad gamete-producing organ- ovary or testis or embryonic rudiment of either.

Gonapophysis (pl. **gonapophyses)** long, slender genital process of coxopodite, arising from masal angle; the two gonapophyses in female genital chamber are associated in forming shaft of ovipositor whereas in males these embrace the penis. Present only in some thysanuran species.

Goniodæ small family of Ischnocera (Mallophaga: Psocopteroidea) that includes bird-lice, about 3 mm long, having broad head with a horn-like projection on either side - *Goniodes pavonis* (Linnæus) and *G. parviceps* Piaget are peacock-lice whereas *G. dissilis* Nitzsch and *G. gigas* Taschenberg are parasitic on domestic fowls and *Lipeurus* Geoffroy is turkeylouse.

Goniodidæ Mjoberg 1910 family of bird-lice (Mallophaga) in an old classification system, now placed under family Philopteridæ (Ischnocera: Mallophaga).

Gonopodium (Gonopod) claspers: paired copulation appendages on IX abdominal segment of male insect; used for grasping the female while mating.

Gonopore external opening of genital duct.

Gonosomite genital segment, specifically IX abdominal segment in male insects.

Gonostyli styli of IX abdominal segment, when present, generally modified to form clasping organs — harpes or harpagones.

Gorytidæ a small family of Sphecoidea (Aculeata: Aprocrita); large norrow-waisted solitary wasps having black coloured body with narrow yellow stripes; second cubital cell not triangular; anterolateral angle of prodoeum rounded; abdomen sessile. Feeds on froghopper larvæ, draggling the larveæ from their protective cuckoo-split and stocking the same in its nest — *Gorytes mystaceus* et al.

Grade evaluation of insect based on numerical or visual rating.

Gram pod borer *Helicoverpa armigera* (Noctuidæ: Lepidoptera), medium-sized (wing spam 32 to 42 mm), stout light brown moth, forewings olive-green to pale brown in colour with a dark brown circular spot in the centre, hindwings smoky-white with a broad blackish outer border. Major pest of pulses, peas (*Lathyrus* spp., Papilionaceae) and tomato (*Lycopersicon* spp., Solanaceæ), specially gram (*Cicer arietinum* Linnæus).

Grannom green-tailed caddisfly, *Brachycentrus subnubilis* (Sericostomatidæ: Limnephiloidea), of Europe having smoky gray wings. Female is often seen with bright green cluster of eggs beneath its abdomen.

Granule coarse particle of inert material impregnated or mixed with a toxin or pesticide.

Graphosomatidæ old and obsolete term for the heteropteran family Podopidæ (Scutelleroidea: Geocorisa) that include terrestrial turtle bugs.

Grasserie a polyhedrosis disease of silkworm characterised by spotty yellowing of the skin and internal liquefaction.

Grasshopper straight-winged, hemimetabolous, orthopteran insect, head bent at right angle to the body; mouthparts biting and chewing thickened in form horny tegmina which cover the membranous hind-wings; some species are wingless, Stricking feature in their power of stridulation, cocophony they makes by rubbing hind femur against the tegmen or rubbing the two tegmina together.

Grasshopper, Great green *Tettigonia viridissima* (Tettigoniidæ: Tettigonioidea) of Eurasia and North Africa. Found in thick vegetation, active hunting other insect; the loud and almost continuous penetrating noise of the males is very common specially during summer night.

Grasshopper, Long horned grasshoppers (Testtigonidæ: Ensifera) also called katydids have antennæ as long as or longer than their body — *Neoconocephatus palutris*.

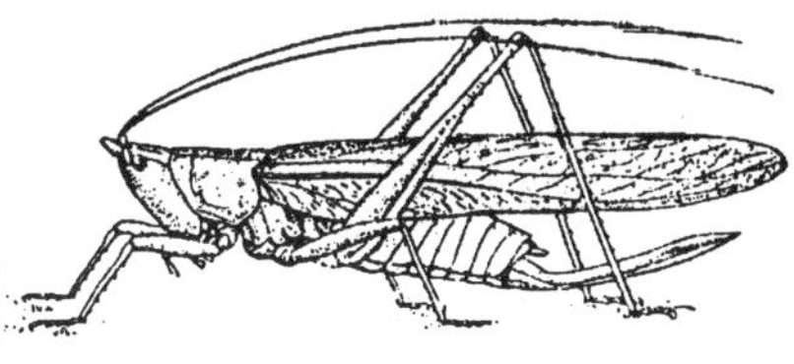

Neoconocephalus palutris

Grasshopper, Meadow small European species *Chorthippus parallelus,* (Acrididæ: Acridoidea); brownish-purple in colour with a whitish stripe down the back; head the elytra green; elytra short in male and vestigial in female and wings in both sexes reduced to more stumps; being wingless, these are unable to fly and move by hopping.

Grasshopper, Short-horned grasshopper (Acrididæ: Cælifera) having very short antennæ. Locusts are also short-horned grasshopper, but all short-horned grasshopper are not locusts - *Atractomorpha crenulata* Fabricius.

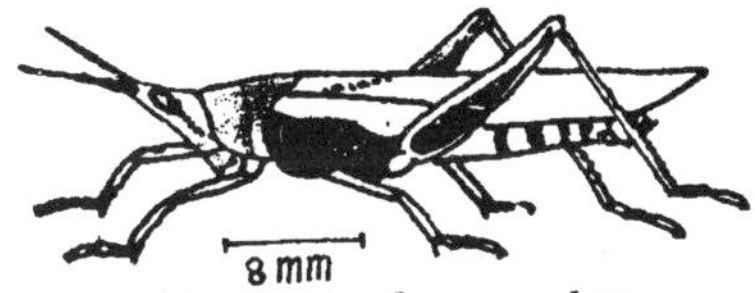

Atractomorpha crenulata Fabricius

Gravid female insect carrying fertilised eggs.

Grease-band adhesive material applied as band round tree trunk to trap ascending wingless insect — used on mango trees (*Mangifera indica* Linnæus) against mango mealybugs, *Drosicha mangiferæ* (Green).

Gregarious parasite parasitoid whose several (many) individual develop simultaneously on or within the body of single host species.

Gregarious parasitism development of several individual of a parasite species in or an single host — *Apanteles flavipes* (Cameron) is a gregarious parasite of sorghum stem borer *Chilo partellus* (Swinhoe) (Crambidæ: Pyralidoidea).

Gripopterygidæ fillipalpian family of stoneflies (Plecoptera: Orthopteroidea) related of Austroperlidæ (Leptoperloidea) from which it differs only in arrangement of anal veins; found only in Australasia and South America (120 species) — species of *Dinotaperla, Leptoperla, Trinotaperla* et al.

Groundnut earwig *Euborellia annulipes* (Locas) (Labidridæ: Demaptera); medium sized, slender, black shining insect, without elytra and no ovipositor in females; body length 10 to 14 mm, forceps 2 to 4 mm. Borers into tender pods of groundnut (*Arachis hypogæa* Linnæus) in the soil and feed on kernels; also attacks cabbage (*Brassica oleracea capitata* Linnæus) and onion (*Allium cepa* Linnæus).

Grub larva of celopteran insect having fleshy, wrinkled, curved body well developed head, apodous or having three pairs of thoracic legs but no abdominal legs and mouthparts similar to adult (biting and chewing type).

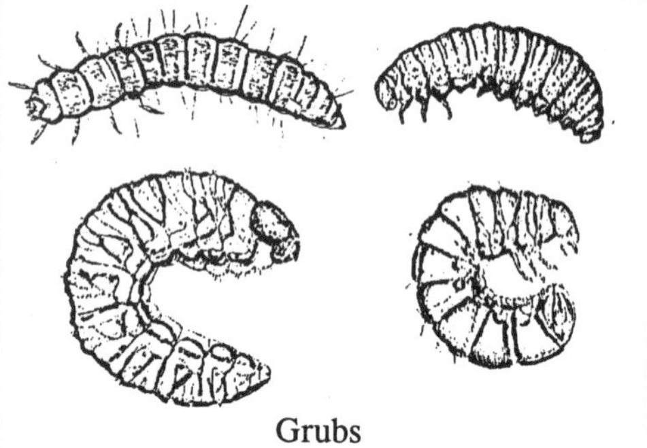

Grubs

Gryllidæ Saussure 1894 a family of crickets (Grylloidea: Ensifera); body depressed tarsi 3-segmented; fore tibiæ normal usually with an auditory organs on each side; ovipositor well developed, spear-shaped or awl-shaped-ant-crickets, desert crickets, house crickets, tree-crickets etc.

Grylloblattodea Brues and Melander 1932 (Notoptera Crampton 1915) monofamilial primitve order of medium-sized, light-coloured, secondary apterous insects that show orthopteran affinities (formerly a family under Orthoptera); head flattened and prognathous, antennæ long, filiform and multi-segmented; compound eyes vestigial or wanting; ocelli absent; mouthparts ectoganthous biting and chewing type; hypopharynx well developed; thorax weakly developed; coxæ large, tarsi 5-segmented; ovipositor long and exserted; male genitalia asymmetrical; cerci long and 8-segmented. Metomorphosis incomplete-egg, nymphal and preoviposition period last for one, five and one year respectively.

Gryllobiattide Walker 1914 only family of Grylloblattodea; found at altitudes of 600 to 2200 metres or in cooler regions (around 1 °C); not found in the Indian subcontinent. About 16 species placed under three genera — *Grilloblatta* (9 in Western North America), *Galloisiana* (6 from Japan) and *Grlloblattina* (one - *G. djakonovi* occurs in Siberia). *Grylloblatt a campodeiformis* Walker was the first species discovered in 1914 in Canadian mountains.

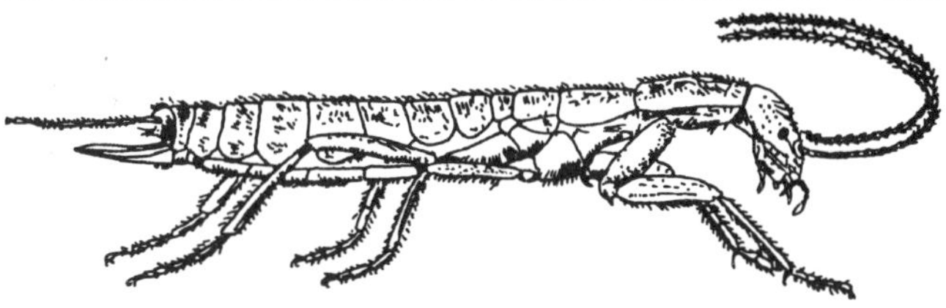

Grylloblatta campodeiformis Walker

Grylloidea recently added superfamily of Ensifera (Orthoptera: Orthopteroidea); comprising of only two families — Gryllidæ and Gryllotalpidæ — these were earlier placed under superfamily Tettigonioidea (Ensifera: Orthoptera).

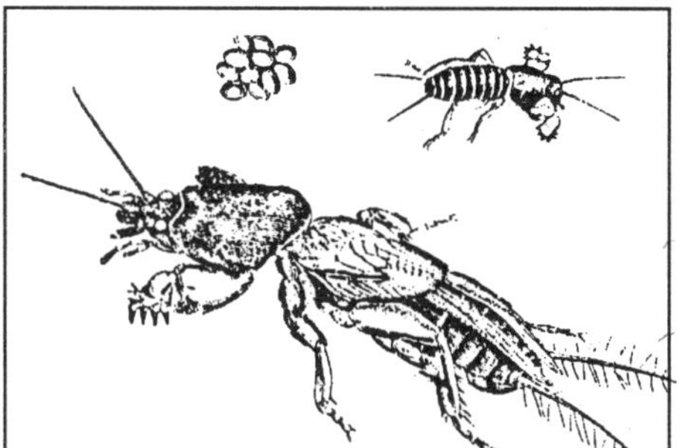

Gryllotalpa gryllotapa Linnæus

Gryllotalpidæ Brunner 1882 a family of Grylloidea (Ensifera: Othoptera); large insect, 18 to 20 mm long, densely clothed with shorts hairs; antennæ filiform and many segmented; two large ocelli present; fore tibiæ greatly enlarged or fossorial and hind greatly enlarged or fossorial and hind femora nor enlarged for jumping; ovipositor not exposed. Burrow in the soil — mole cricket *Gryllotalpha gryllotalpa* (Linnæus), *G. africana* Beauvios, *G. coarctata* Walker, *G. siamensis* Giebel, *G. ornata* Walker, *G hexadactyla* Pertty, *G unispina* Saussure, et al.

Gula median ventral plate or head in some prognathous insects (particularly coleopterous).

Gundhi bug *Polychrysa (Leptocorisa) acuta* (Thunberg) (Coeridæ: Heteroptera); active. yellowish-green, elongated bug with long legs; a major pest of paddy; appears at grains (milky) stage, sucks juice from soft grains, the affected grains shrivel turn yellow and become chaffy; the bugs also emit a peculiar pungent smell when distributed — hence the name. Also known as paddy or rice bug.

Gurdaspur borer *Bissetia steniellus* Hampson (Crambidæ: Lepidoptera); medium-sized dusky-brown moth; male having serrate and ciliated antennæ whereas female has simple, Caterpillar creamy-white with four narrow pinkish longitudinal stripes; attack grown up crop of sugarcane (*Saccharum officinarum*), causing drying up entire whorl of top leaves instead of deadheart.

Gymnocerata equivalent name for Hydrocorisæ, a subdivision of Heteroptera comprising of water-bugs.

Gyne(s) queen(s) in ant-colonies — large individuals with well developed reproductive organs; antennæ and legs relatively shorts; winged, dialated or apterous. The larger species are known as **macrogynes** and dwarfed ones as **microgynes**; in certain dimorphic forms, queen that are apterous and resemble worker are called **ergatogynes**.

Gynecoid(s) individual members of ant colony which if sufficiently fed and cared for, are capable of lying fertile eggs and replacing queens.

Gypsy moth *Lymantria dispar* Linnæus (Lymnatridæ: Noctuoidea); a tussock moth introduced accidentally from Europe to North America where it is now a serious pest of forestry. Minor pest of apple trees in India.

Gypsy moth, Indian *Lymentria abfuscata* Walker (Lymantriidæ: Noctuoidea); a tussock moth; adult have bulky body and feebly developed (non-functional) wings; males winged. Caterpillar gregarious and nocturnal in habit, defoliate the fruit completely, resulting failure of fruit formation.

Gyrindæ Leach 1815, 1817 a family of Gyrinoides (Adephaga: Coleoptera); comprising of whirligigs; antennæ shorts, thick and 11-segmented; eyes divided into dorsal portion adapted for aerial vision and ventral portion to see under the surface of water; forelegs long, prehensile; mid and hind legs short and paddle-like. Gergarious beetles that swim swiftly round the round on the surface of quite water. Grubs aquatic, having elongate paired segmental appendage on first seven or more abdominal segment, functioning as tracheal gills. Both grubs and adults are predaceous — *Enhydrus* spp., *Cyrinus* spp., *Porrhorrhynchus* spp., *Orectochilus* spp. et al.

Gyropidæ Kellogg 1908 a small Neotropical family of Amblycera (Mallophaga: Psocopteroida); that include lice infesting mammals specially rodents; beside having concealed 4-segmented capitate antennæ and horizontal mandibles, these insects have tarsi with only one claw — *Gliricola procellis* (Linnæus) and *Gyropus ovalis* Nitzsch are parasitic in guinea-pigs (Rodentia: Eutheria).

H

Habitable fit to live in can be inhabited.

Habitant an inhabitant.

Habital place or environment in which an animal lives : normal abode or locality of animal or plant species.

Hadda betle(s) *Epilachna dodecastigma* Mulsant and *Henosepilachna vignitiotopunctata* (Fabricius) (Coccinellideæ: Coleoptera); adults spherical in shape, 6 to 8 mm log radish to brownish-yellow in colour; *E.dodecastigma* having 6 black spots and *H. vigniticotopunctala* 14 on each elytron. Polyphagous pest specially of vegetable crop; both grubs and adults feed voraciously by scrapping the chlorophyll and skeletonise the leaves which percent lace-like appearance and later die away.

Epilachna dodecastigma Mulsant

Henosepilachna vignitioctopunctata (Fabricius)

Hæm(o)- denoting blood.

Hæmatomyzidæ an aberrant family of Rhychophtherina (Mallophaga: Psocopteroidea); minute, flattened, apterous lice having head prolonged into rostrum with mandibles at tis apex; pro-, meso- and metathorax fused together. Only two species, both blood-sucking ectroparasites on mammals-elephant-louse, *Hæmatomyus elephantis* Piaget on elephant, both Indian (*Elephas* spp.) and African (*Loxodonta* spp.) and hog-louse, *H. hopkins* Clay on wart-hogs.

Hæmatopinidæ Enderlein 1904 family of blood-sucking lice (Siphunculata); body flattened dorsoventrally, head retracted into thorax; eyes vestigial or absent; abdominal cuticle wrinkled; infest domestic animal (cattle, dogs, horses, pigs, rodent etc.) – dog louse *Linognathus setosus* (Olfers), hours/zebra louse *Hæmatopinus asini* (Linæus), pig louse *H. suis* (Linæus) et al.

Hæmocoel portion of body cavity of insects filled with hæmolymph and functioning as circulatory system: blood spaces of embryo.

Hæmocoelus viviparity phenomenon that occur in certain insects having no oviducts: eggs escape into the hemocoel, hatch there an freshly formed larvæ escape out through secondary openings in the body wall–most strepisteran and some cecidomyiids.

Hæmocyanin bluish respiratory pigment found in insect's hæmolymph; serves to transport oxygen to body tissue.

Hæmocyte(s) blood cells specially those found in invertebrates; their main function is plagocytosis; beside, in insects, these help in defence reaction by devouring foreign particles and also by taking fat and glycogen. There are various types of hæmocytes found in insects : –

— **Proleucocytes (prohæmocytes)** – small, rounded, basophil cells, produced by mitotoc division; generally found in all insects.

— **Phagocytes** (plasmatocytes) – large cells with hyaline cytoplasm; common in all insects.

— **Graanular leukocytes** – granular cytoplasm frequently functioning as phagocytes; present in all insects.

— **Openocytoides** – large cells with spherical or fund forms, having homogenous cytoplasm; occur in Homoptera, Heteroptera, Diptera, Lepidoptera and Coloepetra but not in Orthoptera and Phasmida.

— **Cystrocytes** – speciallised granular hæmocytes, found in most insects.

Adipohæmocytes – hæmocytes containing fat.

Hæmoglobin red pigment of vertebrate blood; found very rarely in insects–aquatic larvæ of chironomid–gnats (Chironomideæ: Culicoidea).

Hæmolymph blood plasma or liquid portion of blood : colourless blood of insects, filling the entire body cavity.

Hair filament or thread–like outgrowth from epidermis (skin) of animal or plant. Insect's hair also originate form the epidermis and cuticle and these may be:-

— **Macrotrichia** – setæ: single or branched hairs; fin or thickened, arise form a pit or socket and are longer than microtrichia.

— **Microtrichia** – minute, non–motile hairs formed form cuticle.

— **Spine(s)** – immovable, sharp–pointed, multicellular external processes of body wall.

— **Glandular hair** - setæ of certain caterpillars that secrete irritant fluid.

— **Scales** - flattened and coloured hair similar to setæ found in some mosquitoes and butterflies.

— **Sensary hair** - setæ with nerve cells leading form their bases, these act as auditory, tactile or even as chemical receptors.

Hairfly alternate name for the flies of family Bibionidæ (Nematocera; Diptera) including St. Mark's fly *Birbio marci* Linnæus, garden hairy fly *B. hortulans* Linnæus and fever fly *Dilophus ferbrilis.*

Hairstreaks(s) group of butterflies (Lycænidæ: Papillonoidea) distinguished by thin white streak on undersurface of wings and hind-wings with tail-like prolongation-*Callophyrs rubi, Thecla crysalus, T.halestis, T.quercus* et al.

Hairy dragonfly a medium-sized dragonfly, *Brachytron prachense* (Anisoptera : Odonata) having dark brown thorax marked with pale green and yellow and covered with find velvety haris. Male has blue eyes and bluish-black abdomen; female has brown eyes and yellowish-black abdomen.

Half-insect(s) obsolete term, earlier used for proturan insects, as these are most primitive, minute, colourless and without antennæ and eyes.

Haliplidæ Kirby 1837 small family of crawling aquatic water beetles (Caraboidea : Adephaga); small brown or yellow beetles that crawl about on submerged vegetation in shallow waters; legs not adapted for swimmings. Grubs feed on algæ in running and standing water (200 species) -*Haliplus* spp., *Peltodytes* spp., et al.

Haltere(s) balancers: reduced (modified) hind-wings of Diptera; dumbel-shaped projections that vibrate and act as gyroscopic stabilisers.

Halticinæ (Alticinæ) Tropical subfamily of Chrysomelidæ (Chrysomeloidea: Heteromera) closely allied to Galerucinæ (both now regarded as independent families); body covered with long stout upright spines. Larvæ leaf-miners-flea beetles, *Phyllotreta chotanica* Duvivier and allied species are pests of Cruciferous crop *Atica cyanea* (Weber) *Padagria bowringi* Baly, *Chætocnema basalis* Baly are also pests of vegetable crops.

Hamula retinaculum : hook-like structure found in collembolan insects (springtails); along with fercula, it enables the insect to leap forwards into the air.

Hamulus (pl. **hamuli**) frenal hooks : series of minute hooks present on anterior margin of hind–wing that catch in a fold on hind margin of forewing; found hymenopteran insects specially the bees.

Hand sprayer small pneumatic sprayer that can carry 0.5 to 1.5 litres of spraying material; a few strokes of pump–piston develops pressure in the small tank and by pressing the lever, nozzle gives a continuous find mist spray; used for spraying potted plants and experimental plants in laboratory; against household pests as also small kitchen gardens; not practicable and uneconomical for field–work.

Hand-sprayer

Hand–picking method of pest control that involves collection of insects (eggs, larvæ/nymphs, pupæ and adults) from plants or soil by hand and destroying/killing the same.

Hanging fly common name for several species of scorpionfilies (Mecoptera) that hang from vegetation by their forelegs and use the other legs for catching and seizing smaller insects flying past.

Harpagone(s) (harpes) claspette: pair of movable periphallic genital lobes of some pterygote insects, when present, always pertain to annulus of IX abdominal segment and serve as copulatory claspers, as in mosquitoes (Culicidæ: Diptera) etc.

Hatch to bring out from the egg : emergence of larva or nymph from the egg.

Head uppermost or foremost part of an animal's body: head of an insect is formed from six embryonic segments fused together and contained in a cuticular capsule, strengthened by endoskeleton. It bears a pair of antennæ, usually two compound eyes and often serval simple eyes (ocelli) also; the mouthparts comprise of labrum, mandibles, maxillæ and labium and vary greatly in structure and function.

Head capsule circular covering of head on an insect composed of number of plates or sclerites fused together. Its main regions are epicranium (dorsal plate), from (frontal plate), cylpeus to which labrum is attached, genæ (two lateral plates) and gula (median vertical plate).

Heart cardiac part of dorsal blood vessel divided into series of chamber by successive constrictions; it acts as pumping organ.

Heart chamber(s) distinct symmetrical segmented swellings of heart.

Heart and Club a noctuid moth, *Agrotis corticea* (Notuoidea: Ditrysla), akin to heart and dart moth, but having a dark club-shaped mark instead of the dart on its forewings.

Heart and Dart grayish-brown noctuid moth, *Argyria exclamationis* (Noctuoidea : Ditrysia); having two conspicuous black marking one on each forewing, one resembling heart and the other narrow steak or dart. Caterpillars feed on turnips (*Brassica rapa* Linnæus) and other vegetables and pupate underground.

Heleomyzidæ family of Anthomyzoidea (Schizophora: Diptera); medium-sized files, postvertical bristles convergent, vibrissæ present; costa with spines and ædeagus often elongate. Maggots scavengers in fungi, excrement, nest or birds and also in caves (300 species).

Helicopsychidæ family of Limpehiloidea (Integripalpa : Trichoptera), ocelli wanting, dorsal part of tentorium reduced, males having 5-segmented maxillary palpi.

Halictophagidæ Perkins 1905 a family of Strepsiptera (Oligoneuropeta: Pterygota); males having 4 or 7-segmented antennæ with 2 or 4 appendages; tarsi 3-segmented with no claws; females remain partly within the bodies of host insects and have body and head with apical tubercles.

Heliodinidæ Wocke 1877 family of sun-moths (Yponomeutoidea : Ditrysia); very small often beautifully coloured moths; forewing with open discal cell hind-wing narrow, venation much reduced; hind tibia heavily spined, tarsi with whorls of bristles at apices of segments, hind legs elevated when at rest. Caterpillars minute, either herbivorous or predaceous- *Eretmocera impactella* Walker, pest of amaranthus (*Amaranthus* spp. and *Oedematopoda* spp. prey upon coccids.

Eretmocera impactella Wallker

Heliothripinæ subfamily of Thripidæ (Terebrantia: Thysanoptera); includes thrips characterised by having antennæ 7-segmented with terminal segment long, thin and needle-like, dorsum of body deeply reticulate with polygonal areas-grapevine leaf thrips, *Rhipphorothrips cruentatus* Hood and *Retithrips syriacus* (Mayet), greenhouse thrips *Heliothrips heæmorrhoidalis* (Bouche) et al.

Heliotropism denotes attraction of insects to the Sun (misnomer).

Heliozelidæ family of Incurvarioidea (Monotrysia : Lepidoptera); distinguished by the scales of head depressed as in higher Tineoids; hind-wing without discal cell and vein Rs branching form M near apex of the wing. Caterpillars apodous; and with two ocelli on each side of head; leaf-miners (100 species).

Hellgrammite nickname of a large predatory aquatic larva of dobsonfly, *Carydalis* species (Corydalidæ: Megaloptera: Neuroptera) of North America; wing expanse 150 mm; also know as arnly or conniption bugs and used as bait for catching freshwater fish.

Helminthidæ old and obsolete term for coleopteran family Elmidæ (Dryopoidea; Polyphaga).

Helodidæ family of small terrestrial, thin, soft-bodied beetles (Eucinetoidea: Polyphaga); body covered with numerous find hairs; hind coxæ narrower than their mesosternum and the postgena not keeled. Found near water, where their grub live (380 species)- *Scirtes* spp., et al.

Hemelytron (hemi-elytron) (pl. **hemelytra)** forewing of heteropteran but that exhibit marked difference in the consistency of its texture; proximal portion is well sclerotised while small distal area remains membranous; the former comprises of two parts-clavus, the narrow area next to scutellum and corium, the remaining broad region. In certain bugs (Anthocoridæ and Dipsocoridæ), narrow strip of corium bordering on costa is demarcated as embolium while in Miridæ and Velocipedidæ a triangular apical portion of corium is differentiated as coneus.

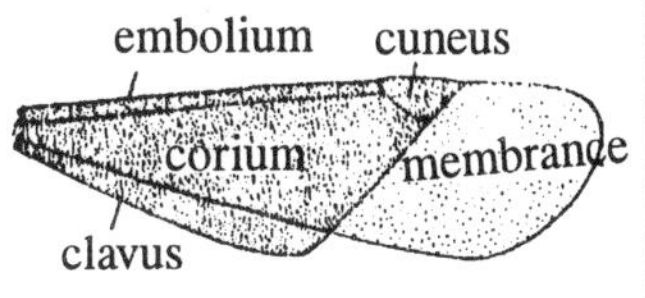

Hemelytron (forewing)

Hemepteriod (orders) an anatomical group of five exopterygote order, also known as Paraneoptera; these include, Psocopetra, Mallophaga, Siphunculata, (Anoplura), Hemiptera (now Homoptera and Heteroptera) and Thysanoptera.

Hemerobiidæ family of Hemerobioidea (Planipennia : Neuroptera); antennæ moniliform, trichosors present; wings oval or narrow. Larva with short hairs, empodium not trumpet-shaped. Adults predaceous on same small insects as the larvæ.

Hemerobioidea superfamily of Planipennia (Neuroptera: Oligoneoptera); common in Temperate regions; antennæ larger than forewing or else vein Sc and R connected by cross-vein. Larvaæ predatory on aphids-Hemerobiidæ, Chrysopidæ and Psychopside.

Hemi–half.

Hemicephalous dipteran having its head incomplete posteriorly and party embedded in protborax. Also called jaw–capsule. acephalous.

Hemileucinæ subfamily of Saturniidæ (Bombycoidea: Ditrysia) having antennæ bipectinate in both sexes and frenulum wanting–Bull's eye moth (*Automeris to* Fabricius) of North America is yellow in colour with two large black circles on hind–wings; *Aglia tau* is reddish-brown moth of Europe having a white nail on each forewing. Both have wing spread upto 80 mm.

Hemimeridæ Krauss 1900 small monogeneric family of Hemimerina (Dermaptera: Demapteroidea); apterous, viviparous earwigs, eyes absent, cerci long feebly sclerotised, unjointed hairy appendages. Then species of *Hemimerus,* all parasitic on pouched rat (*Cricetomys* spp.) of tropical Africa. *H.tapoides* Walker, *H.hanseni* Sharp, *H. deceptus* Rehn, *H. vatus* Deoras, et al.

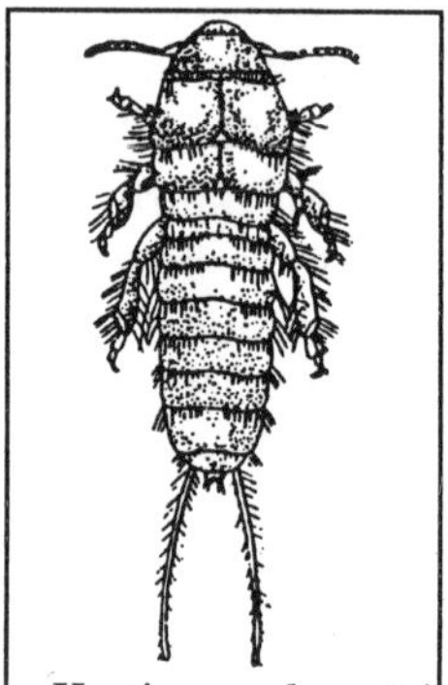

Hemimerus hanseni Sharp

Hemimerina (= Hemimeroidea Heymons 1915) suborder of Dermaptera (Polyneoptera; Pterygota); wingless earwigs; devoid of eyes, cerci long, straight and unsegmented. Ectoparasites of rats; confined to Tropical Africa–only one family (Hemimeridæ) and one genus (*Hemimerus*).

Hemimetabola (hemimetabolous) direct, simple or incomplete metamorphosis–egg, naiad and adult–as exhibited by some exopterygotes. The immature stage of insect (naiad) differs form the adult in general body form and mode of life, besides lacking wings and genitalia–Palæopteran orders (Ephemeroptera and Odonata).

Hemiparasite organism which derives part of its sustenance from other organism(s).

Hemipneustic insects having one or more pairs of spiracles, non-functional.

Hemiptera Linnaæus 1735 an insect order erected to include insects with wings but without wing covers–species of *Cimex, Notonecta, Gryllus, Lampyrus, Formica etc.,* now belonging to four different orders.

Hemiptera Linnaæus 1758 an insect order established by Linnæus in 1735 to include insects having wings but without wingcovers. In

1758, Linnæus restricted the order to only bugs and thrips. Later, Latreille (1810) grouped all the bugs into two suborders Homoptera and Heteroptera that were subsequently raised to order rank, while thrips were placed under the order Thysanoptera Haliday 1836.

Hemipteroidea Geoffroy 1763 superorder of Paraneoptera (Pterygota: Insceta equivalent to original order Hemiptera Linnæus 1758. Divided into two orders–Heteroptera and Homoptera.

Hepialidæ Stephens 1829 a family of Hepialoidea (Monotrysia : Lepidoptera); comprising of swift moths that are robust and hairy; head short and hairy; antennæ short, moniliform and often pectinate in males; ocelli rudimentary or absent, mouthparts vestigial; thorax well developed and hairy; wings long and narrow, hind pair much smaller, primitive, jugum slender, frenulum absent. About 300 species–*Sahyadrassus malabaricus* M., is pest of tea in Nilgirl hills. Australian species include *Charagia nobilis* Turner, C. *mirabilis* Rothschild and Leto *stacyi* (Scott) while *Hapialus humili* (Linnaæus) and *H.hecata* (Linnæus), the European species and *H.gracilis* Grote of North America attack root of ferns.

Heptachlor $C_{10}H_5Cl_7$; cyclodiene insecticide, that is nerve–poison (highly toxic); light–tan, soft waxy solid, insoluble in water, but dissolves in various solvents like acetone, benzene, ethyl–alcohol, kerosine, xylene etc. Acute oral and dermal LD_{50} value for rat being 40 and 200 to 250 mg/kg respectively; mainly used against household pests and as soil insecticide to control beetle-grubs, termites and other soil–borne insects–Drinox®, Velsicol® etc.

Heptageniidæ Bengtsson 1913 (Ecdyuridæ Klapalek 1909) a family of Hetagenioidea (Ephemeroptea); mayflies characterised by having wings with many veins and tow pairs of intercalaries between Cu_1 and Cu_2. Naiads have body and appendages depressed, head large and gills flat and leaf–like, found clinging to the undersurfaces of stones and gills flat and leaf–like, found clinging to the undersurface of stones in fast–flowing streams–*Epeorus assimilis* Eaton and species of *Atopopus, Ecdyurus, Heptagenia, Rhithrogena* etc.

Herbicide substance used for killing weeds or preventing their germination–acrolein (for submerged and floating vegetation), diuron (for selective control in field crops) etc.

Herbivorous feeding or living on grasses or herbage.

Hermaphrodite bisexual : having ability to function both as male as well as female; either simultaneously or in succession. Common in plants and some lower invertebrates (earthworms etc.) but rare in insects–fluted scale, *Icerya purchasi* Maskell (Margarodidæ: Coccoidea) are self–fertilising, males are rare and true females unknown, apparent females being functional hermaphrodite.

Herpetologist person having sound knowledge of herpetology.

Herpetology branch of zoology that deals with study of reptiles.

Hesperiidæ Leach 1815, 1819 large, cosmopolitan family Hesperioidea (Ditrysia: Lepidoptera); includes skippers, so called due to their erratic darting flight which is different from other butterflies; antennæ widely separated at their bases, apices prolonged beyond the club to form a small recurved point–Hesperiinæ (small skippers), Megathyminæ (giant skippers).

Hesperiinæ subfamily of Hesperiidæ (Hesperioidea : Ditrysia); includes small skippers; head large and attached to strongly constricted collar. Caterpillars stout, tapering towards both ends, live concealed in webs our galleries.

Hesperinidæ a dipteran family (Protophthalma: Nematocera); adult files often appear in great numbers, early in Spring–*Hesperinus* spp., et al.

Hesperioidea Wallengren 1853 superfamily of Dirtrysia (Lepidoptera: Mecopetroidea) having antennæ dilated apically to form a gradual club which often ends in a hook; labial palpi rough–haired, maxillary palpi lacking; forewing with Cu_2 absent and M_2 arising from below middle of transverse vein; hind–wing without frenulum and Cu_2-skippers; Australian regent skipper *Euschemom raffesia* MacLeay; dusky wing *Thorybes pylades* Scudder and gray skipper *Pyrgues communis* Grote (North America).

Heter(o)- different; other.

Heterocera one of the two suborders of Lepidoptera; in old classification, that included all moths. *Cf.* Rhopalocera (butterflies).

Heteroceridæ family of variegated mud–loving small beetles (Dryopoidea: Polyphaga); antennæ short and serrated, adapted for digging; labrum and mandibles conspicuous and projecting; tarsi 4–jointed; live in galleries that they excavate in mud near ponds and streams (150 species).

Heterochromous having different or varying colours.

Heterochrony divergence form normal time sequence in development.

Heterodactyla Brauer 1882 one of the two series of suborder Brachycera (Diptera : Oligoneoptera) in the old classification; differs from Homoedactyla (Brachycera: Diptera) in having feet, each with two lobes and a bristle or hair in between and not an empodium.

Heteroecism parasitism upon different hosts at different stages of life-cycle.

Heterogeneidæ Meyrick 1885 (= Cochlidiidæ Hubner 1822) old and equivalent name for family Eucleidæ Comstock 1895 (= Limacodidæ Walker 1855) (Zygænoidea: Ditrysia).

Heterogony allometric growth; disproportioned growth of different part of organs of body as compared to the organism (insect as a whole)-male stage beetles (Lucanidæ: Scarabæoidea) grow comparatively large mandibles as in *Lucanus* spp.

Heterogynidæ extremely small family of moths (Zygænoidea: Ditrysia); confined to South Europe; females apterous, vermiform and degenerate like those of Psychidæ, but remain in cocoons and lay their eggs there. Caterpillars without any case-*Heterogysin* spp.

Heteromera one of the two sections of Cucujoidea (Polyphaga: Coleoptera) equivalent to superfamily Tenebrionoidea; nocturnal beetles having vestigial or no wings and characterised by first and middle tarsi 5-segmented and hind-tarsi 4-segmented; comprising of 24 families, the most important ones being Tenebrionidæ and Meloidæ.

Heterometabola (heterometabolous) incomplete or direct metamorphosis; as exhibited by exopterygotes; young ones (naiads and nymphs) differ from the adults in absence of completely developed wings and genitalia; there is no pupal stage as in holometabola. This is further divided in hemimetabola (naiads) and paurometabola (nymphs)-all palæopteran, orthopteroid and hemipteroid orders.

Heteromophoses type of metamorphic phenomenon found in meloid beetles and Strepsiptera etc.-the egg hatches as an active triungulin larva, lies among the flower lobes and clings to the visiting insect; it becomes endoparasitic, loosing its own limbs and becoming saccular. Later, the larva differentiates an anterior cephalothoracic region.

Heteromyzidæ a very small family to Tanypezoidea (Schizophora: Diptera) earlier placed under family Heleomyzidæ (Anthomyzoidea: Schizophora): small two-winged files; maggots saprophage-*Tephrochlamyks* spp..

Heteroneura Tillyard 1918 (= Fernatæ Comstock, 1892) one of the two suborders of Lepidoptera in old classification insects having well developed sucking mouthparts and dissimilar ventation in fore-and hind-wings. It included all butterflies and majority of moths except Micropterygidæ (Zeguloptera) and Hepialidæ (Monostrysia); no more in usage.

Heteroptera Latreille 1810 order of Hemipteroidea (Hemiptera of Linnæus, 1758); comprises of all true bugs; variable size; head porrect, gular region sclerotised; mouthparts modified to form a beak or rostrum; forewing sclerotised basally and membranous apically (hemelytra); pronotum large. Metamosphosis gradual or incomplete (paurometabolous). Over 48,000 species subdivided into three series-Geocorsiæ (Cimicomorpha, Pentatomorpha), Amphibicorisæ and Hydrocorisæ.

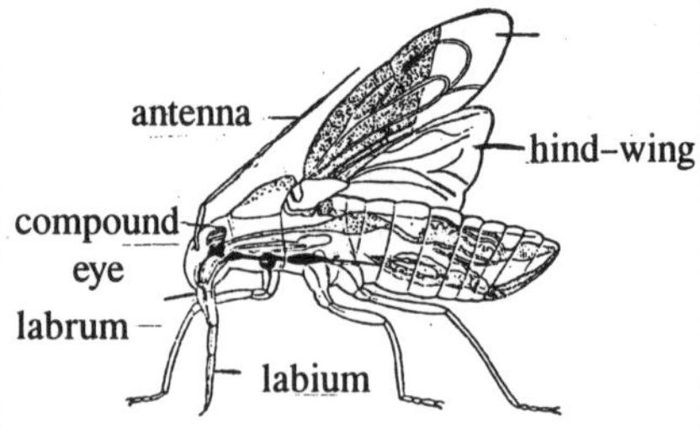

Typical bug

Heterothripidæ Bagnall 1912 subtropical family of Thripoidea (Terebrantia : Thysanoptera); distinguished by antennæ 9-or 10-segmented with short triangular sense-comes-*Heterothrips azaleæ* transmits spores of fungus causing azalea flower spot.

Hetrogenidæ old and equivalent name for family Eucleidæ (=Limacodidæ, Cochlidiidæ) (Zygænoidea: Ditrysia).

Hexapoda Latreille 1825 (Insecta Linnæus, 1785) biggest class of phylum Arthropoda (now placed under phylum Uniramia).; segmented body comprising of head, thorax and abdomen. Head bears a pair of antennæ (except proturans); thorax carries three pairs of legs (without exception) and usually two pairs of wings except apterogotes (wingless) and dipterans (only one pair). Breath by air-tubes (spiracles) and reproductive organs open near hind end of abdomen (937 000 species) classified into two suborders-Apterygota (ametabolous) and Pterygota (metabolour); later are further grouped into Exopterygota (heterometabolous) and Endopterygota holometabolous).

Hibernaculum Winter-shelter for dormant animals (and plants).

Hibernate to pass Winter in resting stage.

Hibernation Winter-sleep: to pass Winter in resting stage.

Hibernation, Deep profound decrease in metabolic rate physiological function during Winter.

Hierarchy system of ranks in animal/plant classification indicating categorical levels of various taxa.

Himantopteridæ (Thymaridæ) a family of Zygænoidea (Diptera: Lepidoptera): including small bright coloured moths with narrow hind-wings, ribbon-like tails and long hair covering the body and wings. Earlier teated as a subfamily Himantopterinæ under family Zygænidæ-*Himantopterus* species, et al.

Hippoboscidæ Leach 1817 family of Muscoidea (Schizophora : Diptera); louse files having head sunk into an emargination of thorax; antennaæ one segmented with or without terminal bristle or long hairs inserted into a depression; eyes round or oval, ocelli present or lacking; legs short and stout, claws strong and often toothed-forest-fly (*Hippobosca equina* Linnæus) is ectoparasitic on camels, cattle, asses, horses and mules in the Old World and *Lipopterna cervi* (Linnæus) on deer in Europe.

Hispinæ (Leach 1815) Samouelle 1819 large family of Histeroidea (Polyphaga; Coleoptera); hard compact shining beetles having geniculate and strongly clubbed antennæ; elytra truncated behind leaving the last two abdominal segments exposed (2500 species)-*Hister unicolor* and allied species frequent carrion and dung; *Hololepta* spp. and *Platysoma* spp. live beneath the bark.

Histeroidea superfamily of Polyphaga (Coleoptera: Oligoneopetra), head without Y-shaped impressed line, exoskeleton hard, black and shining; antennæ geniculate, last three segments, forming a compact club. Comprises of two small and one big family-Sohærittidæ, Syntellidæ and Histeridæ.

Histo tissue.

Histoblast tissue wall.

Histochemistry science of distribution and activities of chemical components in tissues.

Histogenesis (histogeny) origin, development and differentiation of tissues of an organism.

Histologist person having especialised in histology.

Histology science concerning chemical composition, structure and form of organic tissues of animal or plant, as related to their function.

Histolysis breakdown or disintegration of organic tissues.

Histophysiology technique for talking X-ray photograph of cells, tissues etc.

Histotome microtome used for cutting ultra-fine tissues sections for microscopic studies.

Hodotermitidæ Sjostedt 1925 family of harvestor termites, mostly subterranean or wood-inhabiting; ocelli wanting, pronotum saddle-shaped but narrower than head; anterior wing scales short; found feeding on roots or grasses as also *Excalyptus* trees- *Anacanthotermes (Hodotermes) macrocephalus* (Desneux), *A. viarum* (Koenig), *Hodotermes koenigi* Holmgren.

Holdback to retard or prevent the normal development of the immature stage of the insects (or any other organism) by manipulating the temperature etc.

Hollow-cone spray jet with a core of air breaking to give drops in an annular pattern.

Holocephalous inspect larvæ with head fully developed. *Cf.* acephalous; hemicephalous.

Holometabola (holometabolous) complete or indirect metamorphosis; young ones (larvæ) differ markedly form their adults in form, shape and mode of living; between the larval and adult stage, a quiescent stage (pupa is always present-all oligoneopteran orders.

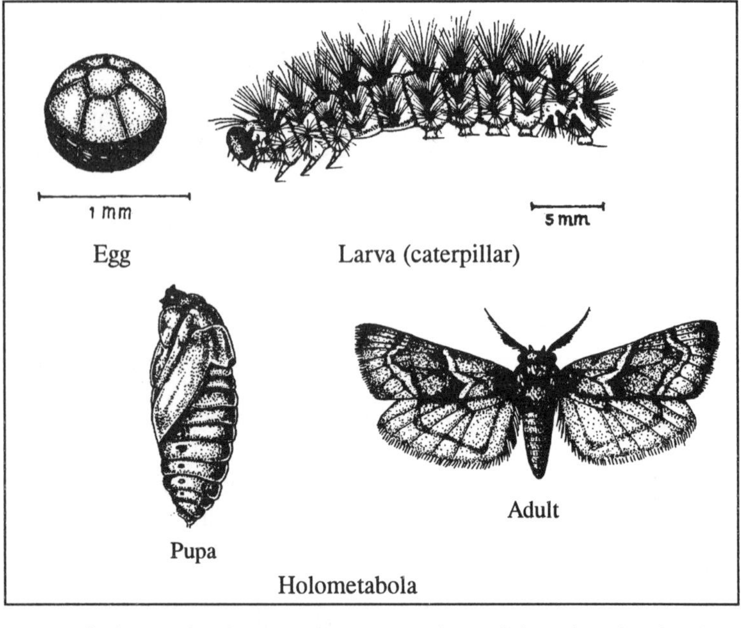

Holometabola

Holopneustic insect having maximum number of functional spiracles; a majority of adults have 10 pairs, two on thorax and 8 on abdomen; some primitive diplurans have more than 11 pair of spiracles, 4 thoracic and 7 abdominal.

Holoptic (files) dipteran files having both their compound eyes, of the two sides contiguous.

Homedactyla Brauer 1883 group (series) of short-horned files (Brachycera: Diptera); having feet with three lobes, central one empodium) being smaller than those on either side (pulvilli)- gallflies and horse-flies (Tabanidæ), snipeflies (Rhagionidæ), soldier-flies (Stratiomyidæ) etc. *Cf.* heterodactyla.

Homo same, like or similar. *Homo* is also generic name-human beings (*H.sapiens* Linnæus)

Homodynamic insects insects that are active all the year round, do not undergo any æstivation, diapause or hibernation etc.

Homoecious having one host for all stages of life-cycle.

Homoecious replacement of a damaged appendage by different appendage *Dixippus* species (Phasmida).

Homoithermic warm-blooded animals that maintain practically constant body temperature independent of environments-all mammals and birds (Aves).

Homoeura Tilyard 1918 (= Jugatæ Comstockj 1892) one of the two suborders of Lepidoptera in old classification, that included moths having mouthparts mandibutate or rudimentary and almost identical venation in both the pairs of wings-microlepidopteran biting moths (Micropterygidæ) and large primitive moths (Hepialidæ); the former now constitutes a separate suborder, Zeugloptera while the later has been placed under suborder Monotrysia.

Homonym one of two or more, identical but independently proposed names in binomial nomencalture for same or different taxa. Name identical in orthography with another but based on a different type- *Noctua variegata* is scientific name of a bird as well as an insect. Bread and bred, dear and deer, fair and fare, etc. are also homonyms.

Homonymy existance of two or more name based on different types- *Noctua variegata* (Noctuidæ : Lepidoptera) is an insect (moth) and *Noctua variegata* (Strigidæ: Striformes: Aves) is a bird.

Homoptera Latreille 1817 (Leach 1815) order comprising of diverse group of sucking insects (Paraneoptera: Pterygota); head deflexed, rostrum attached to back of underside of head; wings when present of uniform texture, forewing of harder consistency than hind-wing; apterous forms frequent. Metamorphosis incomplete except in male scale insects that undergo complete metamorphosis and whiteflies

(Aleyrodidæ) whose last nymphal instar (puparium) differs from other instars and undergoes considerable metamorphosis. Over 58000 species grouped into three subordes–Coleorrhyncha (not in India) Auchenorryncha and Stenorryhncha.

Homotype taxonomic type for specimen that has been compared with holotype by another person than the author of the species and determined to be conspecific.

Homotype protective device based on resemblance of shape to the background.

Honey sweet viscous fluid elaborated by honey bees from nectar of flowers and stored as food in beehive, to be used by the bees during Winter or in time of need. Term of enderment.

Honey, Virgin honey that flows of itself from honeycomb.

Honey Wild honey made by wild bee (*Apis dorsata* Fabricius and *A. florea* Fabricius).

Honey–bag (honey–sac) thin-walled enlargement of alimentary canal of worker honey–bees (crop in general); nectar gathered from flowers is held in this bag, them regurgitated into the cells of the comb (beehive); also called honey-stomach.

Honey–guide(s) dots or lines or petals of flower that guide the visiting insect(s) to the nectaries of those flowers.

Honey-moth *Achroia grisella* Fabricius (Galleridæ: Pyraloidea) is specific enemy of honeybees. Caterpillars feed on beeswax and dead bees specially in neglected beehives, often compelling the bees to abandon the hive.

Honey–stomach thin-walled crop in alimentary canal of Apoidea (Aculeata: Apocrita); dilation of oesophagus in the interior portion of abdomen and lined with a sclerotised membrane; herein the nectar collected from flowers is mixed with enzymes from the salivary glands and converted to honey before being disgorged.

Honey-stopper value at the posterior end of honey-stomach (bee's crop), speciallised for preventing the nectar from going into the stomach.

Honey–tube(s) misnomer for cornicles (siphunculi) of aphids, as these were earlier believed to excerte honeydew whereas actually these secrete wax, used by the aphids as protection against their enemies (predators).

Honeybee social bees, *Apis* spp. (Apidæ: Apocrita); live in organised communities; each colony (beehive) comprising of a few hundred drones (males) whose sole function is to fertilise the female and die soon-after, queen (fertilised female), only one, for egg-laying purpose and workers (sterile females) 20000, that construct the nest (hive), collect pollens, feed the queen and tend the larvæ in their hive. Most valuable insects, as besides providing honey and beeswax (products of commercial value) they also help in pollination-Indian bee (*Apis indica* Fabricius), European bee (*A. mellifera* Linnæus), little bee (*A. florea* Fabricius), rock bee (*A. dorsata* Fabricius) etc.

Honeycomb group of six-sided (hexagonal) wax-cells, constructed by honeybees, usually around the top or sides of their beehive; used for storage of honey and rearing their brood; these are capped with wax.

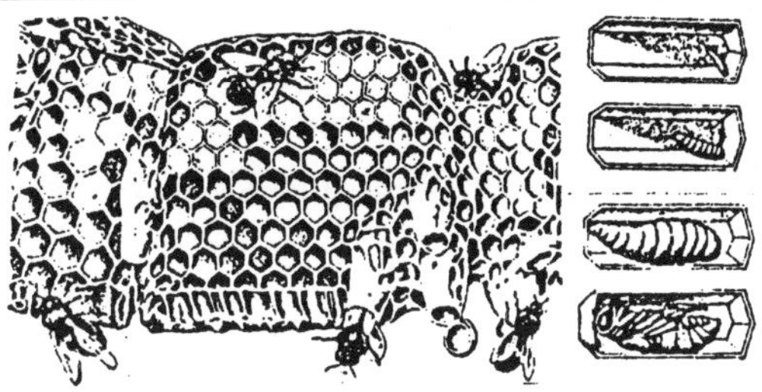

Honeycomb

Honeydew a sugary waste substance exuded by various homopteran insects-aphids, scales, whiteflies etc. These insects suck almost continuously the plant juices, talking in far more sugar and water than they need; the excess is passed out thought he anus as honeydew. It is much sought after by various species of ants (*Camponotus* spp., *Cremastogaster* spp., *Cataulacus* spp. etc.). Besides, it also accelerates the development of certain fungi (*Capnodium mangiferum* Coocke and Brown, *Meliola mangiferæ* Earle etc.).

Hoplopleuridæ largest family of Siphunculata (Paraneoptea: Pterygota), blood-sucking lice (ectoparasites) characterised by having tergal and sternal plates distinct and paratergal plates projecting apically-*Hæmodipsus ventricosus* (Deny), *H.setoni* Ewing and *H.lyriocephalus* (Burmeister) infests hares and rabbits, *Haplopleura* spp. on rodents and *Polyplax* spp. on shrews etc.

Hoppers large group of small sap-sucking bugs (Homoptera: Paraneoptera), closely related to aphids, very variable, but all have powerful hind-legs adapted for jumping-leafhoppers (Cicadellidæ), froghoppers (Membracidæ) cause considerable damage to cultivated crops. Nymphs of locusts (Acrididæ: Orthoptera) seen marching in big bands are also called hoppers.

Horizontal resistance general resistance expressed equally against all biotypes of pest species, having no gene–to gene relationship.

Hormone natural organic substance, secreted by certain glands; it stimulates various organs of the body or regulates and controls specific processes like growth, metabolism and other functions of the body.

Hornet large social wasp (Vespidæ: Vespoidea), most common species, *Vespa crabro* Linnæus grows upto 35 mm in length and nests in hollow trees; voracious predator and also sucks fruit juices White-faced hornet, *Vespula* (*Dolichovespula*) species found in Europe and North America make nest of chewed wood fibre in papery form; *Vespa ducalis* is Himalayan species about 40 mm long with wing span of 80 mm. All hornets are generally docile, having powerful sting, but less ready to use it.

Hornworm caterpillars with a dorsal spine or horn on last abdominal segment as in Sphingidæ (Ditrysia : Lepidoptera).

Horotelic organism that shows moderate rate of evolution.

Hose (pipe) a flexible pipe or tube for coveying liquid or gas : delivery pipe, often fitted with a lanse and nozzle at one end for spraying purpose.

Hoseman fireman who directs the stream of water.

Host organism (plant or animal) that harbours or entertains a pest or parasite. Dominant partner of a symbiotic or commensal pair.

Host, Primary the host in which parasitic insect remains for most of its adult life.

Host, Secondary the host in which parasitic insect remains for most of its a adult life.

Host evasion pseudoresistance where the cultivars escape insect injury by passing the susceptible stage quickly (or delaying), as in case of early–sown or early maturing varieties.

Host plant resistance method of pest control, using cultivars as are resistant or tolerant to target pests.

Host specificity degree of restriction of number of different plant or animal species that can serve as food source for herbivorous or carnivorous insects.

Hot spot site in the field or plantation where natural infestation by a particular species of insect is high. Site in a gene at which there is unusually high frequency of mutation.

Hot-bedbug *Xylocoris galactinus* (Anthocoridæ: Cimicoidea) a close relative of bedbug (Cimicidæ), this small, elongate-oval, flattened, predatory but inhabits manure heaps feeding on small arthropods; also known as piratebug.

HOWLETT, H.B. - appointed as first Pathological Entomologist in 1907 at IARI, Pusa (Bihar); the designation was later changed to Second Imperial Entomologist. Assisted H.M. Lefroy in writing the book, **Indian Insect Life** (1909). Left for England during the first World War, to serve as Entomologist to Hygiene Department of Royal Army Medical Corps (1916-17), returned to India by end of 1917 and rejoined as Entomologist. Died in 1920.

Humectant material which when added to spray solution, delays evaporation of water carrier.

Humeral plate anterior preaxillary sclerite of wing base supporting costal vein, greatly enlarged in Odonata.

Humeral suture (basal suture) line weakness along which shedding of wings takes place after swarming, present in almost all termites except *Mastotermes* spp. Also called fracture-line.

Humidity amount of water vapour in air; expressed in terms of percentage.

HUSAIN, Dr. (Mian) Mohammad Afzal - born March 1889, started his career at College of Agriculture, Lyallpur (Flaslabad); obtained Ph. D. (1919), University of Cambridge (UK); back to his alma mater as Professor of Entomology (1919-33) and Principal (1933-38). Besides teaching, conducted research on locusts and economic entomology. Vice-Chancellor, Punjab University, Lahore (1938-44; 1954-65); first Plant Protection Advisor to Government of India (1946); Chairman Pakistan Public Service Commission (1948-52), Pakistan Sugar Commission (1957-58) and Pakistan Agricultural Council (1961-65). Died November 1, 1970.

Hyaline glassy, colourless and transparent.

Hybrid offspring of genetically dissimilar parents : person or animal of mixed origins : heterogenous.

Hybrid sterility inability of a hybrid to form functional gametes due to disturbances in sex-cell development caused by incompatible genetic constituents.

Hybrid vigour increase in size, yield and performance of hybrid.

Hybridise to interbreed : to cause to interbreed.

Hydænidæ widely spread family of Hydrophiloidea (Polyphaga : Coleopetra); beetles having antennæ 5-segmented, pubescent club preceded by a capsule and 6 or 7 visible abdominal sternites; live in water (running or stagnant); feeding on algæ (300 species)-*Hydraœna* spp., *Ochthebius* spp. et al of Europe and North America.

Hydromenidæ (Lerentiidæ Buler 1874) family of Lepidopteran insects equivalent to Larentiidæ regarded as subfamily of Geometridæ (Geometridæ: Ditrysia); greenish-gray moths with several brown transverse line on forewings. Caterpillars brown and gray striped, feed on willow (*Salix* spp.), hazel poplar (*Populus* spp.) etc-*Hydriomena furcata* et al.

Hydro water.

Hydrocorisæ Latreille 1802 (= Cryptocerata) subdivision (series) of Heteroptera that includes aquatic and semiaquatic bugs having small concealed antennæ and no ejaculatory bulb in males. *Cf.* Geocorisæ.

Hydrofuge (hydrophobe) (insects) having waxy surface that repels water-cuticle and cuticular hairs of many aquatic insects, these enable the insects to breath under water by taking down a bubble of water trapped by these hairs.

Hydrometridæ Stephens 1829, Kirby 1837 family of Gerroidea (Amphibicorisæ: Heteroptera); delicate rod-like insects with long head, 4-or 5-segmented antennæ and 3-segmented rostrum, legs thin and still-like, tarsi 3-segmented, claws apical (80 species)-*Hydrometra stagnorum* is found crawling slowly over the surface of stagnant water.

Hydrophiliæ (Leach 1815) Samouelle 1819 large family of Hydrophiloidea (Ployphaga: Coleoptera); considerable number of terrestrial form (2000 species)-species of *Cercyon* and *Sphœidum* are found in dung. *Helophourous* species cause damage to root crops, while *Hdrophilus* and allied species are predaceous (great black water beetle, *Hydrous piceus* Linnæus of Europe, is the largest species being > 45 mm long).

Hydrophilis (hydrophilous) water-loving.

Hydrophiloidea Leng 1920 superfamily of Ployphaga (Coleopetera : Oligoneoptera); having elongate maxillary palpi that also perform tactile function, as antennaæ are involved in respiration; majority of species are aquatic placed under families-Hydrophilidæ, Hydænidæ, Spercheidæ and Georyssidæ.

Hydrophobia aversion to water inability to swallow water.

Hydropsychidæ Curtis 1835 very small family of Hydropsychoidea (Annulipalpia : Trichoptera); includes a few species of seine-making caddisflies-species of *Hydropsyche* Picte, *Macronema* Picte and *Hydropsychodes* Ulmer (all cosmopolitan).

Hydropsychoidea very small superfamily of Annulipalpia (Trichoptera: Oligoneoptera), caddisflis having last segment of maxillary palpi annulated-Dispseudopsidæ and Hydropsychidæ.

Hydroptilidæ Stephen 1836 family of Rhyacophiloide (Integripalpa : Trichoptera); small hairy; narrow winged caddisflies. Larva with a form of hypermetamorphosis, early instars free living, later ones modified in shape and living in tube-like case.

Hydropyle specialised region of egg-shell through which water can enter.

Hydroscphidæ small family of Sphærioidea (Myxophaga: Coleoptera); small oval, aquatic beetles, found near wet and marshy places near streams. Grubs having a pair of spiracular gills on thorax and fist eight abdominal segments.

Hydrotropism turning of an organism towards (positive) or away from (negative) moisture.

Hygrobiidæ (Pelobiidæ) very small, monogeneric family of Caravoidea (Adephaga: Coleoptera); most primitive beetles having greatly enlarged head and thorax with narrow linear abdomen terminated by two very long cerci and a median process. Both grubs and imagines carnivorous and predaceous-*Hygrobia* (*Pelobius*) *tardus*, *H. hermanni* et al.

Hylesininæ subfamily of Scolytidæ (Curculionoidea: Polyphaga); cylindrical, ash bark weevils adapted for burrowing in the wood; snout very short even wanting; as such, often called beetles. Grubs gray or black in colour, <5 mm long; burrow galleries beneath the bark of the trees-common ash-bark beetle [*Hylesinus* (*Lepersins*) *fraxini*], large ash-bark beetle (*H.crenatus,*) lesser ash-bark beetle (*H.oleriperda*).

Hylicidæ a very small cicadelloid family (Coleorrhyncha : Homoptera) of the Old World, that includes about 30 species of bugs.

Hymenopodidæ small family of mantids (Memtodea : Dictyoptera); head having a process on vertex; pronotum often expanded laterally; inner ventral spines on fore-femur alternately long and short, outer spines document; females sometimes brachypterous. Found in African and the Oriental regions-species of *Acromantis, Hymenopus, Oxypilus* etc.

Hymenoptera Linnæus 1758 order of Hymenopteroidea (Oligoneoptera: Pterygota); mouthparts either chewing type (Symphyta) or sucking and lapping (most Apocrita); wings with present, membranous with few cross-veins; abdomen basally constricted and its first segment fused with metathorax; ovipositor often longer than body and modified for piercing, sawing or stinging. Metamorphosis complete (holometabolous), over 135000 species grouped into two suborders-Symphyta (horntails, sawflies etc.) and Apocrita (ants, bees and wasps).

Hymenopteroidea Handirsch 1908 superorder of Oligoneoptera (Pterygota; Insecta); adults with two pairs of membranous wings with venation greatly reduced hind wings smaller than forewings and interlocked with the latter by means of hooklets. Larvæ apodous with well developed head, rarely eruciform with locomotory appendages. Pupæ adecticus, exarate (rarely obtect) and cocoon generally present. Comprises of single order Hymenoptera.

Hyper-above: more than normal.

Hypermetamorphosis more than complete or complex metamorphosis, in which different larval instars represent tow or more type of larvæ. Found in various species of most panorpoid orders-Neuroptera (Mantispidæ), Coleoptera (Meloidæ, Carabidæ, Rhipiphoridæ, Staphlinidæ), Diptera (Acroceratideæ, Bombyliidæ, Nemestrinidæ, Tachinidæ) and Hymenoptera (Ichneumonidæ, Perilampidæ, Pteromalidæ etc.).

Hyperparasite parasite that parasitises, another parasite. *Mesochorus nigriceps* and *Pezomachus slevenii* have been recorded as hyperparasites of *Bathyplectes anurus* and B. *curculionis* (Thomsom); *Euryltoma albotidbialis* is hyperparasite of Perisierola *nephantids* and *E.oallidiscpus* of *Holcoera pulver.*

Hyperparasite, Indirect secondary parasitoid that searches out and lays its eggs in body of an unparasitised non-parasitic host, these eggs usually do not hatch.

Hypodermatidæ (Rondani 1856) Towsnend 1916 cyclorhaphous dipteran family of warble files (Calyptratæ: Schizophora); large, robust bee-like files; facial grooves broadly separated; arista bare; apical part of scutellum bare or hairy; palpi small or lacking; wings large permitting rapid flight; apical cell open-*Capra ibex, Hipoderma bovis, H. lineata, Portschinskia neugebaueri* et al.

Hypodermis epidermis : outermost single layer of ectodermal cell of body-wall.

Hypoglottis sclerite often present at the base of tongue, between mentum and labium of an insect; mostly found in some beetles.

Hypognathous head of adult insect with facial area projected forward, mandibles, maxillæ and labium directed downwards and labrum hanging as free lobe before the mouth.

Hypomere ventral process of phallobase.

Hyponeustic oligopneustic; having reduced number of functional spiracles as in some coccids (Coccoidea), thrips (Thysanoptera) etc.

Hyponomeutidæ an equivalent name for family Yponomeutidæ (Yponomeutoidea: Ditrysia); includes ash-bud moths, ermine moths etc.

Hypopharyx median postoral lobe, chitinous sclerite arising from floor of the mouth of an insect, bearing usually salivary apertures. In blood-sucking insects (mosquitoes etc), it is enormously elongated and forms part of proboscis.

Hypopygium terminal segments of abdomen of an insect, often modified for copulation; term mostly applied to males of dipteran files.

Hyposmocomidæ very small family of Gelechiodea (Ditrysia; Lepidoptera); includes small cryptically coloured moths having mandibles functionless or wanting, palpi well developed and forewings narrower than hind-wings.

Hypostoma part of each subgenal area of cranium behind mandibles: anterioventral part of head in dipteran files. Projection surrounding oral aperture in many coelenterata polyps.

Hyppotype specimen of a species which though not a member of the original type series is known from published description.

Hypside family of Noctuoidea (Ditrysia : Lepidoptera); medium-sized moth having well developed proboscis and minute maxillary palpi. Caterpillars thickly covered with long hairs. Found in Africa and The Orient-*Argina argus, A.cribraria* (Clerk), *Hypsa alciphron* (Cramer), *H.ficus* Fabricius etc.

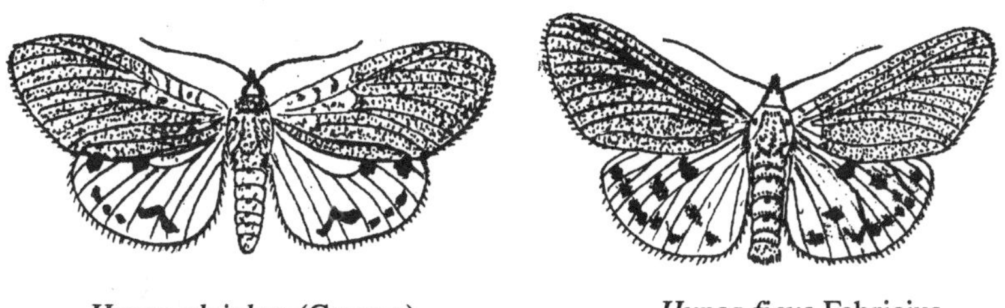

Hypsa alciphro (Cramer) *Hupsa ficus* Fabricius

Hystrichopsyllide Tirboschi 1904 family of Siphonaptera (Oligoneoptera: Pterygota) characterised by metanotum lacking marginal spinelets; includes very large mole fleas-*Hystrichopsylla talpæ* et al.

Hystricothripidæ small family of phytophagous thrips (Phloeothripoidea: Tubulifera); conspicuously characterised by the last abdominal segment being greatly enlarged and nearly as long as all the remaining abdominal segments together-*jamun* thrips, *Leeuwenia ramarkrishnæ* Ananthakrishnan (= *L. karnyi* Ramakrishna).

I

Iapygidæ Lubbock 1873 alternative spelling of Japygigæ (Dicellurata: Diplura).

Ibaliidæ small family of Cynipoidea (Apocrita: Hymenoptera); comprising of very large wasps (>10 mm long) having forewing with cell R_1 closed and at least nine times as long as brand; hind-basitarous twice as long as all other segments out together; ovipositor long and saw-like, normally coiled up underneath within a sheath. Endoparasites of Sircidæ (Siricoidea: Symphyta), female used its long sensitive antennæ to search out tunnels of wood-wasps, then extend its ovipositor to lay eggs in the body of its host. — *Ibalia* spp., widely distributed in Europe and North America, *I. leucospoides* being more common.

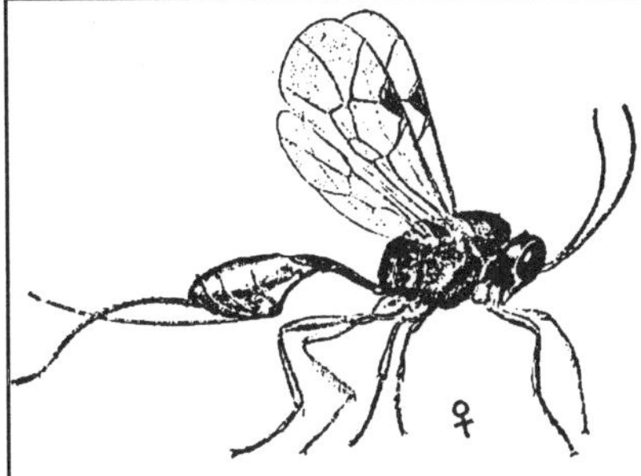

Ichneumon fly

Ichneumon (flies) any species belonging to family Ichenumonidæ. These wasps are mostly parasites and a few hyperparasites, by and large useful as they check the population of insect pests of economics crops.

Ichneumonidæ Leach 1817 family of Ichneumonoidea (Apocrita: Hymenopetera); large, slender, blackish-yellow or reddish-yellow wasps, antennæ having 16 or more segments; forewing with costal cell narrow, cross-vein *2m-cu* present; hind-wing with cross-vein *r-m* meeting Rs after it leaves Sc + R; trochanter 2-segmented; ovipositor arises anterior to tip of the abdomen and remains extruded permanently. Larva endoparasitic, pupates in sliken cocoon. Majority are parasites or less frequently hyperparasites — *Goryphus nersei* (Cameron) parasites *Sylepta derogata* Fabricius and *Earias* spp., *Isotima dammermani* Rohwer and *I. javensis* Rohwer on *Tryporyza nivella* (Fabricius), *Xnathopimpla* spp. on *Chilo partellus* (Swinhoe).

Icneumonoidea Konow 1897 Largest superfamily of parasitic wasps (Apocrita: Hymenoptera); characterised by long antennæ, presence of distinct pterostigma and narrow costal cell on forewing, absence of anal lobe on hind-wing and hind femora with a trochantellus. Without exception, all are parasitic.

Idarninæ subfamily of Torymidæ (Chalcidoidea: Apocrita); having males often apterous; abdomen short and hot tubulary lengthened or broadened at the apex; found associated with fig-insect either as inquilines or parasites; they also attack seeds and fig — *Apocrypta westwood*, *Eokoebelia brevitarus* and *Idarnes testacca* have been reported as minor pests of cluster fig (*Ficus glomerata* Roxburgh) in India.

Idogastra small group of parasitic hymenopteran insect, a relic of an ancient group regarded as a suborder of Hymenoptera, intermediate between Apocrita and Symphyta; but now considered as a superfamily Orussoidea Rohwar under suborder Symphta.

Idiostolidæ family of Lygæoidea (Pentatomomorpha: Heteroptera); plant-bugs characterised by presence of ocelli, prosternum, seutellum and thoracic sterna not punctured; tarsi 3-segmented; abdomen with many trichobothria. Comprises of two monotypic genera confined to Southren hemisphere - species of *Idostolus* are found in Chile and those of *Tricecus* have been recorded from Tasmania.

Idolothridæ Bagnall 1908 a family of phantom thrips (Philoeothripoidea: Tubulifera); head produced in front of eyes; vertex conical and conspicuously overhanging the base of antennæ; usually with strong bristle in front of the eye - giant thrips *Acanthinothrips spectrum* (Haliday), is the largest thrip (12 to 14 mm long) discovered in Australia by Charles Darwin in 1836, feeding upon the dead and moist leave of *Eucalyptus* on ground.

Ileum anterior part of small intestine between ventriculus and colon.

Imaginal buds (discs) cluster of cells that undergo rapid division to form rudiment of future organs during the metamorphosis of an insect.

Immigration movement of individual (insect) into a population.

IMMS, Dr Angustus Daniel — born August 24, 1880; B.Sc. (London) 1903 and D.Sc. (Cantab.) 1907. Demonstrator in Zoology, Birmingham University 1903; Professors of Biology, Allahabad University 1907-11; Forest Zoologist, Government of India at Dehradun 1911-13; Reader in Agricultural Entomology, Manchester

University 1913-18; Chief Entomologist Rothamsted Experimental Station, Harpenden 1918-31, Reader in entomology, Cambridge University 1931-45. Died April 3, 1949. His, **General Textbook of Entomology** (1925) is the most authenic book on insect classification; ninth edition revised by O.W. Richards and R.G Davies (1977) was later (1988) re-revised, enlarged and published by these authors in two volumes (1354 pp.) as Imms' **General Textbook of Entomology.**

Immune not liable to catch infection: cultivar that a specific insect will not feed upon or injure under any known condition.

Incompatibility inability of different pesticides (acaricides, fungicides and insecticides) to mix with each other, without having any adverse effect; use of incompatible chemical often results in undesirable effect on plants, when applied. The incompatibility may be:

— **Chemical** — synthetic organic compounds react with alkaline material and form new compound and such mixture lose their effectiveness.

— **Physical** — chemical when mixed together, change their physical form to one that is unstable and harmful for application.

— **Phytotoxic** — two or more chemical, each non-injurious to plants when mixed, the mixture proves to be phytotoxic.

Incubidæ (= Aphidiidæ) a small family of Ichneumonoidea (Parasitica: Apocrita); active, slender, ant-like, minute parastic wasps; formerly included in the family Braconidæ (Ichneumonoidea) as subfamily Aphidnæ; later separated as an independent family Aphidiidæ; the name subsequently changed to Incublidæ. Adult 1.5 to 2.5 mm long have narrow wings with reduced venation and are parastic on aphids (Aphidoidea: Homoptera). Eggs are laid directly into the bodies of aphids; larva feeds and pupates withing the host and emerges out as an adult; only one parasite comes out from each parasitised aphid — species of *Epherus, Incubus (Aphidius), Monoctonus, Praon, Toxures* et al.

Incurvariidæ family of Incurvarioidea (Monotrysia: Lepidoptera); metallic coloured, tiny moths, active in sunshine; male having antennæ longer than those of female and eyes larger, approximated dorsally. Caterpillars feed on vegetable refuse — *Adela* spp., *Nemaphora (Nemotois)* spp., *Tegticula yuccasella,* et al.

Incurvarioidea Forbes 1923 superfamily of Monotrysia (Lepidoptera: Mecopteroidea); moths characterised by having forewing with closed discal cell; male having frenulum, female with long cloacal duct and sclerotised ovipositor — Adelidæ, Incurvariidæ and Prodoxidæ.

Index list of record surrogated in order of some attributes.

Indigo caterpillar *Spodoptera exigua* (Hübner) (Noctuidæ: Lepidoptera); also known as *ragi* (finger millet) cutworm; medium-sized brown moth with white hind-wing having brown lines. Caterpillar brownish-green with dark wavy lines or dorsum and yellow stripes on the sides. Nocturnal and polyphagous; more harmful in nurseries and to the young crops.

Indirect metamorphosis holometabolous: complete metamorphosis, comprising of four stages — egg, larva. pupa and adult — all panorpoid (endopterygota) order.

Indotermitidæ Roonwal and Sen-Sarma 1958 monogeneric and monospecific family of termites (Isoptera: Biattopteroidea); soldier having profused hairy body; antennæ long and slender; mandibles large and robust; pronotum saddle-shaped, legs long and slender, tarsi 3-segmented; a pair of 2-segmented cerci on X sternum and no styli. — *Indotermes maymensis* Roonwal and Sarma, found only in Myanmar (Burma).

Induced resistance pseudoresistance: temporary resistance acquired by a plant due to change in environmental conditions, like climate, soil fertility etc.

Indusium extra-embryonic covering: in most insect during egg development two distinct embryonic covering are met with — outer (**amnion**) and inner (**serosa**); but in many orthopteran insects an additional third envelops also appear, this is called indusium.

Industrial melanism occurrence of dark forms in industrial district shown to be due of natural selection by the predators of moths against a background of blackened vegetation.

Inert material no biological action.

Infect to contaminate: enter and establish pathogenic relationship with host; introduce pathogenic organism into the host.

Infertile barren: not productive.

Infest be present (usually in large number) and cause inconvenience and/or injury to the host.

Ingest food and water intake in alimentary canal by an individual by swallowing.

Ingesta material ingested or taken into the body.

Ingestion process of taking food and water.

Ingredient component, part of mixture.

Ingredient, Active (a.i) toxic component of formulated pesticide.

Inheritance sum of all characteristics transmitted by germ cell from parent to offspring.

Injector with a needle-like projection for injecting pesticides into tree trunk; or (mostly nematocides) below soil surface, hence also called soil-injector.

Inocellidæ Navis 1916 a small family of Raphidioidea (Megaloptera: Neuroptera); includes snakeflies, differentiated by lack of ocellit and pterostigma not bisected by a veinlet. Only two genera, *Inocellia* Schneider (Nearctic and Palæarstic) and *fibla* Navas (European) — *I. crassicornis* Schummel of Japanese species.

Inopeplidæ release repeated releases of laboratory reared insect with a view to establishing the exotic entomophagous insect in a particular locality.

Inopeplidæ Crowson newly erected very small family of Heteromera (Cucujoidea: Polyphaga), earlier a part of Salpingidæ (Heteromera: Cucujoidea) — *Inopeplus* spp.

Inuiline an animals living in the abode of another and sharing its food — insect living as guest in nests of ants and termites. Some larvæ of Cecidonyidæ (Mycetophiloidea: Nematocera) live in leaf-mines made by the leaf-miners; even some small gall-wasps live in the galls made by other bigger gall-wasps.

Insect a highly specialised tracheate arthoopod (Insecta: Uniramia); adult insect is bilaterally symmetrical, with segmented body covered by Chitinised exoskeleton and having three distinct regions — head, thorax and abdomen — head carries a pair of antennæ (except proturans); thorax bears three pairs of legs (no exception) and usually two pairs of wings (some have only one pair on are wingless). *Phyneilla præcursor,* the oldest known insect (Collembola), lived during Devonian period (extinct since long). The term is also loosely used for other small invertebrates, such as mites, spiders etc., that look like insect; or for an insignificant or contemptible person.

Insecta Linnæus 1758 old and equivalent name for class Hexapoda Latreille 1825 (Uniramia: Arthropoda); includes all insect.

Insect(s), Fossil insect that lived during Palæozoic and Mesozonic eras and are all extinct since long. First insect, on record, arose during Devonian period and the oldest fossil available are collembolan period there lived in France an immense dragonfly *Meganeura* *monyi* (Meganeuridæ: Protodonata) having wing span of 750 mm and a huge cockroach (Archimlacridæ: Blattaria) in Germany and a grasshopper *Oedishia wiiilamsoni* (Oedischiidæ: Protelytroptera) that is considered as a link between cockroaches and earwigs.

Insect number Entomologist guesstimate that there are about 2 million living species of the insect in the world. Of these, about half of the species have been described so far (937000); besides as many as 13000 fossil forms have been also been described. According to Z.P Metacalf (*Entomological News,* Philadelphia, **51**: 219-222, 1940) approximately 1500000 species have been described during 1758 to 1940, but this is based on a hypothetical calculation.

Insect origin it is believed that both myriapods and insect (now both placed under Uniramia) have descended from a common (hypothetical) ancestor Proptaptera, which gave rise to Opisthogoneata and Progoneata; in due course of time Opisthogoneata splitted into Chilopoda and Insecta. Earliest known insects were collombolan that came into being during lower Devoniam period (more than 400 million year ago) followed by proturans, which then got divided into entotrophic Diplura and ectotrophic Thysanura; the latter again gave rise to fixed-winged form Palæoptera — Ephemeroptera and Odonata — subsequent development enabled the insect to fold their wings and these are called Neoptera that includes Poluneoptera, Paraneoptera and Oligoneoptera.

Insect rearing management (IRM) efficient utilisation of resources and insect rearing techniques for the mass production of an insect species in the laboratory (insectory).

Insect size though limited, it varies considerably; fairyfly *Alaptus magnanimus* Annandale (Mymaridæ: Hymenoptera) is 0.02 mm long and fungus beetle *Nanosella fungi* (Ptikkudæ: Coleoptera) is 0.25 mm long white giant birdwing butterfly *Triodes* species (Papilionideæ) have wing spread of 300 mm, hercules moth *Coscinoscera hercules* Miskin (Saturnidæ) 275 to 290 mm, black witch *Erebus agrippina* (Noctuidæ) 280 mm and Atlas moth *Attacus atlas* (Linnæus) and *A. edwardi* White (saturnidæ) 260 to 280 mm; largest living dragonflies are *Tetracanthagyne plagtata* and *Megaloprepus cœulatus* (Drury) having wing expanse of 190 mm to 210 mm; stick-insect (Bacterioidea: Phasmida) *Pharmaica serratipes* (♀) is 320 mm long and *Palohus titan* Sjostedt 260 mm; while male hercules beetle *Dynastes hercules* (Linnæus)(Dynastidæ: Coleoptera) attains length of 155 mm. Among the true flies

(Diptera), Chinese *Tipuka brobdignagia* Westwood is the largest species with wing expanse of over 100 mm, smallest being *Lepoconops caleri* Hoffman < 2 mm long and apterous show-gnat *chionea nivicola* Doane of North America < 4 mm long.

Insect net large hand-net, 250-350 mm in diameter and 500-600 mm deep, close at one end and fitted on a circular frame with 700-800 mm long handle; used for catching adult (flying) insects.

Insectarium vivarium for insects.

Insectary laboratory for rearing and maintaining insects.

Insecticide substance (toxin) effective against insects, it may be contact poison, stomach poison, systemic of fumigant. Use of insecticide dates back to 200 B.C. when fumes of boiling bitumen (asphalt) were blown through grapevines toward off insect pest infestation; use of sulphur started in 100 B.C. and assenic from middle of first century A.D; Paris-green was first used agianst Colorado beetle, *Lepinotarsa decemlineata* (Say) in 1867 followed by other organic chemical and a few botanical. Use of synthetic organochlorine compounds started in 1939 with DDT, followed by BHC (HCH) in 1941. Since then phenomenal progress has been made in the development of insecticides — organiphosphates, carbamates etc. and of late synthetic pyrethroids.

Insectiform insect-like.

Insectifugs substance that drives away insects.

Insectistasis use of phenomones to trap, confuse or inhibit the insect in order to hold their population density below the economic-injury-level.

Insectivore organism (plant or animal) that eats or feed upon insect — butterwort (*Pingicula alpina*), bladderwort (*Utricularia flexuosa*), pitcher plant (*Nepenthes khasiana* Hook) and sundew (*Drosera* spp.) are some insectivorous plants; musk-shrew (*Crocidura maurina*) and pangolins or scaly-ant-eater (*Manis* spp.) are mammals that eat insects.

Insectorubin a brown pigment, usually incorporated with melanin, in this epidermis and cuticle of insects, formerly considered to be a mixture of a group of pigments called ommochromes.

Insecloverdin a green pigment present in the bodies of some insect, usually a mixture of blue-green waste product (biliverdin) with yellow carotinoid substance derived from the food.

Instar form of insect between two successive moults; first instar is the stage between hatching of egg and the first moult. The number of larval instar varies from 3 to 30 in different species.

Instinct inborn impulse or inherited response: institution: an involuntary prompting to specific set up or pattern of behaviour.

Integrated pest control ecologically based system of pest management to keep the density of pest population below the economic-injury-level by adopting number of compatible pest control techniques.

Integricipita Oudemans 1909 one of the two suborder of Siphonaptera (Siphonapteroidea: Oligoneoptera) that included unbroken headed fleas, but being of little systematic value, is no more is usage. *Cf.* Eracticipita.

Integument body-wall of insect; consists of three parts — cuticle, hypodermis and basement membrane.

Intercaste(s) types or variation, rarely found insect, specially in termite (Isoptera) individual that are morphologically intermediate between the castes. Soldier with wing-pads as also forms interbetween worker and soldier are known, but worker-sexual intercastes have been reported. Infection with parasite is an apparent cause of intercaste formation; the other but very rare cause being the postembryonic differentiation of castes under the influence of morphogenetic pheromones.

Intercellular between cells.

Internal parasites endoparasites.

Intersex individual that show both female and male feature. A strong female mated with a weak male may produce male offspring having female characteritics — gypsy moth *Lymantria dispar* (Linnæus) (Lymantriidæ: Ditrysia).

Intestinal myiasis disease of man or other higher animal (vertebrate) caused by the maggots of certain dipteran flies that lives as internal parasite in the intestine of their host.

Intestine proctodæum or part of proctdæum beyond pylorus.

Intima inner chitinous lining of foregut and hind-gut in alimentary canal of an insect; it is selected by a layer of epithlial cells.

Intracellular within the cell.

Inundative release mass production and large scale release of the parasitoids against a particular pest population — *Trichogramma* species reared and released against sugarcane borers.

Iodofenphos organophosphate insecticide, a derivative of thiophosphoric acid: colourless crystalline substance with mild smell; unstable in acidic and alkaline media; contact and stomach poison with prolonged

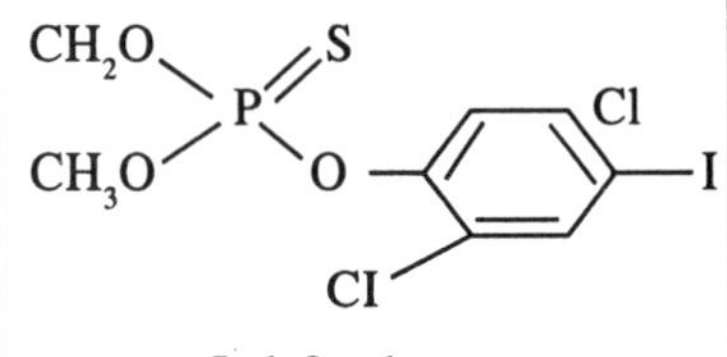

Iodofenphos

action; toxic to bees. Acute oral and dermal LD_{50} values for rats are 2100 and 2700 mg/kg respectively — Nuvanol®.

Ipinæ small subfamily of Scolytidæ (Curcukionoidea: Polyphaga), bark beetle having cylindrical-shaped body with head concealed by long prothorax. Destructive pest of forest trees, in which they make numerous zigzag tunnels, between the bark and the wood; those tunnels are later invaded by fungi, causing fungal diseases. Some species of these beetles, even cultivate fungi on the walls of the tunnels, to provide food for their grubs.

Ipsviciidæ extinct family of plant-bugs of Triassic period is among the earliest fossil Heteroptera; it does not however contributes much to theories of heteropteran phylogeny.

Iridescence production of brilliant colour by optical interference as in the bodies of dragonflies, the wings of some butterflies and the elytra of certain beetles. The effect is due to the presence, in the cuticle, of numerous thin transparent lanellæ separated by material having a different refractive index.

Iris dark pigment surrounding dioptric apparatus of eye (not ocellus). A plant genus under family Iridaceæ.

Iris cell(s) pigmented cells surrounding each ommatidium is a compound eye. In bright light these cells spread round each crysatlline cone, isolating it form the neighbouring ones, as a result the vision is mosaic. In dim light, the process is reversed; iris cells retract, the light into an ommitadium merges with that of neighbouring ommatidia, causing a continuous, though very dim, vision.

Iris tapetum network of fine tracheæ surrounding the ommatidia in some compound eyes, especially those of Noctuid and some other nocuturnal insects.

Irritability ability to respond to a simulus; a fundamental property of protoplasm.

Ischiopodite third segment of generalised appendage of Arthropoda.

Ischnocera Kellogg 1896 suborder of Mallophaga (Psocopteroidea: Paraneoptera) that include biting lice having antennæ filiform and 3-segmented, mandibles vertical and without maxillary palpt, labial palpt single-segmented; meso-and metathorax fused. Further divided into five families - Bovicolidæ, Dasyomygidæ, Trichodectidæ, Philopteridæ and Trichophilopteridæ.

Ischnopsyllidæ family of Siphonptera (Siphonapteroidea: Oligonoptera), that include bat-fleas, distinguished by presence of a peroral comb of two stout bristles on each side.

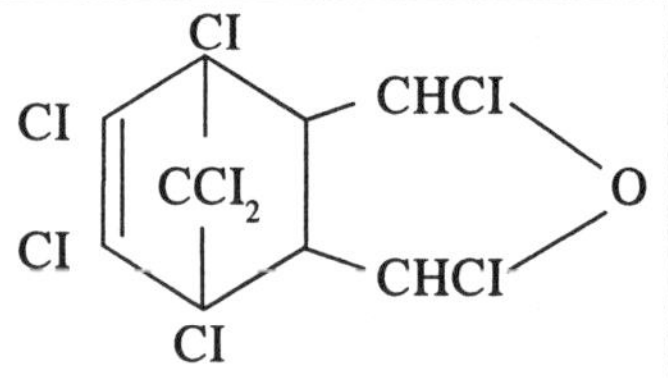

Isoenzan

Isobenzan $C_3H_2CI_8O$; cyclodiene insecticide that act as contact poison. Highly toxic having acute oral and dermal LD_{50} values for rat 5 to 10 and 5 to 30 mg/kg respectively — Telodrin® etc.

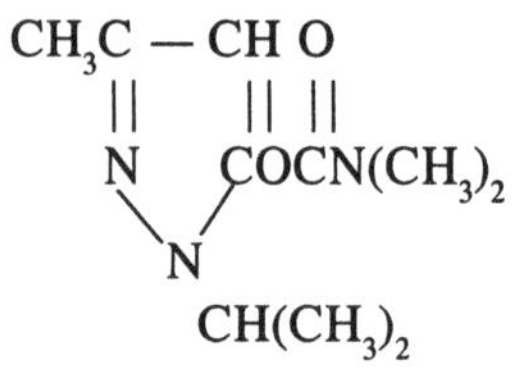

Isolan

Isolan $C_{10}H_{17}N_3O_2$; heterocyclic carbamate insecticide; colourless liquid, soluble in water and organic solvents; contact, stomach and fumigant in action. Acute oral and dermal LD_{50} values for rat are 12 and 40 to 60 mg/kg respectively.

Isopneustic primitive arrangement of species — two pairs thoracic and eight pairs abdominal, all situated in the inter segmental memranes — persists only in some embryos; most insect have some degree of specialisation, may be reduction in number of change position.

Isoptera (Brulle 1832) Comstock and Comstock 1895 order of social polymorphic insect (Blattopteroidea: Polyneoptera); live in communities (in termitaria) composed of reproductive forms (kings and queen) together with numerous apterous sterile soldier and worker. Soft-bodied, having strong mandibulate mouthparts, Ligula 4-lobed; wings when present, equal, elongate and membranous, anterior veins strongly sclerotised, regular cross-veins wanting and an arch edictyon often present; tarsi 4-segmented; creci short, genitalia rudimentary or wanting. Metamorphosis slight or absent. Over 2800 species, grouped in nine families — Hodtermitidæ, Kalotermitidæ, Mastotermitidæ, Termopsidæ, Rhinotermitidæ, Termitidæ, Stylotermitidæ, Indotermitidæ and Serritermitidæ.

Isomidæ family of Entomobryoidea (Arthopleona: Collembola); body setose without scales; postantennal organ present, legs without trochanteral organs, furca well developed, III and IV abdominal segment almost equal - *Isotoma cinerca, I. palustris* (Muller), *I. virdis* (Bourlet) et al.

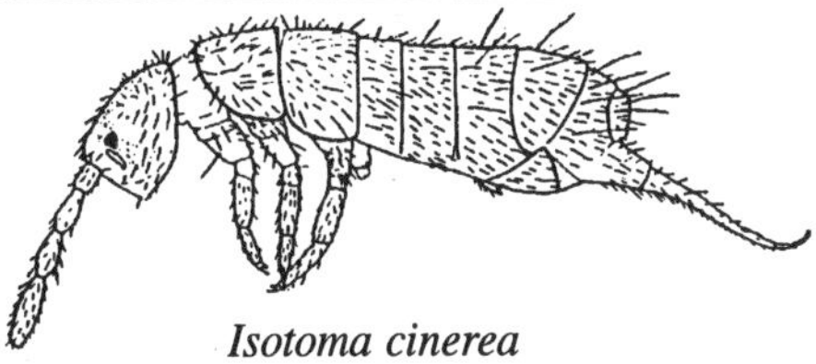

Isotoma cinerea

Isotype duplicate speciman of hototype of a texon, collected at the same time and from place as helotype but deposited usually at some other place.

Inssidæ large family of Fulgoroidea (Achenorrhyncha: Homoptera) plant-bugs having second segment of hind tarsus with one apical spine on each side; tegmina reduced, often brachypterous; hind-wings often absent. Some species have squat beetle-like facies (1100 species) — *Sarina nigrochypeata* infests sandal-wood (*Santalum album* Linnæus) trees in South India.

Ithonidæ Tittyard 1919 family of Osmyloides (Planipennia: Neuroptera). large, soft-bodied insect (moth lacewing) and small head and no ocelli, wing span 50 to 70 mm and venation much branched; larvæ soft, whitish and blind with small mandible and stout maxilla; active runner, found in craks and crevices. Five genera embracing 14 species — *lthone* spp., (Australasia) *Oliarces clara* Banks (USA) and *Rapisma* spp. (Himalayas).

Itonididæ old and equivalent term form gall-midge family Cecidomyidæ (Mycetophiloidea: Nematocera), now no more in usage as it has been ruled invalid.

Isoidicide any chemical that is able to kill ticks (Argasidæ®, Isodidæ®: Acarina) — toxaphene or chlorinated camphene.

J

***Jamun* leaf miner(s)** *Acrocercops phæsopora* Meyrick and *A. telestis* Meyrick (Gracillariiæ: Lepidoptera); microlepidopterous species, species pests of *jamun* (*Syzygium cumini* (L.) sheels and *S.jambos* Linnæus); the caterpillar mine the leave and feed within, causing blister-like swellings on dorsal surface of leaves; the swellings later dry and drop down, leaving big holes in lamina.

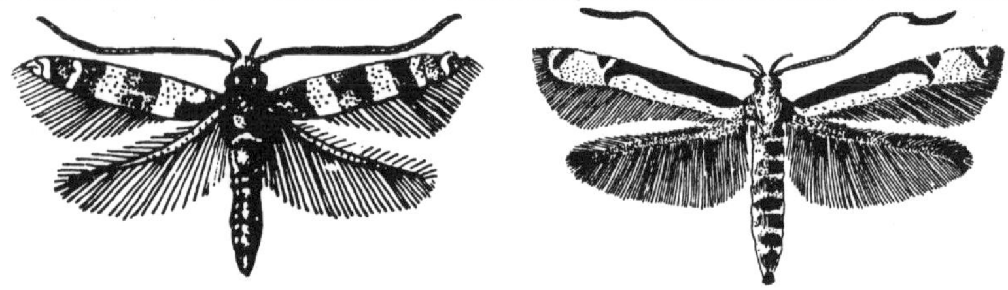

A *Acrocercops phæsopora* Meyrick   *B* *Acrocercops telestis* Meyrick

Japanese oak silkworm *Antheræ yamamai* Guerin (Saturniidæ: Saturnioidea); an oak-feeding, wild and semidomesticated species, common in the forests of Japan; caterpillar are watched on the native tree day and night, till the cocoons are collected; the silk is of very fine quality.

Japygidæ (Iapyigdæ) Lubbock 1873 family of two-pronged bristletails (Japygoidea: Dicellurata); primitive wingless insects having first abdominal segment with styli, cerci modified into sclerotised forceps — *Japyx americanus* Mac Gillivray *J. solifugus* Hailday, *I. subterraneous* Packard et al. *Heterojapyx souliei* (Bouvier) of Tibet is the largest japygid (about 50 mm long).

Japygoidea superfamily of Dicellurata (Diplura: Diplurata); fragile, slender primitive insect having whitish body and a pair of short unsegmented terminal forceps instead of cerci. Considered as evolutionary link between insects and millipedes. Comprises of two families — Japygidæ and Parajapygidæ.

Jassid common name leafhopper (Cicadellidæ: Cicadomorpha: Auchenorrhyncha); both nymphs and adults feed on almost all kind

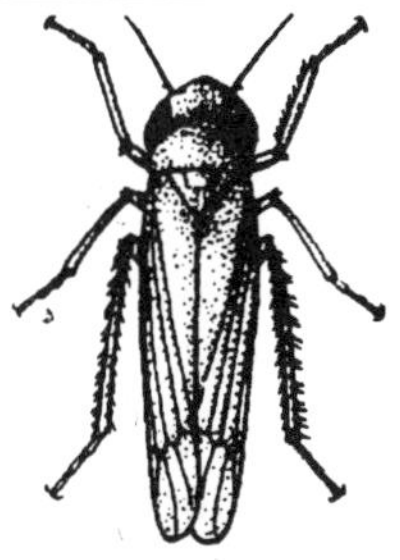

Jassid

of vegetation — crops, grasses, herbs, shrubs and tree; suck sap from leaves, flower stalks and tender shoots, some even transmit virus disease; most species are host specific.

Jassid, Cotton *Amrasca biguttula biguttula* Ishida (= *Empoasca devastans* Distant) (Cicadellidæ: Homoptera); adults wedge-shaped, about 2 mm long, pale green in colour with black dot on posterior portion of each forewing; nymphs whitish-pale-green and wingless, walk diagonally. Both nymphs and adults suck sap from ventral surface of leaves and inject their toxic saliva into plant tissues; affected leaves turn yellowish, curl, then become dark brick red in colour and finally crumple and die away; growth is stunted and yield adversely affected.

Green Jassid
Nephotettix
Species

Jassid(s), Green *Nephotettix virescene* (Distant) and *N. nigropictus* (Stal.) (Cicadellidæ: Homoptera), small 5 mm long, elongated wedge-shaped active leafhoppers; *N. nigropictus* adults have a black submarginal band on the crown. Both nynphs and adults suck sap from the leaves which turn yellow and start drying from tip downwards, the hoppers also transmit Tungro and other virus diseases of rice.

Jassid, Spotted *Nephotettix bipunctatus* Motschulsky (Cicadellidæ: Homoptera), larger is size than green jassid (*Nephotettix virescens* Distant), uniformly pale white in colour with four black spots on the head.

Jassidæ (Jassoidea) alternate name for Cicadellidæ (Cicadelloidea: Homoptera); family of agile plant-bugs or leafhoppers that infest trees and leap from leaf to leaf with remarkable speed and alertness.

Jaws mouth-structure for biting, chewing, gripping, cutting etc., typical insect has two pairs of horizontally moving jaws — mandible (upper pair) and maxillæ (lower pair).

Jet nozzle: stream or spout emitted from nozzle.

Jigger Chigger or Chigoe: a flea found in tropical Africa and South America *Tunga penetrans* (Trungidæ: Pullcoidea); female burrows into skin of human being feet and stay feeding under the toe-nails; male feeds normally as an external blood-sucking parasite.

Johnatonia organ sense organ located in second segment (pedicel) of antennæ of nearly all insect; enables the insect to perceive movement of its antennæ and become aware of air-vibration either as sound or as air current.

Joint worm larva of chalcid wasp *Tetrameca* (*Harmolita*) *tritici* (Eurytomidæ: Chalcidoidea); a major pest of wheat (*Triticum æstivum* Linnæus), also found of barley (*Hordeum vulgare* Linnæus), rye (*Secale cereale* Linnæus) and various grasses (Poaceæ) in North America; it tunnels the stems and produces galls.

Jopyeicidæ a very small of Reduvioidea (Cimicomorpha: Geocorisæ) comprising of small plant-bugs *Joppeicus paradoxus* that carry the rostrum in a forwardly directed position and are predaceous on other insect specially those found on *Ficus* species in Asia Minor.

JOTWANI, Dr. Mulchand Girdharlal — born May 31, 1926; graduated from Karachi (Bombay University) in 1947 and migrated to India; obtained Associate IARI in 1950 and Ph.D. in 1965. Started his career as demonstrator (Entomology), at Central Collage of Agriculture New Delhi; shifted to Directorate of Plant Protection in 1951 and back to Division of Entomology IARI, in 1953 . Worked of economics entomology (toxicology) and pest control for 30 years; was also Head of Division of Entomology for little over one year and died with his boots on, June 4, 1983. One of the most leading insect Toxicologist and World authority on insect of sorghum; his books, **Control of Sorghum Shootfly** (1972) and **Insect in Vegetable** (1984) are still great demand.

Jowl lower portion of the genæ below the eyes, found in head of dipteran flies.

Jugal region (neala) small posterior: basal area of wing set off from vannal region by jugal fold (plica jugallis) and bearing one or two similar.

Jugum small lobe near base of hind margin of forewing that extend backward towards hind-wing, present in some moths, effective in producing coordinated movements of wings. Also known as jugal lobe.

Jumping beans phenomenon caused by larvæ of Mexican moth, *Laspeyresia saltitans* (Westwood) (Olethreutidæ: Torticoidea), which live and develops within the seeds of beans; when the inside of a seed is completely eaten away leaving only the thin-walled outer coat, the jerking movement of the caterpillar within, makes the bean hop up and jump.

Jumping plant-lice minute plant-lice (Psyllidæ: Psylloidea), having long antennæ, long legs and transparent wings; suck plant juices and exude honeydew — *Psyllia mali* Scmidberger (apple-sucker), *P. pregrina* (Hawthorn-sucker), *P.pyri* (Linnæus) (pear sucker) et al.

Juventile immature stage of an organism, usually different from adult form and incapable of reproduction: young stages of various insects.

Juvenile hormone neotenin hormone secreted by endochrine glands (corpora allata) associated with brain, it maintain juvenile character of larval form preventing it from becoming adult.

K

Kairomone(s) allelochemical released by host or prey and used by the entomophagous insect in host/prey location — bee wolf *Philanthus triangulum* a predatory sphecid wasp utilises olfaction to recognise its host, honeybee (*Apis mellifera* Linnæus) Chemical emitted by an organism that induces response in individual of another species and is beneficial to the receiving species — plant scent insect pests.

Kaltotermitidæ Banks 1920 family of dry-wood termites (Isoptera: Polyneoptera) having no worker caste; ocelli present. pronotum flat and broader than head; anterior wing scales long and overlapping posterior wing scales — *Kaleotermes* spp., are common household pests; *Glyptotermes dilatatus* Bugnion is pest of banana (*Mussa paradisica* Linnæus) and rubber (*Fisus elastica* Roxberg) trees while *Nevtermes bosei* Snyder and *N. magiferæ* Roonwal and Sen-Sarma attack mango tree (*Mangifera indica* Linnæus).

Karbi, *Trifona carbonaria* Smith (Meiliponidæ: Apoidea); a small, 4 mm long. stingless bee; common and confined to Australia.

Karunmiidæ small family of Cantharoidea (Polyphaga: Coleoptera); beetles distinguished by an unmodified X abdominal tergite that does not form part of the genitalia. Both larvæ and adults predatory. Only six species found in Central and South America living in termitaria.

Katepimeron infa-epimeron: lower part of epimeron of an insect's leg when it is divided by a transverse suture.

Katepisternum infa-episternum: lower part of epsiternum or anterior sclerite of subcoxa at the base of an insect's leg.

Katydid(s) colloquial name for a large perdominently tropical group of nocturnal noisy, long-horned grasshopper, belonging to family Tettifoniidæ (Tettigonioidea: Ensifera); variously coloured brown, pink or green and having very long and filamentous antennæ (7000 species) — *Amblycorypha oblongifolia* is the largest (40 to 45 mm long) katydid, found in America. *Phasgonura viridissima* (Linnæus) is the Old World species while *Gampsocleis inflata* Uvarov and *G. gratiosa infuscata* Uvarov and Chinese.

Katydid, Florida *Lea floridensis* (Beutenmitter) (Tettigoniidæ); pale yellowish-green species commonly found in the oaks (*Quwecus* spp.) and cabbage palms [*Roystonea olveracea* (Mart) Cook] of Florida (USA).

Ked aptergous, blood-sucking dipteran fly, *Melophagus ovinus* (Linnæus) (Hippoboscideæ: Muscoidea) spend its entire life on sheep (*Ovis ammon*) Larvæ born viviparously, get attached to the animals by a stickly secretion; pupate there and the emerging adults continue to suck blood of their victim.

Kelp-flies small, flattened, bristly flies (Coelopidæ: Schiomyzoidea); larvæ live and feed on rotting sea weed (kelp) on the shore above high-tide mark where the adults often occur in great number — *Coelope frigdis* et al.

Kermidæ old and equivalent form homopteran Eriococcidæ (Coccoidea: Sternorrhyncha); includes scale insect found feeding on oak (*Quercus* spp.) trees.

Key-pests serious, regularly persistent species which if not checked, their population density will exceed economic-injury-level.

King chung (golden bell) a large (about 20 mm long), black, tree-cricket *Homoœogryllus japopnicus* Haan (Gryllidæ: Ensifera) found in India but more common in China and Japan; cacophony of males sound like notes of tiny bells.

Kinetopause arrestration of insect activity without arrestation of its development. Some insect exhibit long period of motionless inactivity interrupted by burst of energy — rest, death-feigning etc.

King sclerotised, macropterous reproductive caste in a termite colony, having lost its wings after founding the new colony. A monarch.

Kindom one of the primary division of living organisms; till recently only two kingdoms were recognised — Animals (all animals including unicellular Protozoa) and Plantæ (all plants, algæ, bacteria and fungi). Of late three more kingdom have been erected, thus Animalia now comprises at only higher animals (Metazoa or multicellular) and plantæ cover only green plants; then there are Monera (bacteria and blue-green algæ), Protista (Protozoa and algæ) and Mycota (fungi).

Kinnaridæ a very small family of Fulgoroidea (Fulgoromorpha: Auchenorrhyncha) with some sixty species recorded feeding on leave of cocoa (*Thevbroma cacao* Linnæus) in Trinidad. Earlier under family Cixilldæ — *Bytrois nemoralis* et al.

Klinokinesia avoiding-action manifested by certain insects which normally fly in straight line under normal or favourable environment the insect immediately change to rapid turning movements.

Klinotaxis movement of an insect in a definite direction in relation to a stimulus — towards or away from light.

Knee joint between tibla and femur in leg of an insect; named by analogy with knee if higher animals (vertebrates).

Knob-horn(s) club-horns: group of sawflies (Symphyta: Hymenoptera) whose antennæ end in a knob of swelling — *Climbex* spp.

Knot-horn(s) colloquial mane for certain pyralid moths (Ditrysia: Lepidoptera) that have swelling at the based of their antennæ.

Knowiellidæ old and obsolete name for hymenopteran family Plumariidæ (Vespoidea: Apocrita).

Kwo-iou the yellowish tree-cricket, *Oecanthus rufescens* Serville (Gryllidæ: Grylloidea: Ensifera) that is popular in China; *O. indicus* Saussure, a pale white of greenish species is common in India.

Kynurenine oxidation product of amino-acid tryptophane that gives rise to some of the brown and red pigment in insects.

L

Labellum one of the two oral lobes, present at distal end of proboscis of true flies (Diptera).

Labial pertaining to labium.

Labial gland(s) salivary glands of insect opening on hypopharynx or between base of hypopharynx and labium.

Labial hooks spines present on labial palpi of naiads of dragonflies (Odonata).

Labial kidney(s) excretory glands located in the head of an insect with tubules opening near the base of labium as in some bristletail (Thysanura) and springtails (Collembola).

Labial palpi (palps) pair of segmented sensory organs or feelers borne on labium of an insect.

Labiduridæ Verhoeff 1902 family of Forficuloidea (Forficulina : Dermaptera); medium-sized, alate earwigs distinguished by male having paired ædeagus, one penis lobe directed forward and one backward; telson smaller than pygidium-*Farcipata quadrispinosa, Labidura riparia* (Pallas), *Nala liridipes, Eudorellia annulipes* (Lucas), et al.

Labiidæ Burr 1909 family of Labioidea (Forficulina: Demaptera); small, winged earwigs, second tarsal segment unlobed and not dilated; male having ædeagus with a single median penis lob-*Labia minor* (Linnaæus), *Marava archidis, Prolabia archidis* (Yersin) et al.

Labioidea superfamily of Forficlina (Dermaptera: Dermapteroidea); body not pubescent; male forces (cerci) usually symmetrical penis lobes variable, single or with only one directed forward-Carciniphoridæ and Labiidæ.

Labium lower lip: posterior median appendage of insect-head formed by union of a pair of gnathal appendages; consists of a fixed mentum and movable prementum to which are attached laterally two labial palpi and distally a ligula.

Labrum upper lip; flat, peroral lobe of head, suspended form clypeus.

Lac-insect(s) highly specialised group coccids (Lecciferidæ: Coccoidea); basically beneficial as they yield stick-lack of great commercial value, from which shellac is prepared. But until unless insects are reared for its lac-production, their presence is harmful as they suck the cell sap and devitalise the trees, affecting adversely the growth and fruit bearing capacity -*Kerria lacca* (Kerremann) and allied species, *Paratachardina lobata* (Green) et al.

Lacciferidæ Chamberlin 1925 (Tachardiidæ Green 1896) family of Coccoidea (Sternorrhyncha: Homoptera): apterous females are highly degenerate having irregularly globular body with vestigial antennaæ and without legs, enclosed in a dense resinous mass secreted by lac-resin glands, found practically at over the dermis. The encrustation covers the twigs of host trees to a thickness of 5 to 12 mm - *Kerri communis* (Mahdihassan), *Paratachardina minuta* (Morrison) and allied species; *Gascardia* species of Malagasy yield inferior lac (gum-lac) containing more of wax.

Lacewings group of neuropteroids (Hemerobioidea: Planipennia); having almost subequal membranous wings with primitive venation and many accessory veins. Larvæ have short broad body with well developed sickle-like mandibles and maxillæ; feed on aphids, coccids, psyllids, thrips etc.

Lecewings, Brown *Kimminisa subnebulosa* and allied species (Hemerobiide: Hemerobioidea); small and more hairy than better known green lace-wing flies; common in woodlands. Larvæ are purple-brown having sharp jaws, feed on aphids and pupate in white silken cocoons. Some sympherobiids (Sympherobiidæ) are also called brown lacewings-*Sympherobins angustatus* Banks (about 6mm long) is the most common species in North America.

Lecewings, Gaint *Osmylus fulvicephalus* (Osmylidæ) is one of the largest European Neuropteran (neuropteroidea: Oligoneoptera) having slender dark-gray or brown body and two pairs of large lace-like wings, wing-expanse about 60 mm; found near streams. Larvæ living among wet moss and feeding on other small insect. *O.chrysops* (Linnæus) of Great Britain has wing span of 50 to 55 mm.

Lacewings, Green chyrsopids having green or yellowish-brown body; antennæ filiform and longer than body; compound eyes iridescent; wing venation complicated with more branchings and cross-veins, veins green in colour tinged with brown, orange or red-*Chrysopa oculata* Say, C. *vittata* Wesmæl, *C.vulgaris* (Linnæus), *C.notata* Walker, C. *flava* (Scopoli) et al.

Lachesillidæ (= Pterodelidæ) family of Psocomorpha (Paraneoptera: Homoptera) that includes aphids and plant-lice often placed under family Aphididæ-*Cinara maritomæ* Dufour, *Lachnus fasciatus* Burmeister, *Tuberolachnus salignus* Gmellin et al.

Lacinia inner endite lobe of maxilla.

Lagriide family of Heteromera (Cucujoidea: Polyphaga), beetles having trochanters of heteromerous type; front coxæ projecting, prosternal intercoxal process narrow; first three visible abdominal sternites connate (400 species) -*Lagra hirla* et al.

LaL, Dr. Krishna Behari-born 1904; M.Sc. Ph.D. (Edinburgh); served as Entomologist to Government of Uttar Pradesh till 1954 where he introduced and established the Plant Protection service. IN 1954, appointed as Plant Protection Advisor, Government of Indian and retired from that post in 1962. Immediately appointed by FAO, as Regional Officer of Plant Protection at Cairo (Egypt). Died May 24, 1986. An Entomologist of rare excellence, who specialised in plant protection.

LALL, Dr. Bankey Shwar (alias **Bankebabu**)-Born January 1, 1914; after matriculation (1932), joined Bihar Agricultural College, Sabour as daily-paid-worker; appointed as Fieldman in January 1940 and Research Assistant in July 1945. After the partition of India (1947), sailed for USA; obtained B.S. (1950) from Cornel University; M.S. (1952) and Ph.D. (1954) form Maryland University-all the while earning his own living. One return to India, became Assistant Entomologist (April 1955) at Sabour, promoted as Professor of Entomology in January 1956 and given additional charge of Vice-Principal of the College form November 1962. Retired January 1, 1973, Reappointed as Professor Emeritus under Bihar University and worked on insect pests of leetchee (*Litchi chinensis* Sonnerat) at Muzaffarpur. Died October 1984.

Lambdacyhalothrin $C_{23}H_{19}ClF_3NO_3$; synthetic pyre-throid; white, odourless substance that is photostable insecticide with contact, residual stomach acting and repellency properties; effective against

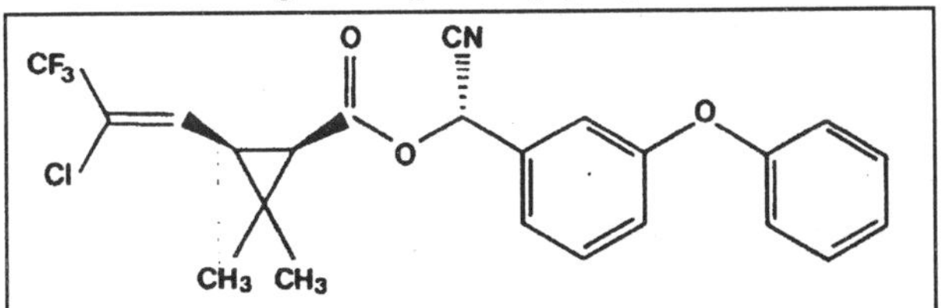

Lambdacyhalothrin

major chewing, sucking and fruit-boring pests of plantation crops; used prophylactically to give persistent control and as quick-acting curative treatment-Karate®.

Lamellicornia equivalent name for superfamily Scarabæiode (Polyphaga: Coleoptera) that includes leaf-horn beetles.

Lamingtoniidæ Sen-gupta and Crowson 1969 very small family of Clavicornia (Cucujoide: Polyphagy); includes one Australian monotypic genus with affinities to languriids.

Lampyridæ Leach 1917 family of glow-works or fireflies (Cantharoidea: Ployphága); nocturnal beetles; usually sexually dimorphic; head concealed by pronotum, eyes large in male much reduced in female; male winged and female often apterous; laminous organs in VI and VII abdominal segments in male but only in VII segment in female. Grubs and adults predaceous on other insects, slugs snails, earthworms etc. (1700 species) - species of *Diaphanes, Lamprophorus* and *Lucicola* are more common in India.

Languriidæ family of lizard-beetles (Calvicornis: Cucujoidea), small, narrow, elongate and parallel sided, 6 to 10 mm long having reddish pronotum and deep-blue, green or black elytra; coxal cavities of forelegs open behind, claws simple. Adults can stridulate by rubbing anterior margin of pronotum against a file on vertex. Phytophagous, adult feed on leaves and flowers (400 species) - *Anadastus parvulus* bore into stems of Italian millets (*Selaria italica*).

Lap to take or swallow by licking up - thrips.

Laphriini tripe (subfamily) of robber-flies (Asilidæ: Adiloidea) closely resembling the bees.

Lappet moth moth, whose caterpillar has fleshy lappets on either side of the body -*Gastropacha quercifolia* and allied species (Lasiocampidæ: Calliduloidea).

Larentiidæ Butler 1874 family of Geometroidea (Ditrysia: Lepidoptera); that includes small pale coloured moths having forewings with one or two accessory cells and areole usually present; found all over all the World-carpet moths, pug moths etc.

Larentiinæ subfamily of carpet moths (Larentiidæ: Geometroidea), small moths which rest with their wings spread out, partially exposing the hind-wings. Caterpillar rest in arched position-speices of *Chloroclysla, Colostygia, Larentia, Xamthorhoe etc.*

Lariidæ Bedel 1891 an equivalent name for Coleopteran family Bruchidæ (Chrysomeloidea: Polyphaga) that includes pulse beetles.

Larridæ Stephens 1829 family of Shepcoidea (Apocrita : Hymenoptera); includes sand-loving wasps; earlier regarded as subfamily (Larrinæ) of Sphecidæ; these wasps nest in burrows in sandy places and provision their nests with bugs, crickets, grasshoppers, spiders etc.—*Tachyles erythropoda, T. mometaritus* are common Indian species.

Larva an independent immature, often vermiform stage of insect that develops from the fertilised egg and must undergo series of changes in form, shape and size before becoming an adult or pupa.

— **Oligopoda** - well developed with efficient thoracic legs, sometimes much reduced; body segmentation distinct.

— **Thysanuriform** - active, flattened, chitinous; free-living and mostly predaceous:-

— **Campodeiform** - elongated, with long or short sickle shaped mandibles and well developed antennæ and cerci - species of Ephemeroptera, Odonata, Plecoptera and Neuroptera.

— **Caraboid** - similar to campodeiform but more chitinised with stronger mandibles, short antennæ and with or without cerci - certain coleopteran families: Cicindelidæ, Carabidæ, Dytiscidæ Hydrophilidæ, Staphylinidæ.

— **Triunguloid** - minute active spiny first instar larvæ of certain predaceous beetles and stylops.

— **Naupliiform** - minute first instar larvæ of Platygaster (Hymenoptera) with branching tails.

— **Planidiiform** - active first state larve or planidia of certain dipterous parasitic flies and hymenopterous perilampids and chalcids.

— **Eruciform** - sluggish, robust, cylindrical; thoracic legs well developed or vestigial, sometimes fossorial; mostly herbivorous and secluded, subterranean or living in burrows in wood, in seed or in fruits : -

— **Scarabæoid** - large wrinkled hairy species, often fossorial and living on roots of plants and on decayed vegetation and dung - Scarabæoidea.

— **Cerambycoid** - large or small; straight; sometimes flattened or cylindrical; smooth or naked, distinctly segmented; mostly herbivorous living in wood or in the soil - Cerambycidæ, Buprestidæ, Elasteridæ.

— **Polypoda** - larvæ akin to eruciform type but having abdominal legs or pseudopods - herbivorous caterpillar (Lepidoptera and Tenthredinioidea/Hymenoptera) and predaceous larvæ of Mecoptera.

— **Protopoda** - primitive parasitic larvæ that emerge early in their embryonic development and have internal system poorly developed and no segmentation of abdomen–Platygastridæ and other parasitic Hymenoptera.

— **Apodous** - robust; crescent- or spindle–shaped; with or without well developed head and legless, some may have minute fleshy lobes or hairy pads in place of legs; herbivorous, parasitic or saprophagous.

— **Curculionoid** - crescent–shaped with well developed head; herbivorous; aerial and subterranean, plant and seed feeders–Brenthidæ, Bruchidæ (except first instar).

— **Muscoid** - headless and spindle–shaped, or cylindrical, truncate maggots, phytophagous, saprophagous or parasitic–most Diptera.

— **Apoid** - well developed head, cared for by dally feeding or lodged in provisioned cells - ants, bees and wasps (Hymemoptera).

— **Lepidopterous** larvæ are known as caterpillars (loopers and semiloopers) while those of Coleoptera and Diptera are called **grubs** and **maggots** respectively. **Nymphs** are the larvæ of paurometalbolous insects whereas **naiads** are aquatic larvæ whose adults are aerial–Ephemeroptera, Odonata, Plecoptera.

Larvævoridæ old and obsolete term for dipteran family Tachinidæ (Musocidea: Calyptratæ).

Larvarium sheller or abode made by insect larvæ, often made of pieces of leaf, pine–needles, soil particles etc. woven together.

Larviparous female insects that deposit larvæ instead of laying eggs - tsetse–flies *Glossina* spp. (Glossinidæ), forest files *Hippobosca* spp. (Hippoboscidæ), sheep ked *Melophagus ovinus* (Linnæus) (Hippoboscidæ) etc

Lasicoampidæ (Harris 1841) Waterhouse 1882 a large family of Bombycoidea (Ditrysia: Lelpidoptera); that includes stout–bodied, hairy and scaly moths, closely related to silkworms (Bombyucidæ); antennæ bipectinate, eyes hairy, ocelli greatly reduced or absent; prodoscis atrophied, maxillary palpi vestigial or wanting, labial palpi well developed; legs very hair tibial spurs short; wings large frenulum wanting. About 1600 species - *Lenodera vittata, Malacosoma indica, Streblote siva* (Lefevre), *Trabala vishnou* et al.

Lauxanidæ (Saprimyzidæ) family of Lauxanoidea (Schizophora: Cyclorrhapha); small soft–bodied flies, shining blackish–blue or yellow; head as broad as thorax, vibrissæ absent, tibia with preapical bristles; wings often clouded with vein Sc complete; ovipositor membranous. Maggost scavengers feed on decaying matter (1200 species) -*Lonachœa* spp., *Sapromyza* spp., et al.

Lauxanoidea superfamily of Schizophora (Cyclorrhapha: Diptera); files having antennaæ with second segment not cleft; postvertical bristles convergent; costa complete (unbroken) and anal vein not reaching wing-margin; comprises of four families-Lauxaniidæ, Celyphiidæ, Chamæmyiidæ and Braulidæ.

LC_{50} lethal concentration of toxicant required to kill 50 percent of a large group of individuals of one species.

LD_{50} lethal does of toxicant required to kill 50 percent of a large group of individuals of an species.

Lead arsenate $PbHASO_4$ (acid orthoarsenate); stomach poison, first used as insecticide in 1892 in USA. It is relatively weak poison; acute oral and dermal values for rate being 10-100 and 2400 mg/kg respectively. In India it was widely used during first half of this century, as spray against *Achæa* spp., *Epilachna* spp., *Aulacophora* spp., *Papillio* spp. etc, as also in poison-baits for controlling grasshoppers and locusts. With the inception DDT, BHC etc., its use was discouraged and by not it is not more in usage.

Leaf-insect flattened, leaf-like insect, *Phyllium* spp. (Phylliidæ: Phasmida); head progathous with compound eyes; mouthparts ectognathous biting and chewing type; legs with lamellate expansions; forewing modified to simulate veins of leaf; hind-wings membranous; found in tropical Asia.

Leaf-miner insect that mines a leaf and lives within feeding upon the tissues between upper and lower epidermis of leaf-*Acrocercops syngramma* Meyrick (Gracilaridæ: Lepidoptera), *Chromatomyia horticola* Goureau (Agromyzidæ: Diptera), *Platypria adrewsesi* (Weise) (Hispidæ: Coleoptera) etc.

Leafhopper(s) homopteran plant-sucking bugs (Cicdelloidea: Auchenorrhyncha); small (4 to 6 mm long), elongated body, can jump and fly with great speed; feed on almost all kind of vegetation, suck cell-sap and inject toxin into plant tissue; some even transmit fungal and viral diseases among the plants-*Cicadella spectra* (Distant), *Nephotettix nigropictus* (Stal), *N. virescens* (Distant), *Recilia dorsalis* Motschulsky, *Tettigoniella albida* Signoret are some of serious pests of paddy (*Orzya sativa* Linnæus) in India.

Leathery jacket(s) maggots of crane–flies (Tipulidæ: Nematocera); fleshy, cylindrical 11 to 12–segmented, having tough leathery integument, often rugose and pigmented or transparent; head deeply invaginated, antennæ well developed, labium and hypopharynx large, with or without pseudopods; spiracles on thorax and abdomen (amphipneutic) or only on anal segments (metapneustic); terrestrial, semi–aquatic or aquatic forms.

LEFROY, Harold Maxwell - born January 20, 1877; educated at King's College Cambridge, Natural Science Tripos 1898; Entromologist Imperial Department of Agriculture, West Indies 1989–1903; Imperial Entomologist, India 1903–12; worked at Surat till 1905, then shifted to Pusa (Bihar; Professor Entomology Imperial College of Science and Technology, London 1912 and Honrary Curator Insect house, Zoological gardens, 1913. Temporary Lieutenant Colonel, Mesopotamia E. Force 1916. Died October 14, 1925. His book, **Indian insect Life** (1909) is still considered as on of the best reference book on entomology.

Leg appendage or limb, used for support and/or locomotion. An insect's true leg is a segmented appendage born on the thorax and there are always three pairs of legs, one each on pro-, meso-and metathorax. Prolegs and pseudopoda that develop usually on abdomen, are not considered as true legs. A typical insect–leg consists of coxa, trochanter, femur, tibia and tarsus terminating in pretarsus (claw). Besides, in primitive insects there is one more segment (subcoxa). The lege are often modified, according to their use and function:–

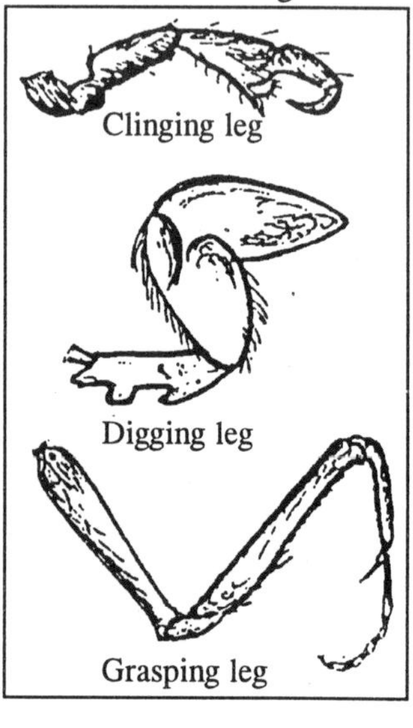

Carrying legs – highly modified legs for carrying out various functions. Hind–legs have a pollen basket (corbiculæ) on outer surface of tibia for carrying pollens and a pollen–comb on inner surface of tarsus for removing pollen from the body. Middle legs with spur of scraper on apex of tibia, to scrape pollen form the basket and antennal-comb on forelegs for removing pollen from then antennæ–honeybees.

Clinging Leg – scansorial, tibia being stout, bearing at one end thumb-like process with a distal spine and tarsus single segmented terminating in a curved pretarsal claw-body louse.

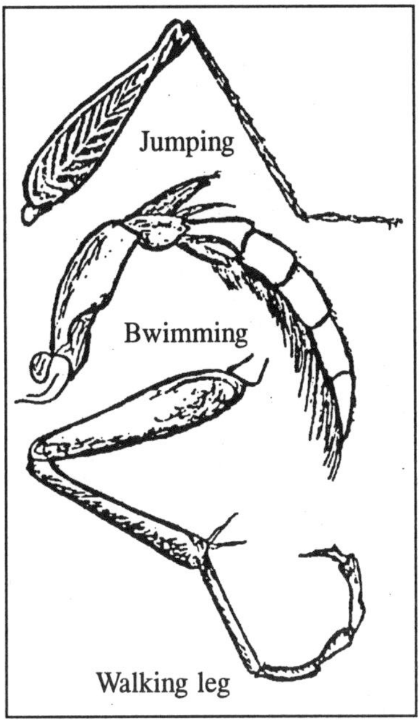

Digging legs - fossorial; stout, toothed forelegs with tibia being spade or rake-like adapted for digging - mole crickets.

Grasping legs - forelegs raptorial with very long coxæ and spiny femur and tibia - mantids.

Jumping legs - hind-legs saltorial, having very large and strong femur - grasshopper and locusts.

Swimming legs - natatorial; femur, tibia and the first foretarsal segment are broad and flat with dense flat setæ that serves as oars-waterbugs and aquatic beetles.

Walking legs - ambulatorial; simple legs with 2-segmented trochanter-wasps etc.

Legume pod borer *Maruca testulalia* (Geyer) (Pyraustidæ: Lepidoptera) also know inflorescence and nearby leaves and developing pods and feed within; affected buds and flower do not yield any pods and the affected pods become malformed.

Leiodidæ (Liodidæ) a cosmopolitan family of Staphylinoidea (Polyphaga: Coleoptera) the round carrion beetles having antennæ with 5-segmented club; found usually under decaying bark-*Liodes* spp. breed on subterranean fungi.

Lemoniidæ (= Erycinidæ, Nemeobiidæ, Rhiodinidæ), family of small butterflies (Bombycoidea: Ditrysia); forewing short and broad with M_1 and Rs on common stalk; no frenulum; males having brush-like, clawless non-functional forelegs while in female forelegs are smaller but perfectly functional (1100 species) - *Dodonia* spp., *Lemonia* spp., *Libythea cltis, Nemeobius lucina* et al.

Lens lenticular outer part of a compound eye, typically biconvex, may be plano-convex an d sometimes rounded on either surface.

Leotichidæ small family of bugs (Leptopodoidea: Pentatomomorpha);

earlier placed under superfamily Saldoidea; only two species of *Leotichius,* confined to bat-inhabited limestone caves in South-east Asia.

Lepiceridæ small family of Sphærioidea (Myxophaga: Coleoptera); very small hornbeetles, found in wet places near stream; includes only two species form Central America.

Lepid(o)-pertaining to scales.

Lepidillidæ Enderlein 1911 family of Trogiomorpha (Psocoptera: Paraneoptera); tarsi triarticulate.

Lepidillidæ Enderlein 1911 Tropical family of Trogiomorpha (Psocoptera: Paraneoptera); claws with a preapical tooth; found under bark and leaf-litter (100 species)-*Lepidopsocus* spp. et al.

Lepidoptera Linnæus 1758 one of the biggest insect order (Oligoneoptera: Pterygota); antennæ variable in length and structure; mandibles vestigial or absent, mouthparts represented by suctorial proboscis formed by maxillæ, compound eyes large and globular, ocelli two; four membranous wings with few cross-veins; body, wings and legs covered with pigmented overlapping scales that provide colour pattern characteristic of the species. Metamorphosis complete. Larva (caterpillar) eruciform and peripneustic with mandibulate mouthparts. Pupa adecticous and obtect, generally enclosed in cocoon or earthen cells. Over 160000 species grouped in four suborders-Zeugloptera, Decnonypha, Monotrysia and Ditrysia. Earliest known Lepidopterous insect appeared during Cretaceous period (Mesozoic era).

Lepidopterist student or scholar devoted to study or butterflies and/ or moths.

Lepidosapthinæ very primitive family of Lepismatoidea (Zygentoma: Thysanura); body not scaled, eyes and ocelli present, abdomenal sterna and coxites well developed with styli present on II to IX segment and overside vesicles on II to VI segments. Only one species found beneath bark of rotten logs of Douglas fit in California (USA) - *Tricholepidion gerischi.*

Lepismatidæ Lubbock 1873 family of Lepismatoidea (Zygentoma: Thysamura); minute primitive wingless insects having compound eyes small or absent, no ocelli; thoracic styli absent, abdominal styli on VII to IX segments or only or VII and IX segment-silverfish (*Lepisma saccharina* Linnæus and *Clenolelpisma longicaudata* Escherich), firebrat (*Thermobia domestica* (*Packard*)] et al.

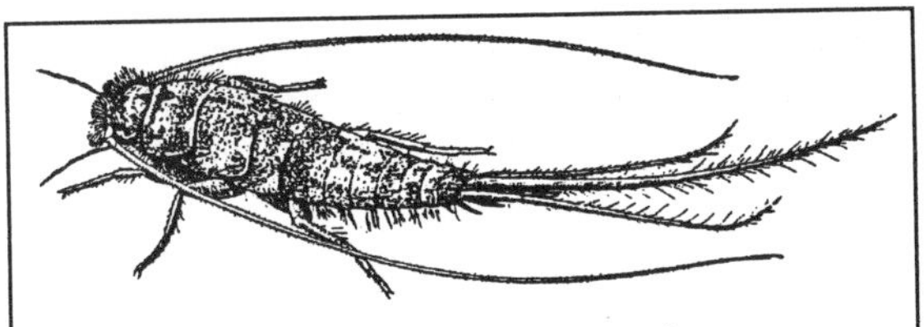

Thermobia domestica (Packard)

Lepismatoidea superfamily of Zygentoma (Thysanura: Apterygota); comprising of 3-pronged bristletails like silverfish and firebrats, divided into three families-Lepidotrichidæ, Lepismatidæ and Nicoletiidæ.

Lepidostomatinæ subfamily of Sericostomatidæ (Limephiloidea: Apocrita), Trichoptera); characterised by the larvæ having tracheal gills.

Leptaleinæ (= Pseudomyrmecinæ) subfamily of Formicidæ (Scollioidea: Apocrtia), male have 2-segmented pedicel and large eyes. Head of larva is surrounded by hood-like thorax; just behind the head is a pocket or surrounded, food received from worker ants is kept in this pouch and gradually drawn into the mouth and swallowed. Usually found in hollow twigs or the galls in plants.

Leptinidæ Le Conte 1866 a small family of mammal nest beetles (Staphylinoidea: Polyphaga) having head behind forming a keel that overlaps the pronotum; eyes rudimentary or wanting; antennaæ long 11-segmented; forecoxæ small and rounded-*Platypsullus castoris* and *Leptinillus* species are found feeding on beaver in Europe and North America while *Leptinus testaceus* feeds on small rodents in Europe.

Leptoceratidæ a small dipteran family of Acalyptratæ (Schizophora).

Leptoceridæ Leach 1817 a large family of caddisflies (Limnephiloidea: Untēgripalpia); body densely hairy, ocelli wanting, dorsal part of tentorium reduced; venation in forewing normal-*Leptocerus ancylus* Vorchies, *Mystacides sepulchrais* Walker, *Setodes argentalis* et al.

Leptoceratidæ (Pelecinellidæ) family of chalcid wasps (Chalcidoidea : Apocrita found on deadwood in Tropical America and Australia.

Leptoperlidæ family of stoneflies (Leptoperloidea : Filipalpia) often combined with Gripopterygidæ, hind-wing with six or less anal veins, no transverse vein connecting R to Cu; confined to Australia - *Leptoperla* species et al.

Leptoperloidea superfamily of Filipalpia (Plecoptera : Polyneoptera); archaic heterogenous group of stoneflies having numerous cross-veins (including distal cross-veins). Nymphs with tubular anal gills. About 170 species distributed in four families -Austroperlidæ (= Penturoperlidæ), Scopuridæ, Pelteoperlidæ and Gripopterygidæ (including Leptooplidæ).

Leptophlebiidæ Klapalek 1909 family of mayflies (Leptophlebioidea: Ephemeroptera); small slender delicate insects having four movable tarsi on hind-legs, fifth fused with tibia; forewing with Cu_2 curved gently to hind margin. Nymph (naiad) aquatic and herbivorous-species of *Leptophlebia, Habrophlebia, Paraleptophlebia, Choroterpes* et al.

Leptophos $C_{13}H_{10}BRCl_2O_2PS$; phosphonothionate insecticide; white microcystalline waxy solid. Acute oral and dermal LD_{50} values for rat being 90 and > 800 mg/kg respectively-Phosvel® etc.

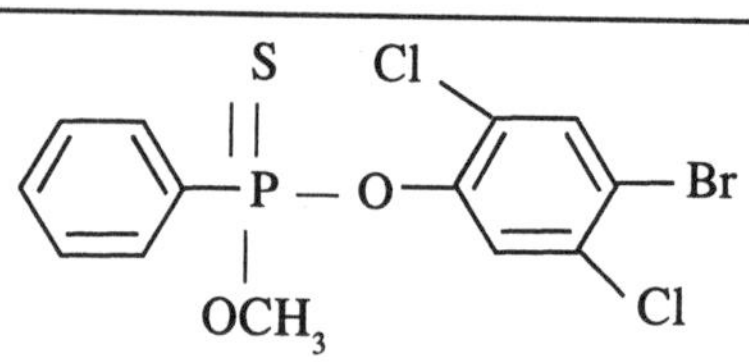

Leptophos

Leptodidæ family of heteropteram bugs (Leptopodoidea: Geocroisæ); distinguished by spiny body, very short rostrum, ocelli on a pedunculate tubercle; all tarsi 3-segmented; found mainly in Tropics and subtropics of the Old World-*Leptopus marmoratus* et al.

Leptopodoidea a superfamily of Pentatomomorpha (Geocorisæ: Heterroptera).

Leptopsyllidæ a small family of fleas (Ceratophylloidea : Siphonaptera); distinguished by presence of a genal comb and upper seta of the ocular now being above the eye - parasitic on some birds and mammals.

Lestidæ Jacobson and Bianchi 1905, family of damselflies (Lestinoidea : Zygoptera); moderately robust species of variable size; having wings distinctly petiolated and less than 5 antenodals; wings held upwards and backwards when at rest-*Archilestes californica* McLachlan *Lestes stultus* Hagen, *Sympecma fusca, S.pædisca, L.vigilax* Hagen, et al.

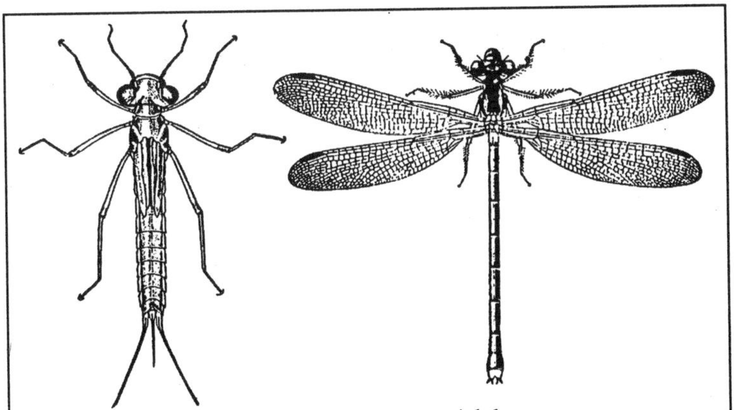

Archilestes californica McLachian

Lestinoidea superfamly of Zygoptera (Odonata: Odonatoptera); damelflies with varied wing-venation; distinguished by hamulli of secondary male copulatory organ elongate-Perilestidæ, Chlorolestidæ and Lestidæ.

Leucospi(di)æ family of parasitic wasps (Chalcidoidea; Apocrita); black or yellow, stout-bodied insects; hind femur swollen with rows of short teeth; wings folded longitudinally when at rest; ovipositor long and bent forward over long and bent forward over abdomen; usually parasitic on bees-*Epexochlænoides pyriformis, Polistromorpha indica* et al.

Leuctridæ family of filipallpain stoneflies (Nemouroidea: Plecoptera); comparatively small insects with rolled wings; found in Northern Hemisphere (170 species-*Leuctra* spp. (females carry their eggs on their back before depositign the same at a suitable place).

Libellulidæ Stephens 1836 widely distributed family of darters (Libelluloidea: Anisoptera); large, fat-bodied, broad-winged and brightly coloured dragonflies characterised by wings having one or more black spots near anterior edges and a brown or black triangular patch near the base-species of *Libellula, Leucorrhinia, Orthetrum, Pantala, Sympetrum* et al.

Libellulides Leach 1815 old and obsolete name for insect order Odonata Fabricius 1793; no more in usage.

Libelluloidea Selys 1840 large dominant superfamily of Anisoptera (Odonata: Odonatoptera); includes all dragonflies; eyes meeting or confluent; wings not petiolate; abdomen variable, fusiform or depressed; ovipositor reduced. Nymphs (naiads) having labium spoon–shaped and setose; four families–Synthemidæ, Corduliidæ, Macrodiplactidæ and Libellulidæ.

Libytheidæ small family of butterflies (Papilionoidea: Ditrysia); often regarded as subfamily of Riodinidæ; characterised by long palpi looking like a beak Caterpillars resemble those of Pieridæ with each segment divided into annulets and numerous secondary setæ present. Pupæ short, smooth and suspended perpendicularly–*Libythea celtis* found in Western Asia, South Europe and North Africa has all the four wings brown with large orange marks and two with spots on each forewing. *L.geoffroyi* of Australasia has forewings lilac coloured with broad brown margin, hind[a]wings half brown and half mauve.

Life existence : state of being alive: unique complex organisation of molecules expressing itself through chemical reaction that lead to growth, development, responsiveness, adaption and reproduction; ending ultimately in death.

Life cycle definite pattern of birth, growth, reproduction, old age and death.

Life history life–time: longevity: age of organism at death.

Life table separation of pest population into different developmental stages (egg, larva, pupa adult).

Lifeless dead: inanimate.

Light arches medium–sized noctuid moths, *Apamea lihoxylæa* having yellowish-while wings with gray streaks and a row of tiny dark dots near the edges. Caterpillars feed on gasses. *A. monoglypha* is commonly called black arches.

Life emerald geometrid moth, *Campæ* (*Metrocmpa*) *margaritaria* [Sellidosemi(at)idæ: Geometroidea] having white body and pale emerald-green wings; forewings with two transverse dark green and white lines and hind–wings with only one such line. Caterpillar feed on foliage of various trees.

Light trap contrivance deviced for catching and collecting insects. Consists of light source (petromax or electricity) that attracts the insects at (dark) night and mechanism that traps these insects, live or dead. Once used as mechanical pest control method.

Ligula terminal lobe (distal part) of labium; typically comprises of a pair of inner lobes or glossæ and a pair of outer lobes, paraglossæ.

Limacodidæ Walker 1855 (= Cochlidiidæ Hubner 1822, Eucleidæ Comstock 1895, Heterogenidæ) small family of Zygænoidea

(Ditrysia: Lepidoptera); mediumsized, stout-bodied moths, wings heavily and loosely scaled; sexual dimorphism evident in some species. Larvæ (slug-caterpillars) having no prolegs but sucking discs and poisonous spines on the body that are irritating to touch. Pupa enclosed in hard round cocoon with an operculum (40 species) -*Parasa lepida* (Cramer), *Spatulicraspeda castaneiceps* Hampson, *Thosea aperiens* Walker, et al.

Limicolous mud dwelling; living in mud.

Limivorous mud-eating: feeding on mud for the organic matter contained in the mud.

Limnadidæ (= Danaidæ, Euploeiinæ) family of nymphalid butterflies (Papilionoidea: Ditrysia) often regarded as subfamily of Nymphalidæ; antennæ without scales and often with small antennla club; fore feet in female terminate in corrugate knob; adults have developed acrid disagreeable odour and taste which coupled with leathery consistency of their body, protects the imagines from insectivorous enemies - *Danaus plexippus, D. menippe* (Hubner), *Euploea gadarti, E. core* Cramer et al. Often considered as a synonym of Danaidæ Hampson 1918 (Papilionoidea: Ditrysia) or Danainæ (Nymphalidæ Papilionoidea).

Limnebelidæ a family of minute moss beetles (Hydrophiloidea: Polyphaga).

Limnephilidæ Kolenati 1859 family of Limnephiloidea (Integripalpia: Trichoptera); caddisflies characterised by presence of ocelli, male having maxillary palpi 3- or 4-segmented and forewing with vien M missing-*Limnephilus amureniis* and *L. correptus* are minor pests of paddy (*Oryzae sativa* Linnæus) in East China, Japan and Korea while *L. flovicornis* Fabricius is pest of watercress in England.

Limnephiloidea superfamily of Integripalpia (Trichoptera : Mecopteroidea); ocelli if present then male maxillary palpi 3 or 4-segmented and if ocelli absent, maxillary palpi 5-segmented - Phryganeidæ and Limnephilidæ have ocelli whereas Sericostomatidæ, Philanisidæ, Helicopsychidæ and Leptoceridæ have no ocelli.

Limnichidæ cosmopolitan family of minute false water beetles (Dryopoidea: Polyphaga); antennæ more or less filiform; front coxæ not projecting, mid coxæ widely separated and hind coxæ contiguous (70 species).

Limnobate(s) alternative name for water-measures (Hydrometridæ: Gerroidea), agile predatory water bugs that move on surface on water-*Hydrometra* spp.

Limnobiidæ family of long-legged wingless craneflies (Tipuloidea : Nematocera); regarded by some as subfamily of Tipulidæ; larvæ vary greatly in habit; those of *Limnobia* species are fungivorous, those of *Dicranomyia* are mostly aquatic or semiaquatic, while some species have been recorded from snows of Carpathians (South-east Europe).

Limnologist person having sound knowledge of limnology.

Limonology branch of zoology that deals with study of life and the conditions for life in lakes, ponds and streams.

Limulodidæ family of Staphylinoidea (Polyphaga: Coleoptera); characterised by antennæ with 3-segmented club and retractile into grooves beneath the head; eyes lacking. About 40 species of myrmecophilous beetles grouped in 5 genera.

Lindane $C_6H_6Cl_6$; name proposed by German chemist Van der Linden in 1949 for an insecticide containing not less than 99 percent gamma isomers of BHC. Colourless or slightly whitish, odourless crystals, insoluble in water slightly soluble in petrolium oils and readily soluble in acetone, aromatic and chlorinated solvents. State to air light heat and carbondioxide but unstable in presence of alkali. It is strong stomach poison and exhibits persistent contact action as also possesses some fumigant properties. Safer than HCH (BHC) on plants but more toxic to insects; harmful to bees, fish and livestock. Acute oral LD_{50} values for male and female rats are 90 and 80 mg/kg respectively.

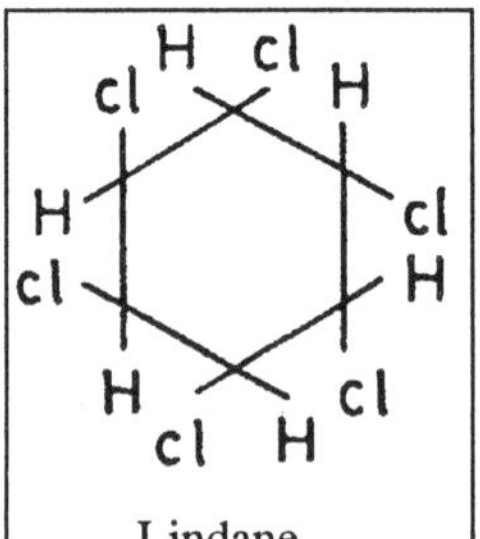

Lindane

Lingua tongue or tongue-like organ; term often incorrectly used for elongate proboscis of certain bees (Hymenoptera) and butterflies (Lepidoptera).

LINNÆUS, Carolus A. (Car von Linne) Dutch biologist, born May 23, 1707 at Rashult (Sweeden); studied at University of Lund and Upsala; qualified first in medicine and later in botany. Married Sara Moræa in 1739. Appointed to the chair of medicine at Upsala in 1741, exchanged for the chair of botany in 1742. First to propone binomial nomenclature of animals and plants;

Carolus A. Linnæus

listed about 3000 insect species (including 12 Indian) under Insecta, which he divided on the basis of form and structure of wings (alary system) into seven classis (orders) - Aptera (all wingless insects as also spiders, terrestrial Crustacea and Myriapoda), Angiosptera (Neuroptera, Ephemeroptera, Angiosptera Embioptera, Isoptera, Psocoptera and Mecoptera), Ryngota (Hemiptera, Siphonaptera and Thysanoptera), Elentherata (Coleoptera and Dermaptera), Glossata (Lepidoptera), Antliata (Diptera, Anoplura, Mallophaga as also some Acarina and Crustacea) and Piezata (Hymenoptera). Died January 10, 1778 at Upsala. Published 180 works including **Systema Naturæ** (12 editions, 1735–1767), **Genera Plantarum** (1737), **Philosophica Botanica** (1751) and **Species Plantarum** (1753); most of his botanical and zoological collection now belong to Linnæan Society of London while his statue stands in Humledgard park at Stockholm (Sweden).

Linognathidæ family of Siphunculata (Psocopteroidea: Pterygota); sucking lice having abdomen membranous, abdominal paraterygal plates lacking and only six pairs of abdominal spiracles–*Lingnathus* spp. on goats and sheep.

Liopteridæ family of large wasps (Cymipoidea: Apocrita); characterised by forewings with cell R_1 closed not even twice as long as broad; gaster attached to top of propodeum. Confined to Africa and South America.

Lipari(di)dæ old and equivalent name for family Lymantriidæ (Noctuoidea: Ditrysia).

Lipeuridæ Mjoberg 1910 family of order Mallophaga (Psocopteroidea: Paraneoptera) in an old system of classification, now forms a part of family Philopteridæ (Ischnocera: Mallophaga).

Lipid(s) esters of fatty acids; neutral fats (triglycerides), phosphatides, cerebrosides, fat soluble vitamins and steroids–these comprise the major food reserves of insects.

Lipoptera Shipley (1904) suggested name for the insect order to include bird lice but not accepted; Mallophaga Nitzsch 1818 has precedence over this name.

Liposeclidæ Enderlein 1911 small family of booklice (Troctomorpha: Pscoptera); body small, flattened and oval; antennaæ short; alate or apterous, latter with

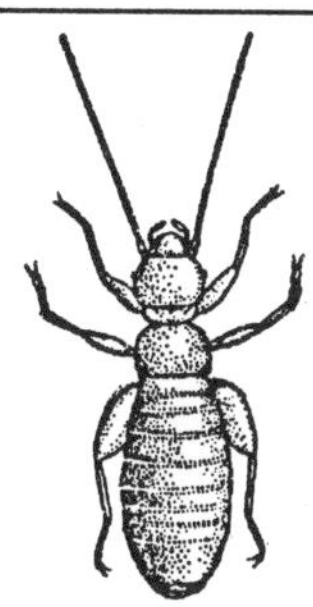

Liposcelis divinatorious (Muller)

reduced eyes, no ocelli and fused meso-and metanotum. Confined to Africa and South America (80 species) - *Embidopsocus* spp. are found under the bark; *Liposcelis divinatiorius* (Muller) and allied species are apterous often found in nests of ants and birds.

Lithosiidæ Stephens 1829 old and equivalent name for Arctiidæ (Noctuoidea: Lepidoptera), a family of tiger moths etc., new considered as subfamily (Lithosiinæ) of Arctiidæ.

Lithosiinæ subfamily of Arctiidæ (Noctuoidea Ditrysia); includes such moths as are called Footman; crepuscular or diurnal in habit; forewings long and very narrow. Caterpillars hairy, feed upon lichens growing on tree trunks-*Lithosia lurideola* is gray coloured having a yellowish band along the front edges of forewings and yellow hind-wings.

Little emerald small geometrid moth, *Lodis lactearia* (Geometroidea: Ditrysia) of Europe, having pale brown wings crossed by two white lines; wings expanse about 25 mm.

Living fossil a living species belonging to an ancient stock otherwise known as fossils.

Locust(s) the most notorious insect pets in the World, closely allied to short-horned grasshoppers (Acrididæ: Cælifera); both having short antennæ, 3-segmented tarsi, auditory organs one on each side of first abdominal segment and short ovipositor. Stridulation is femoroalary, rubbing hind femora against lower edge of tegmen. Locusts are no more than large grasshoppers differing principally in their tendency to periodically migrate in million form country to country, crossing all barriers. Grasshoppers generally feed on specific crops only but locusts devour almost all vegetation and even feed on specific crops only but locusts devour almost all vegetation and even feed upon dead and dying locusts. Unlike grasshoppers which are solitary, locusts exits in two phases-solitary (*phasis solitoria*) and gregarious (*phasis gregarious*), intermediate forms, *phasis transiens* are also observed. India has been subjected form times immemorial to locust invasions which cause immense demage to crops, sometimes resulting in severe famines and substantial loss

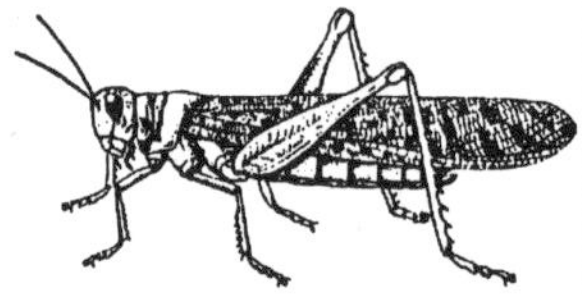

Swarming

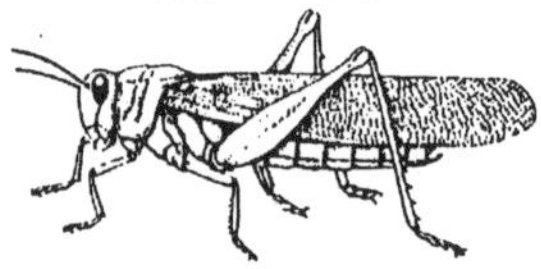

Solitary Locusts

of livestock. There are 11 main species of locusts; of which three are commonly found in India - Bombay locust [*Patanga succincta* (Linnaeus)], Migratory locust (*Locusta migratoria* Linnæus) and Desert locust (*Schistocerca gregaria* Forskal); other important species are Red locust (*Nomadacris septemfasciata* Serville) with outbreaks area in East Africa; South American (*S. paranensis* Burneister) and North American Rocky mountain locust (*Melanoplus spretus* Ulher); Australian (*Chortoicetes terminifera*) and Mediterranian (*Dociostaurus macroccanus* Thumberg). Young ones of locust are known as hoppers; these move in bands, devouring every vegetation that comes in their way. The term is also applied (erroneously to 17-year cicada).

Locustidæ Burmeister 1838 old and obsolete term for family Tettigoniidæ (Ensifera: Orthoptera) as it does not include any locust. Locusts come under family Acrididæ (Cælifera: Orthoptera).

Locustologist entomologist who has specialised in the study of locusts including their behaviour (swarming) and control. Sir B.P. Uvarov was the World famous locustologist while the renowned Indian locustologists include Y. Ramachandra Rao (1885-1972) and Hen Singh Pruthi (1897-1969).

Locustology branch of entomology that deals with the detailed study of locusts.

Lonchæidæ dipteran family of Lonchæoidea (Schizophora: Cyclorrhapha); arista present, postvertical bristles not divergent, vibrissæ present or lacking; male holoptic, female with an oviscapt. Maggots scavengers in dung or rotting vegetation (300 species)-*Lonchæa aristella* is pest of fig (*Ficus carica* Linnæus) in Mediterranean region.

Lonchæoidea superfamily of Schizophora (Cyclorrhapha: Diptera); characterised by antennæ with 2nd segment cleft; only on proclinate orbital bristle present; frons with dense setulæ. Comprises of two families-Lonchæidæ and Cryptochetidæ.

Lonchopteridæ (Musidoridæ) a dipteran family of pointed-winged flied (Lonchopteroidea: Aschiza); very small, slender bristly flies having 3rd antennal segment rounded or globular with a long arista; wings pointed at the apex and having nor cross-veins; ampodium lacing; found along the banks of shady streams-*Lonchoptera* spp. et al.

Lonchopteroidea superfamily of Aschiza (Cyclorrhapha: Diptera); characterised by pointed wings, vein Sc free and Rs with its two branches closely approximated at the wing–apex. Larvæ amphipneutic, flattened and having long anterior and posterior setæ–Lanchopteridæ.

Longevity length of life.

Lophopidæ Old World family of Fulgoroidea (Auchenorrhyncha: Homoptera); distinguished by hind trochanters directed backwards and moderately long basal segment in hind–tarsi (120 species) – *Pyrilla perpusilla* Walker, *P. aberrans* (Kirby) are most destructive pests of sugarcane (*Sccharum officinarum* Linnæus) and other crops; *Elasmoscelis platypoda* and *Lophops carinatus* et al.

Lora the mandibular plates that are parts of modified hypoharynx in homopteran bugs.

Loricifera Kristensen 1983 recently described phylum of Pseudocoelomate (Eumetazoa) comprising a single species *Nanaloricus mysticus,* tiny (0.25 mm long) marine animalcule that looks like a cross between a rotifer and kinorhynch.

Lorum submentum: a cuticular plate or sclerite of the head, having variety of forms and functions in different families of insects–honeybees (Apoidea) here V–shaped lorum supporting their elongated proboscis.

Louse (pl. **lice**) small apterous sucking insects (Pediculidæ: Siphunuculata) that feed on blood or primates including human beings.

Louse, Body *Pediculus humanus corporis* de Geer; about 4 mm long, usually found in body hairs of human beings. Egg to adult development takes 2 weeks and adult longevity is 4 to 5 weeks.

Louse, Crab *Pthirus pubis* Linnæus; about 2 mm long, found mainly in hair around sexual organs.

Louse, Dog *Linognathus setosus* (Olfers); 1.5-2.0 mm long infests domestic logs as also ferrets, foxes and rabbits.

Louse Goat *Linognathus stenopisis* (Burmereister); 2 to 2.5 mm long infests goats and sheep; whereas L. *pedails* (Osborn) is found on legs and feet of sheep.

Louse, Head *Pediculus humanus humanus* Linnæus; about 3 mm long, grayish in colour, found in hair or human head.

Louse, Hog *Hæmatopinus suis* (Linnæus); 3-6 mm long, cosmopolitan in distribution, infests pigs and swines.

Louse, Horse *Hæmatopinus asini* (Linnæus); about 3 mm long; found all over the World on asses, horses, mules and zebras.

Louse Mouse *Polyplax serrata* (Burmeister); about one mm long is confined to the Old World, infesting mice and rats; an allied species

P. spinulosa (Burmeister), known as rate louse is cosmopolitan in distribution and found on rats and cattle as well.

Lousy watchman nickname of Dor beetles, *Geotrupes stercorarius* and allied species (Geotrupidæ: Scarabæcoidea); large convex insects of corophilus habits, burrows several centimetres deep in the earth below a patch of dung and is often infested with mites.

Low volume concentrate (LVC) spraying technique for application of pesticides in concentrated form, undiluted with less than one litre to 3-4 litres per hectare. Also called waterless spraying; cheaper and quicker than conventional power spraying or aerial sprayings.

Lucanidæ (Leach 1815, 1817) small, cosmopolitan family of stag beetles (Scarabæoidea: Polyphaga), large elongate-oval, smooth, rough, straite or punctate, black and brown beetles; antennæ 11-segmented, usually elbowed (geniculate) with loose club, rarely straight; eyes large; labrum indistinct, mentum entire and completely covering the maxillæ; mandibles of males often enormously elongated (1000 species)-*Lucanus lunifer* Hope, *Hemisordoreus nepalensis, Figulus* spp., et al.

Lucifegous (insects) that shun light.

Luciferase oxidising enzyme which acting on protein luciferin causes luminosity in glow-worms (Lampyridæ: Cantharoidea).

Luciola colloquial name of Mediterranean firefly, *Lampyris nactiluca* (Linnæus) (Lampyridæ: Cantharoidea); which exhibits distinct sexual dimorphism-males winged with greately developed eyes, females (larviform) wingless and without well developed eyes.

Lunule crescent shaped structure or spot : crescent-shaped scar between the eyes and above the bases of antennæ in many insects.

Lurker nickame of a cicada *Tibicina hæmatodes* Scopoli (Cicadidæ: Cicadoidea); having black body and wing with brilliant red wings; wing span of 70 to 75 mm; commonly found in Southern Europe. Larva lives in soil for about four years, feeding on roots of various grasses etc.

Lycænidæ Leach 1815 family of small, brightly coloured delicate butterflies (Papilionoidea: Ditrysia); clubbed antennæ and eyes ringed with white scales; forelegs of males short tarsi reduced, one-segmented with one or both claws absent, those of female normal with both claws present; upper surface of wings usually metallic-blue and tail-like prolongation on hind-wings. Caterpillars short, flattened, chiton-shaped or slug-like; smooth or pollinose; some caterpillars secrete fluid, sought by ants; few strictly myrmecophagous, other predaceous on coccids but majority phytophagous-*Catochrysops* (*Euchrysops*) *cnejus* (Fabricius), *Chilades laius* (Cramer), *Deudorix epijabas* (Moore), *Freyeri putli* (Kollar), *Terucus theophrastus* (Febricius), *Virachola isocrates* Fabricius et al.

Catochrysops cnejus (Febricius)

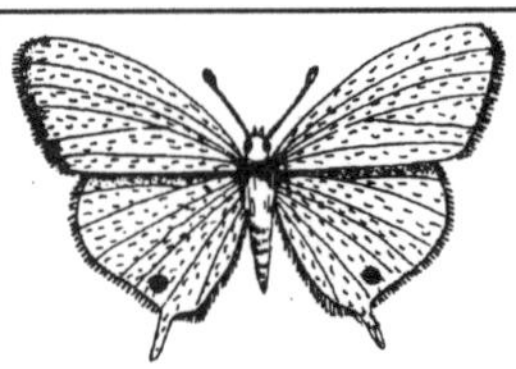

Virachola isocrates Fabricius

Terucus theophratus (Fabricius)

Lycidæ Lacordaire 1857 large family of Cantharoidea (Polypohaga: Coleoptera); brightly coloured diurnal beetles with antennal sockets closely approximated; facing anterio-dorsally; long trochanters, transversely joined to femora; luminous organs absent (3000 species) - *Lycus loripes, Duliticola* spp. et al.

Lyctidæ Le Conte 1862 cosmopolitan family of long-winged powderpost beetles (Bostrichoidea: Polyphaga); small wood-boring insects, body flattened dorsoventrally and head prominent but not covered by pronotum; antennæ short 11-segmented with 2-segmented club; eyes large; thorax well developed; legs slender, fore-coxæ tarsi 5-segmented; elytra and wings normal-*Lychus africana* Lanse, *L. malayanus* Lense and allied species.

Lygæidæ (Schiller 1829) (= Mycodochidæ) family of chinch bugs (Lygæiodea: Pentatomomorpha); head short and broad, deeply inserted in to pronotum; antennæ 4-segmented; eyes large, ocelli present; legs short, tarsi 3-segmented; hemelytra without distinct cuneus and membrane with only 4 or 5 longitudinal viens - *Chauliops nigrescens* Distant, *Nysius minor* Distant, *Oxycarrenus lætus* Kirby, *Spilostethus minor* Distant et al.

Lygæoidea superfamily of Pentatomomorpha (Geocorisæ: Heteroptera); antennæ 4-setmented; rostrum 4-segmented; ocelli present in alate forms, hemelytra with clavus longer than scutellum; pretarsal claws with pulvilli-Lygæidæ, Berytidæ, Colobathristidæ, Picsmatidæ, Thaumastocoridæ and Idiostolidæ.

Lymantriidæ Hampson 1892 (Lipari(di)dæ, Oencriidæ) family of Noctuoidea (Ditrysia: Lepidoptera); moderate-sized brilliantly coloured, nocturnal tussock moths, clothed with hair-like scales and highly developed pectinate antennæ in male; no ocelli, proboscis atrophied; hind-wing with Sc + R_1 separate eggs. Caterpillar covered with thick hair or thick tuffs of hair - *Beara dichromella* Walker, *Euproctis fraterna* (Moore), *Lymantaria concolor* Walker, *Perina nuda* (Fabricius), *Porthesia* (*Euproctis*) *scintillans* (Walker) et al.

Europctis fratema
(Moore)

Lymantaria concolor
Walker

Porthesia scintillans
(Walker)

Lymexylidæ small family of wood-boring beetles (Lymexloidea : Polyphaga); having elongated body with soft integument; antennæ short and serrate tarsi 5-jointed and without membranous lobes; anterior coxæ without trochantin, posterior coxæ prominent but not sulcate. Grubs cylindrical, bore horizontal galleries in dead and dying trees specially oak and palm trees - *Atractocerus reversus* Walker.

Lymexloidea Leng 1920 superfamily of Polyphaga (Coleoptera; Coleopteroidea); comprising of soft-bodied elongated wood-boring beetles, distributed in three small families - Lymexylidæ–, Micromalthidæ and Telegeusidæ.

Lyontetiidæ Stainton 1854 family of Tinæoidea (Ditrysia : Lepidoptera); brightly colored minute moths having very long antennæ and deeply fringed wings. Minute flattened caterpillars burrow into tissues between upper and lower epidermis of the leaf - *Lyonetia speculella* mines into apple leaves; *Asymplecta phorbiophora* is common in forest trees (*Myristica* spp., *Terminalia* spp. etc.), *Decadarchis inculta* mines leaves of *Gonystylus macrophyllus*.

Lysolecithin very toxic substance found in venom of bees and wasps; it breaks down the cells of its victim and sets free histanine.

M

Machærotidæ Kirkaldy small family of tube-forming corcopids (Cerocopoidea: Auchenorrhyncha); bugs having head narrower than pronotum; face strongly inflated, scutellum acuminate apically or with a strongly curved spine, forewings membranous apically with one or two claval veins. Nymphs inhabiting fixed calcareous tubes found mainly in the Old World (100 species) — *Machærota guttigera* Westwood, *Pectinariophyes* spp., el al.

Machilidæ Grassi 1888 family of Machioidea (Microcorphia: Thysanura); comprising of three-pronged bristletails, characterised by having large compound eyes, two ocelli; thoracic styli present on middle and hind coxæ and abdominal styli on II and IX segments — *Machilis maritima* Leach, *M. polyphoda Pætrobius maritimus* et al.

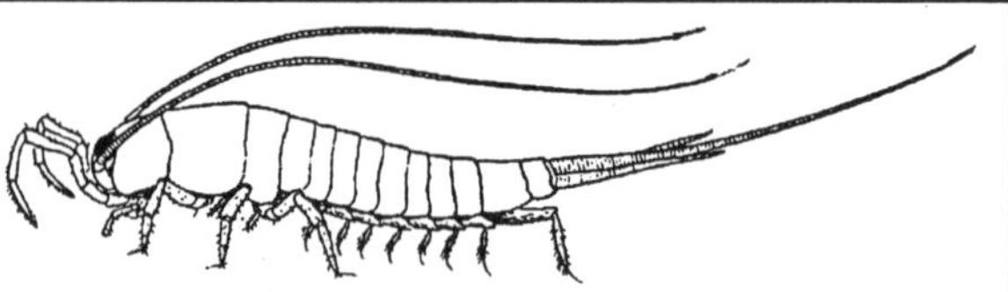

Machilis maritima Leach

Machiloidea superfamily of Microcoryphia (Thysanura: Apterygota); primitive insects having almost same characters as of the suborder, comprises of two families — Machilidæ and Meinertellidæ.

Macraner unusually large male ant (Formicidæ: Apocrita).

Macro - large: long.

Macrobian long-lived.

Macrolepidoptera small family of Libellulidea (Anisoptera: Odonata); large dragonflies found mainly in the Old World; earlier these species were placed under family Corullidæ of Libellulidæ.

Macrolepidotera one of the two main division of Lepidoptera in an old classification (other group being Microlepidoptera) founded mainly upon size criteria. No more in usage.

Macroparastie(s) parasite as are big enough to be seen with naked eye (without help of microscope) — lice etc.

Macropi(di)dæ family of Apoidea (Aculeata: Apocrita); small bees having labrum smaller than clypeus; posterior metatarus equal in width to posterior tibla; basal vein of forewing straight and feebly arcuate.

Macropterous large or long winged — ants, termites etc.

Macrotermitinæ once a superfamily of termites regarded as family Termitidæ, includes about 75 per cent of the isopteran species.

Macrotrichia long hairs or setæ consisting of extended cells surrounded by cuticle. These may be :—

Simple or branched — hairs covering most of the body.

Sensory setæ — those having nervous connection and resopond to smell, taste, touch etc.

Irritant setæ — these secrete fluid which is irritant — some caterpillar.

Pigmented scales — those found on wings of butterflies and bodies of mosquitoes.

Macrovellidæ very small family of semi-aquatic plant bugs (Gerroidea: Amphibicorisæ) living on vegetation and debris near edges of ponds and streams.

Maggot vermiform, acephalous, apodous, eruciform larva of true flies (Diptera). There are five distinct forms in parastics diptera :—

Caudate — having paired caudal appendages.

Microtype — small -sized without extensive cuticular armature and mouth-parts greatly reduced.

Mucoidiform — plain maggot without adoptive characters.

Planidium — more or less elongate form, often highly sclerotised and bearing heavy spines.

Vesiculate — vesicle is in form of large plate occupying main portion of ventral face of last segment.

Malacodermata group of soft-bodies beetle, in some old system of classification of Coleoptera. It included glow-worms (Lamexylidæ), click beetle (Elateridæ) and other species. No more is usage.

Malathion $C_{10}H_{19}O_6PS_2$; organophosphatic insecticide (Phosphorothiolothionate), brownish liquid with unpleasant odour; insoluble in water, slightly soluble in mineral oils and readily soluble in most organic solvents. It is contact and stomach poison (non-systemic) and one of the safest chemicals. Acute oral and dermal LD_{50}value for rat being 1400-1900 and 4000 mg/kg respectively.

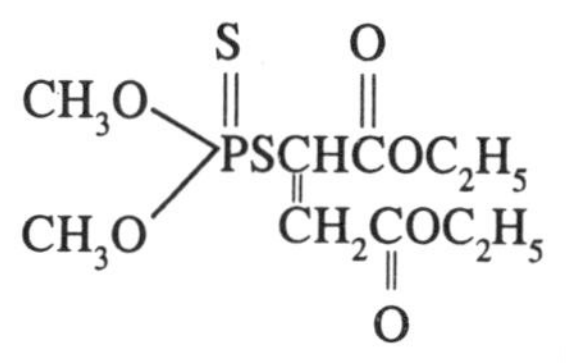

Malathion

Malaxation to make soft and pulpy by chewing with mandible as is done by certain mandibulate insect like ants, bees, wasps etc.

Male (♂) organism of masculine gender: sex (donor) that fertilises ovum of female. *Cf.* female (♀).

Male sterility the condition in which male gametes are absent, deficient in number or non-functional.

Mallophaga Nitzch 1818 order of bird lice or biting lice (Psocopteroidea: Pterygota); small, 0.5 to 10 mm, wingless ectoparastic insect; head flattened, eyes reduced, ocelli lacking, mouthparts biting and chewing prothorax evident and free; meso and metathorax imperfectly separated; thoracic spiracles ventral; cerci absent. Feed by biting off particle of feathers, haris and skin of some birds, less frequently on mammals. Metamorphosis slight. About 4000 species in three suborder — Amblycera, Ischocera and Rhynchopthirina.

Malpighian tubules main excretory organs of insect and most other arthropods; terrestrial narrow, glandular tubules in large number, opening into anterior end of proctoæum.

Mammalia class of phylum Chordata; higher vertebrates, that are warm-blooded and have hair on their skin; they nourish their young ones with secreted from special glands (mammæ) of the mother. It comprises of three distinct subclasses — Protheria (duck-billed platypus etc.) Metatheria (kangaroos, opossum etc.) and Eutheria (whale, bat, elephant, camal, cat, man etc.).

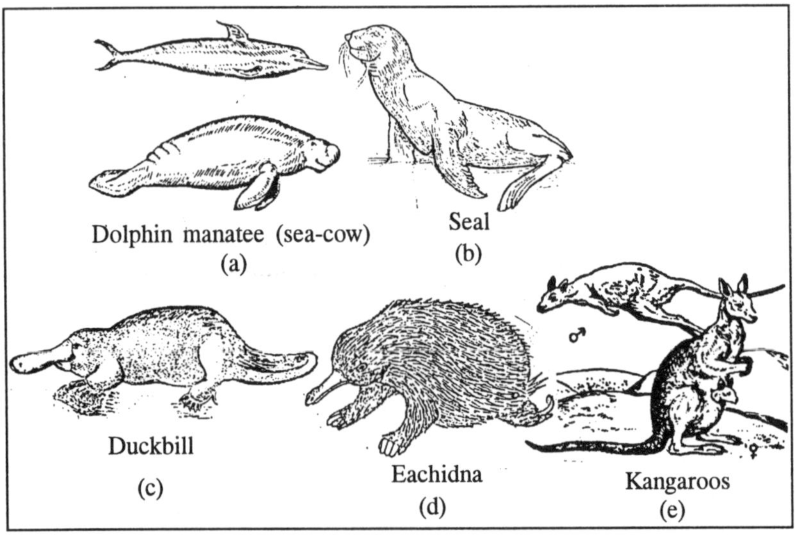

Flying-squirrel
(f)

Flyying-fox
(g)

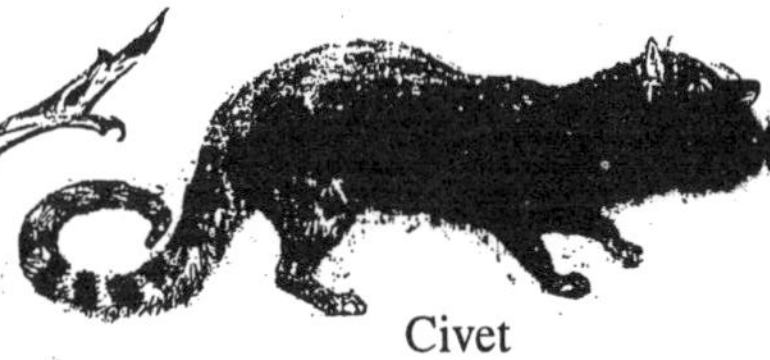

Civet
(h)

Pangolin
(i)

Opossum
(j)

Camel
(k)

Chimpanzee
(l)

gorila
(m)

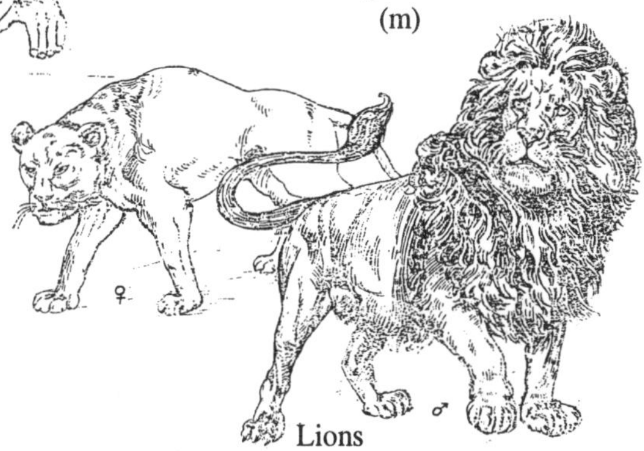

Lions
(n)

Mandible(s) pair of biting law-like organs, being appendage of first gnathal segment of head in all mandibular arthropods.

Mandibular glands pair of glands associated with mandible opening on each side of hypophyrynx near bases of mandible, present in certain insects only.

Mango blackfly (aleyrodid) *Aceurocanthus mangiferæ* Quaintance and Baker (Aleyrodidæ: Homopteran); a whitely with all four wings smoky black colour; specific and minor pest of mango (*Mangifera indica* Linnæus) confined to indian subcontinent.

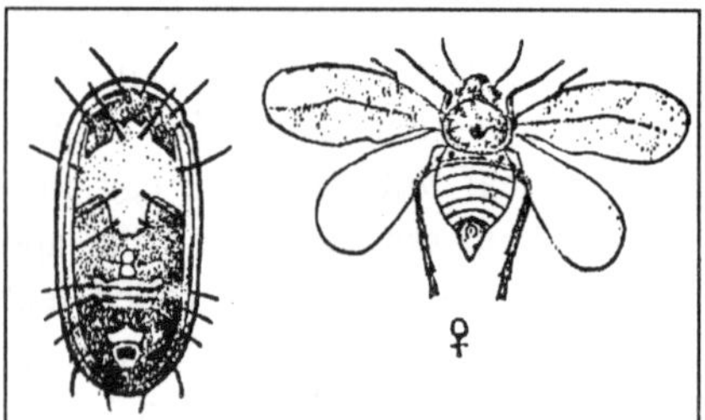

Aleurocanthus mangiferæ Quaintance & Baker

Mango buprestid *Belionota prasina* Thunberg (Buprestidæ: Coloeptera); flat-headed wood-borer; adult oblong elongated beetle, 20 to 30 mm long; head bronze coloured, elytra dark green with coppery-blue reflection and rounded at the shoulder. Feeding on inner side of the bark and boring inside the sapwood, cause minor damages to mango tree (*Mangifera indica*). Also known as mango jewel beetle.

Belionota prasina Thunberg

Mango hopper(s) *Idioscopus clypealis* (Lethierry), *Amritodus atkinsoni* (Lethierry) and allied species (Cicadellidæ: Homoptera); adult wedge-shaped, greenish-brown in colour; forewing thicker than hind-wing, bronzy subhyaline with a distinct white line along costal margin forming characteristic mid-longitudinal line when the insect is at rest. Both nymphs and adult suck cell sap and excrete sweety-sticky flquid—honeydew.

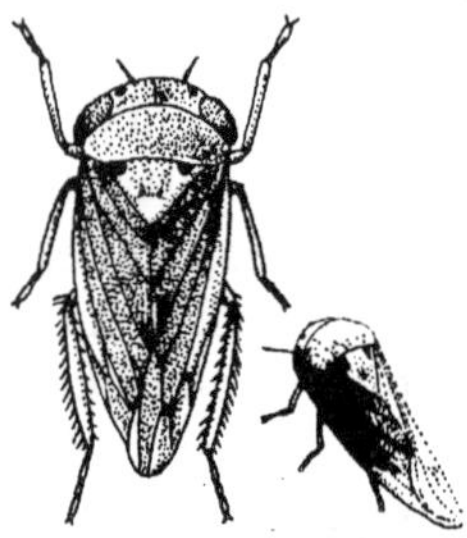

Idioscopus clypealis (Lethierry)

Mango leaf cutting weevli *Deporaus marginatus* (Pascal) (Curcullionidæ: Coleoptera); small 6 to 10 mm long, grayish-brown having long brown snout and shiny elytra. Grubs mine the leaves between two epidermal layer and feed within no mesophyll; pupate in the soil; adults eat away leaf lamina specific pest of mango trees confined to Indian subcontinent.

Mango leaf twisting weevil *Apoderus transquebaricus* (Fabricuis) (Cuculionidæ: Coloeptera); medium-sized reddish-brown weevil with long neck and snout. Soon after oviposition, each leaf with a single egg inside is rolled tip downward and the apodous grub feeds within the rolled leaf and pupates therein, adult come out by making hole in the rolled leaf.

Deporaus maginatus (Pascal)

Mango mealy scale *Chloropulvinaria polygonata* (Cockerell) (Coccidæ: Homoptera); a soft-scale, antennæ and legs in adult females highly degenerated, making the insect almost immovable. Both nymphs and adult female suck plant sap; a single insect can hardly cause any loss but their enormous number collectively cause substantial loss of vitality, affecting the growth and fruit setting capacity of the tree.

Apoderus transquebaricus (Fab.)

Mango mealybug(s) *Drosicha mangiferæ* (Green) and allied species (Margarodidæ: Heteroptera); adult female oval in shape; about 6 mm long, wingless and heavily coated with wax; males crimson-coloured have two dark brownish-black wings. Clusters of nymphs and adult females suck cell sap from young shoots and panicles; as a result leaves start curling and drying, entire shoot looks deformed, ultimately growth and fruit bearing capacity of tree is adversely affected.

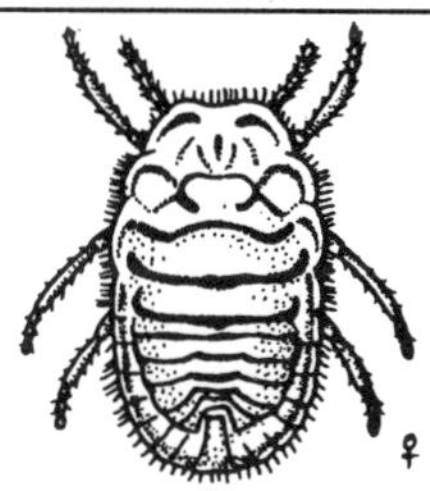

Drosicha mangiferæ (Green)

Mango (gall) psyllid *Apsylla cisellata* (Bluckton) (Psyllidæ: Sternorrhyncha); adult small, 3 to 4 mm long insect with head and thorax black and abdomen light brown. Nymphs feed on vegetative and reproductive buds causing hard conical green galls, from which the adult emerge; the buds obiviously do not blossom or yield and fruit.

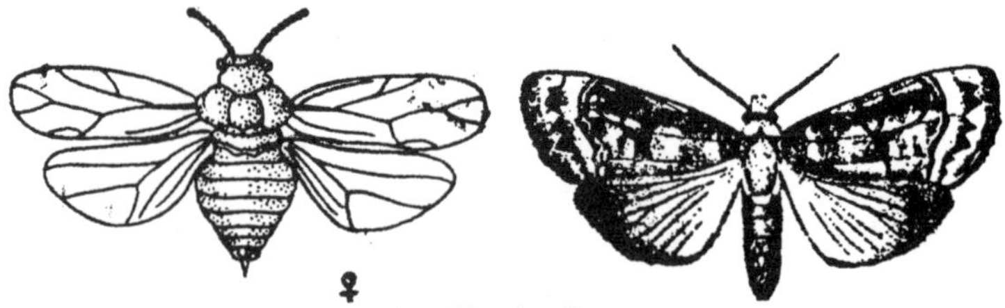

Apsylla cistellata
(Buckton)

Mango shoot borer(s) *Chlumetia tranversa* (Walker) and *C. alternans* Moore (Noctuidæ: Lepidoptera); medium-sized moths having thorax and abdomen clothed with rufous fuscous and gray scale; forewing dark gray, beautifully patterned with wavy design, hindwings fuscous, apical side being darker than proximal side; wing expanse 15 to 20 mm. Caterpillar bore into top shoots, tunnelling downward and throwing their excreta out of the entrance hole; when full grown, these come out and pupate.

Mango stem borer(s) *Batocera rufomaculata* (de geer) and allied species (Cerambycidæ: Coleoptera); stout dark brown, 50 to 55 mm long, longicorn beetle with yellowish-green pubescence; elytra irrorated with small light oranage spot. Grubs make zigzag burrow beneath the bark and tunnel into the trunk and main stems, moving upward and feeding in integral tissue till they reach sapwood, attacked stem die and wither away.

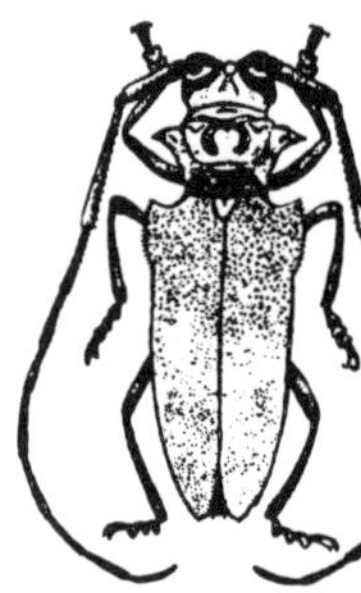

Batocera rufomaculata
(de Geer)

Mantid large carnivorous insect (Mantidæ: Mantodea) having long (60 to 80 mm) and narrow (<8 mm) body; broad head which moves freely upon thorax; prothorax long; forelegs reporial provided with spines for grasping the pray; wings well developed, hind wing larger than

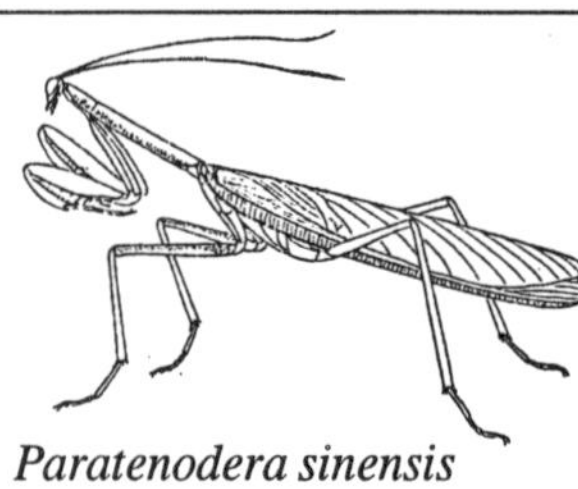

Paratenodera sinensis
(Saussure)

forewings and remain folded in plaits inder forewing on back of the abdomen — *Paratenodera sinensis* (Saussure) et al.

Manitdæ Sausaure 1869 family of cosmopolitan, carnivorous insects (Mantodea: Dictyoptera); head extremely mobile, connected with thorax by slender neck; eyes very large, three dilate ocelli present, prothorax very long, mid-and hind-legs weakly developed, pair of short, joined cerci and the males have pair of anal tufts - mantis [*Mantis religiosa* Linnæus, *Paratenodera sinensis* (Saussure) and *Stagmomantis carolina* (Linnæus)are respectively European, Chinese and North American species].

Mantispidæ Westwood 1840 a family of Mantispoidea (Planipennia: Neuroptera); that includes false mantids (insect greatly resembling mantid); head wider than long with short antennæ, large hemishperical eyes and no ocelli; prothorax greatly elongate and large raptorial foreleg inserted at the frcnt; wings longs, narrow and similar. About 200 species, cosmopolitan in distribution — *Eulimacia flavicincta* Stitz (wing-span 55 to 60 mm), *M. brunnea* Say (22 mm) while *Mantispilla pygmæ*a (8 to 12 mm) is the smallest.

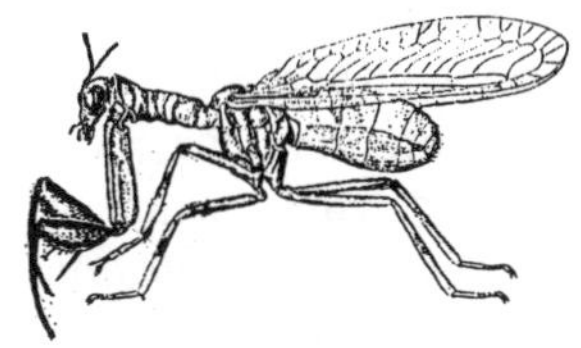

Mantispa brunnea Say

Mantispoidea superfamily of Planipennia (Neuroptreoidea); having elongated prothorax, memberanous wings with nygmata absent but trichosors usually present, less than 5 cross-veins between R_1 and R_S — Dilaridæ, Berothidæ, Polystoechotidæ, Slsyridæ and Mantispidæ.

Mantodea Handlirsch 1903 an equivalent name for insect order Mantodea Burmeister; which is now regarded as a suborder of Dictyoptera and icludes all mandits.

Margarodidæ (Newstead, 1901) Morrison 1927 small family of giant coccids (Coccoidea: Sterorrhyncha); abdominal spiracles present in 11 stages of development antennæ 10-segmented, simple of pectinate; males lack compound eyes, female large and rotund with distinctly segemented body, often hidden away in waxy secretion — *Drosicha*

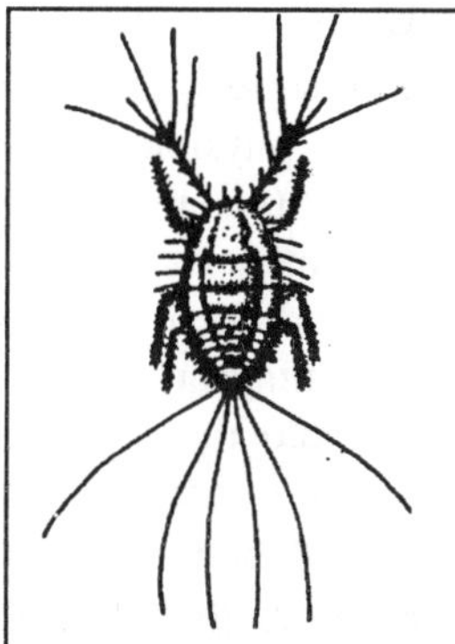

Icerya purchasi Maskell

dalbergiæ (Green), *Hemiaspidoprotus ciner*æ (Green), *Icerya ægyptiaca* (Douglas), *I. purchase* Maskell, *I. purchasi* Maskell *Perissopncumon tamarindus* (Green) et al.

Massidæ Latreille 1802 Leach 1817 family of Vespoidea (Aculeata: Apocrita); not so common wasps, popularly known as Masarid wasps or shinning wasps; found in North America, South America, Africa and Australia; distinguished by thier knobbed antennæ. They build their nests in the stem or mud cells, provision the same with mixture of nectar and pollens except *Pseudomasaris* species which are insectivorous and store caterpillar in their mud cells — *Pseudomasaris vespoidea* (Cresson), *Ceramius* spp., *Masaris* spp., et al.

Mastoteritidæ Silvestri 1909 very small family of termites (Isoptera: Biattopteroidea); distinguished by having 5-segmented tarsi and hind-wings with well developed anal lobe; confined to North Australia — *Mastotermes darwiniensis* Froggatt.

MATHUR Dr. Ram Nath — born July 13, 1903; M.Sc (zoology) Punjab University in 1925; started his career as Asisstant Imperial Entomology at Pusa (Bihar) (1925-27), shifted to Forest Research Institute and Collage, Dehradun from where he retired in 1961 as Forest Entomologist and Head, Division of Forest Protection. He will be remembered for his monograph **Psyllidæ of Indian subcontinent** (1975).

Mating meeting or coming together of individual of opposite sex for sexual intercourse.

Mating flight some insect do not male is captivity or confined space; these days always mate while flying, such flights are called nuptial flight as is seen in ants, bees etc.

Mating strain group of organism of a species that do not interbreed with each other but are able to breed with member of other group of same species that are morphologically identical but genetically and physiologically different.

Maxilla (pl. **maxillæ**) second pair of appendages (jaws) of gnatha segment of head implanted on pleural region of head just behind mandible; composed of number of distinct or partly fused segment forming organs for mastication and for holding food and carrying it into the mouth.

Maxillary glands small inconspicuous glands present in certain insect, opening mesally at bases of maxillæ.

Maxillary palp(i) joined sensory appendages or feeler attached to the maxilla of an insect. The various segment (upto seven) may bear different types of sensory receptors, in some insect of palpi, may be enough to form most obvious part of maxillæ, these help in examining, holding and conveying food into the moth.

Maxillulæ first pair of maxillæ in Crustacea (not in insects).

Maximum dosage largest amount of chemical (pesticide) that can safely be used without producing excessive residues or ill-effects.

Maximum residue limit maximum concentration (ppm or mg/kg) of a residue that legally permitted in or on food at harvest or at a specified post-harvest processing stage upto the point of consumption.

Mayfly ephemerid: not a true fly but a delicate slender-bodied insect (Ephemeroptera: Palæoptera) having eyes conspicuous, mouthparts vestigial rest, wings are held vertically above the body, two (often three) long, many jointed cerci at the end of abdomen. Nymph (nalad) aquatic and long-lived, while adult lives only for a few hours, being unable to feed. Many fossil have been found, which suggest that three were more abdundant in the earlier days than they are now.

Meadow brown large nymphalid butterfly, *Maniola jurtina* (Satyridæ: Papilioidea); that is reluctant to take to wings; found in Eurasia and North America; it has dark brown wings, forewings partly golden-brown, each with one white-centered black spot. Caterpillar bright green, found feeding on grasses; overwinter and pupate early in Summer. Green pupæ, marked with dark brown wing-cases hang grass stems or leaves.

Mealybug(s) group of cosmopolitan polyphagous scale insect (Pseudococcideæ: Coccoidea); female elongate-oval in shape having distinctly segmented body covered with waxy secretion or mealy coating; legs and antennæ well developed. Males winged having compound eyes. Active throughout Winter, more so on bright and sunny days — brinjal mealybug [*Centrococcus insolitus* [Green], citrus mealybug [*Pseudococcus citri* (Risso)], mango mealybug [*Realtrococcus iceyoides* (Green)], rice mealbug [*Heterococcus rehi* (Lindinger)], sugarcane mealybug [*Phenacoccus saccharifollii* (Green)] et al.

Mealybug giant margarodid, *Drosicha* spp. (Margarodidæ: coccoidea); most destructive pest of mango tree, found feeding on over 60 other host plants including fruits and forest tree — *D.mangiferæ* (Green), *D. contrahens* (Walker), *D.dalberiæ* (Green), *D. stebbingi* (Green).

Mealy White-tailed *Ferrisa virgata* (Cockwell) (Pseudococcidæ: Coccoidea); widely distributed in Tropic and subtropics on a large number of host plants. Female apterous with two long prominent waxy filament at posterior end and number of waxy hair over the body covered with waxy powder.

Mealywings conipterygids: small predatory lacewing flies *Coniopteryx* and allied species (Planipennia: Neuroptera); body covered with powdery wax-like secretion; long filiform antennæ and wings fragile covered with waxy powder — *C.pusanus* Withcombe and *Nimboa basipunctata* Withcombe are predaceous on eggs of *Pyrilla* spp. (major pest of sugarcane).

Macaptera Packard 1886 old spelling of the order Mecoptera (Scorpion flies), no more in usage.

Mechano-receptors sensory receptors as a butterfly emerge from its chrysallis; comprises etc., found anywhere on insect body specially on antennæ, base of halteres and cerci.

Meconium dark brown fluid exerted as a butterfly emerge from its chrysalis; comprises of waste product of pupal metabolism. First fæace of a newly emerged insect imago or a new-born child.

Mecoptera (Packard 1886) Comstock and Comstock 1895 small order of slender-bodied scorpionflies (Mecopteroidea: Oligoneoptera); linking Hymenoptera with Trichoptera. Head prolonged and bent downward forming rostrum and bearing chewing type of mouthparts; mostly carnivorous; antennæ filiform, legs long and slender; two pairs of long, narrow, subequal membranous wings with primitive venation; abdomen elongated with terminal segment raised like scorpion; cerci short; male genitalia prominent. Metamorphosis complete. Larva eruciform with biting mouthparts and three pairs of thoracic legs. Pupa exarate and decticous (600 species). Two suborder — Protomecoptera and Eumecoptera.

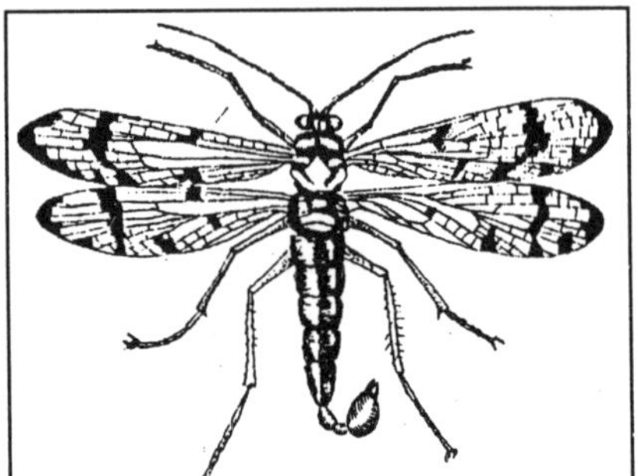

Scorpionfly

Mecopteroidea (Petanoptera) Brauer 1885 superorder of Oligoneoptera (Pterygota: Insecta); comprises of four endoptertgote order — Mecoptera, Trichoptera, Lepidoptera and Diptera.

Media (M) fourth of insect wing that usually into two main branches — media enterior (MA) and media posterior (MP). These further subdivide into two (MA_1, MA_2) and four (MP_1, MP_2, MP_3, MP_4) terminal branches.

Meenoplidæ a very small of Fulgoroidea (Auchenorrhyncha: Homoptera), includes about 50 species, associated with Cixidæ (Fulgoroidea); confined to Eastern hemisphere.

Mega macro- or large.

Megabacterium very large-sized bacterium.

Megachilidæ (Latreille 1802) Kirby 1839 cosmopolitan family of leaf-cutting bees (Apoidea: Apocrita); stout-bodied, densely hairy, solitary bees, nest in cavities of wood or cut stems; these cavities are lined with neatly cut out of pieces of leaves; a few are parasitic — *Megachile anthracina* Smith, *M. disjuncta* Fabricius, et al.

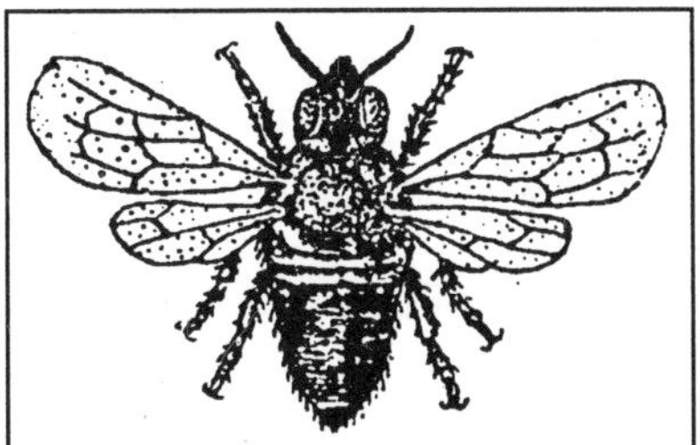

Magachile antracina Smith

Megalodontidæ family of web-spinning sawflies (Megalodntoidea: Symphyta); stout bodied, having flabellate antennæ; mouthparts as long as head capsule; abdomen broadly joined to thorax; ovipositor short. Larva live gregariously in the webs spun on herbaceous plants.

Megalodontoidea superfamily of Symphyta (Hymenoptera: Oligonoptera); having antennæ with 3rd segment not very long, pronotum with straight hind margin; fore-tibla having two apical spurs; forewing with cross-vein 2r present; abdomen flattened and ovipositor not exserted. Two families — Pamphilidæ and Megalodontidæ.

Megaloptera Latreille 1802 suborder of Neuroptera (Neuropteroidea: Oligoneoptera); having elongate prothorax or branches of wing-vein without any tendency to bifurcate at wing-margin and ovipositor relatively long, Larvæ with biting mouthpart. Two superflies — Sialoidea (alderflies) and Raphidioidea (snakeflies).

Megalopygidæ Berg 1882 (Lagoidæ) family of Zygænoidea (Ditrysia: Lepidoptera); distinguished by having vein Sc + R_1 coincident with cell to middle or beyond it — *Lagoa crispata* (Packard) a wooly yellowish moth marked with brown broken bands on forewings.

Megalothripinæ subfamily of Phlæothripidæ (Tubulifera: Thysanopetra) also regarded as family Megalothripidæ Karny 1913 (Phiæthripoidea:

Tubulifera); having 3rd antennal segment broadly attached to 2nd segment; maxillary stylet thickened, band-like and as broad and or broader than labial palpi; include bulk of mycophagous species — *Elaphrothrips productus* Priesner, *Kelothrips gigans* Schmutz, *Tiarothrips subramanii* (Ramakrishna) et al.

Megalyridæ very small of Icheumonoidea (Apocrita: Hymenoptera); having sternites of gaster evenly sclerotised, mesoscutum with a central furrow but no notauli; forewing with costal cell broad; ovipositor long — *Megalyra* spp. found in Australia are parasitic on wood-boring beetle.

Megalyridæ a family of fossil insect belonging to order Proctelytroptera of Permian period (Paleozoic era), the entire order is extinct by now.

Megamerinidæ small dipteran family of Sciomyzoidea (Schizophora: Cyclorrhapha); small, slender flies distinguished by postocellar bristle convergent of absent; hind femora thickened with minute spines and abdomen clavate — three genera and a few Palæarcitic and Oriental species.

Meganeura fossil dragonfly of Carboniferous and Permian periods; largest known insect with wing expanse of over 700 mm.

Meganeuridæ odonate family of Caboniferous period that included gigantic dragonflies — *Meganeura* spp. et al.

Meganisoptera suborder of Odonata in old classification, formerly included in order Protodonta and includes *Erasipteran larischi* of Carboniferous period.

Magathripdæ Tillyard 1893 family of Coenagrioidea (Zygoptera: Odonata); group of rare species, moslty Neotropical and Oriental having affinities with Lestidæ (Lestinoidea: Zygoptera).

Megasclere large sclerite.

Megascoptera Bronghiart 1893 Palæozoic fossil order of Palæodictyoptera (Palæcoptera), small, pterygote insects having biting and chewing mouthparts; wings homonomous, narrowed at the base; tarsi pentamerous; carnivorous (all extinct).

Megathripidæ Karny 1913 family of large thrips (Phoseothripoidea: Tubulifera) having head not produced anteriorly in front of eyes and vertex not sharply conical; characterised by male having stout tubular projection of each side of abdominal segment VI and segment VIII without peg-like projection.

Megathymiinæ Tropical subfamily of Hesperiidæ (Heperioidea: Ditrysia); giant skipper having apex of antennæ devoid of recurved point and male with wing-veins greatly strengthened and peculiarly specicalised - *Ochlodes venata* et al.

Meinertellidæ family of Machiloidea (Microcoryphia: Thysanura); anntenæ and legs without scales, wings wanting; sterna small, never more than one pair of abdominal vesides on abdominal vesides on abdominal segments; found mainly in Southern Hemisphere — *Allomachis* spp., *Nesomachmilis* spp., et al.

Melandryidæ (= Serropalpidæ) family of false darkling beetle (Heteromera: Cucujoidea); head not sharply constricted at neck; pronotum with well-marked side border, tibial spurs simple or serrate (600 species) — *Orchesia* spp., *Osphya* spp. found in woodlands of temperate regions.

Melanin insoluble black pigment found in the cuticle of many insect, formed by oxidation of amino-acid tyrosine in the presence of specific melanogenic enzymes.

Melanothripidæ Bagnali 1926 family of Æolothripoidea (Terebrantia: Thysanoptera); often placed under family Æolothripidæ. Black thrips having all antennal segments usually freely movable; maxillary palpi 3-segmented and labial palpi 2-segmented — *Melanothrips* spp. are pests of cruciferous crops.

Melectidæ Schmiedeknecht 1882 family of parasitic bees (Apoidea: Apocrita); having posterior tibiæ of female without pollen brush; anterior wing sculpture highly differentiated with broad band of minute papillæ along apical margin and nervellus of hind-wing elongate-oblique, often sinuate or curved — *Melecta punctata* preys on *Anthophora* spp.

Meliponidæ (Ashmead 1900), Schwarz 1936 often regarded as subfamily of Apidæ; family of mosquito-bees or stingless bees (Apoidea: Apocrita); very small (<3 mm) bees having vestigial stings; nest in hollows in trees, rocks or walls (2500 species) — *Trigona lutea, T. pallida* (Latreille), *T.carbonaria* Smith, et al.

Melittidæ small and rather diverse family of rare insect (Apoidea: Apocrita); small dark coloured bees that nest in the ground and have labial palpi cylindrical with all segment similar; glossæ acute often elongate — *Macropis* spp., *Melitta* spp., (not to be confused with *Melit*æa spp. and *Melittica* spp. which are lepidopterous) et al.

Melittin protein-like poisonous substance present in venom of bees.

Meloidæ Thomson 1859 family of blister beetle or oil-beetle (Heteromera: Cucujoidea); soft-bodied, with head strongly deflexed (hypognathous), neck distinct but narrow; tarsal claws appendiculate and usually serrate; wings mostly developed loosely covering the body; some with wings vestigial or even apterous — *Mylabris pustulata* (Thunberg), *M. triffensis* Billberg, *Cylindrothorax tenuicollis* Pallas et al.

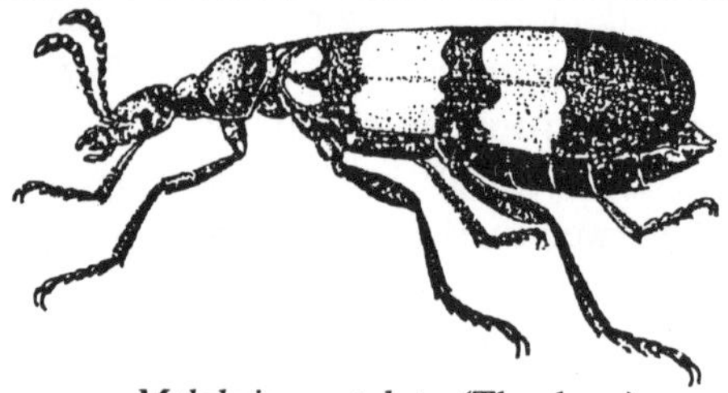

Mylabris pustulata (Thunberg)

Melolonthidæ Leach 1817 Macleay 1819 family of cockchafer, commonly known as May beetle or June beetle (Scarabæoidea: Polyphaga); mouthparts strong, labrum and mandible hidden beneath clypeus, mentum large and quadrate; abdominal spiracles in two rows, each forming a single straight line. Grubs typically scarabæoid, mainly subterranean, feeding on roots; adults nocturnal and leaf-feeder (9000 species) *Holotrichia longipennis* (Blanchard), *Melolontha melolontha* (linnæus), *M. indica* (Hope), *Leucopholis conceophora* Burnmeister, et al.

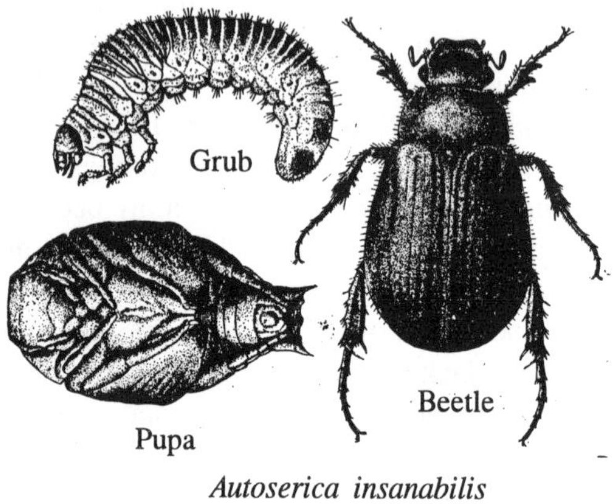

Autoserica insanabilis Brenske

Membracidæ (Germar 1821) family of cow-bugs or treehoppers (Cicadelloidea: Homoptera); head modified with crown vertical and face horizontal; pronotum extending backward over abdomen and queerly ornamented with spines and hooks; wings membranous; tibiæ angular, posterior coxæ transverse (2500 species) — *Anchon Pilosoma* Walker, *Ceresa bubalus* (Fabricius), *Leptocentrus obliquis* Walker, *Ceresa bubalus* (Fabricius), *Leptocentrus obliquis* Walker, *Trichentrus bicolor* Distant, *Venduzea arduata* et al.

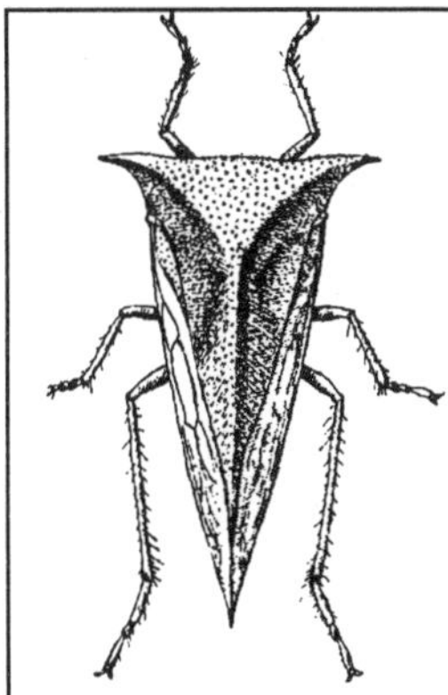

Ceresa bubalus (Fabricius)

Mengenillidæ (Mengeidæ Pierce 1908) a family of Mengenilloidea (Strepistera: Coleopteroidea); most primitive stylops readily distinguished from other by male having 5-segmented tarsi with paired claws, female possessing functional antennæ (6- or 7-segmented), eyes and legs — species of *Eoxenos*, Peyerimhoff, Mengenilla, Hofeneder, *Triozocera* Pierce et al.

Mengenilloidea super family of Strepsiptera (Coleopteroidea: Pterygota); that includes free living females having a single posterior genital opening and the males with a strong vein MA_1.

Mengeoidea a superfamily of Strepsiptera (Coleopteroidea: Pterygota); comprising of a single fossil male in Baltic Amber; that also has five tarsal segment but differs from Mengenilloidea in being without vein MA_1.

MENON, Dr. Mannoore Gopala Ramdas — born January 1, 1913; Ph.D (Zoology) Bombay University, 1940; jointed IARI, New Delhi as Assistant to Imperial Entomologist. Deputed to study Pscoptera collection at U.S National Museum, Washington and British Museum (Natural History), London, 1947-49; back to IARI form where he retired as Senior Systematic Entomologist on December 31,1972. Emeritus professor , ICAR till 1978 and consultant to Kerala Agricultural University, Trichur, till death (February 28, 1990). He was instrumental in developing National Pusa Collection as also Insect Identification Service and laid a solid foundation for study of insect taxonomy, at IARI., New Delhi.

Menoponidæ Mjobeg 1910 family of Amblycera (Mallophaga: Pscopteroidea); biting or bird lice having head broadly triangular, antennæ 4- or 5-segmented and maxillary palpi 4-segmented separate; mid- and hind-legs each with two tarsal claws — *Actornithophilus* spp. and *Austromenopon* spp. on Charadriiformes,

Menacanthus spp. and *Myrsida* spp. on Galliformes and Passeriformes and *Menopon* spp. on Galliformes — common fowl louse (*Menopan gallinæ* Linnæus) (= *palidum* Nitzsch), large poultry louse (*M. stramineum* Nitzsch), glant bird louse [*Ancistrona vagelli* (Fabricius)] et al.

Menotaxis movement in a direction that makes a constant angle with the source of light, as in case of ants etc., it maintains a constant visual pattern.

Mentum distal plate of postlabium between prementum and submentum.

Mermithaner(s) aners or male ant (Formicoidea: Apocrita) that are parasitised by nematodes (Nematoda: Aschelminthes).

Meroblastic division type of egg cleavage in which only nucleus and nuclear cytoplasm are divided.

Meron lateral postarticular area of base of coxa, proximal to basicostal suture.

Meropodite fourth segment or generalised leg of Arthopoda; femur in insects.

Merothripidæ only family of Merothripoidea (Telebrantia: Thysanoptera: Thysanopteroidea); having moniliform antennæ 8- or 9-segmented, style lacking, 3rd and 4th antennal segment with tympanum-like sensory area without sense cone; fore- and hind-femora greatly enlarged; abdomen blunt, ovipositor very weak or vestigial — Merothripidæ.

Mesadenia the accessory glands present in certain insect; these arise from mesodermal lining of the seminal vesicles.

Mesentron mid-gut or mid-intestine, endodermal stomach, middle portion of alimentary canal.

Mesocerebrum deutocerebrum: middle portion of insect's brain, including antennal lobes.

Mesoderm germ layer between ectoderm and endoderm in triploblastic animals: mesoblastic tissue.

Mesonotum dorsal cuticular sclerite or notum covering the middle segment of insect's thorax.

Mesoscutellum scutellum of shield-like plate covering the hind part of insect's thorax.

Mesoscutum middle and the largest of three sclerite that forms the cuticular covering of the middle segment of insect's thorax.

Mesosternum sternum or ventral sclerite under middle segment or mesothorax of insect's thorax.

Mesothorax middle or second of thorax bearing a pair of (mid) legs and in most pterygote (winged) insects, the pair of (fore-) wings.

Mesoveliidæ small family of Gerroidea (Amphibicorisæ: Heteroptera); long-legged, carnivorous bugs, having antennæ 4-segmented, rostrum 3-segmented, claws terminal mass invertebrates; having outer layer of ciliated cells enclosing an internal mass of either sperm or egg cell (50 species) — *Mesovelia* spp., live among the floating vegetation feeding on small aquatic organism. *Phrynovelia pupua* is terrestrial (exception), living among fallen leave in forest.

Mesozoa minor phylum of Acoelomates (Eumetazoa) comprising of endoparasites of marine invertebrates; having outer layer of ciliated cells enclosing an internal mass of either sperm or egg cell (50 species); two classes — Rhombozoa inhabiting excretory organs of cuttlefish, octopods, squids etc. while Orthonectida are found in bivalve molluscs, brittle-stars, flatworm etc.

Metabola metamorphosis or embryological development during growth and development — heterometabola (hemimetabola and paurometabola) and holometabola (as also hypermetabola). *cf.* ametabola.

Metabolism sum total of chemical reaction, both anabolic and catabolic which occur in living cells and organisms.

Metacerebrum tritocerebrum: hind portion of insect's brain, comprising of fused ganglia of embryonic third somite.

Metamere embryonic somite; primary body segment of embryo.

Metamorphosis marked structural change of abrupt transition of one stage of life-cycle into another, that some living-beings undergo during course of their growth and development as in case of frogs and metabolous insect — in insect it may be direct or incomplete (egg, nymph and adult), indirect or complete (egg, larva, pups and adult) or complex (larval stage undergo different forms).

Metamorphosis, Incomplete unique phenomenon wherein only a few individuals that hatch from eggs as nymphs develop into adults — termites. The worker termites remain as perpetual adolescents, having lost the capacity to reach adulthood.

Metanotum notum or dorsal cuticular covering of insect's metathorax.

Metapleuron lateral cuticular plate of insect's metathorax.

Metapneustic (insects) having spiracles only on posterior end of abdomen mosquito larva etc.

Metarbelidæ (= Arbelidæ Moore 1829) family of bark borers (Ditrysia: Lepidoptera) nocturnal moths having antennæ bipectinated; palpi minutes and porrect; no proboscis and no frenulum. Caterpillar feed on bark, then bore into tree trunk and main branches covering entry holes with silken webs — *Indarbela dea* (Swinhoe), *I. theivora* (Hampson) and allied species.

Metascutellum scutellum or dorsal cuticular shield that covers the hind-part of insect's metathorax.

Mestacutum largest part of dorsal cuticular covering with metathorax.

Metasternum sternum or ventral sclerite below insect's metathorax.

Metasternum ventral portion of metathorax in insect.

Metatarsus first segment of tarsus of an insect's leg; more correctly called first tarsomere or first tarsal segment.

Metathetely delay in appearance of adult character due to excess of juvenile hormone, may result in appearance of an insect that is intermediate in form between nymph or pupa and adult. *Cf.* Prothetely.

Metathorax posterior or third segment of thorax, bearing a pair of (hind) legs and second pair of (hind) wings in most pterygote (winged) insects.

Metecdysis period after a moult, in insect or other arthropods, when new cuticle is still soft.

Methamidophos $C_2H_8NO_2PS$; organphosphatic insecticide (phosphoramidothioate) slightly yellowish coloured crystals, readily soluble in water, alcohol ketone etc., but less soluble in ether and insoluble in petroleum ether. It is contact and stomach poison with good systemic action. Acute oral and dermal LD_{50} values for rat being 30 and 110 mg/kg respectively — Tamaron® etc.

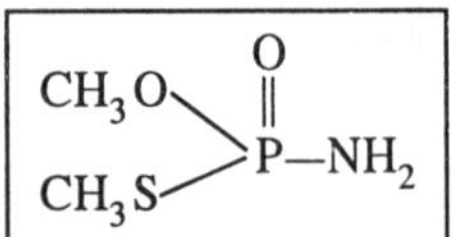

Methamidophos

Methidathion $C_6H_{11}N_2O_4PS_3$; organophosphatic insecticide having contact action with good penetration; less toxic to bees and fishes. Insoluble in water but readily soluble in acetone and benzene and methyl-alcohol. Acute oral and dermal LD_{50} values for rat are 20 to 48 and upto 400 mg/kg respectively— Supracide® etc.

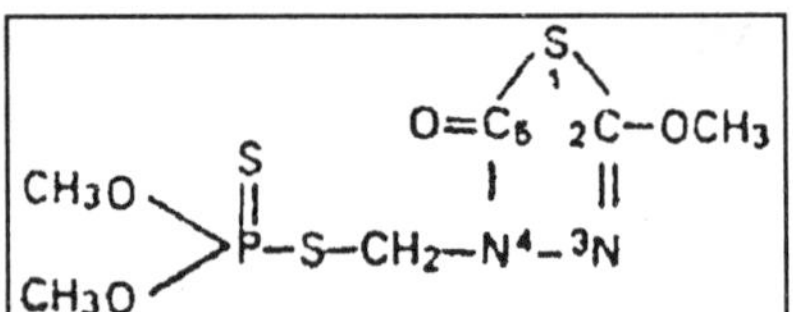

Methidathion

Mevinphos oranophosphatic systemic insecticide, contact poison having strong fumigant action; pale yellow liquid, miscible in water, acetone and xylene, rapidly decomposed

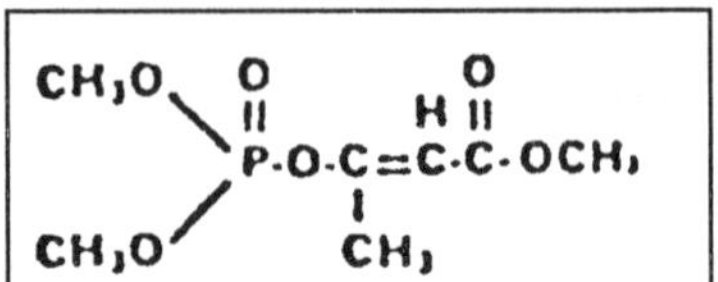

Mevinphos

by alkalies, hence incompatible with alkaline pesticides; toxic to bees, fish, poultry and livestock. Acute oral and dermal LD_{50} value for rat being 3 to 5 and 90 mg/kg respectively — Phosdrin®.

Micr(o)- very small: one millionth (10^{-6}). *Cf.* macro- (mega-).

Micraner very small (dwarf) male ant (Formicidæ: Apocrita).

Micrergate unusually dwarf worker ant (Formicidæ: Apocrita).

Micro-aerophile(s) micro-organism that can grow in extremely low concentration of oxygen; these often spoil food-stuffs stored in air-tight container when only traces of oxygen are left in the container.

Microanatomy histology.

Microbe (micro-organism) very minute organism, visible only under microscope — bacterium, mould, virus etc.

Microbial (microbic) concerning or pertaining to microbes.

Microbial pesticide insecticide composed of pathogenic microbes or their toxic products. It included bacterial insecticides, viral insecticides and entomogenous fungi etc.

Microbicide toxin that can kill microbes.

Microbiologist person that has sound knowledge of microbiology.

Microbiology scientific study of micro-organisms—structure, function, physiology etc. of micro-organisms.

Microbion obsolete term for microbe or micro-organism.

Microcoryphia suborder of Thysanura (Zygoentomata: Apterygota); includes bristletails, having compressed body, well developed compound eyes, median are paired lateral ocelli; mandible with only one articulation, eversible vesicles on II and VII abdominal sterna and no gonangulum; only one superfamily — Machiloidea.

Microgaster alternative name for small braconid fly (*Apanteles* species) whose larvæ are parasitic on caterpillar of cabbage butterflies etc.

Microgasteridæ a small family of Chalcidoidea, often included in the family Parliampidæ (Chalcidoidea: Parastica); the small wasps are parasitic on the caterpillar of Lepidoptera — *Microgaster connexus* parasitises the tussock moth caterpillar *Euprotis similis* and allied species (Lymantriidæ: Noctuoidea) whereas *M. indicus* is a parasite of teak skeletoniser *Hapalia machæralis* (Pyraustidæ: Pyralidoidea).

Microlepidoptera one of the two groups (suborders) of Lepidoptera in old classification, based only upon size criterion. These small-sized moths are now placed under suborder Zeugloptera,

Dacnonypha, Monotrysia and superfamilies Tortricoidea, Pyrallidoidea and Tineoidea of Ditrysia.

Micromalthidæ small family of micromalthid beetles (Cupesoidea: Archostermata), none of the species have been recorded from India so far.

Microparasite parasitic micro-organism.

Micropezide family of micropezoidea (Schizophora: Diptera) having arista dorsal; tibia with preapical bristles; legs long and stilt-like; costa unbroken and Sc complete.

Micropezoidea superfamily of Schizophora (Cyclorrphapha: Diptera), characterised by second antennal segment not cleft, two superior orbital bristles present and inferior orbital bristles lacking; spiracles 2 to 5 in the membrane; cerci not linked to lower telomere — Micropezidæ, Neriidæ, Cypselosomatidæ and Pseudopomyzidæ.

Microphysidæ very small family of Cimicoidea (Cimicomorpha: Geocorisæ); rostrum 4-segmented; tarsi 2-segemented; female brachypterous with swollen abdomen and no ocelli; predaceous, some myrmecophile, a few found among fallen leaves, moss, lichens etc. (20 species).

Micropterous having small or vestigial wings.

Micropterygidæ Comstock 1893 (= Eriocephalidæ) family of small, most primitive moths (Micropterygoidea: Zeugloptera); mouthparts well developed; ocelli present, galea free, distinct from each other, do not form the sucktorial tube; tibia spurs present; wing -coupling jugofrenate type; wing venation of fore-and and hind-wing almost identical — *Micropterix calthella* (found in Britain), *Microparadalis* spp. and *Sabatinca* spp. (confined to New Zealand) etc.

Micropterygoidea Dyar 1902 superfamily of suborder Zeugloptera or old Homoneura (Lepidoptera: Oligoneoptera); small-sized moths having mandibulate mouthparts, withour probocis; wing venation very primitive and no frenulum; comprises of such families as, Eriocraniidæ, Micropterygidæ, Mnesarchæidæ and Neopseutidæ,

Micropyle minute or group of pores in egg chorion that permit entrance of spermatozoa; this is necessitated because the egg-shell is secreted before fertilisation.

Microscope optical instrument for producing magnified images of small object by use of single lens or system of lenses. Galilei Galileo (1564-1642) was first to examine insect with a simple magnifying glass in 1610.

Microscope, Simple fitted with one simple convex lens only; magnifying power limited.

Microscope, Compound having two converging lenses—eye-piece and objective, for greater magnification. Invented by Zacharias Janssen of the Netherlands in 1590.

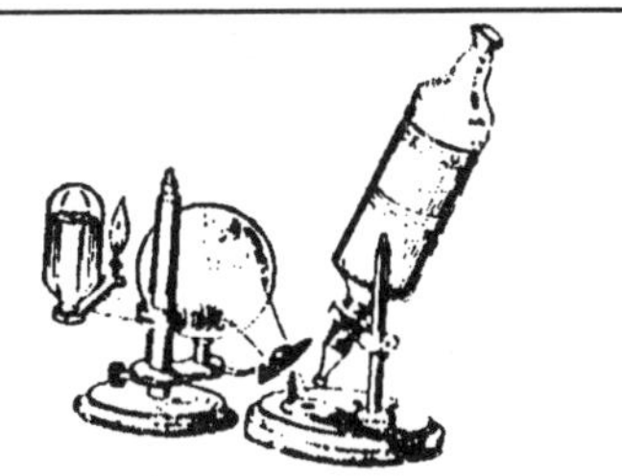
Compound microscope designed by Robert Hooke (1665)

Microscope, Steroscopic having two eye-pieces and two objective, with independent light paths, thereby it gives a three dimentional image.

Microscope, Binocular having two eye-pieces; it may be simple, compound or steroscopic. Cherubin d'Orleans designed the first binocular microscope in 1677.

Microscope, Electron using parallel beam of electron, to achieve magnification upto 250000 times. M. Knolly and E. Ruska constructed first magnetic electron microscope in 1932 followed by Vladimir Kosme Sworykin (USSR) in 1939.

Microsope, Ultra-violet using ultra-violet rays to obtain magnification beyond 1500 times; quartz lenses are used and image is recorded photographically.

Microscopic pertaining to microscope: Object hardly visible without the aid of microscope. *Cf.* macroscopic.

Microscopist person expert in handling and using various microscopes.

Microscopy study of minute object with the help of microscopes.

Microthorax thin neck-like prothorax as in dragonflies (Odonata).

Microtome apparatus for cutting very fine (thin) section of material, for mounting and microscopical examination. Invented by Withelm His (1831-1940) in 1870.

Microtomy technique of cutting animals or plant tissues into minute slices (sections) and preparing the same for histological studies.

Microtrichia minute, non-cellular cuticular processes or outgrowths, characterised by absense of membranous articulation, often found on wings of certain insect. *Cf.* macrotrichia.

Midge common name of many gnat-like flies (Nematocera: Diptera) — biting-midges, *Culicoides* spp. (Ceratopogonidæ), non biting gnats, *Chironmous* spp. (Chironomidæ), gall-midge, many species of Cecidomyids (Cecidomylidæ).

Midgut mesenteron part of alimentary canal lined by endodermal tissue, where most digestion and absorption normally takes place.

Migration seasonal movement of large number of individual (same species) as a regular habit to obtain advantage of favourable environmental conditions offered at the other place; common in certain species of birds (migratory) and like army-ants, butterflies, locusts etc. — *Danaus menippe* (Hübner) (= *plexippus* Fabricius) (Danaidæ: Papilionoidea) breeds in Summer in Northern USA, migrates to breed during Winter at Mexico.

Mimicry adaptive resemblance of certain insect to other insect species; it does not include cryptic colouration.

Mimicry, Batesian vulnerable species copying another species that is unpalatable or other wise defended by warning colouration — *Papilio* spp. (butterflies) and certain dipteran flies that resemble bees, wasps etc.

Mimicry, Mullerian two or more will defined unpalatable species having developed same pattern or warning colouration.

Mimicry, Parasitic parasite resembling its host.

Miridæ Hahn 1831 (Capsidæ) family of leaf-bugs (Cimicoidea: Geocorisæ); flattened, fragile, brilliantly coloured, active insect; rostrum long and 4-segmented; ocelli absent; hemelytra opaque or hyaline with clavus, corium, cuneus, membrane and embolium. Mostly herbivorous, a few predaceous (5000 species) — *Calocoris angustatus* Lethierry, *Cyrtopeltis crassicornis* Distant, *Prodromus subviridis* Distant, *Psallus ambiguus* (Fallen), *Ragmus flavomaculatus* Ballard, et al.

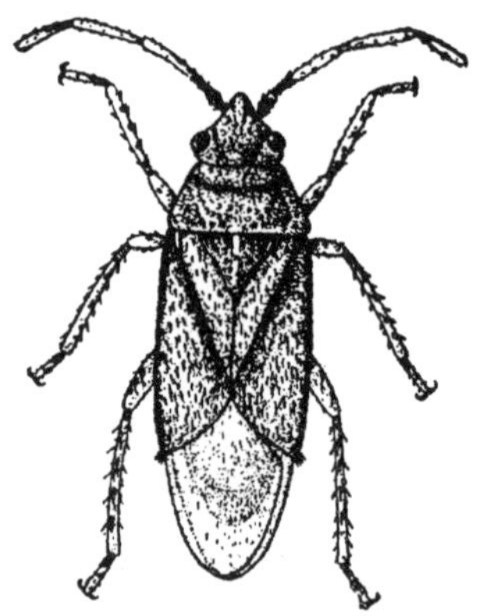

Psallus ambiguus (Fallen)

Miscible that can mixed: (liquid) capable of being mixed to form homogenous mixture.

Miscible liquid (m.l) solution or formation in which technical product is dissolved in organic solvent that when diluted dissolves in water carrier.

Miscogasteridæ family of Chalcidoidea (Parasitica: Apocrita); often regarded as subfamily of Pteromaildæ (Chalcidoidea); very small moths that are parasitic on a variety of insect — *Anysis alcocki* and *A. saissetiæ* are parasites of scale insect (Coccoidea: Sternorrhyncha) *Bruchobium laticeps* parasitises *Bruchus* species (Bruchidæ: Chrysomeloidea), *Ecrizotomorpha taskiri* is hyperparasite of secondary parasite of linseed gall midge *Dasineura lini* (Cecidomyiidæ: Nematocera) whereas *Dinarmus sauteri* parasites *Hypolixus truncatulus* (Curculionidæ: Curcolionoidea).

Mist blower spraying equipment that produces a fine air carried spray, usually power-driven, carried on the back by the operator.

Mite very small acaridan Arachnid (Chelicerata: Arthropoda); body composed of cephalothorax (head and thorax fused together) and abdomen; four pairs in adult stage but only three pair in larval stage; mostly phytophagous, a few parasitic.

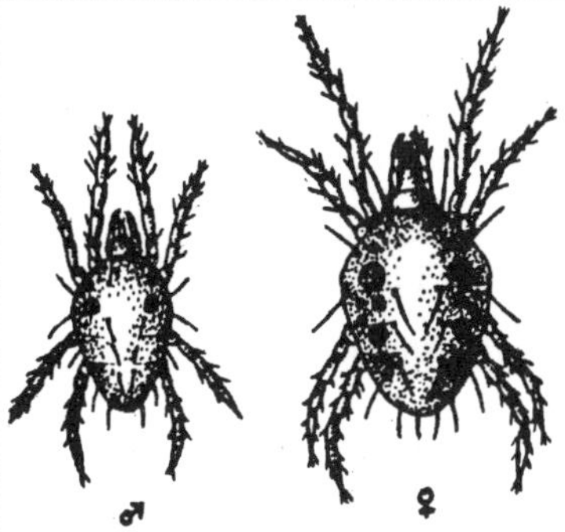

Mite (*Tetranychus sp.*)

Miticide acaricide: Toxin that can kill mites (Acarina).

Mixture heterogenous association of two or more substance mixed together; may be solid or gaseous; the constituents can be separated by suitable physical means.

Mnesarchæidæ monogeneric family of Eriocranioidea (Dacnonypha: Lepidoptera); having ocelli and no mandibles; maxillary palpi 3-segmented, laciniæ absent and galea elongated to form proboscis used as sucking-organ — *Mnesarchæa* spp., confined to New Zealand.

Molannidæ Wallengren 1891 family of Limenephiloidea (Integeripalpa: Trichoptera); caddisflies having long antennæ and very hairy wings. Caterpillar use sand particle to make tubular cases to live in, found in ponds, lakes or stream etc.

Mollusca phylum of non-segmented, soft-bodied animals (Triploblastica: Non-Chordata); having muscular foot and mantle (hard shell), terrestrial, fresh water and marine forms, descendents of ancient marine polychæte annelids (90000 species) — Amphineura (Apalcophora, Polyplacophora), Monoplacophora, Gastropoda, Scaphopoda, Pelecypoda and Cephalopoda.

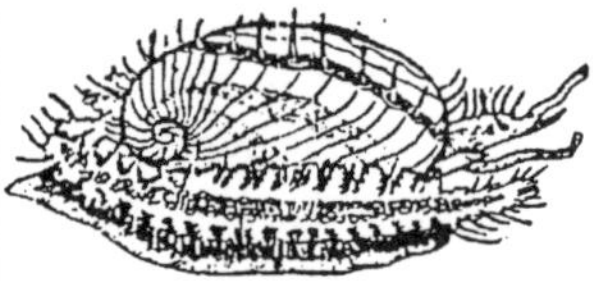

Abalone
(*a*)

Cowry
(*b*)

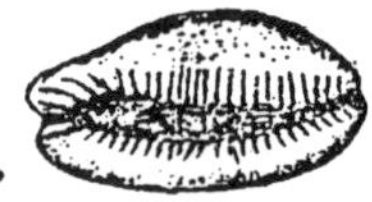

Cowry sheel
(*c*)

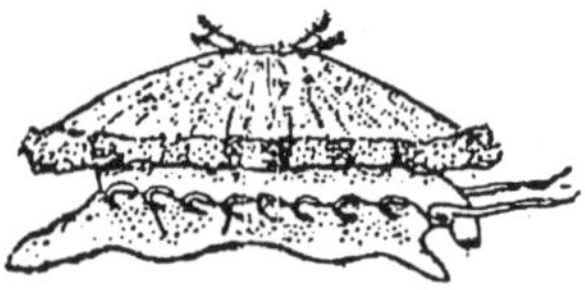

Limpet
(*d*)

Pond–snaill
(*e*)

Garden snail
(*f*)

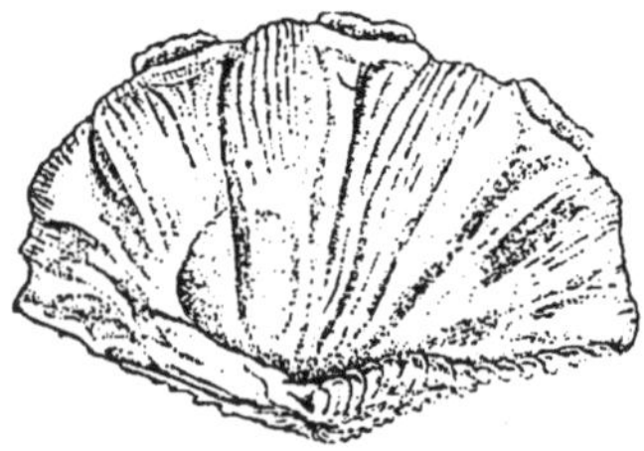

oyster
(*g*)

pilla
(*h*)

Turrilite
(*i*)

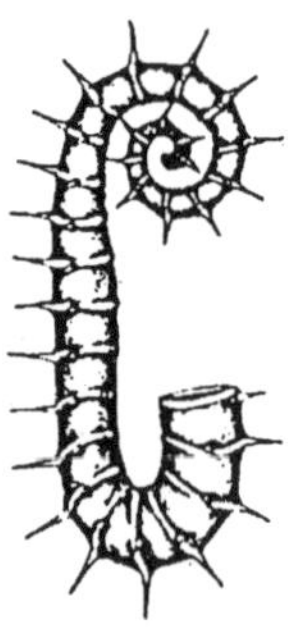

ancylocerus
(*j*)

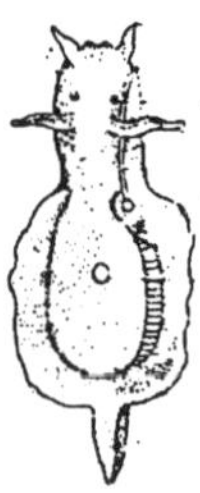

sea-hare
(*k*)

sea-slugs
(*l*)

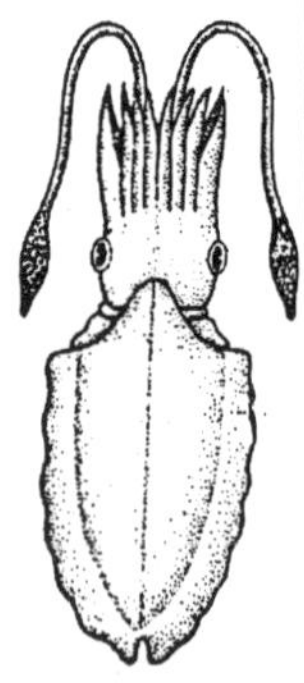

cuttle-fish
(*m*)

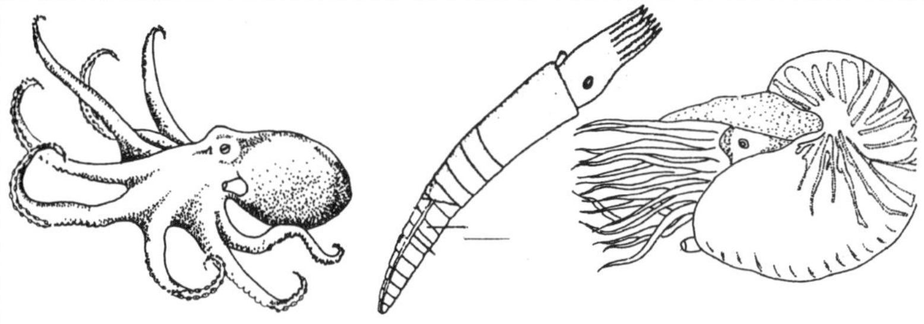

Octopus (*n*) Nautiloid Cone (*o*) Nautilus (*p*)

Molluscicide toxin can kill molluscs specially slugs and snails — Metaldehyde (metacetaldehyde) etc.

Monocondylic joint with a single point of articulation between adjacent leg segments.

Monocrotophos $C_7H_{14}O_5NP$; fast acting organophosphatic insecticide and miticide with systemic, stomach and contact action. Crystalline substance miscible with water, soluble acetone, and ethanol, insoluble in petroleum oil. Toxic to bees, fish and livestock and corrosive to brass and iron, not aluminium and glass. Acute oral LD_{50} value for rats 20 mg/kg — Nuvacron®, Phoskill, Milphos etc.

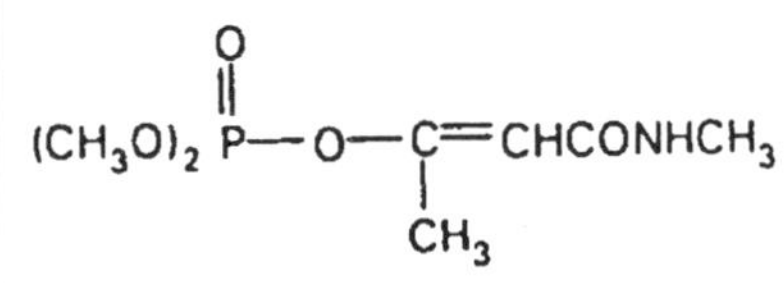

Monocrotophos

Monoecious having male and female flowers on same plant.

Monomachidæ very small and insignificant family of Proctotrupoidea (Apocrita: Hymenoptera); include small black wasps, parasitic on the immature stages of other insects.

Monomm(at)idæ family of Heteromera (Cucujoidea: Polyphaha); beetles characterised by propleuron grooved to receive antennæ; fore-coxæ small and rounded, separated by a broad posternal process, their cavities widely open behind; found in warmer parts of the World (specially Africa) breeding in rotten wood (200 species) — *Monomma indicum* et al.

Mononychidæ old name for family Gelastocoridæ (Ochteroidea: Amphibicorisa) that includes toad-bugs.

Monophagous (insect) that feed on a single plant species — pink bollworm larvæ (*Pectinophora gossypeilla* Saunders), feed only on cotton-boils (*Gossypium* spp.). A parasite parasitic only on one specific host.

Monophagy restriction of animals (insect) to consumption of the one species of host.

Monothalamous galls galls with single cavity, harbouring only one insect larva — oak marble galls, hard spherical galls formed by cynipid wasp, *Andicus Kollari*.

Monotrophic monophagaous: subsisting on a single kind of food.

Monotrysia small suborder of Lepidoptera (Oligoneoptera: Pterygota); having functional mandible, maxillæ without lacinia; wing usually aculeate; female moth with two genital opening and an anus or coloaca opening on or behind IX sternite; male without a saccus on IX sternite. Three superfamilies — Hapialoidea, Incurvarioidea and Nepticuloidea.

Monotype a holotype based on a single specimen: a single type of organism that constitutues a species or genus.

Monotypic taxonomic rank only one immediately subordinate rank or taxon — order with only one family or family comprising of only one genus of genus with single species.

Monterey pine engraver barkbeetle, *Ips radiatæ* (Scolytidæ: Curculionidea); both grub and adult tunnel under the bark of monterey and other pine trees in California (USA) making large number of radiating galleries that often attract fungus and viruses.

Mordellidæ (Leach 1815) large family of bristle or tumbling flower-beetles (Heteromera: Cocujoidea); small boat-shaped insect (<5 mm long) having silky body, tibial spurs pubescent, pronotum with distinct side border as wide as elytra at shoulder; claws serrate having a bristle-like lobe beneath; abdomen extended to a point at anal end. When distributed, it leaps away quickly (900 species) — *Mordelia aculeata, Tomoxia* spp. et al.

Mordelloidea Leng 1920 a super family of Polyphaga (Coleoptera: Coleopteroidea) in an old classification system; now merged with superfamily Cucujoidea (Polyphaga: Coleoptera) under section Heteromera.

Moribund (insect) knocked down and showing no sign of recovery: dying.

Morphoidæ (Biosduval 1836) Kollar 1850 (Argidæ, Amathusiidæ) a family of large, brilliant metallic butterflies of Northern India and Tropical America (Papilionoidea: Ditrysia) having upper surface of wings often blue and remarkably iridescent; hind-wing having a long depression on the anal margin in which the abdomen rests. Caterpillar gregarious, often look like bird dropping on tree trunks — *Amathusia* species, *Morpha achilles* (Linnæus), *M. menelaus* (Linnæus), *M. sulkowskyi* Kollar et al.

Morphology study of forms and feature of different parts of an organism.

Morphometrics study of body measurement.

Morphopathology branch of pathology dealing with the morbid change occurring in the structure of cells, tissue and organs. *Cf.* physiopathology.

Morphotype a selected specimen of the second or later form of a dimorphic or polymorphic species.

Mortality death-rate: frequency or number of deaths in proportion to population.

Morula globular mass of cells or blastomeres formed by holoblastic cleavage of egg.

Moth(s) all lepidopterous adult insect except butterflies and skippers; mostly nocturnal; antennæ not clubbed but usually hairy, branched or feather-like; possess a mechanism for hooking the fore- and hind-wing white flying.

Moth, Acorn *Melissopus latiferreanus* (Walsingham) and allied species (Olethreutidæ: Tortricuoidea), found in North America. Caterpillar feed on acorns.

Moth, Atlas giant saturniid moth of Tropical Asia, *Attacus atlas* (Linnæus) (Bombycoidea: Ditrysia); one of the largest insect of the World; female wing span upto 280 mm; brown to blackish-red in colour with one transparent spot on each wing. *A. edwarsi* white and *Cosinoscera hercules* Miskin are other similar and equally large species found in Australia.

Moth, Burnet group of diurnal, slow-flying moths (Zygænidæ: Zygænoidea); having metallic-blue or greenish-black forewing with scarlet marking which act as warning colouration as these insect are distasteful to the predators — five-spotted (*Zygæna loniceræ* and *Z. trifolii*) and six-spotted (*Z. filipendulæ*) are pests of clover (*Trifolium* spp, Papilionaceae) in Britain and Europe; while *Heterusia magnifica* Butler and allied species are pest of tea (*Camellia sinensis* Linnæus) in India.

Moth, Fox *Macrothylacia rubi* (Lasiocampidæ: Bombycoidea); medium-sized light brown coloured moth, antennæ bipectinate; female greyer than male; two pale oblique lines crossing forewing.

Caterpillar dark brown and hairy feed on bramble, billberries, heather etc. and pupate in brown hairy cocoons.

Moth, Goat *Cossus cossus* (Linnæus) (Cossidæ: Cossoidea); large, widespread, nocturnal moth. Caterpillar bore inside tree trunk of forest, fruit and shade trees, make large tunnels into solid wood, feed for 3 to 4 years and emit repulsive smell suggestive of a goat.

Moth, Honey lesser waxmoth, *Achoria grisella* (Fabricius) (Galleriidæ: Pyraloidea); pest of beehives particularly neglected ones may cause the bees to leave the hive; outside the hive, the caterpillar feed on dead insect and dried fruits.

Moth, Honeycomb waxmoth or waxworm, *Galleria mellonella* (Linnæus) (Galleriidæ: Pyraloidea); common pest of beehives, wing span about 30 mm, male smaller and paler than female. Caterpillar feed on wax and in doing so, spoil the honey as well; spin tough silk cocoons, clustered together in the comb wherein the hibernate during Winter and then pupate.

Moth, Indian-meal *Plodia interpunctella* Hübner (Phycitidæ: Pyraloidea); small, delicate moth with elongate forewings and broad hind-wing; cosmopolitan in distribution. Caterpillar damage maize (*Zea mays* Linnæus), fig *(Ficus carica* Linnæus*)* and seeds of various grain:adult are found in flour, cereals shelled nuts and dried fruits.

Moth, Leopard *Zeuzera pyrina* (Linnæus) (= *æsculi* Linnæus) (Zeuzeridæ: Cossoidea); nocturnal moth, white in colour having beautifully patterned wings. It is destructive pest of temperate fruit trees.

Moth, Lily *Polytela glorios*æ Fabricius (Noctuidæ: Noctuoidea); stout bodied moths; forewing having mosaic pattern of yellow, red and black hind-wings uniformly blackish; caterpillar, specific pest of lilies (*Lilium* spp.), are smooth with beautiful black, white and red spots; defoliate the leaves completely.

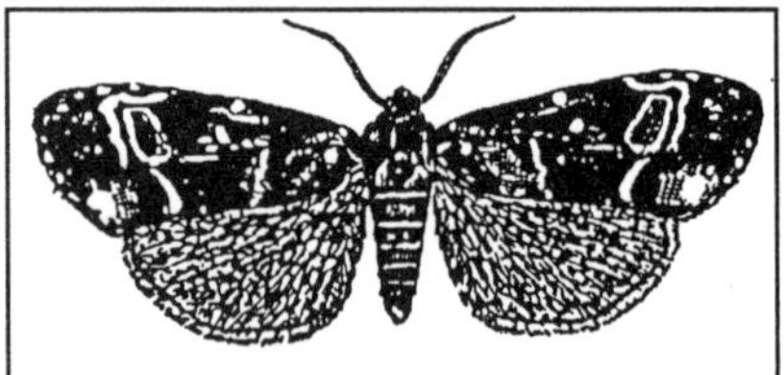

Polytela gloriosæ Fabricius

Moth, Lunar hornet *Sphecia bembeciformis* (Seaiidæ: Yponomeutoidea) closely resembles hornet; head and thorax brown with distinct yellow collar; wings transparent; abdomen yellow with two brown bands. Caterpillar whitish in colour, bore into trunks of willows (*Salix* spp., Salicacese) and poplars (*Populus* spp., Salicaceae).

Moth, Lunar underwing *Omphaloscelis lunosa* noctuid moth (Ditrysia: Lepidoptera); light brown with black dots and patches. Caterpillar brown, found feeding on grasses.

Moth, Monk shark-moth(s) group of noctuid moths; grayish-brown in colour having tuft of long hairs on thorax projecting forward and resembling monk's cowl. Caterpillar bright yellow having large number of conspicuous black spots; very specific in their feeding habit — *Cucullia companulæ, C. scrophulariæ, C. umbratica* and allied species.

Moth, Moon (Luna) delicate apple-green coloured moths (Saturniidæ: Bombycoidea); having two large slender trails on hind-wings and moon-like marking on forewing; wing expanse varies between 140 and 180 mm; female being bigger than male. Caterpillar are defoliators of various temperate fruit tree like apple (*Malus pumila* Miller), cherry (*Prunus avium* Linnæus, *P. cerasus* Linnæus), pear (*Pyrus communis* Linnæus), walnut (*Juglans regia* linnæus) etc. — Indian moon moth, *Actias selene* (Hübner), produces poor silk as its cocoons are thin, irregularly shaped and closed; American moon moth is *A. luna*.

Moth, Mouse noctuid moth, *Amphipyra tragopaginis*, brown moth with 2 or 3 small black spots on each forewing when disturbed it runs away like a mouse, hence the name.

Moth, Nail-mark *Aglia tau* (Saturniidæ: Bombycoidea); large brown moth of Europe; having white nail-shaped mark on each forewing and average wing expanse of 80 mm; males smaller and darker than females. Caterpillar feed on beech (*Fagus sylvatic* Linnæus) leaves.

Moth, Silk (Silkworm) caterpillar of various species of Bombycidæ and Saturniidæ (Bombycoidea: Ditrysia); make dense silken cocoons for their protection and silk is produced commercially from these cocoons. Chinese silkworm, *Bombyx mori* Linnæus (Bombycidæ) is the most common species, feeding on mulberry leaves, a single cocoons can produce continuos 300 metres of silk thread and 50000 cocoons are required to produce one kg of silk. Saturnid moth used commercially for silk production include, eri silkworm (*Attacus ricini*), muga silkworm (*Anthraræa assamensis* Helfer) and tasar silkworm [*A. paphia* (Linnæua)] from India and oak silkworm (*A. roylei* Moore from Himalayas, *A. pernyi* Guerin from China and *A. yamamai* Guerin form Japan). There are many other species also that can be used for silk production, but the cost involved is too high.

Moth, V small grayish moth, *Itame wauaria* (Geometridæ: Geometroidea) with distinct black V-shaped mark on each forewing. Its yellowish-gray caterpillar are pests of currant (*Ribes nigrum* Linnæus, *R. rubrum* Linnæus etc.) and gooseberries (*R. grossularia* Linnæus etc.) in Britain.

Moth, Winter geometrid moths, *Operophtera brunata* and *O. fragata* (Geometroidea: Ditrysia); adults abundant during Winter in orchards and woods; female having rudimentary (non-functional) wings. Caterpillar feed voraciously on foliage of various fruit and forest trees during Spring.

Moth flies minute (4 mm long), fragile, dipteran flies (Psychoidæ: Psychodoidea); having body and wings covered with hair or scales, so as to resemble moths; no ocelli. Maggots feed on decaying vegetation — *Maruina californiensis* Kellogg, *Pericoma funebris* Hudson *Psychoda alternata* Say and allied species.

Moult to cast or shed outer covering — cuticle, scales or feathers — a common practise in insects, birds, snakes etc.

Moulting ecdysis: process of loosening or discarding the cuticula periodically; it enable the insect to grow rapidly after each moult before the new cuticula hardens.

Mouth anterior opening of alimentary canal; it leads to buccal cavity and helps in ingestion of food.

Mouth hooks chitinous hooks found in certain larvæ and maggots; used for tearing the plant or animals material on which they feed; well developed in leaf-miners

Mouthparts feeding tools: mouthparts of an insect usually comprise of an upper lip or labrum, lower lip or labium and between these, two pairs of jaws-mandibles and maxillæ. The labium and maxillæ bear sensory appendages or palpi. Mouthparts are variously modified according to mode of feeding; these may be biting and chewing (cockroach), chewing and lapping (honeybee), piercing and sucking (mosquito, bedbug), rasping and sucking (thrips) or siphoning (butterfly).

Movement active spatial displacement.

Mulberry bug *Halys dentatus* Fabricius (Pentatomidæ: Heteroptera); polyphaga pest; swarms of nymphs and adult suck juice form stems and tree trunks during Summer; their deceptive colouration, which closely resemble the bark of the tree makes it difficult to spot them.

Multiparasitism simultaneously parasitisation of a single individual host by two or more different species of primary parasites — eggs of various lepidopterous borer are parasitised at the same time by the egg-parasites *Telenomus* species and *Trichogramma* species.

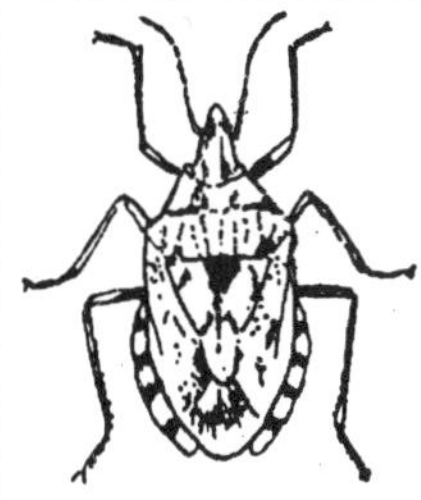

Halys dentatus Fabricius

Multiple manifold: repeated several times.

Multiple parasitism more than one parasitoid species feeding and developing simultaneously on or within the same host.

Multiple resistance simultaneous resistance to several stresses, such as diseases, insect pests, mites, nematodes, draught etc.

Muscidæ (Fallen 1810) Leach 1815, 1819 family of Mucsoidea (Schizoptera:Cyclorrhapha); small to medium-sized flies, having fleshy proboscis, oral vibrissæ present and bristles on mesonotum. Wing large and strong, vein Cu_1+ 1A not reaching margin of the wing; halters small. Abdomen hairy, short and with constricted base; basal are without macrochætæ. Larvæ are typical maggots having pseudopodia, metapneustic and hemispherical stigmata — *Musca domestica* Linnæus, *M. nebluo* Wiedemann are common houseflies and vectors of various diseases.

Musturd sawfly *Athatia lugens proxima* (Klug) (Tenthredinidæ: Hymenoptera); small, 8 to 10 mm long flies (not true flies), body orange-coloured with transluscent smoky wing having black veins; females possess strong saw-like ovipositor. Grubs (look this caterpillar) feed tender leaves by cutting small holes in leaf lamina. Polyphagous, but prefer cruciferous crops.

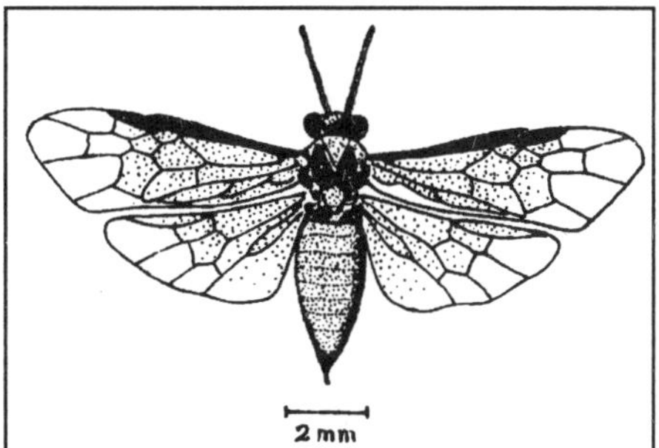

Athalia lugens proxima (Klug)

Mutagen mutagenic agent, capable of causing mutation or genetic changes in living cells — proflavin, cosmic rays etc.

Mutagenic (Mutagenecity) property of a substance (chemical) to produce genetic change in living cells; cause mutation.

Mutant organism, population, chromosome gene etc., that differs distinctly from the original type; the abnormally to be internal (genetic) rather than environmental and capable of being transmitted to offsprings.

Mutillidæ (Latreille, 1802), Stephens 1829 family of Vespoidea (Aculeata: Apocrita); 8 to 30 mm long, brightly coloured wasps, males usually winged and females apterous; antennæ in females 12-segmented and curled, in males 13-segmented and straight; thoracic segments closely fused and legs stout — *Mutilla lilliputiana* Andre (< 3 mm), *Dasymutilla occidentalis* (Linnæus) (upto 30 mm).

Mutualism biotic interaction between two or more organism in which all the interacting partner benefit jointly and none is harmed.

Mycetocyte specialised cells that cluster together to form mycetome; found in insect containing symbiotic bacteria of protozoa.

Mycetome specialised structure in certain insect for holding the endosymbionts.

Mycetophilidæ Macquari 1838 a family of fungus gnats or mushroom flies (Mycetophiloidea: Nematocera); small flies with long antennæ; ocelli present; coxæ elongate and tibia spurred. Maggot vermiform having 8 pairs of spiracles and living gregariously in fungus or decaying vegetable matter. More than 2500 known species — *Arachnocampa luminosa* Skuse, *Ceroplatus sessiodes, Dahlica larviformis* Enderlein, *Macrocera fasciata* Meigen, *Pnyxia scabiei, sciara semilata* et al.

Mycteridæ very small family of Cucujoidea (Polyphaga: Coleoptera); distinguished by having mid coxal closed by sterna and penultimate tarsal segment lobed; found mostly on flowers — *Mycterus* spp., et al.

Mydaidæ (Mydidæ) (Leach 1815) family of Asiloidea (Brachycera: Diptera); very large flies having no bristles; antennæ terminating in a clubbed style; pulvilli large, no empodium and wing-venation complex. Adult predaceous, maggot in decaying wood, ossasionally predaceous on the grubs — *Mydas heros* Perty of Brazil is 52 mm long with wing of 85 mm.

Myiasis the affection caused by dipteran larvæ — the larvæ may be parasitic or saprophagous. Primary myiasis is induced by true parasites whereas secondary mylasis is caused by saprophagous larvæ. According to nature and site of infestation, mylasis is called cutaneous, oral, ocular, nasal, rhinal, gastric, intestinal, urinogenital etc.

Mymaridæ Haliday 1840 a hymenopteran family of Chalcidoidea (Parasitica: Apocrita), minute (< one mm long) black and yellow fairy-flies that fly and jump with agility; have -8 to 13- segmented antennæ without ring joint; legs long and slender, tarsi 4- or 5-segmented; wings narrow and fringed; abdomen sessile subsessile or petiolate Cosmopolitan in distribution, 15 genera embracing about 200 species; many are beneficial egg-parasites — *Alaptus magnanimis* Annadale (0.2 mm long) is probably to smallest insect in the World) and *Metalaptus torquatus* Malenotti is also less than 3 mm. *Anagrus empoascæ* and *Lymænon empoascæ* are egg-parasites of *Amrasca biguttula biguttula* Ishida in India. *Paranagrus obtabilis* Perkins is a very effective egg-parasite of sugarcane leafhoppers (*Pyrilla* species).

Metalaptus torquatus Malenotti

Mydochidæ old and equivalent term for heteropteran family Lygæidæ (Lygæoidea: Pentamomorpha).

Myopsocidæ small family of Psocomorpha (Psocoptera: Psocopteroidea); characterised by 3-segmented tarsi and wings regularly spotted; found mainly in the Old World (50 species).

Myriapoda class of Uniramia (Arthopoda: Coelomata: Triploblastica); small invertebrate, terrestrial animals, trunk divided into head and body; head bears two preoral antennæ (antennules); body long, slender, uniformly segmented and having numerous legs.

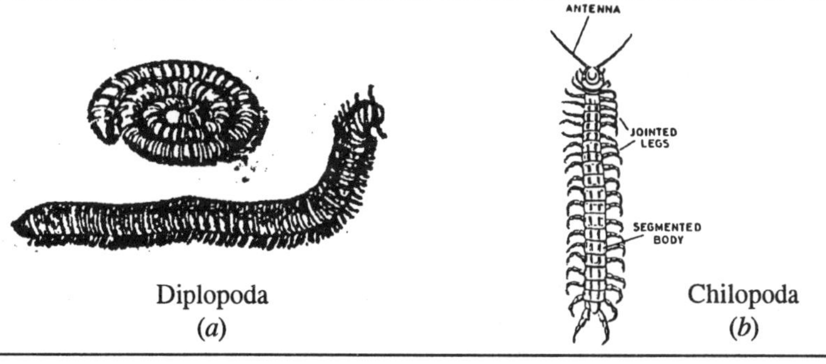

Diplopoda (*a*) Chilopoda (*b*)

Opisthogoneate form or Chilopoda have one pairs of legs on each segment and gonads open behind penultimate segment (centipedes) while progoneate forms include Diplopoda, Pauropoda and Symphysis, having two pairs of legs per segment and gonads opening near anterior and of body (millipede etc.).

Myrientomata Berlese 1909 suggested name for the insect order earlier established as protura Silvestri 1907; no more in usage.

Myrientomata old name for insect order Protura, now regarded as superorder of Protura, under Entognatha: Apterygota.

Myrmeciinæ most primitive subfamily of Formicidæ (Scolloidea: Apocrita); carnivorous ants, with a peculiar habit that often queen also forages for food — bull dog ants, *Myrmecia* spp. of Australia are 20 to 25 mm long, bite and sting very ferociously.

Myrmecoid ant-like.

Myrmecolacidæ Saunders 1872 small family of Stylopoidea (Stresiptera: Coleopteroidea); stylopids very long mandible and base of maxilla not swollen; male having 7-segmented antennæ; tarsi with less than 5 segment and no claws; in female cephalothorax with hood-like projection behind the spiracles — males attack ants and female orthopteroids.

Myrmecocleptic(s) a term coined by W.M Wheeler for the lepismids (Thysanura) which inhabit the nests of ants as guest wherein they steel regurgitated food as it passes from mouth of one ant to that of another — *Atelura formicaria* Heyden (Europe), *A. wheeleri* Escherich (North America), et al.

Myrmecologist person having sound knowledge of myrmecology.

Myrmecology branch of entomology that deals with the study of ants.

Myrmecophaga (myrmecophilous) having a symbiotic relation with ants.

Myrmecophagous feeding on ants.

Myrmecophiles various insect, mites, spiders etc., that inhabit nests of ants either as scavenger or more likely as true guest, fed and tended by the ants.

Myrmecophilinæ Schimmer 1909 subfamily of Gryllidæ (Grylloidea: Ensifera); very small subsphreical apterous cricket that live in association with ants.

Myrmeleontidæ Burmeister 1829 family of Myrmeleontoidea (Planipennia); antennæ short, stout and thickened with apex; wings long, narrow and many-veined, without trichosors but with elongate long hypostigmatic cell; abdomen long and slender. Larvæ (ant-lions) stout having bristles of body directed forwards, mandible strongly developed and curved; they lie concealed in conical or funnel-shaped pits in sand and feed on ants etc. that fall into the pits (1200 species) — *Myrmelcon formicarious* Linnæus (Europe and Japan), *M. contractuss* (India), *M. immaculatus* de Geer (North America) et al.

Myrmeleontoidea superfamily of Planipennia (Neuroptera: Neuropteroidea); body with dolichaster; antennæ short, usually clubbed or thickened; wing long and subequal, trichosors absent except in Nymphidæ — Nymphidæ, Myrmeleontidæ, Ascalaphidæ, Stilbopterygidæ and Nemopteridæ.

Myrmica a common and generic name of a small species of red ant, *M. ruginodis* that nests under stones or in old tree stumps.

Myrmicinæ (Smith 1851) Dalla Torre 1893 largest subfamily of Formicidæ (Scoloidea: Apocrita); carnivorous ants having pliable integument that allows great distention when their crops are gorged with food, as their habit of collecting and storing nectar and honeydew is highly developed; female and worker have large and well developed sting; some species also eject foul-smelling secretions as means of defense — *Holcomyrmex scabriceps* Mayr, feed on wheat grains, [*Ecophylla smaragdina* (Fabricius)] is a nuisance in orchard to all the worker, *Solenopis geminata* Fabricius damage seedlings of egg-plants (*Solanum melongena* Linnæus) and pigeonpea (*Cajanus cajan* Linnæus).

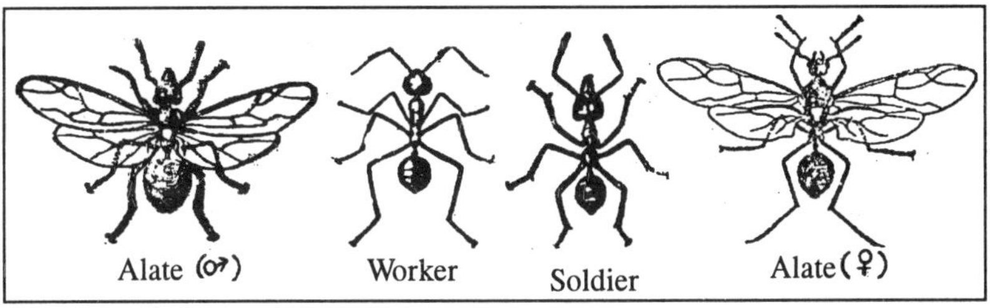

Myrmosinæ subfamily of Tiphiidæ (Scolioidea: Apocrita); characterised by having males winged and females apterous — *Myrmosa* spp. parasitise various ground nesting Aculeata.

Mystax patch of bristles or stiff setæ above the mouth in certain dipteran flies: also called moustache.

Mytilococcus alternative term for mussel-scale, *Lepidosaphes* species (Diaspididæ: Coccoidea); pest of Temperate fruit trees, so named because the scales resemble tiny mussel shells.

Myxophaga fourth suborder of Coleoptera (Coleopteroidea: Oligoneoptera) erected by Crowson in 1955 that includes only one superfamily (Sphærioidea); minute to small-sized beetle, prothorax having notoplural sutures; oblengum more or less distinct, but Rs reduced; found near wet places by the side of streams etc. Grubs of most species have spiracular gills on first eight abdominal segment — Calyptomeridæ (hitherto part of Clambidæ), Lepiceridæ (horn's beetle), Sphæriidæ (bog beetles) and Hydroscaphidæ (skiff beetles).

N

Nabi(di)dæ (Costa, 1852) family of Cimicoidea (Cimicomorpha: Geocorisæ); damsel-bugs having antennæ thin, slender and 4- or 5-segmented, rostrum 4-segmented, forelegs raptoral, tarsi 3-segmented; cuneus absent; mostly precaceous and found on herbags where they attack small phytophagous insect (300 species) — *Allodorrhynchus* spp. *Prostemma* spp. et al.

Naeogæidæ old and obsolete name for heteropteran family Hebridæ (Gerroidea: Amphibicorisæ).

Naiad aquatic gill breathing nymph, usually long-lived years and show varied adoption for different kind of aquatic environment and plecoptera (stoneflies).

Nannochoistidæ Tillyard 1918 very small family of Eumecoptera (Mecoptera: Mecopteroidea); scorpionflies with a distinct tendency for mouthparts of become sucktorial, mandible being vestigial and the labrum forming a sharp projection. Both larvæ and imagine are carnivorous. Confined to Australasia and South America — *Nannochorista* spp. and *Charistella* spp.

Naphthalene $C_{10}H_8$: white crystalline material that vapourises slowly with vapour pressure 0.08 mm Hg at 25 °C; slightly soluble in water but dissolves readily in organic solvents; used in house against clothes moths and carpet beetle and also in museums to protect the preserved specimen against insect pest damages.

Naskemidæ an extinct family of extinct order Paraplecoptera ; these insects; that lived during Carboniferous and Permian period, show venational feature that suggest an ancestor of Plecoptera.

Nasute termite-soldier having a long pointed snout from which they discharge acrid (poisonous) fluid for defending the colony.

Nasutitermitinæ subfamily of Termitidæ (Isoptera: Biattopteroidea); distinguished by nasute soldiers having vestigial mandible coupled with a long median frontal nasus or rostrum, at the end of which the frontal gland opens; builds very large nests of wood-pulp and sand, above the ground — *Nasutitermes costalis* Holmgren, *N. exitiosus* (Hill), et al.

Natality birth-rate: frequency or number of births in proportion in population.

Natural occurring in nature: not artificially prepared or synthesised.

Natural control collective action of environmental factors to maintain the number of a population within certain upper and lower limits over a period of time.

Natural enemies a common name for the parasites, predators and pathogens associated naturally with a specific population of plant or animals. These may or may not able to keep the host population under complete check.

Natural host the plant or animal species on which an insect normally feeds and develops in nature.

Natural population insect population that is naturally present on test plant/tree in the field or orchard.

Naucoridæ Fallen 1814 family of Notonectoidea (Hydrocorisæ : Heteroptera); small to medium-sized, broadly oval and flattened, creeping water-bugs having antenæ 4-gegmented; rostrum 3-segmented; memberane without veins, forelegs strongly raptroid and hind-legs fringed with swimming-hairs (150 species)- *Aphelocheirsu* spp. *Naucoris* spp., *Pelocoris* spp. et al.

NAYAR, Dr. Krishna Pillai Karunakaran-born June 12, 1920; B.Sc. zoology (1939) B.Sc. (Hons. (1941), M.A. Zoology (1943 and Ph. D. (1947), all from erstwhile University of Travancore. Started his career as Lecturer in Government College, Travancore (Kerala) in 1945, subsequently taught at Government colleges, Chittor, Plaghat, Calicut and Trivandrum as Reader, became Professor and Head of Zoology Department, University of Kerala in 1968. Awarded Ford Foundation grant to complete the studies (USA and Japan) on reproductive biology and endocrinology of insects. Elected Commonwealth Visiting Professor and worked at University of East Anglia (UK) and later invited to work as Research Professor at Univversity of California, Berkeley and then as Guest Professor at Albert Einstein College of Medicine, New York. University Grant Commision selected him as National Lecturer in zoology for 1971-72. Just before his death (June 12, 1975), he complete the books, **Elements of Insect Endocrinology** (1973), **General and Applied Entomology** (1976) and **Invertebrate Reproduction** (1992).

Neauridæ family of Poduroidea (Arthroplenona : Collembola); mouthparts adapted for piecing, mandibles without plate or absent; no pesudocell-*Anurida maritima, A. denisi, Neanura muscorum* et al.

Necrosis death of part of living body (plant or animal).

Neeliidæ Folsom 1896 small family of Symophyleona (Collembola: Oligoentomata); minute springtails with antennæ shorter than head, eyes absent and no tracheæ - the most important genera are *Megalothorax* Willem and *Neelus* Folsom.

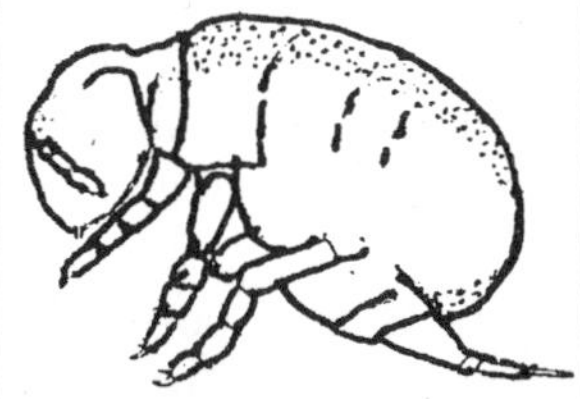

Neelus species

NEGI, Thakur Pratap Singh - born March 16, 1898; obtained M.Sc. (Zoology from Banaras Hindu University in 1925 and joined Indian Lac Association for Reasearch at Ranchi, as Assistant Entomologist in July 1926, later promoted as Entomologist and for about a year as Diretor of the Institute; retired in 1956. An eminent Lac-Entomologist, he died on January 14, 1961.

Neididæ old and equivalent term for heteropteran family Berytidæ (Lygæoidea: Pentatomomorpha) that includes stilt bugs.

Nema(s) nematodes.

Nematicide pesticide (toxin) effective in killing nematodes.

Nematocera (Letreille 1825) Braure 1880 suborder of primitive Diptera (Oligoneoptera: Pterygota); antennæ many-jointed, longer than head and thorax; arista wanting; palpi 4 or 5-segmented and pendulus; forewing without discal cell, cubital cell if present, widely open. Larva with well developed heat and exerted with horizontally bitting mandibles-Anisopodoidea, Tipuloidea, Psychioidea, Culicoidea, Bibionoidea and Mycetophiloidea.

Nematode microscopic, thread-like, unsegmented roundworm (Neamatoda: Aschelminthes); both free-living and parasitic on plants and animals-*Nothotyenchus basiri* Khan, *Tylenchorhynchus acutus* Allen and *Xiphinema basiri* Siddiqi attack mango trees. *Ascrais lumbricoides* is found in the intestine of human beings and pigs.

Nematologist nematode-specialist.

Nematology branch of zoology dealing with study of nematodes.

Nemeobiinæ subfamily of Riodinidæ (Nemeobiidæ) a family of Papillionoidea (Ditrysia: Lepidoptera); includes small butterflies having short broad forewings. Caterpillars exhibit marked diversity of form-*Dodonia* spp. in India, *Hamearis* (*Nemeobius*) *lucina* of Europe.

Nemertinea (Rhychocoela) phylum of bilaterally symmetrical ribbon like unsegmented wormiform lower animals (Acoelomata:

Triploblastica); elongated body having ciliated epidermis, no body cavity, complete digestive tract with mouth and anus: mostly oviparous. About 900 species of bootlace worms approtioned between two classes and four orders–Anopla (Heteronemertea and Palaeonemetea) and Enopla (Bdellongemertea and Hoplenemertea) -*Lineus longissimus* is the longest known animal (worm) its half-unravelled portion measures about 30 metres.

Nemestrinidæ family of large bristleless dipetran flies (Asiloidea : Brachycera); having very long prodoscis, 3rd antennal segment simple with terminal style; pulvilli and arolium pad-like; wing-veins running parallel to hind margin of the wing (200 species) - *Hirmoneura annandale* and *H. bruner* frequent flowers to suck nectar while *Neorhynchocephalus* species parasitise grasshoppers.

Nemonychidæ (Rhinomacridæ) very small family of Curculionoidea (Polyphaga: Coleoptera); having maxillary palpi flexible, labrum distinct and separate; pronotum bordered at sides; fist four abdominal sternites connate (20 species) - *Cimberis* species et al.

Nemopteridæ Burmeister 1839 highly specialised family of Myrmeleontoidea (Planipennia: Neuroptera); head prolonged into a rostrum, antennæ long but not clubbed or thickened at the apex; hind-wing enormously elongated and ribbon-like, seen streaming in the air while the insect fly with a curious up and down motion. Larvæ long-necked, predatory, live in sand places-*Croce filipennis* Westwood (India), *Pterocroce storeyi* Withycombe (Egypt), *Nemoptera bipennis* Illiger (Europe) et al.

Nemuridæ (Selys 1888) Klapalek 1905 largest filipalplian family (Plecoptera : Orthopteridea); small to medium-sized stoneflies found in Holaretic and Palaeretic regions having wing-venation uniform without any distal crossveins; cerci one-segmented or vestigial (350 species)- *Nemoura punjabensis* Jewett, *N.depressa* Banks, *Tæiopteryx fasciata* Burmeister, *T. rivalis* Fitch, et al.

Nemouroidea superfamily of Filipalpia (Plectoptera: Orthopteroidea); homogeneous group of small stoneflies, having reduced cross-veins, anal lobe and cerci. Nymphal gills often on head or thorax and not anal-Tænliopterygidæ, Leuctridæ, Capniidæ, Notonemuridæ, Nemouridæ.

Nemurædes Billberg 1820 old and obsolete term for insect order Plecoptera Burmeister 1809 that includes perlids, salmonflies and stoneflies.

Neoephernidæ newly erected family of large holarctic mayfles (Caemoidea: Ephemeroptera); copious wing venation, more or less similar to Ephemeroidea recognised mainly form nymphs that have no gills in the first abdomial segment.

Neogæidæ old and equivalent name for heteropteran family Herbridæ (Gerroidea: Amphibicorisæ).

Neomale famale becomes masculinsed: ovaries get modified into testis, genita ducts appears as in male and somatic male character develop; such a male can fertilise a female.

Neomecoptera newly errected order (not yet commonly used) for the family Boreidæ (Eumecoptera: Mecoptera); adults having polytrophic ovarioles, no cerci and six rectal glands.

Neoparatype paratype that has been described after the publication of description of original type.

Neoptera Martynov 1923 one of the two major (section) of subclass Pterygota; winged insects with muscular and articular mechasim allowing the wings to be flexed over the abdomen when at rest; it is divided into three infraclasses-Polyneoptera, Paraneoptera and Oligoneoptera.

Neopseustidæ small family of Micropterygoidea (Zeugloptera : Lepidoptera); comprising of tiny moths having reduced mandibles with no teeth.

Neotenin juvenile hormone secreted by cells of the corpus allatum in larvæ (nymphs) of insects. It maintains the juvenile character of the larvæ/nymphs after each moult; as soon as the production of this hormone ceases, the larvæ/nymphs become pupæ or adults, as the case may be.

Neotype nomenclatural type: specimen selected as type after the publication of original description; when the primary type has been lost or destroyed.

Nepidæ (Laterille, 1802) family of water scorpions (Notonectoidea: Hydrocorisæ); predaceous aquatic bugs; antennæ 3-segmented; forelegs raptorial without any claws; hind-legs adapted for walking taris single segmented. Breathing tube formed by cerci at the anal end of abdomen is the characteristic feature of these insects. Inflict painful bit to man, when handled- *Ranatra brevicollis* Montandon, *R. quadridentata, Nepa cinerea* Linnæus, *Laccotrephes maculatus* Fabricius et al.

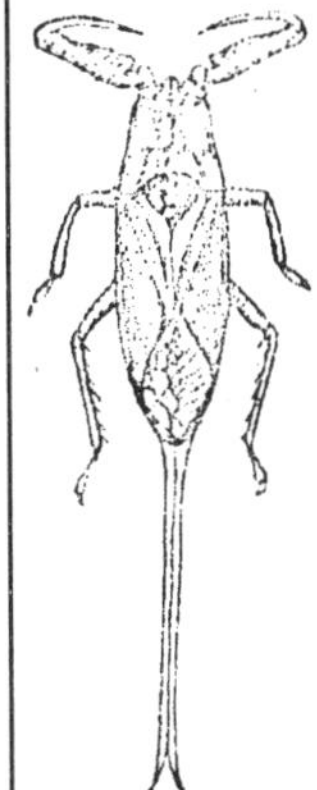

Laccotrephes maculatus Fabricius

Nepticulidæ (Tutt 1899), Spuler 1910 family of Nepticuloidea (Monotrysia: Lepidoptera); distinguished by wings narrowly lanceolate and clothed with aculei; frenulum present in males; in females jugum is present on forewings and a row of hooked spines on hing-wing. Caterpillars mostly leaf-miners - *Nepticula microtheriella* Stainton (wing expanse 3-4 mm), *N. braunella* Jones et al.

Nepticuloidea Forbes 1923 superfamily of Monotrysia (Lepidoptera: Mecopteroidea); forewing without closed discal cell, wing-venation reduced especially in hind-wing; male with frenulum, female with short cloacal duct and fleshy ovipositor-Nepticulidæ and Opostegidæ.

Nereistoxin toxin (cartap hydrochlorie) isolated from marrine annelids (*Lumbrineris brevicirra* and L. *heteropoda*), it causes paralysis in insects by blocking their central nervous system-Padan®, Sanvex® etc.

Nervure any one of the veins in insect's wing.

Net-winged having membranous wings with numerous cross-veins-Ephemeroptera, Odonata, Neuroptera etc.

Neurocyte cell-body of neuron; also called cyton or nerve cell.

Neuron nerve cell with its processes; it consists of cell-body (perikaryon) with a nucleus and one or more protoplasmic projections, the longest projection is known as axon and the short ones are called dendrites; neuron may be unipolar (monopolar), bipolar or multipolar depending upon the number of protoplasmic extensions from the perkaryon. It is a functional unit of nerve system.

Neuron, Internuncial neuron(s) that link sensory neurons with motor neurons within central nervous system through synapses.

Neuron, Motor efferent neuron that brings response from central nervous system to muscles (effector organs).

Neuron, Sensory afferent neuron; concerned with perception of stimuli and conducting them from site of perception to central nervous system.

Neuroptera Linnæus 1758 order of Neuropteroidea (Oligoneoptera: Pterygota); soft-bodied insects having long antennæ, bitting and chewing mouthparts, two pairs of similar long membranous wings with primitive venation, numerous accessory veins and decticous. Metamorphosis complete or inderect. About 6000 species grouped into two suborders - Megaloptera (alderflies, dobsonflies and snakeflies) and Planipennia (ant-lions and lacewings).

Neurothidæ very small family of Osmyloidea (Planipennia: Neuroptera); having no ocelli; wings with numerous costal veinlets and fewer-veins. Larvæ aquatic-*Neurothus faliax* et al.

Neurosecretory cells secretory neurons.

Neurosemidæ (Calopsoidæ) small South American family of book lice, allied to Cæciliidæ (Psocomorpha: Psocoptera) but with many cells in the wing-venation. venation.

Neuter sexless; neither feminine nor masculine: without functional and roecium or gynoecium.

Nicoletiidæ family of Lepismatoidea (Zygentoma: Thysanura); having no eyes, no ocelli and no scales; male gonapophyses long - *Nicoletia phytophila* Gervais is widely distributed whereas *Atelura formicaria* Heyden, a small myrmecophilous species is confined to Central Europe.

Nicomiidæ a very small family of Cicadelloidea (Cicadomorpha: Auchenorrhyncha) allied to Memberacidæ, comprising of only 15 species.

Nicotine $C_{10}H_{14}N_{2;}$ natural organic compound, colourless liquid that darkens on exposure to air and becomes more viscous. It is obtained form leaves and stems of waste tobacco by steam distillation or solvent extraction. A nerve poison, highly toxic to insects and mammals, its acute oral and dermal, LD_{50} values for rat are 70 and 140 mg/kg respectively. The commercial product, used as insecticide is nicotine sulphate which is relatively safe and readily soluble in water-Nicofume.

Nidi group(s) of regenerative cells that often replace worn out secretory cells of midgut epithelium; their number and frequently vary in different species of insects.

NIRULA Dr. Kanwal Krishan born May, 5, 1917; B.Sc. (Hons.) 1938 and M.Sc. (Hons.) 1939, from Punjab University; Associate IARI 1942; Ph.D. (Entomology) in 1955 form Changdigarh University for his work on coconut pests. Entomologist at the Central Plantation Crops Research Institutes Coconut Research Station at Kayangulam (1950–57). Joined Potato Research Institute, Shimla in 1957; was elevated to the post of Nematologist in 1961. In 1969 appointed as Deputy Director (Training) and Head, Central Plant Protection Training Institute, Hyderabad; promoted as Project Director and retired in 1975. Joined ICRISAT as Plant Quarantine Officer; died August 30, 1981 at Rome while on an assignment to FAO.

Nii(s) eggs of louse (Pediculidæ: Siphunculata); usually found glued on to the body hair of the host.

Nitche place or position, both functional and physical, of a species population in an ecosystem.

Nitidulidæ (Leach 1815), Stephens 1830 family of sapbeetles (Cucujoidea: Polyphaga); head relatively large, antennæ short, clavate and 11-segmented, eyes large, elytra truncate, not extending to tip of the abdomen; legs short, front coxæ transverse and tarsi 5-segmented. Both grubs and adults mostly saprophagous, feeding on fruits and garbags dumps- *Carrpophilus dimidiatus* (Fabricius), *C.hemipterus* (Linnæus), *Cybocephalus semiflaus, Meligethes æneus* Fabricius, et al.

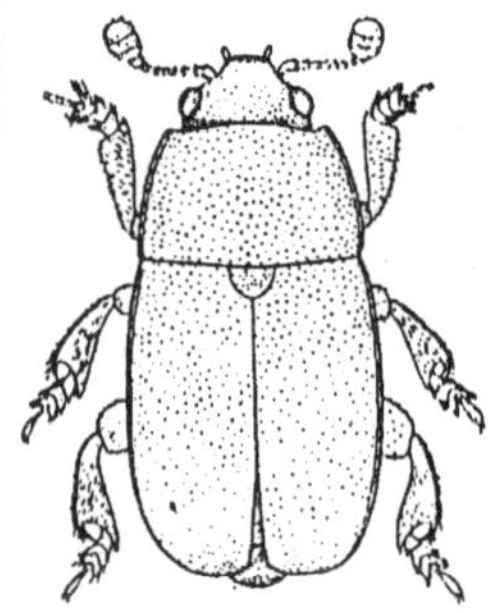

Moligethes æneus Fabricius

Noctuidæ Stephens 1829 (Agrotidæ) very large family of nocturnal moths (Noctuoidea: Ditrysia); cryptic and sombre coloured, proboscis almost always present, labial palpi well developed, frenulum present; hind-wing with Sc + R1 separate from Rs. Caterpillars usually naked or slightly hairy, few with thick tufts or spatulate filaments; mostly polyphgous (22000 species) - *Adisura atkinsoni* Moore, *Agrotis ipsilon* (Hufnagel), *Earias vittella* (Fabricius), *Helicoverpa* (Heliothis) *armigera* (Hübner), *Spodorptera litura* (Fabricius) et al., *Lobocraspis griseifusa* and allied species frequent eyes of cattle, man and other animals.

Noctuoidea Mosher 1916 superfamily of Ditrysia (Lepidoptera : Mecopteroidea); moths having maxillary palpi minute, tympanal organ present in metathorax; forewing with M_2 basally approximated to M_3; Cu_2 absent from both wings. Some important families are- Notodontidæ, Arctiidæ, Noctuidæ, Lymantriidæ Agristidæ, Dioptidæ, Hypsidæ etc.

Nocturnal active during night-most moths.

Nodus slight distinct notch near the middle of costa of each wing of Odonata, there is always a stout cross-vein at this point.

Nogodinidæ family of Fulgoroidea (Auchenorrhyncha: Homoptera); having clypeus with laternal carinæ: second segment of hind tarsus with one apical spine on each side; tegmina well developed; wing venation with supernumerary veins and cross-veins. About 130 species found in America, Africa and Indomalayan region.

Nolidæ small family of Noctuoidea (Ditrysia: Lepidoptera), moths resemble Noctuidæ but are smaller in size, have scale-tufts on wings and in hind-wing Sc + R_1 fusses with Rs from its base to near the middle of the wing. Caterpillars sparesly covered with hair and having 8 pairs of legs. (100 species -*Rhynchopalpus argentalis* Moore, *Selca scripta* Moore, et al.

Nomadidæ (Fallen 1813) Kirby 1837 family of wasp-like hairless, parasitic bees (Apoidea: Apocrita); posterior tibiæ of female without pollen brush: differs from Melectidæ in having anterior wing sculpture uniformly distributed and nervellus of posterior wing short and almost right-angled-*Nomada marshamella, N. ruficornis* and allied species prey on burrowing or mining bees (*Andrena* spp.).

Nomen dublem proposed taxonomic name but invalid as it is not accompanied by description or definition of the said taxon.

Nomen numdum proposed taxonomic name but invalid because accompanied description of taxon is not clear and cannot be interpreted satisfactorily.

Nomenclature terminology; mode or system of naming; for naming the living organisms, (plants and animals) binomial nomenclature as suggested by Carolus A. Linnæus (1707–78) has been universally adopted.

Non-compatibility incompatibility: property of a pesticide, which on mixing with another pesticide undergoes chemical or physical interaction and thereby reduce biological efficacy or increase phytotoxicity.

Non-persistant pesticide that breakdowns quickly and become inactive.

Non preference behaviour of insects avoiding a plant or host species.

No-see-um(s) nickname of minute bitting midges (Ceratopogonidæ: Culicoidea); common in North America-*Leptoconops carteri* Hoffman is $<$one mm long.

Nosodendridæ small but widely distributed family of wounded-tree beetles (Byrrhoidea: Polyphaga); antennæ having distinct 3-segmented club: mentum relatively large and tarsi 5-segemnted (30 species).

Notaulix (p. **notaulices**) longitudinal furrow convergent posteriorly in anterior part of mesonotum.

Notiothaumidæ Brues and Melanders 1932 monogeneric and monospecific family of scorpionflies (Protomecoptera: Mecoptera);

full-venation with more than 4 branches to Rs and many randomly placed cross-veins, male genitalia not swollen-*Notiothaums reedi* McLaclan often characterised as the living fossil.

Notodontidæ Stephens 1829 family of stout-bodied nocturnal moths (Noctuoidea: Ditrysia); having tympanal organs in metathorax; forewings elongated; midtibia with one pair of spurs and hind tibia with two pairs; tarsi short and hairy. Caterpillars without anal prolegs carrying anal somite more or less erect; eject irritant fluid from ventral thoracic gland-*Anticyra combusta* Walker, *Thiacides postica* Walker, *Turnaca acuta* Walker, et al.

Notoligotomidæ Davis 1936 very family of web-spinners (Euembioptera Embioptera); includes a few Asian and Australian species.

Notonectidæ (Leach 1815) family of back-swimmers (Notonectoidea: Hydroorisæ); small aquatic, light-coloured, carnivorous bugs, having body convex dorsally, antennæ very short and 4-segmented; tarsi 2-segmented, fore-tarsi not flattened, posterior tarsi without claws; swims with upside down, propelling with long hairy hind-legs; carry an air bubble beneath their wings for breathing under water. Feed on small crustaceans and other insects; also predaceous on fish tadpoles etc.-*Enitharees templetoni* Kirby, *Notonecta glauca* Linnæus and *Anisops bauvieri* commonly found in ponds etc. *Anisops species* is a small gnat-like diptera fly (Anisopodidæ, Nematocera).

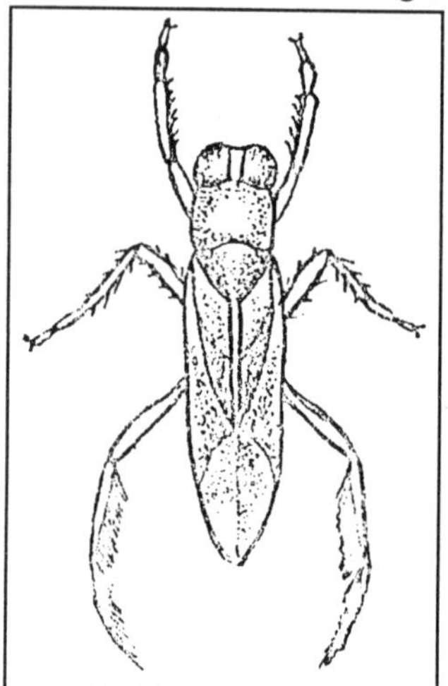

Enithrares templetoni Kirby

Notonectiodea superfamily of Hydrocorisæ (Heteroptera : Hemipterioidea); aquatic species; rosturm more or less elongate; antennæ 3- or 4- segmented and always concealed; ocelli lacking; fore-tarsi not spatulate-Naucoridæ, Belostomatidæ, Nepidæ, Notonectidæ, Pleidæ and Helotrephidæ.

Notonemuridæ family of Nemouroidea (Filipalpia : Plecoptera); includes *Notoneumura* species and its allied, confined to Southern Hemisphere and often placed under family Capniidæ (Nemouroidea: Filipalpia).

Notoptera Crampton 1915 alternative name for a very small primitive orthopteroidean insect order, commonly known as Grylloblattodea Brues and Melander.

Notum tergum or dorsal cuticular plate, covering each of the three thoracic segmented of an insect–pronotum, mesonotum and metanotum.

Nozzle a pointed spout specially one fitted at the end of the lanse or attached to hose or delivery-pipe. There are number of different types to choose one, depending upon the nature of spraying to be done.

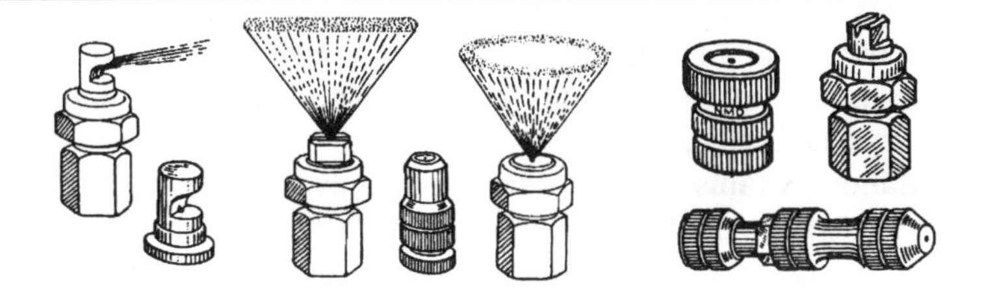

Nozzles

Nucleolus small spherical body located in the metabolic nucleus and composed mainly of protein : also called plasmosome.

Nucleus (pl. **nuclei**) small specialised mass of protoplasm in cell that contains cohromosomes and nucleoli and produces enzymes which stimulate various activities of cytoplasm.

Nun moth *Lymantria monacha* (Linnæus) (Lymantriidæ: Lepidoptera); tussock moth, forewings with a zigzag pattern of transverse black and white lines and hind-wings grayish–a serious pest of coniferous forests in Central Europe and oak woods in North America.

Nurse(s) young worker bee (Apoidea : Apocrita); whose main function is to feed the larvæ in beehive. Person trained for the care of sick or injured.

Nurse cells trophocytes (nutritive cells) of ovary or testis.

Nutrition chemical process invloving various activities and physiological mechanisms by which nutrients are transformed into body materials and energy; act or process of nourshing.

Nycteolinæ subfamily of Arctiidæ (Noctuoidea: Ditrysia) in old system of classification; now regarded as Choleophorinæ (Noctuidæ: Noctuoidea); generally green in colour having thorax and abdomen slender and apex of forewings produced to a point. Caterpillar fleshy having 8 pairs of legs and anal somite tapering to a point–

Earias insulana (Boisduval), *Pseudoips* (*Halias*) *arientalis* Hampson, et al.

Nycteribiidæ small family of Muscoidea (Schizophora: Cychorrhapha); wingless files, having head folded back, when at rest, in a groove on dorsum of thorax; antennæ 2-segmented terminated by a dendriform arista; eyes vestigial with one or two facets or absent; legs long and spider-like. Found in the fur of bats (mammals) sucking their blood (about 100 species)-*Cyclopodia greeffi, Miniopterus schreibersi, Penicillida jenynsi,* et al.

Nymph immature stage of heterometabolous insects that emerges form an egg and is similar to adult in appearance, but with wing and sex organ not fully developed.

Nymphalid any species of butterfly belonging to family Nymphalidæ (Ditrysia: Lepidoptera).

Nymphalidæ Swainson 1827 dominant family of brightly coloured, brush-footed butterflies (Papillionoidea: Ditrysia); antennæ knobbed; palpi large and clavate; forelegs greatly reduced (non-functional), in males brush-like and in females comb-like with 4 or 5 segments and no claws; tibia short and hairy. Wings large and subcostal cell rarely present. Sexual and seasonal dimorphism common. Caterpillarts leaf defoliators (5000 species) - *Egolis merione* Cramer, *Junimia orithya* Swinhoe (Linnæus), *J. almana* (Linæus) et al.

Nymphidæ Tillyard 1926 small neuropteran family (Myrmeleontoidea: Planipennia); distinguished by the presence of trichosors and antennæ not thickened or clubbed. Larvæ live in rubbish. Confined to Australasia-*Myiodactylus* spp., *Nymphes* spp., et al.

Nymphomyiidæ family of Psychoidea (Nematocera : Diptera); small files, mouthparts atrophied; antennæ with three stout segments and a short style; eyes separated dorsally but meeting beneath the head; wings having long fringes and reduced venation. Larvæ aquatic-*Palaeodipteron walkeri* Ide, et al.

Nymphulinæ subfamily of Pyraustidæ (Pyralioidea: Ditrysia) relatively large moths with proboscis vestigial; hindwings with vein Cu_2 fully developed. Caterpillars aquatic with (*Paraponyx* spp.) or without

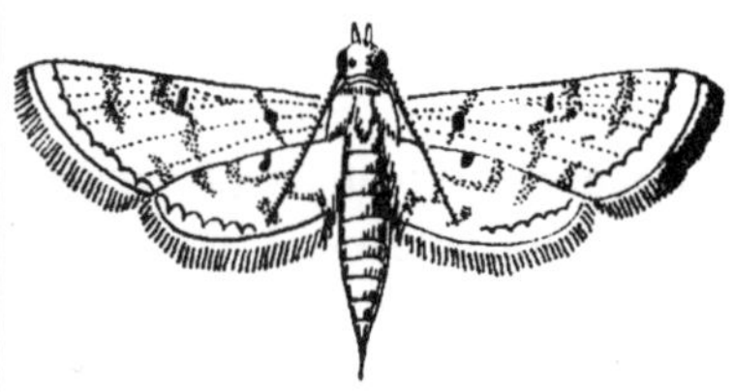

Nymphula depunctalis Guenee

(*Nymphyla* spp.) breathing gills; remain in cases made of leaf pieces and feed by scrapping leaf tissue of aquatic plants–rice caseworms, *Nymphula depunctalis* Guenee and *N.fluctuosalis* Zeller are pests of paddy in India; *N. enixalis* in Malaysia, *N. vittalis* in Japan and *N. nymphæta* (Linnæus) in Italy and Hungary; *N. stagnata* Donovan are semiaquatic, living in silk–lined sacs on water plants in Europe.

Nyssoninæ subfamily of Sphecidæ (Sphecoidea: Apocrita); commonly known as thread–waisted wasps are predaceous insects that nest in shallow burrows in the soil and provision their nests with cicadas, cockroaches, grasshoppers, flies, spiders etc.–*Mellinus arvebsis, Sphex lobatus, S.viduvatus, S.umbrous* and allied species.

O

Obcompressed flattened from front to back.

Oblet almost spherical in shape but slightly flattened at the poles (orange-shaped).

Obligate restricted to specified condition of life.

Obligate parasite parasite incapable of free-living.

Oblong two to three times as long as broad and of uniform width.

Obnoxious offensive: undesirable (said of insect pests).

Obovate egg-shaped with broadest part above the middle and tapering towards the base.

Observation a perception, either quantitative or qualitative, of a set of phenomenon.

Obsolete no more in usage: out of date.

Obtect type of pupa or chrysalis in which the wings and other appendages are glued to the body and most the abdominal segments are immovable-most Lepidoptera (butterflies and moths).

Occasional pest pest that is normally under natural control, but sporadically, its population increase, specially in localised areas, exceeding the economic-injury-level.

Occipital arch horse-shoe shaped band of cranium between occipital and postoccipital sutures.

Occipital foramen opening in the back of an insect's head, through which oesophagus, nerve cord etc. enter the thorax.

Occipital region occiput: dorsal part of occipital arch: back of head, where it joints the thorax: area between vertex and neck

Oceanic inhabiting ocean of deep-sea. Strictly speaking there is no oceanic insect species. There are some water-skaters, *Halobates* spp. (Gerridæ: Heteroptera); found on high seas, but these are semi-aquatic, as they breath air from atmosphere. Others live in under water but only temporarily and by utilising trapped pocket of air.

Ocellus (pl. **ocelli**) simple eye or eye-spot in insects and other lower invertebrates : photoreceptor having single diopteric apparatus. In adult insect, ocelli are dorsal while in larvæ these are lateral and have a different nerve supply. Eye-spot of eye-like pattern on the wings of some lepidopterous insects (moth or butterfly) are also called occellus.

Ochteridæ (Pelogonidæ) family of Ochteroidea (Hydrocorisæ: Heteroptera); includes small, oval, black velvety store-bugs having antennæ short and 4-segmented but not concealed; ocelli and eyes large, rostrum long and 4-setmented; legs similar. Predaceous, found mostly near banks of lakes and ponds (20 species).

Ochteroidea superfamily of Hydrocorisæ (Heteroptera: Hemipteroidea); semi-aquatic unsects having 4-segemented, elongated rostrum, ocelli present and fore-tarsi not spatulate-Gelastocoridæ or Monoychidæ (toad-bugs) and Ochteridæ or Pelogonidæ (shore-bugs).

Ochthiphilidæ Zetterstedt 1890 (Chamaemyiidæ) Hendel 1910) family of minute gray or back spotted dipteran files (Lauxanioidea: Schizophora); vibrissæ absent, vein Sc complete; predacceous on aphids and coccids (1200 species)- *Leucapis griseola* (Fallen) and *L.nigicornis* Egger are predaceous on *Myzus persicæ* (Sulzer) *Aphis gossypii* Glover et al. while *L. luteicornis, L. bella* Loew and *L. bellula* Williston prey upon eggs and young ones of *Pseudococus* spp.

Ocneriidæ old and equivalent name of lepidopteran family Lymantriidæ (Noctuoidea: Ditrysia).

Ocular tubercles supplementary eyes found in addition to compound eyes in many aphids (Aphididæ) and jumping plant lice (*Livia* spp., Psyllidæ).

Odiniidæ family of small dipteran files (Agromyzoidea: Schizophora); characterised by theca not developed at the base of proboscis; wings having costa broken at R_1, cross-veins *r-m* and *im* close together, basal, median and anal cells well defined and squamæ vestigial. Only a few species associated with sap-fluxes of the trees.

Odonata Febricius 1793 (= Crytodontia Laterile 1802, Libnellulides Lech 1815, Libellulina Newman 1834) order of Odonateoptera (Palaeoptera : Pterygota); brilliantly coloured carnivorous insects having long, slender body; head hypognathous and exceptionally mobile, antennæ very short and filiform; eyes very large and conspicuous; mouthparts mandibulate; wings membranous, equal or subequal with numerous veins and cross-veins and prominent stigma. Nymph (naiad) aquatic. Metamorphosis incomplete (6500 species) - Anisoptera (dragonflies), Anisozygoptera and Zygoptera (damelflies).

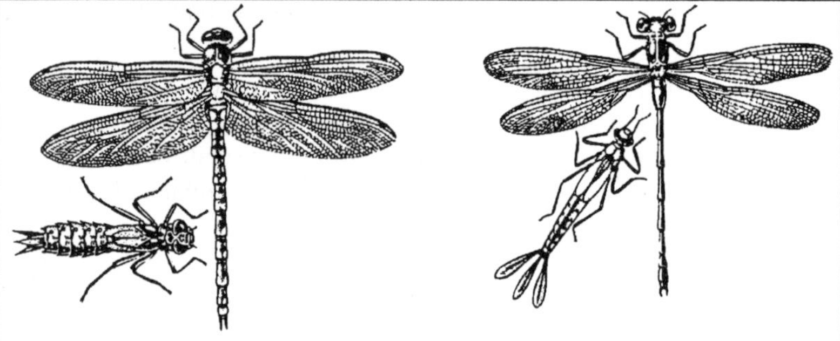

Dragonlly Dameslfly

Odontologist person who has specialised in odontology.

Odontology branch of entomology that deals with dragonflies and damselflies (Odonata: Odonatoptera).

Odonatoptera Martynov 1938 superorder of Palaeoptera (Pterygota: Insecta); that includes only one living order- living order - Odonata - hence occasionally used synonymously with Odonata.

Odontoceridæ Wallengren 1891 small family of Trichoptera; widely distributed all over the World, except Australasia; caddisflies having no ocelli and forewings with veins R_1 and R_2 confluent apically or connected by an apical cross-vein. Larva eruciform having labrum much longer than broad with dorsal surface without any stout bristles; inhabit swift streams - *Psilotreta fontalis* Banks, et al.

Oecanthinæ subfamily of Gryllidæ (Grylloidea: Ensifera); includes large, pale-coloured tree-crickets.

Oecology branch of biology dealing with living organisms' habits, modes of life and relations to their surroundings.

Oecophoridæ (= Depressariidæ) family of small, brownish coloured moths (Gelechiodea: Ditrysia); head scaled, antennæ with basal pectin; maxillary palpi and ocelli vestigial or absent; labial palpi long and porrect; wings apically broad. Caterpillars leaf-rollers and webbers (3500 species) - *Psorosticha zizyphi* (Stainton), *Porthomologa parachina* Meyrick, *Agonopterix heracliana,* et al.

Oedemeridæ Thomson 1859 family of brightly coloured, flower-beetles (Heteromera: Cucujoidea) elongate soft-body with long filamentous antennæ head little deflexed without any neck; eyes enarginate; tarsi with one tow penultimate segments lobed. claws toothed or simple. Adults feed on flower and grubs on dead wood- *Oedemera nobilis* is bright green, thick-legged flower-beetle.

Oedischiidæ an extinct primitive ensiferan family (Orthoptera) of Upper Carboniferous period.

Oenochromatidæ Guenee 1867 family of Geometroidea (Ditrysia; Lepidoptera); that includes delicate, sombre or bright coloured moths having forewings without accessory cells, R_1 fusing with areole of Sc; females of some species with vestigial wings. Cosmopolitan in distribution–cankerworms *Alsophila pometaria* (Harris) et al.

Oencoytes large cells present in various part of insect body, probably producing a hormone, connected with moulting process.

Oenocytoids round acidophil corpuscles in blood of insects; their function in not known.

Oesophageal pertaining to oesophagus.

Oesophageal ganglia pair of small ganglia behind the brain of an insect and joined to it by thin connective nerves. The ganglia rest upon the dorsal surface of oesophagus and are closely connected with corpora allata.

Oesophagus tubular portion of stomodæum between pharynx and crop, in the alimentary canal of an insect.

Oestridæ (Leach 1815, 1817) small family of bot–or warble files (Muscoidea: Schizophora); stout built, head large with small eyes; antennæ short and partially sunk in facial grooves; mouth parts atrophied; postscutellum distinct. Larvæ endoparasitic on mammals–cannel botfly *Cephalopina* (*Cephalopsis*) *titillator* (Clark), sheep botfly *Oesturs ovis* Linnæus, elephant botfly *Pharyngolobus africanus* Brauer, kangaroo botfly *Tracheomyia macropi* Froggatt, et al.

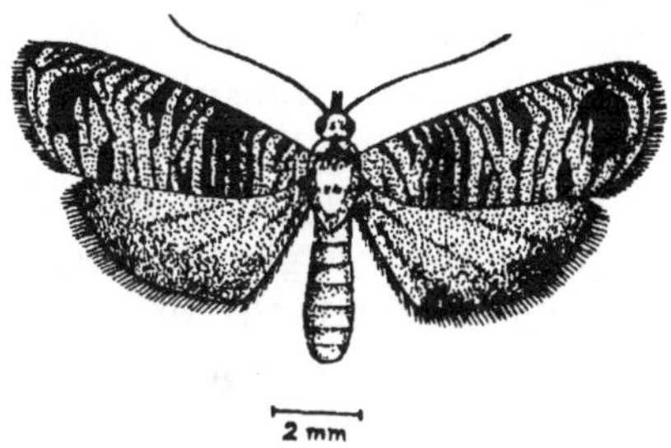

Cydia pomonella (Linnaus)

Olethreutidæ Walsingham 1900, 1913 (= Eucosmidæ) family of Tortricoidea (Ditrysia: Lepidopteera); often included in family Tortricidæ (Tortricoidea); brownish-gray, small moths with wavy pattern on wings; characterised by fringes of long hair on basal part of Cubitus in hind-wings. Caterpillar web or roll up the leaves and bore into fruits–codling moth *Cydia* (*Carpocapsa*) *pomonella* (Linnæus), *Dudua aprobola*

(Meyrick), *Ancylis aromatias* Meyrick, *Laspeyresia saltitans* (Westwood) et al.

Olfactorial pits small pits on the antennæ containing a number of hair-like receptors and nerve connections by which the insect carries out the sense of smell.

Olfactory sense of smell; concerned with smelling.

Oligembiide- Davis often considered synonymous to family Teratembiidæ Krauss (Euembioptera: Embioptera); the largest embiopteran family that includes small enbiids (webspinners) found in Africa and the New World.

Oligoærobe any organism that thrives at low oxygen concentration.

Oligoærobic state of low oxygen concentration, pressure or tension.

Oligoneoptera panorpoid group of neopteran insects with substantial reduction in number of veins in their wings–all holometabolous (endopterygota) orders.

Oligoneura Barner 1904 alternative, obsolete and equivalent name for insect order Embioptera Shipley 1904 which includes embiids and webspinners.

Oligoneuriidæ (Oligoneuriellidæ Ulmer 1920) family of Heptagenioidea (Euphemeroptera: Plaeoptera); having large triangular forewings with reduced venation; legs reduced and non-functional. Nymphs having simple gills-*Oligoneurilla rhenana* el al.

Oligoneustic having reduced number of spiracles.

Oligophagous (organism) feeding upon a few closely related host plants; confined to one family - cabbage-butterfly larva- *Pieris brassicæ* (Linnæus) and allied species (Pieridæ: Papilonoidea) are pests of cole crops only.

Oligotomidæ Enderlein 1909 family of webspinners (Euembioptera : Embioptera); characterised by vein R_4-5 forked and X tergite in males asymmetrical-*Oligotoma greeniana, O. minuscula, O.* saundersii (Westwood) and allied species are Indian while *Haploembia* spp. are Mediterranean.

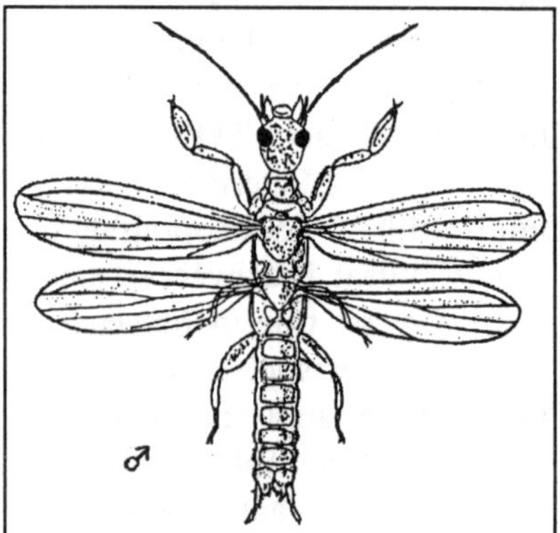

Oligotoma saundersil (Westwood)

Oligotrophic poorly fed : deep water bodies (lake, sea etc.)

having abundant oxygen and small nutrient supply resulting in low production of organic material. *Cf.* entrophic.

Olivaceous olive-green; dusk yellowish green (colour).

Olivarly olive-shaped : oval.

Ombrophilous able to thrive in areas of abundant rainfall.

Ombrophobous unable to live in presence of long continuous rains.

Omethoate $C_5H_{12}NO_4PS$; organophosphatic (phosphorothiolate) systemic acaricide and insecticide; only liquid, slightly yellowish in colour readily soluble in alcohol, acetome and water, but less soluble in ethyl ether and petroleum ether. Toxic to bees. Acute oral and dermal LD_{50} values for rat are 50 and 1400 mg/kg respectively – Folimat® etc.

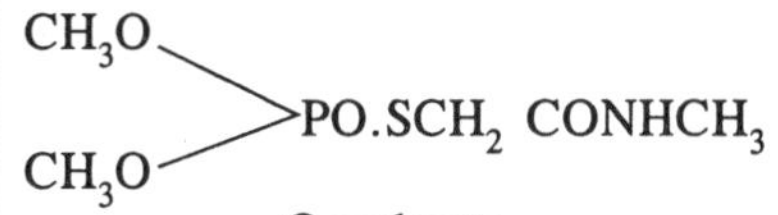

Omethoate

Ommateum (pl. **ommatea**) compound eye; generally hemispherical with all the ommatideia converging inwards towards the optic nerve.

Ommatidium (pl. **ommatidia**) single optical unit of the compound eye of insects and crustaceans; cone-shaped structure having a circular lens, a crystalline cone and retibulæ or sensitive cells at the base.

Ommechrome pigments brown, red or yellow pigments that given the characteristic eye--colours of many insects; sometimes also found in other parts of body: derivaties of amino-acid tryptophane

Ominvore (amnivorous) organism that fees upon almost anything and everything may be plants, animal organic matter etc.–wasps, termites etc.

Omophronidæ a small family of savage beetles (Caraboidea : Adephaga); hitherto regarded as a subfamily of Carabideæ–*Omophron* spp. et al.

Omoptera Leach 1815 old and obsolete term for insect order Homoptera Laterille.

Omphralidæ (= Scenopinidæ) a small family of orthorrhaphous files (Aslioidea: Brachycera); small dark coloured narrow oblong flies devoid of true pubescence or bristles. About 200 species – *Scenopinus* spp. et al.

Omphralidæ family of Entomobyoidea (Arthropleona : Collembola); having scales and long mucro; antennæ long with vesicular sense organ on IV antennal segment-*Oncopodura* spp. et al.

Onion fly *Delia (Hylemya) anitqua* (Meigen) (Anthomyiide: Diptera); slender, grayish flies, 6 mm long, having large wings with an expanse of 12 to 15 mm. White maggots burrow don into underground portion of stem and often the bulb; damage is often followed by attack of fungus (*Bacillus carotovorus*), resulting into soft rot of onion.

Onychium pulvillus: slender or pad-like cushions on ventral tarsal segments or a pair beneath the claws of true flies (Diptera), often having hollow tenant hairs for exuding adhesive secretions to aid the legs in clinging to various surfaces.

Onychiuridæ family Poduroidea (Arthropleona : Collembola); sense-organs on III antennal segment; pseudocelli present on thoracic and abdominal terga;-*Onychiurus armatus* Tullberg *Tetrodontophora bielanesis* et al.

Onychophora (Prototracheata) a minor phylum of primitive, free-living, worm-like terrestrial creatures (Coelomata: Tripoloblastica); having combine features of both annelides and arthropods; cylindrical body, short velvety skin with dermopmuscular body wall devoid of chitin; pair of ringed antennæ and numerous paired laterally placed segmental appendages that end in claws; earlier a class of Arthropoda; includes about 80 species-*Peripatus* spp.

Oocytes egg-cell before maturation usually accompanied by nutritive cells.

Oogenesis development of female gamete or ovum.

Oogonium (pl. **oogonia**) first in differentiation of egg cell from a primary female gamete or ovum.

Oology scientific study of eggs (generally of avian eggs).

Ootheca horny egg-case or capsule formed by hardening of stickly secretion from the colleterial glands of cockroaches.

Opadothripidæ Bagnall 1930 family of thrips (Thysanoptera) that are mostly fossil and not yet placed under amy suborder or superfamily.

Open (blood) circulation a system characteristic of insect and some arthropods wherein the blood is not contained in arteries, veins and capillaries but files the body cavities.

Operculum a lid, flap or valve: cap-like structure of an egg. commonly found in plant bugs (Heteroptera); the cap is lifted by the young insect (nymph) when it emerges form the egg.

Opisthognathous head with the mouthparts placed between the fore-legs and the head directed backwards. *Cf.* hypognathous, prognathous.

Opostegidæ small family of Nepticuloidea (Monotrysia: Lepidoptera); earlier included in family Nepticulidæ as moths of both these families have similar wing venation in their fore-and hind-wings.

-optera (Greek: *opteron* = having wings) suffix of majority of insect orders (living and non-living) denoting the texture or type of wings possessed by the insects of that order–Palæoptera (ancient winged), Neoptera (new or modern wings), Orthoptera (straight winged), Thysanoptera (fringed wings), Coleoptera (sheath winged), Hymenoptera (membranous wings).

Optic lobes lateral expansions of protocerebrum of an insect innervating the compound eyes and ocelli, these lobes are situated immediately above the oesophagus and form the largest region of insect's brain.

Optic nerves nerve fibers emerging form the complex region of optic lobes and leading to the compound eyes.

Opticon inner zone on internal medullary mass of optic lobes.

Orbit(s) parafrontals: the upward continuations of the genæ, along the inner border of eyes as found in dipteran flies.

Order category in taxonomic hierarchy just below the rank of class/ subclass and comprising of one or more related superfamilies/ families. According to modern classification, there are 30 order of living insects, besides eight extinct ones (appendix 1.2) Name of most order end with 'ptera', which means wings-Orthoptera (straight wings), Thysanoptera (fringed wings), Diptera (two wings), Zoraptera (without wings) and so on.

Organ group of tissues that perform one or more particular function(s).

Organism an individual living being–plant or animal–organised body composed of various parts, each bearing structural and functional relationship to every other part.

Organochlorine(s) compounds having carbon (C), hydrogen (H), and chlorine (Cl) as base molecular constituents and may have oxygen (O) and sulphur (S) as well BHC or HCH ($C_6H_6Cl_6$) DDT ($C_{14}H_9Cl_5$), dicofol ($C_{14}H_9Cl_5O$) etc.

Organophosphate(s) chemicals with molecules having one or more atoms of phosphorous, usually derivatives of phosphates (PO_4); Gerhard Schrader, a German Scientist, discovered the insecticidal potency of this chemical group, during the World War II– Chlorfennvinphos ($C_{10}H_{14}NO_5PS$) ethion ($C_9H_{22}O_4P_2S_4$) etc.

Oriental fruit moth *Grapholitha molesta* (Busck) (Eucosmidæ: Lepidoptera); a close relative of codling moth; the caterpillars bore into young shoots and fruits specially of peach *Prunus persica* (Linnæcus) and other stone and pome fruits.

Ormyridæ family of Chalcidoidea (Apocrita: Hymenoptera); often united with family Torymidæ, but it differs from Torymidæ in having indistinct parapsidal sutures and a short ovipositor. Parasitic on gall-making insects.

Orneodiidæ (= Alucitidæ) a small isolated family of many-plume, silvery-white moths [Copromorphoidea (Alucidoidea): Ditrysial]; both pairs of wings cleft into six or more narrow plume-like structures which are fringed densely with hairs along both the margins. Caterpillars burrow into shoots, flower-stalks and buds, causing the galls-*Orneodes hexadactyla, O. pygmæa* et al.

Orothripidæ Bagnall 1926; family of Æolothripoidea (Terebrantia: Thysanoptera); often included in family Æotothripidæ; Nearctic thrips having all segments of antennæ freely movable and distinguished by maxillary palpi being 7-or 8-segmented and labial palpi 3 to 5 segmented; ovipositor saw-like.

Ortheziidæ (Green 1896 as Orthezites) Enderlein 1920 small family of homopteran scale insects (Coccoidea: Sternorrhyncha); having elongate oval, convex and distinctly segmented body with prominent 4-to9-segemented antennæ, one pair of ocellanæ on tubercules, rostrum 2-segmented. Males with well developed compound eyes and 3 ocelli; females covered with hard white waxy plates and carry a large waxy egg-sac-lantana but *Orthezia insignis* (Browne) is a pest mainly of an obnoxious weed, lantana (*Lantana* spp).

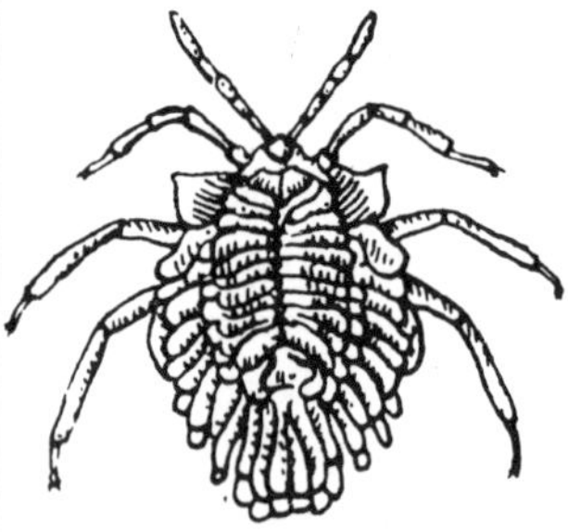

Orthezia insignits
(Browne)

Orthokinesis a simple reaction of insects by which a particular stimulus can cause them to move in straight line.

Orthoptemology branch of entomology that deals with study of only Orthopteran insects.

Orthoptera Olivier 1789 order of polyneopteran insects (Orthopetroidea: Pterygota); mostly terrestrial, having large compound eyes and generally three ocelli (two in long-horned grasshoppers and lacking in apterous forms); mouthparts mandibulate; forewings narrow and leathery (tegmina), hind-wings memberanours; fold longitudinally and are covered by tegmina when at rest; hind-legs enlarged and adapted for jumping; ovipositor well developed; auditory and stridulatory organs usually present. Metamorphosis incomplete or direct (paurometabolous). About 28000 species classified into sub-order-Cælifera (locusts and short-horned grasshoppers) and Ensifera (crickets and long-horned grasshoppers).

Orthopteroidea Handlirsh 1903 old obsolete term for a subclass of Insecta, now regarded as a superorder under Polyneoptera (Neoptera: Pterygota); having wings with well developed jugal areas and numerous malpighian tubules; comprises of four extant orders-Plecoptera, 'Grylloblattodea, Phasmida, Orthoptera. Extinct order Protorthoptera and extant order Embioptera are placed under superorder Embioidea Handlirsch.

Orthorrhapa Braurer 1863 suborder of Diptera in, an old classification, that included more primitive forms of Nematocera and some Brachycera. The adults emerge form pupæ by a straight longitudinal slit.

Orussidæ Haliday 1840 very small family of sawflies (Orussoidea: Symphyta: Hymenoptera); evidently a relic of an ancient group. These so called wood-wasps have wings with reduced venation and no closed submarginal cells in hind-wings. Larva apodous and ectoparasitic on wood-boring beetles (Buprestidæ) *Orussus obietinus* et al.

Orussoidea Rohwer 1911 superfamily of symphyta (Hymenoptera: Hymenopteroidea); antennæ inserted between the eyes but below the clypeus; wing venation weak, hind-wins without any enclosed cells; ovipositor long and when retracted it is lopped within the abdomen-Orussidæ

Oscinidæ Fallen 1820 synonym of dipteran family Chloropidæ Rondani 1856 (Acalyptratæ: Schizophora).

Osmeterium (pl. **osmeteria**) glands found in many caterpillars including those of puss-moths and swallowtail butterflies (Lepidoptera); consists of an eversible pouch or sac wherein repellent scent or other secretion collects; on being disturbed, this pouch is suddenly everted and the scent (secretion) thrown out, acts as efficient defence mechanism.

Osmylidæ Brauer 1868 family of Beautiful giant lacewings (Osmyloidea: Planipennia); ocelli present; maculated wings with numerous cross-veins, vein Sc joined to R_1 at the apex. Larvæ have long, slender stylet-like mandibles and maxiilæ and no gills; live in wet moss or under stones; prey upon dipteran larvæ-*Osmylus fulviicephalus, Spilosmyus flavicournis* et al.

Oxmyloidea superfamily of Planipennia (Neuroptera), membranous wings, venation much branched, nygmata present on hind-wings between two posterior branches of vein Rs.-Ithonidæ, Neurorthidæ and Ismylidæ.

Ostium (pl. **ostia**) one of the paired lateral slit-like openings of heart by which blood flows into heart for pericordial sinus in insects and other arthropods.

Ostomatidæ Gangibaurer 1899 (Ostomidæ) alternative name for family Trogossitidæ (Cleroidea: Polyphaga); includes lignicolous beetles; larva commonly called Cadela (cadelles).

Othnidæ old and equivalent name for coleopteran family Elacatidæ (Heteromera: Cucujoidea).

Otiorrhynchinæ subfamily of Curculionidæ (Curuculionoidea: Polyphaga); weevils having short rostrum and memtum expanded to conceal the maxilæ.

Otitidæ Curran 1934 (Ortalidæ Macqart 1835) a large family of dipteran files (Agromyzoidea: Schizophora); having mottled wings with costa usually unbroken and cell Cu_2 pointed distally. Male having five evident abdominal segments and elongate rolled up ædeagus while in female there are six abdominal segments with a flattened corneous oviscapt. Recently this family has been splitted into seven families - Pyrgotidæ, Tachiniscidæ Phytalmiidæ, Ulidiidæ, Pterocallidæ, Platystomatidæ and Pallopteridæ.

Oubit hairy caterpillar.

Outhbreeding mating of genetically dissimilar and relatively unrelated individuals.

Oval broadly elliptical, rounded at both ends, nearly twice as long as broad, broadest in middle.

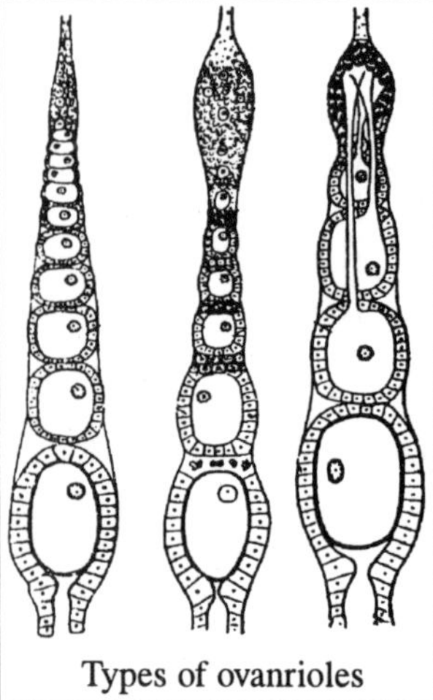

Types of ovanrioles

Ovariole long slender, delicate, transparent and tapering subunit (organ) of ovary, in which developing eggs (oocytes) are disposed in a single chain, the oldest oocyte being nearest the union of oviduct; usually number of ovarioles in an ovary is four (a few may have one or none) but 100 to 200 is not uncommon and some termites (Isoptera) have over 2000 in each ovary. Each ovariole is distinctly marked into three regions–terminal filament, gernarium and vitellarium. There are three types of ovarioles, recognised on the basis of their location and presence or absence of nutritive cells:–

Pansoistic – primitive type; specially differentiated nutritive cells wanting–palæopteroid orders as also Thysanoptera and Siphonaptera.

Meroistic– distinguished by presence of nutritive or nurse–cells. These may be *Polytrophic* wherein nerve–cells alternate with oocytes [Neuroptera, Adephaga (Coleoptera), Lepidoptera, Diptera and Hymenoptera] or *Acrotrophic* having nutritive cells situated at the apices [Polyphaga (Coleoptera) and Homoptera].

Ovary female gonad wherein egg–cells multiply and are nourished; an insect normally has to ovaries, each having its own duct into which eggs from the ovarioles are discharged; the two ducts join together into a common duct (oviduct). Each ovary is composed of a number of ovarioles in which eggs are produced.

Ovate egg–shaped, with broadest part below the middle and tapering towards the top.

Ovicide toxin that can kill eggs, specially of insects and mites.

Oviduct the median duct through which the ova pass, it is formed by joining of two ducts, one from each ovary. In primitive insects, it has an opening (gonopore) in VII abdominal segment; while in others the oviduct is connected to vagina with opening in IX abdominal segment.

Oviductus communis median ectodermal outlet duct of female reproductive system, opening into genital chamber or vagina.

Oviductus lateralis one of the two lateral ducts in female genital system that lead posteriorly form ovaries and discharge into oviductus communis (oviduct).

Oviferous (ovigarous) egg–carrying.

Oviform egg–shaped : ovoid.

Oviparous reproduction by laying eggs, that develop and hatch outside the body of mother: egg-laying animals– insects, birds, fishes and even some species of mammals.

Oviporus posterior opening of vagina in most lepidopterous insects.

Oviposition process of egg–laying.

Ovipositor egg-laying organ in female adult insect.

Ovisac brood–pouch; egg–capsule.

Oviscapt functional ovipositor. in many endopterygote female insects; the distal abdominal segments become attenuated and often telescoped to form a retractile tube or oviscapt; commonly found in most dipteran flies including Cecidomyidæ. Agromyzidæ,

Muscidæ and Trypetidæ as also in most monotrysian and some ditrysian insects.

Oviviviparous (insects) producing eggs that hatch within the mother's body but are not nourished by mother through placenta marsupials as in case of mammals.

Ovogenesis formation of ova : oogenesis.

Ovoid oval in shape.

Ovulation release of ovum from the ovary.

Ovum (pl. **ova**) nature unfertilised egg-cell: non-motile female gamete.

Oxybelidæ small family of spiny digger wasps, earlier included in subfamily Carbroninæ (Sphecidæ: Aprocrita); predaceous wasps having characteristic spines on postscutellum; nest is a burrow in the soil; prey on dipteran flies that are brought in the nest, impaled upon the sting-*Oxybelus uniglumis* et al.

Oxychirotidæ small Indo-Australian family of Pterophoroidea (Ditrysia: Lepidoptera); characterised by wing divided into two plumes. Caterpillar feeds on fallen seeds of mangrove (*Avicennia* species)-*Cenoloba* species et al.

Oxycoryindæ very small family of Curculionoidea (Polyphaga: Coleoptera); comprises of hardly 20 species - *Oxycorynus* species in South American and *Metroxena* species of East Indies.

Oxdemeton-methyl $C_6H_{15}O_4PS_2$; phosphorothioate acaricide and insecticide; systemic in action; odourless yellowish liquid, miscible in water, soluble in most organic solvents and practically insoluble in petroleum ether an unstable in alkalies. Toxic to bees. Acute oral and dermal LD_{50} values for rat 57 and 100 mg/kg respectively-Metasystox-R.

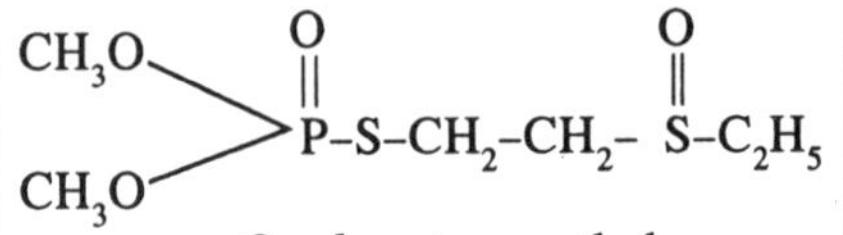

Oxydemeton-methyl

Oxytenidæ small family of Bombycoidea (Ditrysia: Lepidoptera); having short probosic and no frenulum; confined to South America.

Oyster-shell colloquial name of several armoured scale insects (Diaspididæ: Cocoidea); wingless females are covered with tiny shell-like scale formed form cast-off skin-*Oudraspidiotus ostreæformis* (Curtis) and *Lepidosaphes ulmi* (Linnæus), are cosmopolitan species and most omnivorous feeders.

Ozæninæ a subfamily of Carabidæ (Caraboidea: Adephaga), earlier placed under Paussidæ (Caraboidea: Adephaga).

P

Pachyneuridæ small family of dipteran flies (Bibionoidea : Nematocera); antennæ having 16 or 17 elongate segments; eyes holoptic or separate, ocelli present; wings well developed, veins almost of uniform thickness.

Pachytroctidæ small family of booklice (Troctomorpha: Psocoptera); having convex body with long antennæ; winged or wingless; hind femora not flattened and abdomen conical. Found in litter, nests and houses (40 species).

Pædogenesis precocious maturity: production of offsprings (parthenogenetically by an immature animal or larva/pupa. In insects, larvæ of gall-midges, *Heteropeza* spp. and *Miastor* spp. (Cecidomyidæ Diptera) give birth to 25 to 30 daughter larvæ which eat their way out of the mother larva and the process is repeated for several generations. In *Heuria* spp. (gall-midges) the pupæ reproduce pædogenetically, some of the emerging larvæ become normal adults most of other remain as hemipupa with only vestiges of wings and legs. Grubs of *Micromathus* species (Micromalthidæ: Archestemata) also reproduce parthenogenetically.

Painted beauty a small-sized North American butterfly *Vanessa virginiensis* Drury (Nymphalidæ : Papilionoidea); widely distributed in Holarctic region; distinguished by rosy-orange wings with patches of dark drown and irregular white spots near the tip of forewings. Caterpillars feeds on plants of Asteraceæ family.

Painted lady a medium-sized European butterfly *Vanessa cardui* (Linnæus) (Nymphalidæ: Papilionoidea); found all over the World except South America; similar in appearance to painted beauty but bigger in sized and often migrates in enormous swarms. Caterpillars feeds on nettles (*Celtis* spp., Ulmaceæ) and thistles (*Carduus nutans* Linnæus) and allied species, Asteraceæ).

Palæ(o)- ancient: prehistoric fossil (species).

Palæobiochemistry study of chemical processes used by the organisms that lived in geologic past.

Palæodictypoptera Goldenberg 1854 superorder of fossil primitive insects of Carboniferous period, resembling dragonflies. In

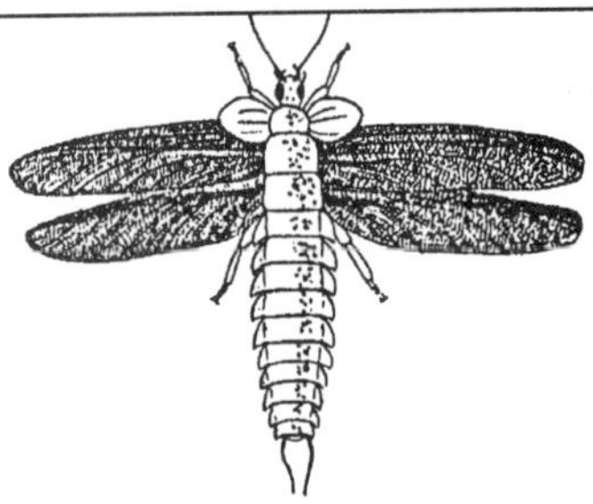

Stenodicya lobata
(Dictyoneuridæ : Eupalæodictyoptera)

addition to two pairs of elongate, broad based wings with highly branched venation, prothorax had a pair of short wing-pads (paranota), these probably helped the insects to glide and all abdominal segments with a pair of lateral lobes. These insects were phytophagous, saprophygous and carnivorous, grouped into three orders (all extinct) – Eupalæodictyoptera, Protohemiptera and Megasecoptera.

Palæoecology ecology of geologic past.

Palæogæa Old World as a biological region.

Palæomorphology study of the form and structure of fossil remains in order to describe the original anatomy of an organism.

Palæontinidæ an extinct family of Mesozoic Lepidoptera; large, stout moths, hind-wings much smaller than forewings; medial cell fully divided in both wings and R_4 branched forewing-*Limæodites mesozoicus* and *Eociada lameeri* lived during Jurassic period.

Palæontography descriptive palæontology.

Palæontology branch zoology that deals with the study of only extinct organism

Palæontology branch of zoology that deals with the study of only extinct organism and their fossil remains.

Palæoptera Martynov 1923 infraclass or section of old primitive pterygote insects (extinct and extant), wings without jugal area, held in primitive position (not foldable), venation simple unspecialised network. Hemimetabolous and polyneophoidia; comprising of three superoders-Palæodictyoptera (extinct), Ephemeroptera and Odonatoptera.

Palæzoologist person well versed in palæozoology.

Palæozoology scientific study of fossil animals including insects.

Palingeniide Klapalek 1909 family of mayflies (Ephemeroidea: Ephemeroptera); having forewing with distal half of Sc concealed in a fold; female with reduced legs and claspers of male 3-segmented. Nymphs with dorsal gills and externally toothed mandibles-*Palingenia longicauta* Olivier (largest European mayfly) is 30-38 mm long.

Paligernesis unaltered recaptilation of ancestral characters by the developing stages of an organism.

Palp (pl. **palpi**) sensitive tactile organ, near the mouth of insects and various other invertebrates (arthropods, molluscs, polychætes etc). In insects these are either on labium (labial palpi) or maxillæ (maxillary palpi) but never on mandibles; often elongate and multisegmented but some may be short and inflated; equipped with obvious olfactory and tactile cells.

Palpicornia alternative name of coleopteran family Hydrophilidæ (Hydrophiloidea: Polyphaga); vegetarian water-beetles, having short antennæ and conspicuous thick palpi as long or longer than antennæ. Adults good swimmers, live upon decomposing vegetable matter while the grubs of most species are predaceous-*Hydrophilus piceus* is about 50 mm long.

Palpifer a small sclerite, situated on insect's maxilla one on either side, and bearing a 5-segmented maxillary palp.

Palpiger a small lobe (sclerite) situated lateral on prementum (labium) on either side, from which arises a 3-segmented labial palp.

Palpus telopodite of gnathal appendage.

Pamphiliidæ Virereck 1916 family of Megalodontoidea (Symphyta: Hymenoptera); stout-bodied, leaf-rolling sawflies having simple antennal flagellum, board wings with primitive venation but Sc absent in forewings and short ovipositor. Larvæ devoid of abdominal appendages; live gregariously in rolled leaves or leaf-webs-*Neurotoma inconspicua* Norton in plum webspinning sawfly and *Pamphilus persicus* MacGillivary is peach leafroller sawfly (both in Eastern North America), whereas *P.sylvaticus* is found on rose plants.

Panchætothripidæ Bagnall 1912 family of bristle thrips (Thripoidea: Terebrantia); found in the Oriental region; characterised by last abdominal segment in females cylindrical and heavily chitinised; segments IX and X with very long, stout, spine-like bristles.

Pamoistic egg-tube an egg-tube in which vitellarium contains only eggs.

Panorpidæ Samonelle 1819 family of scorpionflies (Eumecoptera: Mecoptera); tarsi 2-segmented; wings well developed, legs not to long; female without conspicuous ovi-positor. Found in Europe and North

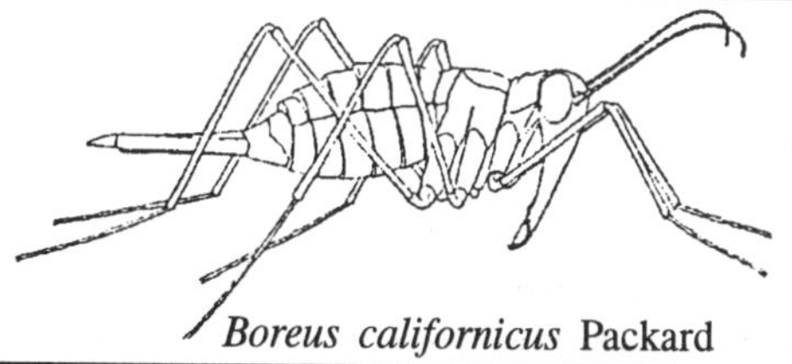

Boreus californicus Packard

America *Boreus californicus* Packard, *Panorpa communis* (Linnæus) et al; *P. klugi* McLachlan is Asiatic species.

Panprotura Crampton 1920 an old and obsolete term for the apteran insect order Protura Silvestri (Myrientomata: Apterygota).

Pantropical (species) widely distributed in Tropical and subtropical region of the World.

Panurgidæ Schmiedeknecht 1882 family of Apoidea (Aculeata: Apocrita); having tongue elongate and acute; marginal cells of anterior wing truncate or subtruncate distally and the apex separated form wing margin; these bees are pollen and nectar gatherers-*Panurginus* species, et al.

Papilionoidæ Leach 1819 family of swallowtail butterflies (Papilionoidea: Ditrysia); large and beautifully patterned, having normal legs; wings generally iridescent black with shades of blue, green, red and yellow; hind-wing with one anal vein and conspicuous tail-like prolongation. Over 600 species maximum species of genus *Papilio*-Indian species include *P. demoteus* Linnæus, *P. helenus* Linnæus, *P. machaon* Linnæus, *P. polytes polytes* Linnæus, *P. p. alphenor* Cramer, *P. p. pammon* Linnæus, *P. porinda* Moore, *P. protenor* Cramer and *P. rummanzorla* Eschscholz, while *P. dardanus* is African. All these are pests of citrus trees, first one being the most destructive.

Papilionoidea Dyar 1902 superfamily of Ditrysia (Lepidoptera : Oligoneoptera that includes all butterflies; formerly classed as Rhopalocera Dumeril 1806 (Lepidoptera). These are distinguished form moths by their clubbed antennæ and absence of frenulum and jugum on the wings; vein Cu_2 is also missing from both wings-Nymphalidæ, Riodinidæ Lycænidæ, Pieridæ and Paliplionidæ. Danaidæ, Satyridæ and Heliconiidæ are considered as subfamilies of Nymphalidæ and Libytheidæ as subfamily of Riodinidæ.

Paraglossæ lateral ligular lobes of labium each with muscle arising in prementum.

Paragnath one of the paired lobes of hypophatrnx in certain insects.

Paramere(s) a pair of lateral processes of phallobase, which together with penis and claspers comprise the external genitalia of male insect.

Paraneoptera Martynov 1923 infraclass of Pterygota; heterometabolous insects similar to Polyneoptera having larger number of veins and no cross-veins, but of much recent evolutionary origin-all hemipteroid orders (Psocoptera, Mallophaga, Siphunculata, Homoptera, Heteroptera and Thysanoptera).

Paraneuroptera old and obsolete term for order Odonata.

Paranotal lobes lateral lobes of pronotum in certain fossil insects: hypothetical lobes of meso-and metanotum supposed to be precursors of wings.

Parapatric species or population that occupy non-overlapping but adjacent geographical areas without interbreeding.

Paraproct(s) a pair of ventral plates just below the anus, found in certain insects; these are reduced sternal lobes of XI abdominal segment.

Parapteron small additional sclerite often present in meso-and metathorax of an insect, just below the wings.

Parasita insect order (Pterygota) propositioned by Latreile (1802) to included both biting and sucking lice; subsequently birdlice (biting lice) were placed under newly erected order Mallophaga and sucking lice grouped as Siphunculata.

Parasite any organism that lives all or part of its life in (endoparasitic) or upon (ectoparasitic) another living organism (host) of different species, deriving nutriment and shelter from its host without rendering any service in return; generally the host is not killed, but occasionally the parasites do cause severe loss of vitality or transmit disease which may eventually result in death of the host.

Parasite, Adult parasite that thrives on nymphal-adult insects-larvæ of *Epipyrops melanoleuca* (Fletcher) suck body juice of nymphs of adults of *Pyrilla* spp.

Parasite, Egg endoparasite that completes its life-cycle within the egg of its host *Trichogramma chilonis* Ishii in eggs of *Corcyra cephalonica* Station (in laboratory) and in eggs of various lepidopterous insects (in nature).

Parasite, Egg-larval parasite that lays its eggs in eggs of its host; the latter hatch into larvæ with parasites still feeding and growing within; ultimately the host larvæ die, as the parasite adults emerge from the same

Parasite, Gregarious parasite whose number of individuals feed and grow together on/in single individual host species-*Apanteles glomertus* (Linnæus) on cabbage worm (*Pieris* spp.) *Cf.* solitary parasite.

Parasite, Larval parasite that attacks larval stage of insects; lays eggs in/on larvæ of its host and pass the entire life-cycle on that host-*Sturmiopsis inferens* Townsend is larval parasite of sugarcane bores.

Parasite, Nymphal parasitoid that lays egg in/on host nymph, passes its entire life-cycle on that host and emerges out as an adult, leaving the host nymph dead–*Chlorodryinus pallidus* Perkins and *Pseudogonatapus pyrillæ* are nymphal parasite of *Pyrilla* spp., while *Amitus aleurolobi* Mani is pupal parasite of sugarcane whitefly.

Parasite, Obligae parasite which cannot live independent of its host.

Parasite, Primary one that feeds and grows on or within non–parasitic host.

Parasite, Pupal (larval–pupal) parasitoid that lays eggs in full grown larvæ (mostly caterpillars), which pupate but instead of host adults, adult parasites emerge form the parasitised pupæ, which ultimately die away–*Xanthopimpla stemmator* Thumberg is pupal parasite of sugarcane bores, *Trichospilus pupivora* is pupal parasite of *Oisine arenosella.*

Parasite, Solitary such parasite that feeds and develops singly in or on its host–*Isotima javensis* Rohwer is solitary ectoparasite of *Tryporyza nivella* Fabricius.

Parastic castration destruction of reproductive organs by parasites.

Parasitica (Terebrantia) one of the two main division of Apocrita (Hymenoptera: Hymenopteroidea); includes parasites as against Acuteata or stinging forms. In Parasitica the ovipositor is adapted for piercing and in Aculeata, it is usually for stinging. However, the distinction between the two divisions is not sharp, as number of Parasitica are phytophagous and many Aculeata are parastic. Superfamilies included are–Ichneumonoidea, Chalcidoidea, Cynipoidea, Evanioidea, Trigonaloidea and Procototrupoidea; the last one often considered as third independent division of Apocrita.

Parasiticide toxin that can kill or destroy the parasites.

Parasitise infest as parasite.

Parasitism temporary or permanant symbiotic relationship between two different species in which one derives the benefits (parasite) while the other way or may not feel the loss or injury caused by the parasite.

Parasitism, Facultative phenomenon adopted by an organism by which it can lead parastic or free–life.

Parasitism, Incidental type of parasitism wherein an insect parasitises an organism, which is not its normal host - *Bracon greeni* Ashmead on *Tryporyza nivella* Fabricius.

Parasitism, Indirect form of secondary parasitism wherein a parasite established itself in a host, in which it is capable of developing, for the sake of primary parasite which the host may or may not harbour.

Parasitism, Obligatory process wherein a parasite is restricted to specified or limited existence-all true parasites.

Parasitism, Social form of parasitism in which the individuals of certain species invade the nests of social insects like ants, bees wasps or termites for parastitisation and for taking up permanent residence there. Not all such interlopers are harmful to the nest community. Some live as scavengers and are consequently tolerated, a few exude certain fluids which either repel the attack or are accepted as sweet bribes by the inmates of host colony.

Parasitism, Polyembryonic a phenomenon wherein an individual host is parasitised by only one larva of the parasitic species but this larva develops into several parasitoids within the host body-*Copidosoma* species on groundnut leafminer *Stomopleryx subsecivella* Zeller (Gelechiidæ) and potato tuber moth *Phthorimæa operculella* (Zeller) (Gelechiidæ).

Parasitism, Solitary development of only one individual parasite in or on an host individual-*Apnteles augaleti* (Viponidæ), *Trioxys indicus* (Incubidæ) *Trissolcus basalis* (Scelionidæ) et al.

Parasitism, Super parasitisation of individual host by more than one larvæ of a single parasitic species, all maturing therein/thereon. Parasitism in which a single host individual is attacked repeatedly by the same species of the parasite and the host is not in a position to support that much of population (suer abundance of parasite) which resulted in into abnormal progeny.

Parasitoid organism which is alternatively free-living and parasitic in its life-history. Insect parasitoids are essentially predator, living within tissues of their hosts in such a subtle way that the hostls death is postponed until the interlopes have completed their larval development.

Parasitoidism systematic feeding by an insect larva on living host tissues to that the host will live until completion of larval development.

Parasitologist person having sound knowledge of parasitology.

Parasitology branch of biology that deals with the study of parasites, nature of parasitism and its effects on host and parasite.

Parasitosis infestation with parasites.

Parathion $C_{10}H_{14}NO_5PS$; phosphorothionate insecticide, contact and stomach poison with some fumigant action. Ethyl parathion is dark brown liquid with unpleasant garlic-odour, slightly soluble in water insoluble in mineral oils and kerosene and readily soluble in

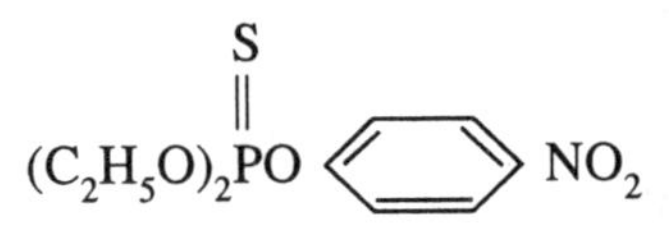
Parathion

benzene, glycols and xylene. Extremely toxic to man and mammals. Acute oral and dermal LD_{50} values for rat being 3–6 and 4–200 mg./kg respectively–Ekatox®, Folidol®, Paramar etc.

Paratype specimen other than holotype available with original author at the time of describing a new species by the author.

Parieta(s) $(CH_3COO)_2Cu.3Cu(AsO_2)_2$; copper acetoarsenite insecticide; emerald green powder, highly phytotoxic; used since 1867 as stomach poison against Colorado potato beetle, *Leptinotarsa decemlineata* (Say); also used in poison-baiting in bran against slugs and snails. Acute oral LD_{50} value for rat 22 mg/kg. With the inception of synthetic organic insecticides (1939) its use has been widely discouraged.

Parmidæ (Leach 1818) Stephens 1929 an old and equivalent name for coleopteran family Dryopidæ Erichson 1847 (Dryopoidea: Polyphaga) that includes subaquatic and aquatic beetles–*Psephenus* spp. occur in North America and Asia.

Parnassidæ Swainon 1840 small family of mountainous butterflies (Papilionoidea: Ditrysia); usually translucent white, partly or completely marked with brownish-gray and circular ping, orange or red spots the wings–Apollo butterfly *Paranassius apollo* Linnæus, Alaskan apollo *P.eversmanni* Menetries, black apollo *P. mnemosyne* (Linnæus) et al.

Parthenogenesis reproduction without mating or fertilisation: development of unfertilised egg into new individual–common in aphids (Aphidoidea) or male (Hymenoptera).

Passalidæ family of scarabæoidea (Polyphaga: Coleoptera); dark brown to black coloured, flattened and parallel sided beetles, with a horn-like process on head; antennæ not geniculate; mentum in males deeply notched; elytra with longitudinal grooves cover the abdomen completely. Produce squeaking sound by rubbing lower surface of wings with upper surface of abdomen, (500 species) – *Epispheus neelgheriensis, Meolontha melolontha* (Linnæus), *Ophrygonius cantori* et al.

Passandridæ a small family of Clavicornia (Cucujoidea: Polyphaga); beetles charcterised by gular sutures confluent tarsi 5-5-5, first segment usually small and no segment lobed beneath. About 100 species, found mostly in warm climate – *Læmotmetus ferrungineus* et al.

Patagium short lobe or wing–like outgrowth on each side of prothorax as in certain noctuid moths. These were well developed in primitive fossil dragonflies (Palæodictyoptera).

Pathodome host population in which all individuals have a given resistance in common.

Pathogen micro–organism that feeds on or within its host causing small to severe (even fatal) injury; disease causing micro–organism.

Pathotype parasitic population in which all individuals have a given parasitic ability in common.

Paurometabola (paurometabolous) incomplete or direct metamorphosis wherein young ones (nymphs) resemble the adults in general body form and mode of life; but do not have completely developed wings and genitalia.

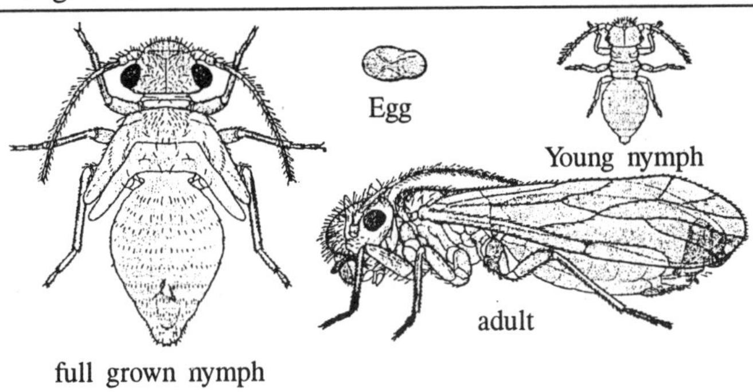

Paurometabola

Paussidæ family of Caraboidea (Adephaga: Coleoptera); now included as subfamily of Carabidæ (Caraboidea), small, reddish–brown to black coloured, flattened and parallel–sided beetles; antennæ thick, broad and expanded; elytra truncate leave the anal end of abdomen exposed; secrete a liquid which is liked and licked by the ants- *Ceratoderus bifasciatus* and *Platyrrhopalus denticrnis* are found in the nests of ants *Pheidole latinoda*.

Peach tree borer *Sphenoptera lafertei* Thomson (Buprestidæ: Buprestoidea); serious pest of stone fruits (mostly temperate, especially of Rosaceæ family) and prefers bright sunshine and old, weak and unhealthy trees; trees less than five years old are generally not attacked.

Peacock the medium-sized European butterfly, *Vanessa io* (Linnæus) (Nymphalidæ: Papilionoidea); beautiful and brightly coloured with large eyespots on the wings; commonly found in Europe.

Pecton comb-like stridulating organs on some insects; row of spines or hair on antennæ of certain moths (Lepidoptera); as also on respiratory tube of mosquito-larvæ (Diptera).

Pedicel second antennal segment between scape and flagellum; it usually carries sense organs (organs of Johnston). Delicate middle portion of haltere (balancer) in dipteran files; it is supported by dialted basal portion (scabellum) and surmounted by knob-like capitellum. Short duct form egg-tube to oviduct in female insects.

Pedicle (pedicel) narrow, between thorax and abdomen in some hymenopterous insects: between cephalothorax and abdomen in certain arachnids (spiders etc).

Pediculidæ Samouelle 1819 family of blood sucking lice (Siphunculata: Psocopteroidea); 3 to 4 mm long, grayish, wingless insects; body flattened dorsoventrally with all abdominal segments unwrinkled and free; infests man and other mammals-*Pediculus humanus humanus* Linnæus and *P.h. corporis* de Geer are found in human head and body hairs respectively.

Pedilidæ very small family of Heteromera (Cucujoidea: Coleoptera); often included in family Anthicidæ (Heteromera); both grubs and adults are found in vegetable refuse in dump places.

Pelecinidæ Haliday 1840 family of Proctrupoidea (Apocrita : Hymenoptera); large pelecinid wasps characterised by first segment of gastor being as long as head and thorax together; hind- and basitarsus much shorter than 2nd segment-*Pelecinus polyurator* Drury common in Temperate North America (not in India); females 55 to 60 mm long and male only 15 to 20 mm, larvæ parasitic on white grubs (Coleoptera).

Pelogonidæ equivalent name for family for family Ochteridæ (Ochteroidea: Hydrocorisæ); that includes about 20 species of predaceous velvety shore-bugs that can fly and jump actively; found near ponds and streams.

Peloridiidæ only family of Coleorrhyncha (Homoptera: Paraneoptera); small primitive bugs, confined to Southern Hemisphere; body flattened; antennæ short, 3-segmented and without terminal arista; rostrum 4-setmented, party ensheathed by propleura; male with simple ædeagus, distinct parameres but no submarginal plates (20 species)-*Oiphysa fuscata* (New Zealand), *Peloridium hammoniorum* (South America).

Peltidæ small family of Tropical beetles, often included in family Torgossitidæ (Cleroidea: Polyphaga); characterised by being broad and rounded with forecoxal cavities open behind-*Peltis* and allied species.

Peltoperlidæ small family of Filipalpian stoneflies (Plecoptera : Orthopteroidea); having uncertain affinities and reported only form Holarctic and Oriental regions (35 species)-*Peltoperla* spp., et al.

Penetrant oil, added to spray-solution so that the solution can penetrate, the waxy cuticle of insects.

Penicillate brush-shaped : tufted.

Penis ædeagus: male intromittant or copulatory organ containing the gonopore from which seminal fluid is discharged; consists of two elongated outgrowth of IX abdominal segment; situated between the two claspers.

Pentastomida uncertain Arthopoda, now considered as an independent phylum; soft, worm-like unsegmented animals, having short cephalothorax and elongated ringed abdomen; tow pairs of hooks surround the mouth; no circulatory, excretory or respiratory organs; sexes separate; endoparasitic in vertebrates-*Sibekia* spp. in crocodiles, *Lingnatual serrata* in goats, horses, rabbits etc. and *Cephalobæna tetrapoda* in snakes.

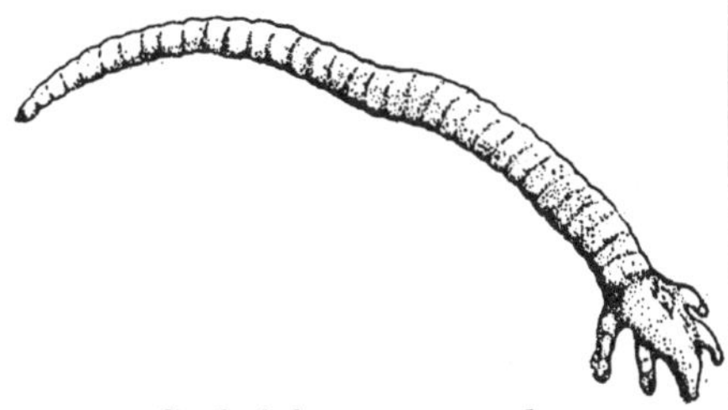

Cephalobæna tetrapoda

Pentatomidæ (Leach 1815) (Pentatomides Leach 1815) family of Pentatomoidea (Pentatomomorpha: Geocorise); brightly coloured, broadly oval, shield-shaped stink bugs; scutellum larger and triangular; hemelytra well developed, membranae with 5 to 12 veins; emit offensive smell and are mostly phytophagous and polyphagous (Pentatominæ), a few predaceous on other insects (Asopinæ, Acanthosominæ) - *Bagrada cruciferaum* (Fabricius), *Carabula biguttala* Fabricius, *Dalpada pillicornis* Stoll,

Bagrada cruciferarum Fabricius

Tessaratoma javanica Thunberg are some crop pests; *Andrallus spinidens* and *Cantheconidia furcellata* are predaceous on various caterpillars (Lepidoptera).

Pentaomoidea superfamily of Pentatomomorpha (Geocorisæ: Heteroptera); strongly sclerotised body, antennæ generally 5–segmented; pronotum six-sides, mesocutellum large; legs with 2 or 3 tarsal segments; claws with pulvill; membrane always with veins. Families included are - Plataspididæ, Cydnidæ, Scutelleridæ Acanthosomatidæ, Pentatomidæ, Thaumastellidæ and Phloeidæ.

Pentatomomorpha section of Geocorise (Heteroptera: Hemipteroidea); includes true–bugs having antennæ 4 or 5 segmented; hemelytra without costal fracture, hind–wings with veins R and M Separate distally. Eggs not operculate and not implanted in substrate. Classified into six supefamilies–Aradoiea, Saldoidea, Lygæoidea, Coreoidea, Pentatomoidea and Pyrrhocoroidea.

Pergidæ (Ashmead 1900) Rohwer 1911 family of Tenthredinoidea (Aporcita: Hymenoptera); antennæ 6-segmented, serrated or clubbed, 3rd segment short; cross–vein 2*r* absent and hind–wings with no enclosed medium or anal cells; abdomen cylindrical. Found in Australia and South America. Larvæ feed on laves of *Eucalyptus* spp. (50 species) –species of *Perga, Phylacteophaga, Xyloperga* et al.

Pericardial cells neptrocytes in heart region of an insect believed to be connected with excretion as these can absorb nitrogenous waste products blood.

Pericardial sinus part of hæmoceol or body cavity surrounding the hart of an insect of other arthropod.

Perientomidæ Enderlein 1927 family of Trogiomorpha (Psocoptera: Psocopteroidea); small, soft-bodied insects; antennæ 20 to 25-segmented; wing scales usually asymmetrical or two sided and hind–wins having a very narrow closed cell between M and Cu, at the base.

Perilampidæ Cameron 1884 small family of Chalcidoidea (Apocrita: Hymenoptera); metallic–green or black, stout–bodied wasps distinguished by coarsely punctate large thorax and small triangular abdomen. Mostly hyperparasites on *Microgaster indicus* and *Apanteles machæralis* which are primary parasites of *Hapalia machæralis. Perilampus hyalinus* Say is hyperparasite on *Olesicambe* species and *Varichæta* species that parasitise the larvæ of *Hypantria* spp.

Periodic release(s) repeated releases of indigenous biotic agents in a particular locality with a view to artificially raising and maintaining the high population level of the parasitoid which is normally unattainable in nature.

Periopticon outer layer of optic lobes, nearest the eye.

Peripneustic having lateral spiracles on all abdominal segments.

Peripsocidæ family of Psocopmorpha (Psocoptera: Psocopteroidea); includes giant cockroach of Malagasy, *Grompnadorrhina protentosa,* a flightless blattid, 85 to 90 mm long. makes snake-like when disturbed.

Perissomatidæ family of few small dipteran flies (Mycetophiloidea : Nematocera); having each compound eye separated into two complete parts and wing-venation very peculiar or rather curious; breed in fungi. Reported only from Chile and Australia.

Peristome ventral marginal rim of epicranium.

Perlaria Handlisch 1903 old and equivalent name for Plecoptera Burmesiter 1939, a polyneopteran order than includes all stoneflies; no more in usage.

Perlarides Leach 1815 old and equivalent name for Plecoptera Burmeister 1939, that includes perlids, samonflies and stoneflies; no more in usage.

Perlidæ (Latreille 1802) Stephens 1829 largest family of Perloidea (Setipalpia: Plecoptera); stoneflies having mandibles reduced or vestigial; anterior coxæ widely separated, 3rd tarsal segment longer than 1st and 2nd together; no archedictyon. Nymphs carnivorous. Found all over the World except Australasia (350 species) - *Acroneuria pacifica* Banks, *Alloperla borealis* Banks, *Neoperla clymene* Newman, *Perla capitata* Pictet, et al.

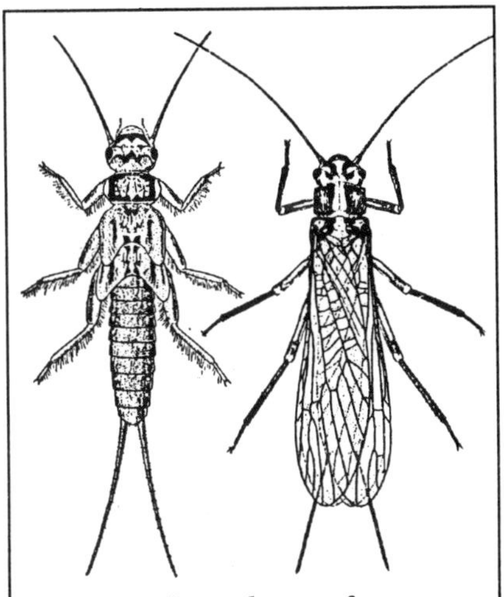

Isoperla consfusa

Perlodide heterogenous family of stoneflies (Setipaplia : Plecoptera), glossæ reduced, mandibles weakly developed and non-functional, males of some species short-winged; found mostly in Northern Hemisphere (over 220 species) – *Diura bicaudata, Isoperla* spp. (nymphs without gills), *Isogenus* spp. (nymphs polyphagous).

Perloidea (= Subulipalpia) superfamily of Setipalpia (Plecoptera: Orthopteroidea); comprises of small stoneflies having glossæ reduced, mandibles small and non–functional; first two tarsal segments very small and furst abdominal sternum small. About 700 species distributed in three families – Perlidæ Perlodiæ and Chloroperlidæ.

Perophoridæ equivalent name for family Locosomidæ (Bombycoidea Ditrysia).

Perothopidæ very small family of coleopteran insects (Elateroidea: Polyphaga); resembles eucnemids (Eucemidæ: Elateroidea) but having mandibles slender; fore–coxe small with concealed trochantins, hind–coxæ contiguous, tarsal claws pectinate and V abdominal sternite free – only three species of *Perothops,* found in North America.

Persistant property of pesticide to remain active for a long period after its application.

Pest any organism (animal or plant) that causes damage, loss or annoyance to man or his property to the extent that warrants its control.

Pest density population level at which a pest species causes economic damage.

Pest management reducing pest population by careful manipulation and adoption of selected control measures.

Pest spectrum complete range of pests attacking a particular crop.

Pest upset upsetting the balance of nature : a phenomenon, wherein the natural pest population is upset by excessive use of pesticides that destroys the natural enemies of the pest, thereby allowing the pest population of increase without much hindrance from the nature.

Pesticide an omnibus term covering all chemicals that can be used for preventing, destroying, repelling or mitigating the pest population– acaricide, avicide, bactericide, fungicide, herbicide, insecticide, ixodicide, larvicide, miticide, molluscicide, nematocide, ovicide, oxyuricide, rodenticide, sylvanicide, tickicide, weedicide etc.

Petalluridæ Tilyard 1926 family of Æshnoidea (Anisoptera: Odonata); large dragonflies, measuring 60 to 80 mm, having notched labia; and angle of hind–wing angularly produced in males, ovipositor well developed - very conspicuous in their distribution; *Petalura* spp. occur in Australia, *Uropetala* spp. in New Zealand, *Phenes* spp. are neotropical, *Tanypteryx* spp. are nearctic and *Tachopteryx* spp. are holarctic.

Petiolata Bingham 1897 old and obsolete name for suborder Apocrita (= Clistogastra Konow 1905) (Hymenoptera).

Petiole narrow stalk: narrow part of abdomen as seen in ants, bees and wasps (Apocrita: Hymenoptera).

Petriidæ very small family of Heteromera (Cucujoidea: Coleoptera); includes a few species of beetles from Central Asia.

pH symbol indicating concentration of hydrogen ions in a solution expressing acidity or alkalinity of the medium. pH value of 7.0 indicate neutral (water), lower values indicate acid and higher alkalies.

Phæomyiinæ subfamily of Sciomyzidæ (Sciomyzoidea: Schizophora); often regarded as separate family, includes small dipteran flies that are predatory on slugs and snails.

Phagocyte(s) blood-cells (hæmocytes) that ingest and digest small solid particles found in the blood.

Phagocytoblast primitive blood-cell that develops into phagocyte.

Phagocytolysis destruction of break down of phagocytes.

Phagocytosis engulfment and devouring of small solid particles, specially foreign ones that are harmful to the body.

Phalacridæ family of shining flower-beetles (Clavicornis: Cucjoidea); small convex, 1 to 3 mm long and dark brown to black in colour; fore-coxal cavities open behind; tarsal claws toothed or appendiculate. Larva having paired urogomphi (500 species) - *Phalacrus* larvæ develop on sugarcane smuts and grasses while those of *Olibrus* are phytophagous feeding on flower-heads specially of Asteraceæ.

Phalænidæ alternative and obsolete name for family Noctuidæ (Agrotidæ Digrysia); comprising of owlet moths.

Phalænoididæ old and equivalent name for family Agaristidæ (Noctuoidea: Ditrysia).

Phallobase proximal part of phallus supporting ædegus.

Phallcrypt pocket or pouch or phallobase or genital chamber wall containing base of ædeagus.

Phallomere(s) secondary phallic-lobes arising form primary phallic-lobes; found in higher insects; the median pair is mesomere and lateral ones are parameres.

Phallotheca distal part of phallobase that forms a fold or tubular sheath which party or wholly encloses the ædeagus.

Phallotreme opening of endophallus located at the distal end of ædeagus.

Phallus median or unpaired intromittent organ, including phallobase, ædegus and any include endophallus also.

Phaloniidæ equivalent but obsolete name for family Cochylidæ (Tortricoidea: Ditrysia).

Phanerognatha sub-division of Curculionidæ (Curculionoidea: Polyphaga); a group of weevils characterised by absence of mandibles but having conspicuous maxillæ not hidden by the mentum. Larva with conically projecting antennæ. *Cf.* adelognatha.

Phantom larva predatory aquatic larvæ of gnat, *Chaoborus crystallinus* (Chaoboridæ: Culicoidea) having transparent body with internal organs clearly visible. Adult non-blood-sucking and harmless. Also known as phantom gnat.

Pharate adult an adult insect that is still within its pupa though fully formed and free from pupal integument.

Pharate instar a stage in development of an insect during which its cuticle has separated form the hypodermis but has not yet been cast off a ruptured.

Pharate pupa an instar in the development of holometabolous insect which is in-between larva and pupa; feeding has cased and the insect has separated from the last larval cuticle but is still within it; often considered as prepupa or perpupal stage.

Pharynx (pl. **pharynges**) part of stomodæum between buccal cavity and oesophagus; saves as passage for food.

Phasmatodea Jacobson and Bianchi 1905 alternative and equivalent name for insect order Phasmida Leach 1815; no more in usage.

Phasmid common name for any species belonging to family Phasmidæ (Phasmida: Orthopteroidea). One of a pair of lateral caudal pores the function as sense organs in certain nematodes (Secernemtea: Nematoda).

Phasmida Leach 1815 order or large-sized, alate or apterous but predominately tropical insects (Orhopteroidea: Polyneoptera); body well sclerotised resembling leaves or twigs, head prognathous, mouthparts ectognathous and mandibulate; prothorax short, meso- and metathorax elongated, if alate, forewings usually small and strongly sclerotised; hind-wings membranous; auditory and stridulatory organs lacking ovipositor small and cerci short.

Metamorphosis incomplete (4000 species) embracing to superfamilies–Phosmoidea and Bacterioidea.

Phasmidæ Brunner 1893 family of long slender, alate or apterous insects (Phasmoidea: Phasmida); antennæ long in both sexes; metanotum longer than wide; abdomen without lateral outgrowths; feed on leaves of *Eucalyptus* spp. – *Argosarchus horridus* White (130 to 160 mm long), *Extatosoma tiaratum* Macleay (120 to 14 cm) and *Podocanthus wilkinsoni* Macleay (90 to 100 mm) are some species found in Australasia destructive to *Eucalyptus* trees.

Phasmodea Burmeister 1838 (= Phosmoidea Handlirsch 1903) alternative names suggested for insect order Phasmida Leach 1815; no more in usage.

Phasmoidea Brues and Melander 1932 superfamily of Phasmida; includes leaf-insects having leaf–like expansions on forefemora; middle and hind tibiæ with sunken arolia on the underside of the apex – Bacillidæ, Phyllidæ and Phasmidæ.

Phaudinæ a small and very aberrant subfamily of Zygænidæ (Zygænoidea: Ditrysia) distinguished by lack of mouthparts.

Phenacolechiidæ monogeneric family of scale insects (Coccoidea : Homoptera) that comprises of only one species–*Phenæcleachia zelandica* found in New Zealand.

Phengodidæ family of Cantharoidea (Polyophaga : Coleoptera); antennæ 12–segmented, antennal sockets widely separated; male abdomen not entirely covered by elytra, female without wings; luminous organs producing light of more than one colour present in both adults and larvæ eve eggs and pupæ may be luminous to some extent. About 50 species, confined to America; feeding on millipedes.

Phenocopy non–hereditary altenation of a phenotype to another type imitating a mutant trait; caused by external conditions during the development.

Phenogenetics study of phenotypic effect of genetic material: physiological genetics.

Phenotype totality of characters of an individual due to interaction between geneotype and environment.

Phenthoate a phosphorothiologhionate acaricide and insecticide; reddish yellow oily liquid with slightly aromatic smell; slightly soluble in water but soluble in other organic solvents like acetone, benzene, ethanol ethyl–ether, xylene etc., phytotoxic to some varieties of apple and peach. Acute oral and dermal LD_{50} values for rat 200 to 300 and 700 to 1400 mg/kg respectively - Cidial® etc.

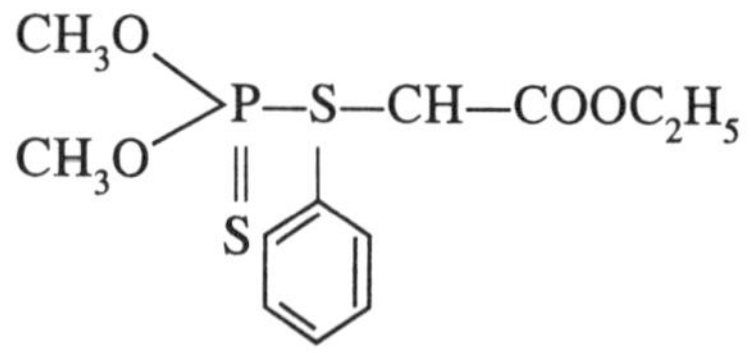

Phenthoate

Phenyl mercury acetelat $C_8H_8O_2Hg$; a white odourless free-flowing powder; soluble in acetone, alcohol and benzene, sparingly soluble in water, used after formulation (as mercurial fungicide) for seed-dressing to protect various cereals and vegetable crops from seed-borne and soil-borne diseases.

Pheromone(s) chemical substances that are secreted by an organism to external environment that elicit behavioural response from other individuals of some species: chemicals used for intraspecific communication - sex pheromones help in locating the opposite sex: alarm pheromones and recruitment pheromones released by social insects are ultilised for collective defence or increased foraging activity.

Pheromones, Aggregation arrestants : chemicals or chemical combinations that cause insects to aggregate. These are released only by one sex of the insect species but elicit response in both sexes of that species-Sulcatol produced by males of ambrosia beetle *Gnathroticus sulcatus;* bark beetle *Dendroctonus brevicomis* releases a mixture of exo-breviconin, frontalin and myrcene.

Pheromones, Alarm substances elaborated by mandibular glands, anal glands or the sting apparatus produced by ants, aphids, honeybees, locusts, wasps etc. to influence the behaviour (flight or aggression) of their fellows-citronellal tridecane, undecane etc. are produced by ants white citral, isoamylacetate are released by bees.

Pheromones, Territorial chemicals secreted by males of many species to mark mating sites and attract both males and females. More common in bees (*Binus terrestris* Linæus, *Melipona* spp., *Xylocopa hirutissina*), ants (*Atta* texana, *Lasius fuligienous, Monomorium pharaonis* Linnæus) and termites (*Nastitermes* spp. *Reticulitermes flavipes* Kollar, *Zootermopsis nevadensis* Hagen).

Phernomens, Trail marking more stable and presistant than other types of pheromones; common in Dictyoptera, Lepidoptera and ensocial Hymenoptera and used to form a chemical trait which is perceived by other individuals, whose behaviour is thereby changed–ant *Formica rufa* Linnæus uses formic acid as a trail marker.

Philanisidæ small family if Limnephiloidea (Integripalpia: Trichoptera); caddisflies having no ocelli, dorsal part of tentorium reduced by male has 5–segmented maxillary palpi. Forewings with normal wing venation. Larva eruciforms, pests of paddy (*Oryza sativa* Linnæus) specially in China, Japan and Korea causing minor damage.

Philanthinæ subfamily of Sphecidæ (Sphecoidea: Apocrita); includes digger–wasps and bee–killer wasps; sting and paralyse their victims (bees, winged ants and beetles) and provision their nests (burrows) with the same to feed their larvæ–*Philanthus ramkrishnæ* is found in South India, while *P.tringulum* is serious pest of bees in Germany where it is called Bienenwolf.

Philopotamidæ small family of Hydropsychoidea (Annulipalpia: Trichoptera); caddisfies characterised by last segment of maxillary palpi annulated. Larvæ compodeoid, live in silken non–portable retreats.

Philopteridæ Brumeister 1938 Nirmidæ Samouelle 1819 (Nirmides Leach 1818) major family of birdlice (Ischnocera: Mallophaga); antennæ 5–segmented, last one curved, front margin of clypeus smoothly curved; legs with two claws. Different species found on different types of birds–*Calumbicola columbæ* (Linnæus) on *G. pavonis* (Linnæus) on peacock, *G. dissimilis* Nitzsch on chicken, *G. damicornis* Nitzsch and *G. minor* Piaget on pigeon, *Lipeurus gallipavonis* Geoffroy on turkey, *Degeeriella* species on birds–of–prey, *Cuclotogaster hetergraphus* on poultry, *Esthiopterum crassicorne* (Scopoli) on divided this family into Eurymetoipidæ Mjoberg 1910, Goniodidæ Mjoberg 1910 and Lipeuridæ Mjoberg 1910.

Philotarsidæ widely distributed family of Psocomorpha (Pscocoptera: Psocopteroidea); having forewings with veins and margin strongly setose; tarsi 3–segmented; areola postica free; female sub–genital plate one–lobed–*Philotarsus* spp. et al.

Phlœthripidæ Uzel 1895 widely distributed family of thrips (Phlæothrioidea: Tubulifera); dark brown to black in colour; often with withish, dusky or mottled wings; head rounded in front antennal cones present, III antennal segment largest; last abdominal segment narrow and relatively short–tamarind thrips *Haplothrips ceylonicus* Schmutz mango thrips *Neoheegeria mangiferæ* (Prisener) lily thrips *Liothrips vaneeckei* Priesner.

Phlebotominæ subfamily of Psychodidæ (Psychodoidea: Nematocera); now regarded as family; minute blood–sucking sandflies; body and wings clothed densely with long coarse hairs. Female suck blood vertebrates including mammals and man, some eve act as vectors and transmit various diseases–*Lutzomyia* spp., *Phlebotomus argentipes* Annandale and Brunnetti, *P. minutus* et. al.

Phlœidæ Leston 1953 very small family of Pentatomoidea (Pentatomomorpha: Geocorisæ); large flats insects with lamellate lateral extentions of head and body; antennæ 3–segmented tarsi 3–segmented and same colouration as the bark of trees where these are commonly found–only three Neotropical species.

Phlœothripoidea Karny 1907 superfamily of Tubulifera (Thysanoptera: Thysanopteroidea); having antennæ usually 8–segemented (rarely seven), maxillary palpi 2–segmented; middle coxæ farther apart than fore and hind–coxæ–Hystricothripidæ Megathripidæ, Phlœothripidæ and others.

Phloiophilidæ very small family of Bostrychoidea (Polyphaga: Coleoptera); characterised by body being without setæ and antennæ having a loose 3–segemented club; tarsi 5–segmented and femoral plate on hind coxæ wanting–*Phloiophilus edwardsi*.

Phobotaxis reflex action, by which an insect turns to avoid an adverse stimulus and by repeated trials and errors eventually moves away from it.

Phonoreceptor sense organ responsive to sound.

Phorate $C_7H_{17}O_2PS_3$; Phosphorothiolothionate insecticide with systemic, contact and fumigant action; clear mobile liquid having low solubility in water but miscible in alcohols, ethers, esters and xylene; gets hydrolysed in presence of moisture and alkali. Acute oral LD_{50} value for rat 2 to 3 mg/kg and dermal 70 to 300 mg/kg - Thimet®, Umet etc.

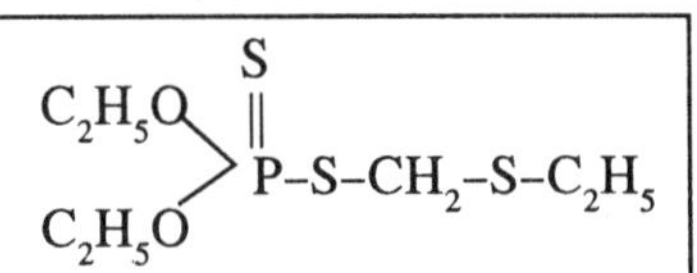

Phorate

Phoresy form of communication (transport) wherein a small insect goes from place to place with the help of another larger species of insect–small flies often ride on back of dung beetles; some oil beetles travel on the bees and wasps while a small chalcid wasp is carried on legs of ants.

Phoridæ Haliday 1851 family of hump-backed flies (Phoroidea: Aschiza); very small grayish-black flies, antenna apparently one large segment which actually conceals the other segments and terminates in long subdorsal arista: thorax conspicuously hunched. Active runners, abundant in decaying vegetation, a few are either parasitic or inquilines in nests of ants and termites- *Aphiochæta xanthina, Chonocephalus felcheri, C. similis, Indoxenia flavescens* et al.

Phosalone $C_{12}H_{15}ClNO_4PS_2$; organo-phosphatic insecticide; white crystalline substance with garlic-like smell, insoluble in water but soluble in most organic solvents; stable in acidic medium but hydrolyses repidily in an alkali. Acute oral LD_{50} value to rat is 135 mg/kg and dermal 1500 mg/kg-Zolone®, Bensophos etc.

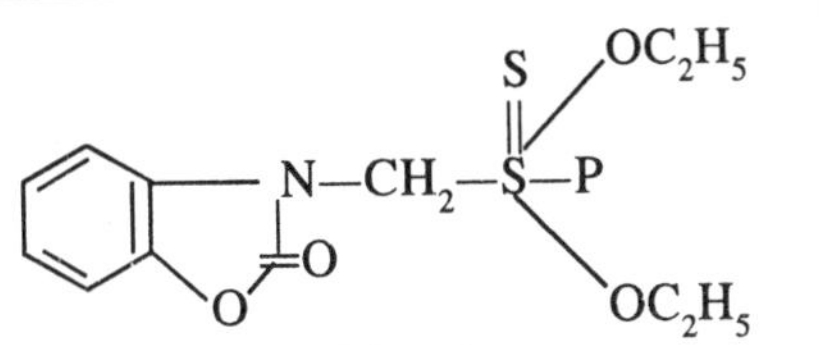
Phosalone

Phosmet $C_{11}H_{12}O_4NPS_2$; organophosphate acaricide and insecticide; white crystalline solid; slightly soluble in water and other organic solvents like acetone, ketone, xylene etc. Acute oral D_{50} value for rat is 113 to 230 mg/kg and dermal for rabbit is 3160 mg/kg-Imidan, Prolate®, etc.

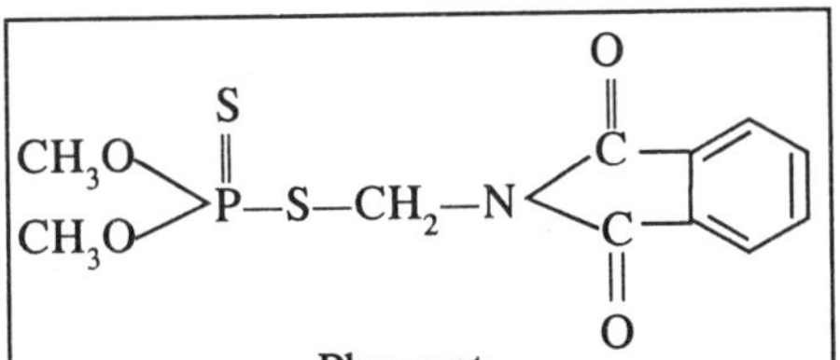
Phosmet

Phosmodidæ very small family of Tettigoniodea (Ensifera: Orthopera); includes a few species of terrestrial insects having long slender rod-like body with vestigial wings, looking superficially like stick-insects-*Phasmodes rantriformes.*

Phosphamidon $C_{10}H_{19}O_5NClP$; organophosphate insecticide, systemic in action and stomach poison, syntheised by E. Beriger or Ciba Ltd., Basle, in 1955. Pale yellow liquid with faint pleasant smell; commercial product is bright violet, very stable and miscible with water and almost all organic solvents except saturated hydrocarbons. Acute oral and dermal LD_{50} values for rat are 15 and 125 mg/kg respectively-Dimecron®, Kinadon etc.

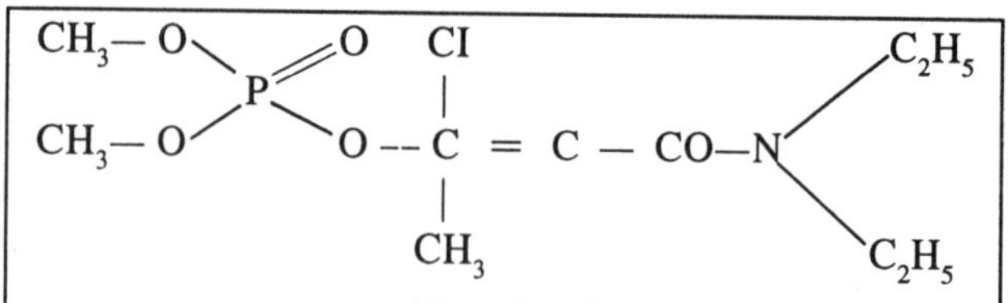

Phosphamidon

Phot(o) pertaining to light.

Photobiology branch of biology that concerns effect of light on various organisms, including insects.

Photocytes cell that emit light, situated just beneath the integument where the cuticle is transparent for permitting the light to pass; oxygen is supplied by tracheoles.

Photogen light producing organs in animals.

Photogenic organs luminous or light emitting organs, consist of large number of photocytes and are found in different stages of various insects, situated very close to body surface behind a window of translucent cuticle and located form head to tip of abdomen.

Photonegative insects that show negative reaction to artificial light being not affected even by exposure to red light – *Trogoderma* Everts sun or avoid artificial light.

Photophobia shrinking from light, as exhibited by termites.

Photophygous thriving in shade.

Photopositive insects that are attracted to artificial light; prefer to gather at violet light–moths of *Amascta moorei* Butler, et. al.

Photoreceptor sense organ responsive to light.

Phototaxis phenomenon by which the insects are either attracted or repelled to artificial light, the reaction may be positive or negative.

Photoropism behavioural response of an animal (including insects) to light stimuli.

Phoxim $C_{12}H_{15}N_2O_3PS$; organophosphatic (Phosphorothionate) insecticide; technical material is reddish-brown, slightly soluble in water, readily soluble in alcohols, ketones, aromatic hydrocarbons etc. comparatively safe, less toxic to man and mammals and having long residual effect; often used against stored-grain pests. Acute oral and dermal LD_{50} values for rat 168 to 184 and 1000 mg/kg respectively–Baythion®, Volaton® etc.

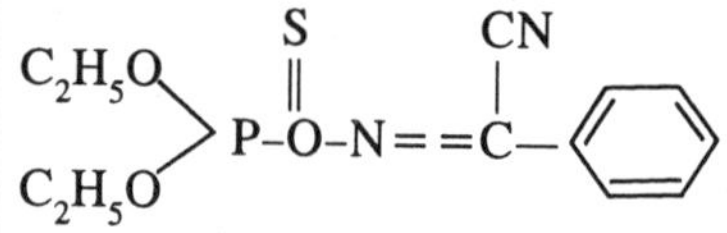

Phoxim

Phryganeidæ Burmeister 1839 family of Limnephiloidea (Integripalpia: Trichoptera); caddisflies having ocelli, maxillary palpi 3 or 4 segmented and forewings with vein M missing Larvæ eruciform with long, cylindrical tube-like cases, found mostly in standing waters-*Phryganea grandis* Linnæus, *Neuronia semifasciata* Say, et al.

Phthiraptera order in old system of classification divided into four suborders-Amblycera, Ischnecera, Rhynchophthirina and Anoplura (Siphunculata); now first three are regarded as suborders of Mallophaga while Siphunculata is treated as an independent order.

Phthiridæ Ewing 1929 family of sucking lice (Siphunculata: Psocopteroidea); small wingless insects having forelegs very slender, legs spread and moved like a crab; abdominal segment 3 to 5 fused and with three pairs of spaircles-crab louse, *Phtthirus pubis* (Linnæus) found mainly in hair around sexual organs. *P. gorillæ* infests gorilla in Africa.

Phthisaners aners or male-ants (Formicoidea: Apocrita) that are parasitised by eucharids (chalcid wasps).

Phycitidæ (Meyrick, 1893), Ragonot 1893 large family of Pyralidoidea (Ditrysia: Lepidoptera); often regarded as subfamily of Pyralididæ small delicate moths having elongated forewing without vein R_5, hind-wings with pectin of hair on dorsal side; frenulem simple in both sexes, males characterised by swollen basal antennal segment (800 species-*Etiella zinckenella* (Treitschke), *Euzophera perticella* Ragonot, *Phycita clientella* Zeller, *Plodia interpunctella* (Hübner) et al.

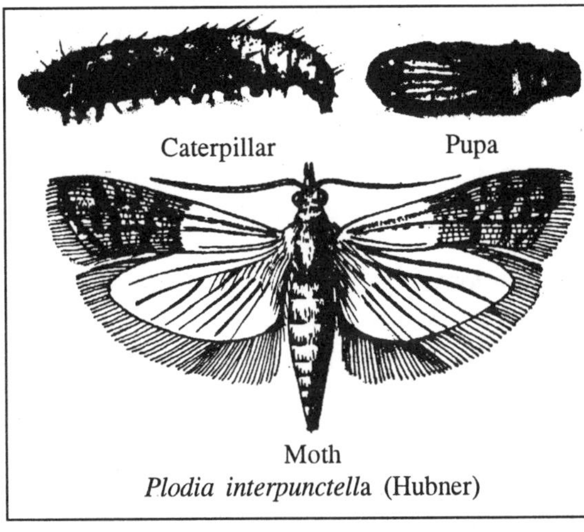

Plodia interpunctella (Hubner)

Phyllidæ Brunner 1893 family of apterous insects (Phasmoidea : Phasmida); antennæ long in males and short in females; tegmina of female large and leaf-like; forewing reduced and strongly sclerotised often resembling leaf-venation or a dead leaf hind-wing membranous; when at rest folded flat over the body - leaf-insects or walking leaves, *Phylium bioculatum* Gray, *Timema californica* Scudder, *Pulchriphyllium scythe* (Gray) et al.

Pulchriphyllium scythe (Gray)

Phyllocinstidæ family of Tineoidea (Ditryasia : Lepidoptera); smallest delicate moths having antennæ with an eyecap and hind tibial having dorsal row of stout bristles. Characteristic feature is that the caterpillars are apodous (50 species) - *Phyllocnistris* spp., et al.

Phyllodromiidæ Brunner 1865 family of Blattaria (Dictyoptera : Blattopteroidea); includes cockroaches, irregular in size and colour; winged forms, like Blattidæ, have large anal area with several folds but the famales are without any valves on seventh sternite-German roach [*Blattella germanica* (Linnæus)], giant drummer [*Blabeus giganteus* (Linnæus)], burrowing roach [*Leucophæa surinamensis* (Linnæus)].

Phylloptera Packard 1883 an equivalent name proposed for Mantodea Burmeister 1838 a suborder of Dictyoptera; often regarded as an independent order by some entomologists.

Phylloxeridæ (Koch 1857) Dreyfus 1889 family of bark/gall aphids (Aphidoidea: Homoptera); confined to deciduous trees and vines; wings held flat over abdomen when at rest, ovipositor vestigial or absent; all females oviparous -*Dactylosphæra* (*Phylloxera*) *vitifoliæ* (Shimer) is a severe pest of grapevines (*Vitis vinifera* Linnæus), causing galls on leaves and gall-like swelling on roots (not recorded in India).

Phylongeny evolutionary pedigree or genealogical history.

Phylum (ph. **phyla**) main division in classification of animals; it is further divided into subphyla, classes, orders, families etc.

Phymatidæ (Laporte 1833) family of Reduvioidea (Cimicomopha: Geocorisæ); predaceous bugs having antennæ with last (4th) segment largest clavate or fusiform; forelegs raptorial, mid-and hind-tarsi 2-segmented (100 species)-*Phymata pennsylvanica, P.*

erosa (Linnæus), *P. monstros* Fabricius, *P. crassipes* Fabricius, et al.

Phymatini Costa 1838 old and obsolete name of the family Phymatidæ (Geocorisæ: Heteroptera), no more in usage.

Phymatites Laporte 1833 another name of the heteropteran family Phymatidæ; not in use.

Physical poison substance that brings about death of an insects by exerting physical effect rather than biological one–asphyxial effect by use of heavy oils or tar oils or by affecting loss of body moisture by inert dusts like aluminium oxide etc.

Physiologist person well versed in the science of physiology.

Physiology biological science concerning functions of various systems of living beings.

Physiopathology study of abnormal functions(s), involving pathological alternation of a bodily function. *Cf.* morphopathology.

Physopoda equivalent term for order Thysanoptera (Thysanoperoidea: Paraneoptera); includes thrips, as these insects have protrusible bladder–like structure at the end of their tarsi.

Physoptilidæ family of microlepidoptera in old system of classification based on sized of moths–*Physoptila scenica* Meyrick is pest of oak trees (*Careye arborea*) in India.

Phytalmiidæ small family of Agromyzoidea (Schizophora: Chclorrhapha); elongate, long legged flies with a neck like a giraffe.

Phytophaga group of beetles (Cerambycoidea Leng 1920); having head not prolonged into a beak with tarsus of all legs 4–segmented and densely pubescent. Includes three families–Cerambycidæ Chrysomelidæ and Bruchidæ (Lariidæ).

Phytophaga Latrielle 1807 (Sessiliventres Haliday 1840) old and equivalent names for suborder Symphyta (Hymenoptera); no more in usage.

Phytophagous herbivourous: animal (insect) that feed and lives on plants.

Phytophagy consumption of plants by animals.

Phytosanitation destruction or removal of infected or infested plant material likely to cause reinfestation or reinfection.

Phytotoxic substance (toxin) capable of damaging or destroying the plants even some pesticides applied for controlling insect–pests or weeds etc. may be phytotoxic to all or certain plant species–HCH on cucurbits.

Pieridæ (Boisduval 1836) Duponcel 1844 family of medium-sized butterflies commonly called Whites (Papilionoidea: Ditrysia); characterised by hind-wing having tow anal veins; legs all alike and normal in both sexes. Butterfly migration is common in certain species; *Pieris* spp. in North India move to cooler regions during March–April and are back again around October. Catepillars are elongate with their segments divided into annulets and body bearing numerous secondary setæ of varying size; feed on plants belonging to families Brassicaceæ, Capparidaceæ and Fabiaceæ - *Pieris brassicæ* Linnæus, *P. canidia* (Sparrman) *P. napi* (Linnæus), *P. oleracea* Harris are pests of cruciferous crops.

Piesmatidæ small family of Lygæoidea (Pentatomomorpha; Geocorisæ); small elongate–oval, polyphagous bugs, formerly placed under Tingidæ; ocelli present in macropterous forms; rostrum 4–segmented; mandibular plates produced into a pair of horn–like structures; scutellum visible. coxæ rotatory; tarsi 2–segemted, pulvilli present.-*Piesma quatratum* Fieber, is pest of beetroot (*Beta vulgaris* Linnæus).

Piezata Linnæus 1758 old, equivalent and obsolete term for insect order Hymenoptera Linnæus 1758.

Pilose covered with hair.

Piophilidæ Macquart 1835 family of small dipteran flies (Agromyzoidea: Schizophora); characterised by costa interrupted well before R_1, and Sc complete; vibrissæ present, postvertical bristles not convergent. Female with an oviscapt. Larva saprophagous - Cheese skipper, *Piophila casei* (Linnæus), et al.

Pipunclidæ (Zetterstedt 1842) Curtis, Walker 1851 distinct family of dipteran flies (Syrphoidea: Aschiza); having very large subhemispherical mobile head, comprising almost entirely of two big eyes; antennæ with long dorsal arista wings much longer than abdomen with cell Rs open; ovipositor horny and exserted - *Pippunculus annulifamur* parasitic on mango hoppers while *P. industrius* Knab, *P. similis* Hugh, *P. subitens* Cresson, *P. uniguiculatus* Cress and *P. vagabundus* Knab have been reared from beet leafhopper *Eutettix tenellus* (Baker) in Western North America.

Pisces superclass of Gnathostomata (Craniata: Chordata); poikilothermal aquatic verterbrates, breathing by means of gills (except Choanichthes or lung–fishes); streamlined body, skin usually covered with scales, pectoral and pelvic fins paired, median fins not paired; no tympanic membrane–all fishes (dolphin, seal and whale are not fishes but mammals).

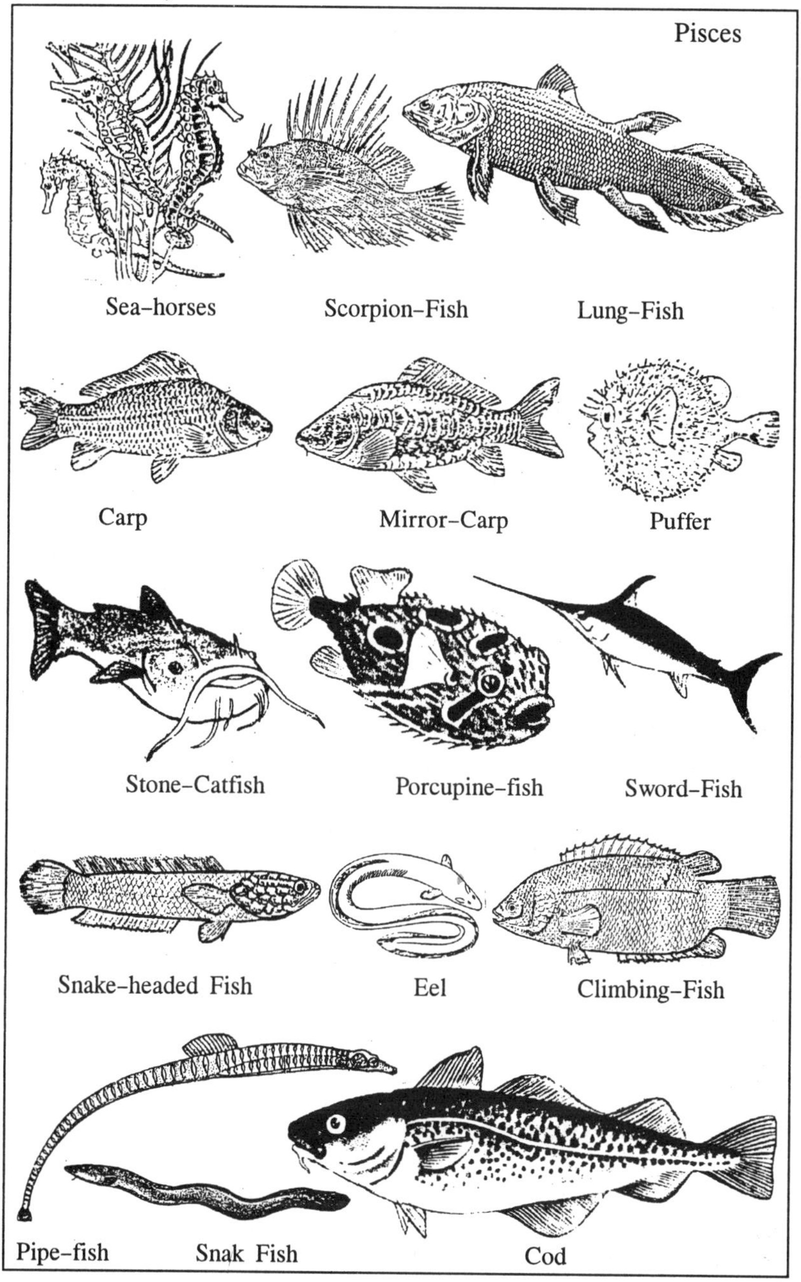

Shark (Bule) Shark (Hammerhead)

Pistol case bearer *Coleophora malivorella* Riley and cigar case bearer, C. *flectcherella* Fernald (Coleophoridæ: Ditrysia) are common pests of apple (*Malus pumila* Miller), cherry (*Prunus avium* Linnæus, *P. cerasus* Linnæus) and pear (*Pyrus communis* Linnæus) in Northern USA and Southern Canada; the nicknames are derived from the shapes of dark larval cases.

Planipennia suborder of Neuroptera (Neuropteroidea: Oligoneoptera); comprising of ant–lions and lacewings characterised by wing venation with branches of veins usually bifurcated at the margin of the wings. Larva Predaceous, having suctorial mouthparts, prey upon aphids and other injurous insects - Coniopterygoidea, Osmyloidea, Mantispoidea, Hemerbioidea and Myrmeleontoidea.

Planta median ventral sclerite of pretarsus, distal to unguitractor plate.

Plantulæ pulvilli; small pads on the lower surface of tarsal segment.

Plasmatocyte(s) type of hæmocytes or blood cells, found suspended in the plasma, these are most abundant, with variable form, basophilic, cytoplasmic and phagocytic function.

Plataspididæ (= Brachypaltidæ; Coptosomatidæ) family of Pentatomidea (Pentatomomorpha: Geocorisæ); strongly convex, broadly rounded, shiny stink bugs; scutellum covering almost entire abdomen, hemelytra long and bent benath the scutellum when a test (500 species) - *Coptosoma cribraria* Fabricius and *C. naziræ* Atkinson are common pests of pulses in India.

Platycnemidæ small heterogenous Old World family Coenagrioidea (Zyygoptera: Odonata); feather–footed or white–legged damselflies having narrow wings with anal vein and Cu_2 normal, discoidal cell elongated; hind legs white, flattened and hairy, resembling feathers - *Platycnemis pennipes* occurs in Europe.

Platygasteriadæ (Westwood 1840) Ashmead 1900 large family of Scelionoidea (Apocrita: Hymenoptera); minute black wasps, antennæ 10–segmented forewing without margin or stigmal veins; parasitic on cecidomyiids and aleyrodids-*Platygaster oryzæ* Cammeron on rice gall–midge *Pachydiplosis oryzæ* (Wood–Mason) and on larvæ of *Elasmus albopictus* Crawford, *Cardiogater secundus* Mani and *Amitus aleurobi* Mani on nymphs of sugarcane whiteflies, *Aleurolobus borodensis* Maskell and *Neomaskellia bergii* Signoret.

Playhelminthes phylum of invertebrate animals (Acoelomata: Triplolastica) said to have been originated from now extinct mobile coelentrates. Soft–bodied, bilaterally symmertical and dorsoventrally flattened roundworms characterised by lack of skeleton, circulator and respiratory systems, body cavity and anus but having a complicated reproductive system (2500 species–Turbellaria (flatworms), Trematoda) (liver flukes and Cestoda (tapeworms).

Platypezidæ Walker 1851 small family of dipteran flies (Aschiza: Cyclorrhapha); 2 to 4 mm long, sombre–coloured, thinly hairy, antennæ 3–segmented with long arista; eyes often holoptic, ocelli present; tarsi elongated and flattened, tibia spurless. Live and breed in fungi specially mushrooms and toabstool (100 species) - *Callimyia amoena* Meigen, *Platypeza agrici* Willard, *P. insignis* (Aldrich), *P. polypori* Willard, et al.

Platypodidæ family of Curcullionoidea (Polyphaga: Coleoptera) often placed as subfamily (Platypodinæ) of Curculionidæ; characterised by cylindrical body. short antennæ and no rostrum; legs and antennæ folded close to the body. Commonly called pinhole borers *Diapus furtivus* (Sampson), *Platypus indicus* Stroyan, *P. pilifrons* Chapius, *P. sordidus* Walker are pests of forest trees.

***Platypus* duckbill** most primitive, semi–aquatic, egg–laying mammal, *Ornithorhychus anatinus* (Minotremata: Protheria); having elongated, leathery, bill–like muzzle and no teeth (Page 274). Nipples lacking milk of lactating females exudes from mammary glands and is lapped up by the yound ones form mother's abdominal fur. Not to be confused with *Platypus* spp. that are tiny beetles, found boring into bark and wood of forest trees.

Platysticitidæ family of Cœnagrioidea (Zygoptera: Odonata); small to medium-sized damselflies from Neotropical and the Oriental regions, having narrow petiolated wings; arculus distal to second antenodal; anal vein and often Cu_2 also greatly reduced; supplementary postcubital cross–veins present basally, joining Cu_2 to hind margin.

Plaumanniidæ small Central American family (Troctomorpha: Psocoptera); characterised by wing being not scaly forefemora simple and tarsi 3–segmented–Psocids.

Plecoptera Burmeister 1939 (Perlaria, Perlarides) order of Orthopteroidea (Polyneoptera: Pterygota); soft-bodied insects; prognathous head with poorly developed biting type mouthparts; elongate setaceous antennæ, lingula held flat over the back while at rest; well developed anal lobes; tarsi 3-segmented; cerci long and multi-articulated; ovipositor wanting. Hemimetabolous. Nymphs aquatic (1800 species) spread in three suborders-Archiperlaria, Filipalpia and Setipalpia.

Plectoptera Packard 1886 old and equivalent name for insect order Ephemeroptera Hæckel 1896 (Palæoptera Martynov 1923), as also for Odontota Latreille 1806, Ephemerida Leach 1817 and Agnatha Meinert 1883.

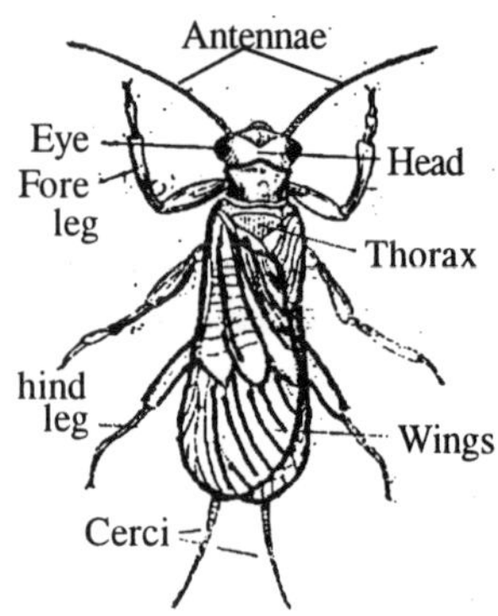

Stonelly

Pleidæ small family of Notonectoidea (Hydrocorisæ : Heteroptera); often included in Notonectidæ; pigmy back-swimmers; minute and stout water bugs, < 2 mm long, body strongly convex dorsally; antennæ and rostrum 3-segmented; forewings uniformly coriaceous, hind-wings reduced, hind tibia and tarsi without conspicuous swimming hairs. Predaceous mainly on fish; common in ponds among dense water weeds (20 species) - *Plea atomeria* (Leach) et al.

Pleocomidæ Arrow 1913 very small family of Scarabæoidea (Polyphaga: Coleoptera); includes a few species of small chafer beetles-*Pleocoma bethrensi* Lec., *P. fimbriata* Lec., et al.

Pleregate(s) honey (pot) ants or repletes that are capable of engorging themselves with liquid food and then hang from the roof of the next as living stores of food for rest of the members of that colony.

Pletyptera old and obsolete term for an insect order that included embiids (Embioptera), termites (Isoptera) and booklice (Psocoptera).

Pleural pertaining to pleuron.

Pleural apophysis internal arm of pleural ridge.

Pleural ridge endopleural ridge formed by pleural suture, bracing pleuron above the leg or between coxal articulation and wing support.

Pleural suture external groove of pleural ridge, separating episternum form epimeron.

Pleural wing process wing support of pleuron at upper end of pleural ridge.

Pleuron (pl. **pleura**) lateral part of cuticle in each segment of an insect or crustacean.

Pleuropoda vestigial leg-like organs on first abdominal segment of embryos of grasshoppers and locusts (Cælifera : Orthoptera) functioning as grandular organs secreting an enzyme that helps to dissolve the egg-shell and facilitate the emergence of the nymph.

Pleurostome portion of each subgenital area of cranium above mandible.

Plica (pl. **plicæ**) fold.

Plica basalis basal fold of the wing; line of flection between base of medio-cubital field and axillary region.

Plica jugalis jugal fold of wing: radial line of folding separating jugal region from vannal region.

Plica vannalis vannal fold of the wing: radial line of folding setting off vannal area form remigial area.

Plicate folded fanwise: plaited (said of insect wing).

Ploiaridæ Dohrn 1863 (= Emesidæ Walker 1873) a small family of thread- legged bugs (Geocorise : Heteroptera); remarkably slender, fragile insects with long bodies and thread-like appendages; antennæ long and slender rostrum 3-segmented; first segment stout and curved; forelegs raptorial, middle and hind legs long and filamentous-*Emesa* species, *Empicoris rubromaculatus* (Blackburn), *Brace annulipes* Stal, *Ploiaria culiciformes* de Geer, *P. huttoni* (Scott) et al.

Ploidy number of complete chromosome sets in a nucleus.

Plumariidæ small family of Bethyloidea (Apocurita: Hymenoptera); characterised by antennæ inserted close to clypeus and in females 13-segmented; fore-tarsus in females not chelate. Parasitic on coleopteran and lepidoteran insects.

Plusiidæ Guenee 1842, 1852 Tropical family of moths (Noctuoidea: Ditrysia); body very slender, forewings broader, vein M_2 of hind-wing well developed. Caterpillers semiloopers-*Plusia chalcites* Esper, *P. nigrisigna* Walker, *P.orchalcea* Fabricius, *P. signata* Fabricius, et al are crop pests.

Plutellidæ (Guenee 1845) Station 1854 family of Yponomeutoidea (Ditrysia : Lepidoptera); often regarded as subfamily of Yponomeutidæ; head smooth; maxillary palpi short and porrect; Sc and R approximate basally in hind-wing. Caterpillars mine into leaves and stems and feed within (200 species)-diamond back moth, *Plutella xylostella* (Linnæus) (= *maculipennis* Linnæus) is a major pests of cruciferous crops all over the World.

Podagrioninæ a subfamily of Torymidæ (Parasitica : Apocrita) often regarded as an independent family of Chalcidoidea; characterised by hind femora having a row of teeth-specific parasites of oothecæ of preying mantids (Mantodea).

Podeon (podeum) narrow stalk or petiole connecting thorax with abdomen in some insects.

Podex pygidium or hind part of insect's abdomen, comprising of last two or three segments, often found fused together or modified to perform some special function.

Podical plates latero-ventral plates formed by modification of X abdominal segment of certain insects; these are on each side of anus and may even bear anal cerci.

Podite promere: segment or part of an appendage of an insect or other arthopod that is independently movable by its own musculature.

Podopidæ (Graphosomtidæ) small family of Terrestrial turtle bugs, Scutelleroidea; (Geocrisæ; Heteroptera); 4 to 10 mm long, having very large scutellum, covering a large part of the abdomen and pronotum bearing a prominent tooth or lobe in front of the humeral angle and another at anterior angle-*Scotinophara lurida* is a pest of barley (*Hordeum vulgare* Linnæus), maize (*Zea mays* Linnæus), paddy (*Oryza sativa* Linnæus), wheat (*Triticum æstivum* Linnæus) sugarcane (*Saccharum officinarum* Linnæus), etc. In Indian subcontinent, China, Japan and South-east Asia whereas *Graphosoma italicum* is European species infesting umbelliferous flowers.

Poduridæ Lubbock 1870 family of Poduroidea (Arthropleona: Collembola); comprising of very small but widely distributed and abundantly represented, springtails and snowfleas. Head hypagnathous, antennæ 4-segmented; eyes often present, psendocelli wanting, scales absent, furca when present arising from IV abdominal segment and extending beyond ventral tube. Feed on decaying vegetable matter and occasionally damage the germinathing seedlings (350 species) - water springtail *Podura aquatica* Linnæus is widespread in Europe and North America;

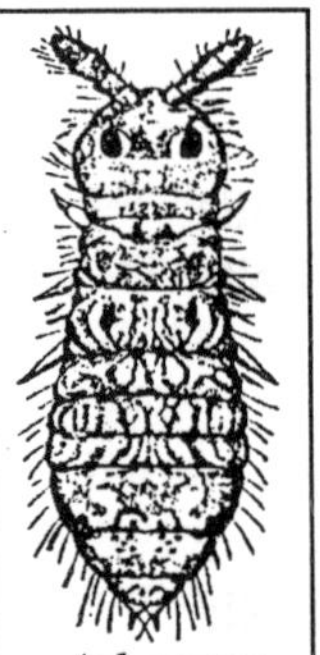

Achorutes armatus (Nioclet)

seashore springtails *Anurida maritima* Lab. and Onychirus cocklei (Folsom) are and *Xenylla maritima* Tullberg are common along seashores of Africa and Europe; and armed springtail *Achorutes armatus* (Nioclet) is pest of mushrooms while snowflea *A. nivicolus* Fitch and *Onychirus Cocklei* (Folsom) are found at high altitudes on snow surface.

Poikilothermic cold-blooded animals; having body temperature varying with external environmental temperature - fishes, frogs, reptiles and all invertebrates, including insects.

Poison substance that causes damage or kills when introduced into living organisms. Certain insects also produce poison to kill or paralyse their prey; some may discharge in weaker connecntration that will repel their enemy. It is introduced in various ways-by stinging (bees and wasps), by squirting (ants), injecting saliva that contains the poison (bugs and mosquitoes), through tiny wound made by their strong mandibles (beetles). Some caterpillars (Arctiidæ) have stinging hairs, other discharge posion from special glands in their skin, in such case are contact will have toxic effect on the victims.

Polistidæ (Lepeletier 1836) Cerstæcker 1863 family of polistes wasps (Vespoidea: Apocrita); large, slender, long-legged social wasps that resemble hornets and yellow jackets (Vespidæ: Vespoidea) in colour pattern but differ in having short and slenderer abdominal petiole - *Polistes aurifer* Saussure, *P. herbæus* Fabricius, *P. stignata* Fabricius and allied species.

Polistes aurifer Saussure

Pollen fertilising powder formed in anthers of flowers.

Pollen basked corbiculum part of hind leg of bee (*A. poidea*: Hymenoptera); specially modified or carrying pollen, tibia is flattened and bordered by long stiff hairs forming a trough into which the pollen can be pushed; in some solitary bees pollen basket is situated beneath the abdomen.

Pollen brush scopa: comb-like structure formed of ring of stiff hairs on first tarsal segment of hind leg of bee; ;used for brushing or removing the pollen, off the body, into pollen basket of corresponding leg on opposite side.

Pollen comb brush-like structure found in joint between tibia and tarsus of foreleg of a bee; used for cleaning or removing the pollen of the head and antennæ.

Pollen compressor flattened plate at the base of concavity of the upper end of first tarsal segment of bee; the pollen is brushed into here and compressed by flexing the leg, then the pollen is pushed into pollen basket.

Polliniferous pollen-bearing.

Poly-many; several.

Polyhedrosis any of the several virus diseases of insect larvæ characterised by breakdown of tissues and presence of polyhedral granules.

Polymitracidæ Klapalek 1909 family of Ephemeroidea (Ephemeroptera: Palæoptera); slender-bodied, amphibiotic insects having mouthparts vestigial; wings transparent in males and opaque in females and hind margins without intercalary veins. Naiads with long tusk-shaped mandibles found burrowing into mud - virgin mayfly *Polymitarcys virgo* (Olivier), *Campsurus* spp., *Excuthyplocia* spp., et al.

Polymorph an individual that exhibits polymorphism.

Polymorphic organism having several different forms/castes.

Polymorphism occurrence of different forms of individuals in a single species: there are as many as 29 different types of individuals among the various castes of ants (Formicoidea: Apocrita). Occurrence of different structural forms in a single individual at different periods of its life-cycle.

Polyneooptera Martynov 1923 infraclass of pterygota (Insecta) that includes insects having wings with large number of veins arranged in elaborate branching pattern and no cross-veins. Spread into three superoders-Orthopteroidea, Dermapteroidea and Blattopteroidea.

Polyphagidæ family of Blattaria (Blattopteroidea: Dictyoptera); both alate and apterous forms; underside of mid and hind femora not spinose; anal aea of hind-wing flat and without fan-like folding when at rest; IX sternite of male rounded posteriorly and with long symmetrical styles-Polyphagine includes Mediterranean species while Attaphiliæ is American and Atticolinæ Neotropical.

Polyphagous one who eats several different kinds of food-locust, San jose scale, various hairy caterpillars etc.

Polyphagy capability to feed on wide range of unrelated hosts.

Polyptera Banks 1892 Old and equivalent name for insect order Anoplura Leach 1815 or Siphuculata Latreille 1825, that includes sucking lice; also considered as synonyms of Pediculina Burmeister 1835, Pediculida Mayer 1876 and Pediculoidea Crampton 1921; no more in usage.

Polythoridæ family of Agrionidea (= Caloptrygoidea) (Zygoptera: Odonata); dameseflies having wings indistinctly stalked basally, antenodal cross-veins numerous; discoidal cell irregular with basal side longer than apical side and connected directly to Rs; arculus absent.

Polytrophic egg-tube egg-tube in which groups of trophocytes accompany oocytes.

Pomegranate (*amar*) butterfly *Virachola isocrates* Fabricius (Lycænidæ: Lepidoptera); medium-sized bluish-brown butterfly, female having orange coloured patch on forewings and black spot on hind-wings. Dirty brown, stout built caterpillar, punctures the fruit, goes inside and feeds within on pulp and immature seeds; affected fruits often fall down prematurely.

Pompilidæ Lech 1815, 1819 large family of wasps (Apocrita: Hymenoptera); relegated as synonyon of Psmmocharidæ Rohwer 1910 on the basis of generic priority. Genus *Psammochares* latreille was established in 1796 while genus *Pompilus* Fabricus was erected in 1798.

Population aggregation of inbreeding organisms of a given species present in a particular habitat; fundamental unit of an ecological system.

Porifera phylum of lower invertebrates (Metazoa : Non-chordata); commonly called sponges; multicellular, aquatic sessile animals, asymmetrical or radially asymmetrical, having single body cavity with numerous pores; mostly marine (500 species)-marine colonial sponges *Leucosolenia* species (Homocoela: Calcarea), bath sponge, *Euspongii* species (Kerotosa: Demopongia) fresh-water sponge *Spongilla fragilis* (Monaxonida: Demospongia), Venus flower *Euplectella aspergillum* (Hexasterophora : Hexactinellida) et al.

Post after. behind.

Post alare lateral extension of postnotum of wing-bearing segment behind wing base, often united with epimeron.

Post coxate postcoxal part of pleuron united with sternum behind coxa.

Post cubitus sixth vein of the wing connected basally with cubitus (Cu - 5th vein).

Post gena area of cranium, posterior to gena: lateral and ventral parts of occipital arch.

Post mentum basal part of labium proximal to stipital region or prementum.

Post notum postscullar, phragma-bearing plate usually present in dorsum of alate segment.

Post occiput extreme posterior rim of cranium behind postocciptial suture.

Posterior rear end of organism or structure.

Postural hairs row of speciallised hairs of satæ situated between head and thorax or between thorax and abdomen.; these carry sensary receptors at their bases, as such a change in their position gives the insect indication about its position.

Potamanthidæ Klapalek 1909 family of Ephermeroidea (Epheneroptera: Palæoptera); soft-bodied, amphibolitic mayflies having forelegs as long as body; wings transparent with numerous intercalary veins specially in hind-wings and Cu_1 of forewing forked. Naiads have crab-like legs with tibia prolonged into flat spine-species of *Patamanthes* Picet, *Potamanthodes* Ulmer, *Rhoenanthus* Eaton, et al.

Potato tuber moth *Phthorinæa* (*Gnorimoschema*) *operculella* (Zeller) (Gelechiidæ: Lepidoptera); a pest mainly of stored potatoes (*Solanum tuberosum*), small, narrow winged (wing span 12 to 16 mm), grayish-brown moth, fly from godown to field and lays eggs in or near the eyes or cracks on the skin of exposed tubers, on hatching the caterpillar enter the tuber, feeds within and is carried to godown along with tuber.

PRADHAN, Dr. Shyamsunder Lal (Srivastava) – born May 13, 1913; D.Sc. in 1938, from Lucknow University. Jointed ICAR in 1938; worked on sugarcane pests for two years before becoming Assistant Entomologist at IARI New Delhi Deputed to work on insect toxicology at Rothamsted Experimental Station (England), 1946–48 and was awarded Ph.D. in 1958, became the first Professor of Entomology of the Post-Graduate faculty of IARI and Head of Division of Entomology in 1962. Died on February 5, 1973. He was internationally recognised as an authority on insect ecology, toxicology and Integrated Pest Control. His books **Insect Pests of Crops** (1969) revised by M.G. Jotwani (1983), **Agricultural Entomology and Pest Control** (1983) still very popular with all concerned.

PRASAD, Dr. Shree Krishna - born December 1, 1923; B.Sc. (Agri) 1948, Associate IARI 1952 and Ph.D. (Minnesota) in 1957. Worked on sugarcane pests at CSRS, Pusa (Bihar) 1948 to 1958; them joined IARI where he was instrumental in establishing the Division of Nematology at I.A.R.I; from where he took voluntary retirement in January 1976 and went as Visiting Professor to the University of Guyana, Georgetown; there he founded the faculty of Agriculture and retired as it Dean in 1986. Returned to India and died July 1,1993.

Pre- anterior: before

Pre-oviposition period time gap between emergence of adult female and its oviposition (egg-laying).

Prealare lateral extension of precutal area of alinotum before wing base.

Precosta small first vein in wings of certain fossil insects.

Precoxale precoxal part of pleuron, anterior to trochantin.

Predaceous (predacious) living by preying: capturing and eating living animals of other species.

Predation biotic interaction involving association between strong and week species of animal.

Predator individual that lives by preying upon individuals of other animal species-dragonflies, praying mantids, various beetles, robber-flies, some scorpion-flies, majority of wasps etc.

Preharvest interval time gap between the last application of pesticide and safe harvesting of edible crops for immediate consumption.

Prehistoric insects fossil insects that lived during Devonian, Carbonifersous and Permian periods of Palæozoic era-*Rhyniella præcursor* (Collembola), *Triassomachilis uralensis* (Thysanura), *Erasipteron larischi* (Odonata), *Protoprosbole stræteni* (Homoptera) et al.

Prelabium movable distal part of labium comprising of prementum, labial palpi and terminal lobes (ligula).

Prementum free distal part of labium of an insect, bearing labial palpi and ligula.

Prepectus anterior marginal sclerite of sternopleural areas of segment set off by transverse suture continuous through sternum and episterna.

Preplanting treatement pesticidal treatment applied in or on the soil before planting the crop, to safeguard the crop against the attack of pest and diseases-application of imidacloprid dust or E.C. before planting sugarcane (*Saccharum officinarum* Linnæus) prevents termite attack.

Prepupa short dormant state just before pupal stage in that development of holomentabolous insects. There is no feeding during this period and often a cacoon in also formed–Tysanoptera and Diptera.

Prescutum anterior of three sclerites which typically form the notum or dorsal cuticular covering over each of the three thoracic segments of an insect.

Prestomal teeth small chitinous teeth round prestomum or outer mouth of certain insects (blowflies etc.) distinct from pseudotracheal teeth which are formed from chitinous rings of pseudotracheæ.

Prestomum pseudomouth to which food channels or pseudotracheæ converge at the base of the two oral lobes; found in blowfly, housefly and similar insects. It is always situated further back at the entrance to the pharynx and is not true mouth.

Pretarsus small distal portion of apical segment of tarsus, usually bearing two claws with a pad or arolium in between : often called 'foot' of the insect.

Preventive applied in anticipation to prevent occurrence or appearance of a pest.

Priapulida a minor phylum of worm–like marine animalcules; peritonum of coelom greatly reduced; body covered with spines and tuberches; mouth surrounded by concentric rows of teeth; proboscis reduced. Only 8 species – *Priapulus* and allied species.

Primitive ancinet: an early underveloped form or species.

Primitive insects insects believed to have been evolved from some wingless arthropods, intermediate between insects and millipedes. The most primitive living insects are Apterygota; these are all primarily wingless and do not undergo any metamorphosis – bristletails and silverfish (Thysanura), campodids and japigids (Diplura), springtails and snowfleas (Collembola) and telson–tails (Protura).

Probnisidæ family of Protorthoptera (Polyneoptera: Pterygota); comprising of extinct insects, found abundant during Carboniferous period in France and USA.

Proboscis tubular organ of varying form and function (mainly feeding found in number of invertebrates including insects, annelids and tapeworms. Lengthened mouthparts of an insect adapted specially for piercing and sucking. There are serveral forms fund in different insects. In bees, labium is lengthened for sucking nectar; in butterflies it is formed by two lengthened, interlocking maxillæ; in houseflies and blowflies it is highly specialised for suction and filteration of food that has been partly digested externally. In most cases, the proboscis, when not in use is folded back under the head. Elongated, flexible trunk of an elephant.

Procephalon part of insect's head that lies anterior to the gnathal segments bearing labrum, mouth, eyes and two pairs of procephic appendages (antennæ and postantennæ).

Procerebrum protocerebrum or the fore-brain of an insect; more strictly protecerebrum and some parts derived from embryonic presegmented ganglia.

Process projection or outgrowth.

Proctiger cone-shaped reduced terminal segment of abdomen that bears the anal opening or anus.

Proctodeum hind-gut or hind-intestine, posterior ectodermal part of alimentary canal.

Proctotrupidæ (Latereillæ 1802) Stephens 1829 (Serphidæ Kieffer 1909), family of Proctotruopoidea (Apocrita : Hymenoptera); very small wasps (2.5 to 10 mm long), having antennæ 13-segemented with one ring joint; mandibles edentate, forewings with distinct stigma hind-wings without distinct venation; abdomen with short petiole. Parasitic mostly on beetles and weevils-*Prototurpes gravidator* (Linnæus) parasitises *Boletophila fusca.*

Prototrupidæ (Laterille 1802) Stephens 1829 hymenopteran family equivalent to family Serphidæ Kieffer 1909; the type genus *Serphus* was erected by Schrank in 1780 whereas genus *Proctotrupes* was established by Latrielle in 1796, hence Serphidæ is the preferred name of the family.

Proctotrupoid group of parasitic wasps under suborder Apocrita (Hymenoptera) that is further divided into superfamilies Ceraphronoidea, Serphoidea (Proctotrupoidea) and Scelionoidea.

Proctotrupoidea (Serphoidea Viereck 1916) superfamily of Apocrita (Hymenoptea: Hymenopteroidea); minute, slender, parasitic insects, mostly pterous or alate with reduced venation and radial cell of forewing normally closed; trochanter 2-jointed; ovipositor issuing from the apex of abdomen-primary and secondary parasites of other insects, a few are hyperparasites and some are inquilines.

Prodoxidæ a very small family of Incurvarioidea (Monotrysia : Lepidoptera) confined to North America; adults small pale coloured moths, caterpillars seed and stem bores the insects show and intimate association with yucca flowers (*Yucca filamentosa* Linnæus) and are often responsible for pollination of these flowers - *Tegeticula yuccasella, Prodoxus* spp., et al.

Proecdysis duration or period of preparation for a moult in an insect or other arthropod.

Proenzyme zymogen; enzyme in inactive form that can be activated by another enzyme.

Progeny offspring; descendant: race.

Prognathous head of adult insect wherein cranium is turned upwards on the neck and mouthparts are projected forward.

Projapygidæ Primitive family of Projapygoidea (Rhabdura: Diplura); bristleails closely approximating to supposed ancestral type of insects, which are believed to be link between insects and millipedes. Antennæ with trichobothria on segmented 4 to 20, style present on first abdominal segment; cerci short, stout, few-segmented and perforated apically. Only three genera-*Anajapyx, Projapyx* and *Symphyurinus,* and five species have been described.

Proleg(s) false largs(s) : short unsegmented appendages found on some abdominal segments of caterpillars etc. Though used mainly for locomotion, these are not true legs. Most caterpillar have five pair situated on II, IV, V VI and X segment; semiloopers (Plusiinæ: Noctuidæ) have 3 to 4 pairs while loopers (Geometridæ) have only two pairs, one each on VI and X segment; those of Micropterygidæ have eight pairs and larvæ of sawflies usually have 6 to 7 pairs. At distal end the prolegs have numerous small hooks (crochets) arranged in definite pattern, which often serves as diagnostic character. There are no crochets on legs of sawfly larvæ.

Prolific multiplying abundantly-aphids are prolific breeders.

Pronotum dorsal cuticular covering of prothorax of an insect; in some species, it is well enlarged to form a shield covering rest of the thorax.

Pronymph newly hatched first instar nymph of dragonfly (Odonata) or other hemimetabolous insect.

Prophalangopsidæ (Haglidæ) very small family of orthopteran insects (Tettigonioidea: Ensifera); characterised by long and many-jointed antennæ; stridulatory organs present on forewings in males-only three species, on Indian (*Prophalagopsis obscura*) and two North American (*Cyphoderris monstrosa* and *Cyphoderris* species).

Prophylactic pesticidal treatment given in anticipation, to prevent the pest attack or pathogen invasion.

Propleuron pleuron or lateral cuticular plate of prothorax in insects.

Propneustic insects having functional spiracles, only in prothorax-a rare condition; found in pupæ of some dipteran flies. *Cf.* Holopneustic, Metapneustic.

Propodeum (propodeon) first abdominal segment fused with the posterior part of thorax and party surrounding the petiole; found in various hymenopterous insects (ants and bees).

Propodite penultimate segment of generalised arthroped leg; tarsus in insect.

Propolis bee-glue; resinuous material of vegetable origin, used by bees in addition to wax, for building and repairing their hives.

Proprietory name particular name, usually registered, given by the manufactures to their own product : trade name-Dimecron® is proprietory name of phosphamidon manufactured by Ciba Ltd., whereas Kinadon is the proprietory name, also of phosphamidon, but manufactured by United Phosphorus Ltd.

Prosechomorpha Enderlein 1936 suborder of Diptera (true flies) in an old system of classification, comprising of a single family-Thaumaleidæ (stream-midges) that has now been placed under suborder Nematocera.

Prosopistomatidæ Lestage 1919 very small family of Prosopistomatoidea (Ephemeroptera: Plaæoptera); mayflies having wing venation without cross-veins; legs reduced and non-functional. Nymphal carapace covers almost entire abdomen-one genus, *Prosopistoma* with its species found in Africa.

Prosopistomatoidea very small superfamily of Ephemeroptera (Palæoptera: Pterygota) related to Cænoidea. Nymphs with characteristic carapace formed by extensions of mesonotum, to cover thorax and half or more than half of abdomen-Bætiscidæ and Prosopistomatidæ.

Protanisoptera an extinct suborder of very big dragonflies, evolved from Meganisoptera, the ancestors of present day Odonata.

Protaptera hypothetical common ancestor of insects, centipedes and millipedes.

Protarsus trasus of first leg (foreleg) of an insect.

Protectant any chemically applied to prevent damage by pest or disease.

Protective colouration colour pattern of various insects that protects them from their enemies, when at rest the colour blend with the background to well that insects become almost invisible-grasshoppers on grass, beetles on bark, bugs and moths on tree trunks, caterpillars on flowers, etc. The camouflage may be effected not only by matching of colours but also by disrupting pattern which disguises the insect's outline.

Protective resemblance resemblance of an animal (insect) to the environment that tends to make it inconspicuous. Many insects are proctected by their form and shape–stick-insects and leaf-insects (Phasmida); catrerpillars (loopers) or Geometrid moths (Lepidoptera) etc.

Protein organic compound with carbon (C), hydrogen (H), Oxygen (O), nitrogen (N) and other elements, easily hydrolysed into mixtures of amino–acids; play an important role in plant and animal bodies.

Protelytridæ family of Protelyroptera, an extinct order of fossil insects that live in Permian period–*Protelytron permianum* recorded form Kansas (USA).

Protelytroptera Tillyard 1931 an extinct order of small Palaeozoic insects found during lower Permian period, related to Blatteria and Dermaptera; antennæ 22–segmented; forewings convex having wide costal area; vein R_2 branched; Cu bifurcate basally into Cu_1 and Cu_2 three simple anal veins present and cross–veins absent–Elytroneouridæ Blattelytridæ Megelytridæ, Protelytridæ and Archelytridæ.

Protembiidæ family of Orthopteroidean insects (Protorthoptera: Polyneoptera); insects reported living in France and North America during Carboniferous period and are al extinct since ages.

Protentomidæ Ewing 1936 one of the two families of Acerentomoidea (Protura: Myrientomata); minute (1.5 to 2 mm long), apterous, blind insects having no tracheæ and spiracles; abdominal segments with a single complete transverse row of dorsal setæ and segment VIII with or without reduced pectines. Only 20 species of *Micrœntomon* Ewing. *Hesperentomon* Ewing. *Protentomon* Ewing and *Proturentomon* Womersley recorded from Europe and North America.

Protephemeroidea Handlirsch 1908 an extinct, monofamilial order of insects (Palæoptera: Pterygota) which flourished during Palæozoic era, reminiscent of Palæodictyoptera and foreunner of order Ephemeroptera. Head relatively large with very bith compound eyes; thorax and abdomen homonomously segmented; both pairs of wings rather oval and nearly equal; an unpaired caudal filament (paracerus) between two long multirticulated cerci–Triplosobidæ–*Triplosoba pulchella* reported from France.

Protephemeroptera fossil insects of Permian period possessing certain affinities with present day mayflies (Ephemeroptera) and primitive fossil dragonflies (Protodonata) in having complete pattern of wing venation with numerous cross-veins forming network or archedictyon.

Protereisma common and generic name of fossil insects of Lower Permian period that looked like present-day mayfly (Ephemeroptera), but had both pairs of wings alike in shape and size–*P. minus* Sellards and *P. latum* Sellards.

Proterhinidæ old and equivalent name for coleopteran family Aglycyderidæ (Curculionoidea: Polyphaga).

Prothetely precocious appearance of adult characters due to insufficient juvenile hormone. *Cf.* metathetely.

Prothoracic gland the gland located in prothorax of certain insects, it secretes moulting hormone, which stimulates the glands in the skin to produce moulting fluid that dissolves away the lower layers of the cuticle to facilitate the moulting.

Prothorax first segment of insect's thorax that is attached to the head; it bars a pairs of prolegs (front legs) but no wings, which if present, are always on meso-and metathorax.

Prothysanura Schepotieff 1909 old and obsolete name of the apteran order Protura Silvestri (Myrientomata).

Protista newly erected kingdom that includes all unicellular organisms and few multicellular ones that are one border line between plants and animals–Protozoa (unicellurlar animals) and Protophyta (algæ and fungi); Mycota has agian been proposed as separate biological kindgom for fungi (90000 described species).

Proto ancestral: primitive.

Protoblattoidea Handlirsch 1908 extinct order of Orthopteroidea (Polyneoptera: Pterygpta); that lived during Carboniferous period as reported form France and USA. Head free, rounded with strong mandibles and long multi-segmented antennæ; prothorax flattened and disc like; abdomen with cerci and also gonapophyses–*Adiphlebia lacbana* (Adiphlebiidæ), *Eucanus ovalis* (Eucænidæ), *Protophasma dumasi* (Protophasmidæ), *Stenoneura fayoli* (Stenoneuridæ), et al.

Protoceredrum forebrain front and the largest part of insect's or arthropod's brain, formed embryonically form used ganglia of fist somite and containing ocular as also associated centres.

Protochordata (Acaniata) phylum/subphylum of Chordata, that includes lower chordates–Urochordata (tunicata), Cephalochordata (lanclets) and Hemichordata (tongue–worms).

Protocoleoptera Tillyard 1924 extinct monofamilial order of Palæozoic beetles represented by only one species, *Protocoleus mitchelli* recorded from New South Wales, lived during Permian period.

Protocucujidæ Crowson 1952 a very small recently erected family of Clavicornia (Cucujoidea: Polyphaga): akin to Sphindidæ (Clavicornia) but differing form it in having the penultimate tarsal segment the smallest and a large bisetose empodium–only one Chilean and one Australian species.

Protodonata Handlirsch 1908 fossil order of Carboniferous (Palæozoic) and Triassic (Mesozoic) periods; wings nearly equal similar and stretched out horizontally; main longitudinal veins touch each other basally, cross–veins straight and regular; eyes large but separate. Three families–Erasipteridæ Meganeuridæ and Paralogidæ (all extinct) –*Meganeura monyi* Brongniart 1884 and *M. brongniarti* Handlirsch 1908 (Meganeuridæ) reported form France had wing nearly 300 mm long with win expance of about 700 mm.

Protohemiptera Heandirsch 1903 an extinct order of Palædictyoptera (Palæoptera: Pterygota); having relatively small head, bearing elongate mouthparts, comprising of labrum and 2 pairs of styles. Prothorax and very large with paranotal expansions; forelegs massive and turned forward with uniarticulate trasus ending in strong paired claws; wing horizontal and homonomous, with complete venation–*Eugereon backingi* Dohrn (Eugereonidæ) was found in Germany during Permain period, with estimated wing span of 160 mm.

Protohymenidæ extinct family of Palæzoic order Megasecoptera, and offshoot of Palædictyoptera, akin to Protodonata but differs by having very narrow or subpetiolate wing with single anal vein and elongate cerci. Found in France, England and North America during Carboniferous period and are all extinct since long.

Protohymenoptera fossil insects found in Permian and Jurassic rocks; these looked like bees and wasps, had two pairs of wings of equal size with enation of generalised type and without coupling apparatus.

Protokollariidæ family of Protorthoptera (Orthopteroidea: Polyneoptera); comprising of extinct insects that lived during Carboniferous period, mostly in France and North America.

Protolepidoptera Laciniata : one of two suborders of Lepidoptera (Mecopteroidea: Oligoneoptera) in an old system of classification, abased only on mouthparts; these insects have biting and chewing mouthparts.

Protomecoptera suborder of Mecoptera (Mecopteroidea: Oligoneoptera); includes mostly fossil insects but a few living ones as well–Chilean *Notiothauma reedi* McLachlan (having ocelli), West Australian, *Austromerope* species (no ocelli), American *Merope tuber* Newman, et al.

Protoneuridæ widely distributed family of damselflies (Coenagrioidea: Zygoptera); having very long abdomen and characterised by wing with Cu_2 reaching the margin before middle of the wing, 1A absent and no supplementary postcubitals.

Protonymph second instar of a mite's nymph (Arachnida); first instar is known as larva and third as deutonymph.

Protoperlaria an extinct order of Palæzoic era (Carboniferous and Permian periods), believed to be ancestors of present day stoneflies (Perlaria or Plecoptera); also showed some resemblance to modern Orthoptera.

Protoplasm living matter of biological cell: physical basis of life. The word was coined by J.E. Purkinje in 1840.

Protopod larva most primitive type of insect larva having unsegmented abdomen and rudimentary appendages. These are endoparasitic and can survive only when the food is abundant, making locomotion unnecessary–larvæ of minute wasps (Proctotrupoidea: Apocrita).

Protopsyllidæ group of extinct psyllids that lived during Permian period, extant Psylloidea (Sternorrhyncha: Homoptera) are its direct discendents.

Protorthoptera Handlirsch extinct insect order, an early offshoot from Palæodictyoptera that gave rise to modern Orthopteroid insects (Polyneoptera : Pterygota); these were abundant during Carboniferous period and disappeared during Triassic period. These Palæozoic Forms had head vertical and free with long antennæ and strong mandibles; wing heteronomous which could be flexed backwards. Over 250 species grouped into 20 or more families–*Cnemidolestes woodwardi* (Cnemidolestidæ), *Gerarus danielsi* (Geraridæ) *Gyrophlebia longicallis* (Spanioderidæ), *Ishnoneura oustaleti* (Ischnoneuridæ), *Oedischia willamsoni* (Oedischiidæ) et al.

Prototype first or original specimen.

Protozoa phylum of simplest and most primitive organisms belonging (till recently) to kingdom Animalia (Non-chordata). Of late placed under newly erected kindom Protista that also includes algæ. About 30,000 species grouped under five classes–Mastigophora (Phytomastigina, Zoomastigina), Sarcodina or Rhizopoda, Sporozoa (Teloporida, Acnidosprida), Ciliata (Protociliata, Euciliata) and Suctoria.

Protozoon (pl. **protozia/protozoons)** microscopic animals consisting of one cell or of colonies of like cells, multiplying by fission–amoeba, paramicium et al.

Protozygoptera an extinct suborder of large dragonflies that included earliest known odonates; it was evolved from Meganisoptera during Carboniferous period; these gave rise to Zygoptera during Permian period and later on to Anizygoptera (Triassic period to recent); both these are now regarded as extant suborders of Odonata.

Protrude stick out: thrust forward.

Protuberance swelling or buldge.

Protura Silvestri 1907 order of minute, slender, white, apterous insects (telson-tails); head pointed without visible antennæ and no eyes; mouthparts degenerated; abdomen 9-segmented in young ones and 12-segmented in adults; first three segments each with a pair of small one–or two–segmented ventral appendages; no cerci. Metamorphosis primitive (slight) or wanting (300 species) two super–families each embracing two families–Eosentomoidea (Eosentomidæ, Sinentomidæ) and Acerentomoidea (Protentomidæ, Acerentomidæ.)

Telson-tails

Proventriculus gizzard or chewing stomach of insects and other arthropods; terminal region of stomodæum between crop and ventriculus, usually thick and muscular with chitinous teeth or ossicles on its inner wall.

Proximal nearest to body or point of attachment. *Cf.* distal.

PRUTHI, Dr. Hem Singh, O.B.E.–born February 23, 1897, started his career at Agricultural College, Lyallpur (1920); obtained Ph.D. (Cantab) 1925; joined the Zoological Survey of India, Calcutta (1925–34) and IARI in 1934 as Imperial Entomologist (first

Indian); became the Director of the Institute in 1944, then Plant Protection Advisor to Government of India and Director Locust Control (1949–54). An Entomologist and Locustologist of international repute; his book, **A Textbook on Agricultural Entomology** (1969) was well received and soon became out of print. Died December 23, 1969.

Psammocharidæ Rohwer 1910 (= Pompilidæ Leach 1815, 1819) family of minute to medium-sized spider-wasps (Apoocrita: Hymenoptera); body shiny smooth or pilose; antennæ straight in males and curved in females; prosternum extending back to base of wings; legs long and slender; tibia of hind legs two-spurred, claws simple; wings well developed, hyalline, infuscated or reddish-*Psammochares ilus* Bingham (28 mm long.) *P. mirandus* (Saussure) (female 5 mm long).

Psammon any organism that lives between sand particles on lake, sea-shore etc.

Pselaphidæ (Leach 1815) large family of small short-winged mould beetles maxillary palpi with last segment very large, tarsi 3-segmented, claws unequal; elytra covering half of abdomen; found under logs and stones or in moss, a few mymecophilous or termitophilous while other are predaceous on mites, spiders etc (5000 species)-*Mastiger abruptus* et al.

Pselaphinæ subfamily of Pselaphidæ (Staphylinoidea: Polyphaga); minute brachypterous beetles having 11-segmented antennæ and well developed maxillary palpi. Widely distributed; some being myrmecophilus, live in ant's nests, other found among moss.

Psephenidæ Le Conte 1862 family of subaquatic beetles (Dryopoidea: Polyphaga), maxillary palpi inserted more laterally than usual; found in great number in Himalayan river and swift rivers. cascades and waterfalls of North America-species of *Psephenus, Psephenoides, Eubria* et al.

Pseudo- false.

Pseudo-imagines supplementary repoductive forms found occasionally in certain termite species. These appear in the colony that has lost its queen, tear off the wings with their mandibles and become sexually functional without leaving the termitarium-*Kalotermes flavicollis, Reticullitermes lucifugus* etc.

Pseudococcidæ (Heymons 1915) large family of mealybugs (Coccoidea: Stermorrhyncha), soft elongate-oval body with distinct segmentation, usually covered with mealy or cottony wax secretion. Females have well developed legs, one pair of ocellanæ as also anal lobes, anal rings and anal ring setæ. Suck plant juices and are destructive pests of various crops-*Ferista virgata* (Cockerell), *Nipæcoccus vestator* (Maskell), *Planococus citri* (Risso), *Pseudococcus nipæ* Maskell et al.

Pseudocoele body cavity not completely lined with mesoderm.

Pseudocone eyes compound eyes in which crystalline cones are imperfectly formed and consist of a mass of liquid; in such eyes cuticular lenses from the main part of optical system–house–flies (Diptera).

Pseudocopulation attempted copulation by male insect with a flower that is mistaken as female insect due to strickingly similar colouration, shape or scent usually results in pollination of flower.

Pseudogamy (gynogenesis) development of ovum into a new individual as a result of stimulation by male gamete; the nucleus of sperm does not fuse with that of ovum and constitutes nothing to the hereditary constitution of the embryo–in certain *Luffia* species (Lepidoptera) the sperm only just penetrates and does not actually fertilise the egg; in *Ptinus* species (Coleoptera) mating occurs between two different species. Also occurs in some nematodes.

Pseudogyne(s) individual ants (Formicoidea: Apocrita); having the size and form of the gaster of the worker combined with the thorax of fertile female.

Pseudolteres reduced or vestigial forewings that look like small club–shaped processes, though these resemble halteres (dipteran hind wings) but are not homologous with the same–stylopids (Strepsiptera).

Pseudolestidæ small family of Agrionoidea (Zygoptera: Odonata); dameslflies with broad, often coloured wings, having 5 or more antenodal cross–veins.

Pseudomyrminæ (Leptaleinæ) subfamily of Formicidæ (Scolioidea: Aporita); males of these ants are characterised by 2–segmented pedicel and a large eyes occupying almost half the side of the head, ocelli present. Found in hollow twigs and on plant galls.

Pseudophyllinæ subfamily of Tettigoniodea: Ensifera); often regarded as independent family, long–horned grasshoppers predominently green in colour, live amidst herbage, simulating the leaves.

Pseudoplacental viviparity phenomenon found in certain insects wherein egg has very little yolk and the embryo is nourished through placenta–like structure in contact with material tissue–some aphids, batbugs, booklice, parasitic earwigs etc.

Pseudopupa larva in quiescent pupa–like condition, actually one or two instars before true pupal stage–Meloidæ (Coleoptera).

Pseudoresistance apparent resistance that results from transitory characters in potentially susceptible host plants; it includes escapes, host–evasion and induced resistance.

Pseudostigmatidæ small family of damselflies (Coenagrionoridea : Zygoptera); characterised by Cu_2 straight; pterostigma absent or indistinct; nodus at or between the basal sixth or seventh wing margin.

Pseudotracheæ food channels consisting of a large number of tubules and resembling tracheæ, found in the membranae covering of labellar lobes in certain dipteran flies–blowflies, house–flies etc.

Pseudotracheal teeth crushing organs formed from the chitinous rings of pseudotracheæ.

Psilidæ Walker 1853 dipteran family of small, shining, slender flies (Acalyptratæ: Schizophora)–carrot rust fly *Psila rosæ,* (Fabricius), maggots feed on roots and crowns of carrots, celery, parsnips and other umbelliferous plants.

Psocid(s) booklice : common name for all the insects belonging to order Psocoptera.

Psocidæ (Leach 1815) Samouelle 1819 dominant and most widely distributed family of Psocomorpha (Psocoptera : Psocopteroidea); antennæ 13–segmented; tarsi 2–segmented; forewings with Cu looped forwards at apex; found feeding on bark of trees, generally in clusters and often in various stages of development and usually covered with silken webs –*Psocus confraternus* Banks, *Amphigerontia bifasciata,* et al.

Psocomorpha suborder of Psocoptera (Psocopteraoidea : Paraneoptera); booklice having antennæ with 13 or less segments, never secondarily annulate; labial palpi 1– or 2–segmented; pterostigma more strongly sclerotised than wing memberane; tarsi 2–or 3–segmented. Comprises of more than dozen families, including–Amphisocidæ (*Kolbea quisquiliarum*), Stenopsocidæ (*Graphopsocus cruciatus*), Psocidæ (*Ptycta distinguenda* Parkins) et al.

Psocoptea Sipley 1904 [Corrodentia (Brumesiter 1939) Comstock & Comstock 1895] order of small paraneopteran insects (psocopteroidea: Pterygota), 3 to 6 mm long, soft–bodied, head orthognathous, mouthparts degenerate biting and chewing type; antennæ multiarticular, long and filiform; compound eyes protruding; wings, when present, have simple venation with few cross-veins;

abdomen globulous, cerci wanting. Metamorphosis incomplete or gradual (3000 species), three suborders–Trogiomorpha, Troctomorpha and Psocomorpha (including Eupsocida).

Psoculidæ monogeneric and monospecific family of Psocomorpha (Psocoptera: Procoptroidea); allied to Mesopsocidæ–*Psoculus negletus* from France and Germany.

Psoquillæ Latreile 1810 old and equivalent name for insect order Psocoptera Shipley 1904.

Psoquillidæ Kolbe 1884 very small family of Trogiomorpha (Psocoptera: Psocopteroidea), about 20 species confined to tropical Old world.

Psychidæ Boisduval 1829 family of bagworms or case–moths (Psychoidea: Ditrysia); cosmopolitan in distribution; sexes distinctly different. Male very active having strongly bipectinate antennæ; tibia with spurs; wings thinly clothed with hair and imperfect scales and frenulum exceptionally large. Female apterous with antennæ, mouthparts and legs degenerated, reproduction sexual as also parthenogentic. Caterpillars have scattered hairs or setæ; strong thoracic legs and prodegs reduced to crochets, live in a bag or case of varying shape, size and material, constructed by the caterpillar (800 species) pests of economic crops–*Arytania obscura* Crawford on mango (*Mangifera indica* Linnæus), *Chaliodes vitrea* Hampson on tamarind (*Tamarindus indica* Linnæus), *Kophere cupea* Moore on banana (*Musa paradisiaca* Linnæus), et al.

Psychodidæ Bigot 1854 family of mothflies or sand–flies (Psychodoidea: Nematocera); small–moth like flies having body and wings densely covered with coarse hair; antennæ long, 15- to 16–segmented; wings lancolate having anterior cross–veins closer to wing–base; bite very vicious. Larvæ aquatic or saprophagous–Phlebotominæ comprises of bloodsucking species; only females suck blood of vertebrates; *Phlebotomus* (*Flebotomus*) *minutus* preys upon lizards and geckos whereas *P. papatasi* suck mammalian blood and even act as vector and transmit various human diseases.

Psychodoidea superfamily of Nematocera (Diptera: Oligoneoptera); mesonotum without V–shaped suture; vein costa continuing around wings margin beyond apex, veins and hind margin very hairy or scaly and wings, at rest, sloping rooflike–Psychodidæ, Tanyderidæ and Nymphomyiidæ.

Psychoidea (Tineoidea in restricted sense) superfamily of Ditrysia (Lepidoptera: Oligoneopotera), antennæ often with eyecap, prodoscis atrophied; maxillary palpi 1–to 5–segmented; Cu_2 present on both fore–and hind–wings and vein M often present in discal cell.

Psychomyiidæ Kolenati 1859 family of Hydropsychoidea (Annulipalpia: Trichoptera); caddisfilies distinguished by larvæ campodeoid type, having compressed body, without abdominal papillæ, tracheal gills wanting but rectal gills often present.

Psychopsidæ Handirsch 1908 family of Hemerobioidea (Planipennia: Neuroptera); predatory insects that suck juices of their victims; distinguished by a broad double vein (rib) in the middle of each wing that strengthens the wing-*Psychopsis* spp. et al.

Psychrophilic micro-organisms that thrive best at low temperature; minute insects found at high altitudes where normal temperature is below 2°C.

Psyllid(s) common name for all plants bugs belonging to superfamily Psylloidea-olive psyllid *Euphyllura olivina* Costa, tomato pysllid *Paratrioza cockereli* (Sulc) et al.

Psyllidæ Latreille 1807 (= Chermidæ Fallen 1841) family of jumping plant-lice, small active psyllids incapable of sustained flight; hind legs large and more muscular adapted for jumping. Nymphs oval with conspicuous large wing-pads. Over 1500 species including many destructive crop-pests-*Apsylla cistellata* (Buckton), on mango (*Mangifera indica* Linnæus), *Diaphorina citri* Kuwayama on citrus, *Psylla pyricoda* (Foerster) on apple (*Malus pumila* Miller) and pear (*Pyrus communis* Linnæus).

Psyllina suborder of Homoptera in certain system of classification; equivalent to superfamily Psylloidea of suborder Stemorrhyncha, comprising of family Psyllidæ.

Psyllipsocidæ Enderlein 1911 small family of Trogiomorpha (Psocoptera: Psocopteroidea); head long and vertical (not transverse), antennæ 27-to 46-segmented; maxillary palpi without sense organs; wings often reduced or showing various degrees of alary polymorphism (20 species) -*Psyllipsocus ramburi* et al.

Psylloidea superfamily of Sternorrhyncha (Homoptera : Hemipteroidea); characterised by having mouthparts and two pairs of wings in both sexes; antennæ 10-segmented; tarsi 2-segmented with paired claws; femora thickened; forewing harder in texture than hind-wing. Only one important family-Psylidæ.

-ptera wings; suffix of majority insect orders denoting the number or type of wings possessed by the insects belonging to that order-Diptera (two-wings), Isoptera (equal wings), Lepidoptera (scaly wings), Trichoptera (hairy wings) and so on.

Pteralia small sclerites at the base of insect's wing which make up the articulation, including the humeral plate of the wing.

Pteregate(s) exceptional form of ants (Formicidæ: Hymenoptera) - workers and soldiers-having rundimentary wings.

Pteridine basic chemicals constituent of pteroic acid and the pterin pigments in insect's wings.

Pterin group of substances occurring as pigments (red, yellow, white) in butterfly body and wings; probably waste products of metabolism.

Pteroic acid constituent of vitamin (folic acid), an essential substance for formation of pterin pigments (pterines) in insects.

Pteromalidæ Walker 1833 largest family of minute chalcids (Chalcidoidea: Apocrita); commonly called jewel wasps; thorax usually not coarsely punctate; pronotum narrow and transverse; hind coxæ not enlarged and femora not toothed; tarsi 5-segmented. Parasites or hyperparsites on species of almost all insect order - *Cardiogaster secundus* Mani is parasitic on *Aleurolobus barodensis* Maskell, *Pachyneuron eros* Girault on *Nipæcoccus vastator* (Maskell), *Aplastomorpha uandinei* Timberlake on rice weevil and *Merisus destructor* (Say) on Hessian fly.

Pteronarcidæ (Jacobson and Bianchi 1905) Enderlein 1909 smallest family of Setipalpain stoneflies (Pteronarcoidea: Plecoptera); mandibles relatively better developed and wings with archedictyon present except on anal lobe of hind wing. Found mostly in East Asia and North America (13 species) -giant stonefly of North America, *Pteronarcys dorsata* Say is 40 top 45 mm long with wing expanse of 50 to 60 mm while California salmonfly, *P. californica* Newport has wing-span of 40 to 46 m.

Pteronarcoidea monofamilial superfamily of Setipalpia (Plecoptera: Orthopteroidea); characterised by wing margin incised at apex of Cu_2 Nymphs with branched gills on thorax and anterior part of abdoment-Pteronarcidæ.

Pterophoridæ Zeller 1841 family of plume-moths (Pterophoroidea: Ditrysia); characterised by lack of maxillary palpi; wings splitted into plumes, forewing normally cleft longitudinally into two, rarely into 3 or 4 parts (plumes) and hind-wing into three plumes. Caterpillars feed exposed on flowers and leaves; includes destructive pests of various economic crops-*Exelastis atomosa* (Walsingham), *Oxyptilus rugulus* Meyrick, *Sphenarches caffer* (Zeller) et al.

Pterophoroidea Tillyard 1926 superfamily of Ditrysia (Lepidoptera : Mecopteroidea); ocelli, chætosemata and tympanal organs absent;

haustellum naked; maxillary palpi reduced and wings deeply fissured into plumes–Pterophoridæ.

Pterostigma wing–spot: small, thickened opaque or dark spot along the costal margin of insect's wing, usually near wing–tip–bees, wasps, flies and some other insects.

Pterothorax the two wing–bearing segments (meso–and meta–) of thorax, closely united with each other.

Pterygogenea old and obsolete term coined by Braure (1885) for subclass Pterygota.

Pterygoid wing–like.

Pterygota subclass of insects (Hexapoda) that includes vast majority of insects, mostly having wings or if wingless, descendent of winged ancestors; terrestrial, semi–aquatic and aquatic forms, found almost every where, from bottom of seas to snow–capped mountains; phytophagous as also parasitic; further divided into four intfraclasses–Palæoptera, Polyneoptera, Paraneoptera and Oligoneoptera.

Pterygote winged.

Ptiliidæ (Trichopterygidæ) family of Staphylinoidea (Polyphaga: Coleoptera); very minute beetles (<2 mm long); elytra variable in length, wings very narrow with margin fringe of along setæ. Found in decaying vegetation fungi etc. (300 species) – some species of *Eurygyne* are parthenogenetic; *Nanosella fungi* is the smallest known beetle (0.25 mm long).

Ptilinum bladder–shaped structure found within lunule on the head on certain dipteran flies; when the fly is ready to emerge from its pupal case, the ptilinum is pushed through the slit to help the fly to escape and is then drawn back into the head.

Ptiloneuridæ little known tropical family of Dryopoidea (Polyphaga: Coleoptera); head with an occipital keel; antennæ not retractile into grooves. Larvæ live in rotten wood or are aquatic or semi–aquatic (170 species).

Ptinoidea Leunis 1860 superfamily of Polyphaga (Coleoptera : Colepteroidea) maxillary palpi not longer than antennæ, ocelli absent; pronotum largely developed and hood like. Comprises of four families–Anobiidæ, Ptinidæ, Bostrychidæ and Lyctidæ–all these families are often placed under superfamily Bostrychoidea (Polyphaga: Coleoptera).

Ptinidæ (Leach 1815, 1819) family of spider–beetles (Bostrychoidea: Polyphaga); earlier placed under superfamily Ptinoidea (Polyphaga:

Coleoptera); small cylindrical beetles having small head, long filiform antennæ, inserted on frons close together; thorax without lateral margin. Fore–and mid coxæ contiguous, hind coxæ transverse, metatarsus longer than second tarsal segment, trochanters distally truncate; abdomen with 5 visible tarsal Mostly scavengers, a few found in storehouses – *Gibbium psylloidea* (Czempinski) (= *scotias* Fabricius) is cosmopolitan pest of stored–products.

Ptychopteridæ small family of Tipulaoidea (Nematocera: Dipera); moderate–sized flies resembling tipulids but having legs not deciduous and wings with characteristic vena spura between R and M. Larvæ aquatic having long breathing tube at hind end similar to that of rat–tailed maggots of dronefly (*Eristalis spp.*, Syrphidæ: Diptera) and are found in humus and muddy water. Adult flies found usually at the edges of streams. Also known as phantom craneflies.

Pubescent covered with addressed or soft hairs.

Pulicidæ Stephens 1829 most widely distributed family of blood–sucking fleas (Pulicoidea: Siphonaptera); body laterally compressed, eyes usually present, hind coxæ with strong internal bristles and the sensilium having 14 pits on each side–human flea *Pulex irritans* Linnæus, rat flea *Xenopsylla cheopis* (Rothschild,) dog flea *Ctenocephalides canis* (Curtis), cat flea C. *felis* (Bouche), et al.

Pulicoidea one of the two superfamilies of Siphonaptera (Siphonapteroidea: Oligoneoptera); differs from Ceratophylloidea by lack of an outer external ridge on mid coxæ, no apical tooth the hind tibia and sensillum having 8 or 14 (not more) pits on each side–Pulicidæ Tungidæ.

Pulsatile organs accessory pumping orangs present sometimes in blood system of an insect at the base of antennæ and wings.

Pulvillus (pl. **pulvilii**) fleshy pad or cushion beneath each claw on terminal segment of trasus of an insect; used as an adhesive organ by means of hairs and sticky secretion; well developed in dipteran flies.

Punctate pitted or beset with punctures.

Pungent caustic : affecting organs of smell or taste with pricking sensation.

Punkie (punky) nickname of various small biting gnats or midges (Ceratopogonidæ: Nematocera).

Pupa (pupal case) fourth nymphal instar of whiteflies (Aleyrodidæ : Homoptera); it differs from first three instars and undergoes considerable metamorphosis before emerging as adult. It also serves as a diagnostic character for taxonomic study of Aleyrodidæ.

Pupa developmental phase, preimaginal stage between larva and adult of insects undergoing complete or indirect metamorphosis : non-motile and non-feeding metamorphic form in holometabolic insects. These may be :-

Exarate pupa with its appendages (antennæ, legs and wings) not glued to its body-most Coleoptera, Hymenoptera, Neuroptera, Trichoptea and some Lepidoptera.

Obtect pupa with antennæ, legs and wing rudiments fastened or glued to the body along its length - Lepidoptera (except Dacnonypha and Zeugloptera), Brachycera (except Stratiomyidæ) and Nematocera (except Cecidomyiidæ).

Coaretate pupation in puparium; last larval moult skin ins retained and its gets hardened to form a protective cover-Cycorrhaphora (Diptera).

Pupa, Adecticous obtect pupa whose mandibles are non-articulaed, often reduced and not used for escaping from the cocoon - most Lepidopteran insects (except Dacnonypha and Zeugloptera).

Pupa, Decticous exarate pupa having powerful mandibles that help the imago to escape form its cocoon-Lepidopteran insects belonging to suborders Dacnonypha and Zeugloptera.

Pupal pertaining to pupa.

Pupal shelter(s) shelter or case constructed under water by the aquatic larvæ of Trichoptera. These are of two distinct types, (*i*) case-bearing larva shortens its habitant (case), constructs a silken wall across either end and the pupa lies free within this case, on cocoon is formed; (*ii*) case-less larva constructs an oval cavern-like structure of sand particles and small stones; the pupa living in such a shelter is enclosed in brownish cocoon.

Puparium (pl. **puparia**) hardened last larval skin that forms a protective covering over the pupa; commonly found in various flies (Diptera).

Puparous female insect that gives birth to full grown larvæ that are about to pupate-Pupipara (Cyclorrhapha: Diptera).

Pupation process of transformation form the last larval instar to pupal stage: a larva becoming pupa.

Pupipara Nitzsch 1818 section (group) of dipteran flies (Schizophora: Cyclorrhapha) degenerated insects with wings mostly reduced but toothed and spined and having powerful claws; larvæ retained in mother's body till full grown, come out and pupate. Ectoparasites, suck blood of birds and mammals (not man) - Hippoboscidæ, Nycteribiidæ and Streblidæ.

Pygidial glands glands that open in pygidium of certain insects and secrete offensive or poisonous fluid for self protection - ground beetles, *Carabus* spp. (Carabidæ: Adophaga) secrete butyric acid.

Pygidicranidæ Verhoeff 1902 primitive, tropical family of Pygidicranioidea (Forficulina: Dermaptera); earwigs having forefemora compressed and keeled; paired ædeagus distinct; targum and telson on XI abdominal segment distinct, telson as large as pygidium-*Cranopygia* spp., *Diplatys gerstæckeri* Dohrn, *Echinosoma* spp., et al.

Pygidicraniodea superfamily of Forficullina (Dermaptera: Dermapteroidea); large earwigs, pubescent or tomentose; usually winged; opisthomeres comprising metapygidium, pygidium and telson, forceps of male often asymmetrical-Pygidicranidæ and Diplatyidæ

Pygidium last dorsal segment of abdomen.

Pygopodia a pair of eversible foot-like processes at the posterior end of abdomen of some insect larvæ; aids in locomotion where true legs are vestigial or wanting.

Pygothripidæ Hood 1915 family of Phlœthripoidea (Tubulifera : Thysanoptera); includes Australian thrips having head not produced anteriorly in front of eyes and characterised by the last abdominal segment not entirely tubular, short, swollen, rounded on sides.

Pyloric cæcum (pl. **cæca**) enteric cæca; digestive glands of an insect; sac-like branched or diverticula from mid-gut just behind the gizzard-cockroach etc.

Plyorus anetrior part of proctodeum, containing pyloric valve; anatomically rear end of stomach.

Pyralididæ Leach 1815, 1819 family of medium-sized somber and cryptic moths with closely appressed scales (Pyralidoidea : Ditrysia); antennæ well developed and tapering to a point; ocelli and proboscis present; labial and maxillary palpi well developed; wings with long frenulum and without pectin; abdominal tympanal organs present. Caterpillars naked with primary setæ arising from small tubercles, mostly phytophagous and pests of various economic crops-*Hypsopygia mauritialis* on wax in wasps' nest, *Pyralis farinalis* (Linnæus) on stored potatoes et al.

Pyralidoidea Mosher 1916 superfamily of small to medium-sized, fragile slender moths (Ditrysia : Llepdoptera); maxilary palpi usually present; vein Cu_2 vestigial or absent in forewing but present in hind-wing; legs long and slender–Thyrididæ, Hyblæidæ, Pyralidæ, Crambidæ, Galleriidæ, Pyralididæ and Pyraustidæ).

Pyraustidæ (Smith 1891) Meyrick 1896 large family of Pyralidoidea (Ditrysia : Lepidoptea); often regarded as subfamily of Pyralididæ; large moths with vestigial proboscis. Caterpillars mostly borers, bore into stems, pods and fruits Comprises of three subfamilies–Schoenobiinæ, Nymphulinæ and Pyraustinæ - all these are often considered as independent families.

Pyraustinæ subfamily of Pyraustidæ, often regarded as an separate family of Pyralidoidea (Ditrysia: Lepidoptera); includes large number of pests of economic importance – *Dichocrocis punctiferalis* (Guenee), *Hellula undalis* Fabricius, *Hymenia recurvalis* (Fabricius), *Sylepta derogata* (Fabricius) et al.

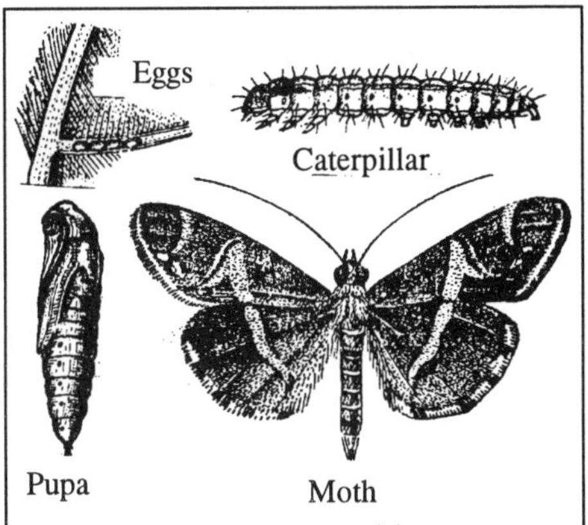

Hymenia recurvalis (Fabricius)

Pyrethroids, Synthetic group of synthetic similar in structure and activity to natural pyrethrums; broad spectrum insecticides having quick knock-down effect low mammalian toxicity and minimum atmospheric pollution -Cypermathrin ($C_{22}H_{19}Cl_2O_3$), Fenvalerate ($C_{25}H_{22}NO_2Cl$), Permathrin ($C_{21}H_{20}Cl_2O_3$) and others.

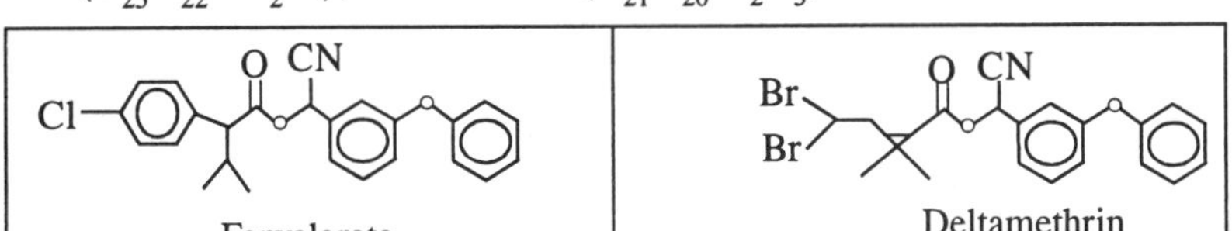

Fenvalerate   Deltamethrin

Pyrethrum dried and powdered flower heads of chrysanthemum (*Chrysanthemum cineraiffolium* Trevisano and *C. concineum* Willdenow); used as insecticide since 1800; it is nerve as also contact poison; not stable in presence of sunlight, air and moisture, hence leaves behind no harmful residues on plants. Acute oral and dermal LD_{50} values for rat are 200 and 1800 mg/kg respectively; not much in use these days.

Pyrilla generic and common name of sugarcane leafhoppers (Lephopidæ: Fulgoroidea); adult is soft-bodied, straw coloured bug, head prominently drawn forward as a snout; females have anal pads bearing boundless of cretaceous threads for covering their eggs; wing-expanse 16 to 20 mm. Nymphs have two characteristic filamentous process at their anal end, almost double in size of their body. Both nymphs and adults suck juices from leaves of sugarcane (*Sccharum offcinarum* Linnæus) and a large number of other economic crops, affecting adversely the yield of cane and sugar recovery-*Pyrilla perpusilla* (Walker) and *P.aberrans* (Kirby) with their numerous geographical subspecies.

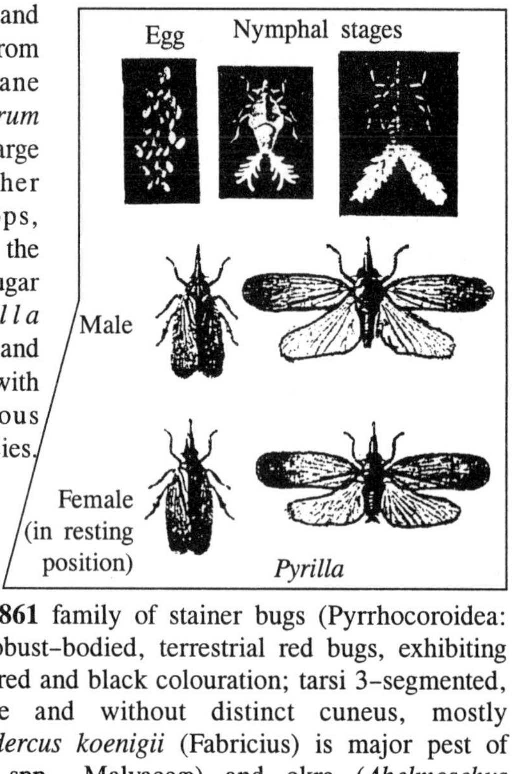

Pyrilla

Pyrrhocoridæ Fieber 1861 family of stainer bugs (Pyrrhocoroidea: Pentomomorpha); robust-bodied, terrestrial red bugs, exhibiting strongly contrasting red and black colouration; tarsi 3-segmented, hemelytra complete and without distinct cuneus, mostly phytophagous -*Dysdercus koenigii* (Fabricius) is major pest of cotton (*Gossypium* spp., Malvaceæ) and okra (*Abelmoschus esculentus* Linnæus) in India and *D.albidiventris* Stal is prevalent in North America, while *Dindymus sangunieus* is carnivorous and prefers termites.

Pythridæ family of small blue and purple beetles (Heteromera : Cucujoidea); fore-coxal-cavities open both externally and internally. Larvæ flattened with cribriform spiracles and broad almost lamelliform, urogomphi; found under bark of conifers-*Pytho* species et al.

Q

Qualitative analysis determination of chemical nature of a substance (insecticide).

Quantitative analysis to determine, by chemical methods, the quality of toxin (actual ingredients) present in an insecticide.

Quantitative characters characters of community used in the ecological description that can be objectively studied; these include population, abundance, density, dominance, frequency etc.

Quarantine methods or operation associated with prevention of exportation or importation of unwanted organisms (pests) form or into a territory.

Quartan fever (malaria) fever caused by *Plasmodium malariæ* and transmitted by female mosquitoes (*Anopheles* spp.).

Queen fertile female of eusocial insects — termites (Isoptera) and ants, honeybees and wasps (Hymenoptera). King's wife: female sovereign.

Queen ant fertile female founder of ant colony, characterised by her large stature and developed reproductive organs; usually larger than the male and workers of the same species; antennæ and legs shorter and stouter than in the male mandible well developed and the gaster large. The fully grown winged female (future queen) leave her colony and flies on her nuptial flight allowed by males; after fertilised, she sheds her wings and starts a new colony; it often attain a great size in certain forms.

Queen bee fertile female and founder of the beehive; longer than worker with large and more pointed abdomen but relatively shorter wings. After being fertilised on her nuptial flight, she return to her hive where she lives for 4 to 5 years laying several thousand eggs. In bumble bees, *Bombus* spp. (Bombidæ: Apoidea), the queen hibernates and is the only member of the colony that survives the Winter and starts a new colony every Spring.

Queen substance 9-oxydeo-trans-2-enoic acid, chemical substance produced by the mandibular glands of queen bee, which inhibits the construction of queen-cells. During her nuptial flight, the scent of this pheromone attracts the drones (males). Later, when the queen is settled in the hive, the worker bees lick this substance and pass it on to each other as a sign of recognising the present of queen in the hive.

Queen termite the only fertile female. After mating and having shed her wings, she stays underground for rest of her life, inside also she hardly moves, become a sort of egg-laying machine, her abdomen full with egg grows many times its normal size and looks like a giant caterpillar. Some species may attain the length of over 100 mm.

Queen wasp fertilised female, larger than males and worker having 12-segmented antennæ and 6-segmented gaster. The colony usually exist for a single season. males and worker perish during Autumn and Winter but the impregnated females (future queen) hibernate and each other found a new colony in next Spring.

Queiscence resting: non-feeding temporary slowing down of metabolism: state of inactivity mostly due to climatic factors.

Quetta borer *Æolesthes sarta* Solsky (Cerambycidæ: Polyphaga: Coleoptera) a destructive beetle causing enormous damage to popular (*Populus alba, P. ciliata, P. euphratica*), elms (*Ulmus villosa, U. wallichiana*) and willow (*Salix alba, S. babylonica, S. caprea*) etc. in entire Baluchistan, Turkestan and Western Tibet. Adults easily distinguished by pubescence thickly coating the elytra; prothorax disc or plate not well defined and more or less rugose; elytra obliquely truncate at the apex; Body length 30 to 45 mm., antennæ of males twice the body length of those of females less than body length.

Æolesthes sarta Solsky

Quinalphos $C_{12}H_{15}O_3H_2PS$; organophosphatic acaricide and insecticide; colourless, white crystalline solid, hardly soluble in water but readily soluble in acetone, alcohols, ethyl ether, benzene, tolune, xylene etc. Acute oral and dermal LD_{50} values for rat 137 and 1400 mg/kg respectively—Ekalux®, Kinalux®, etc.

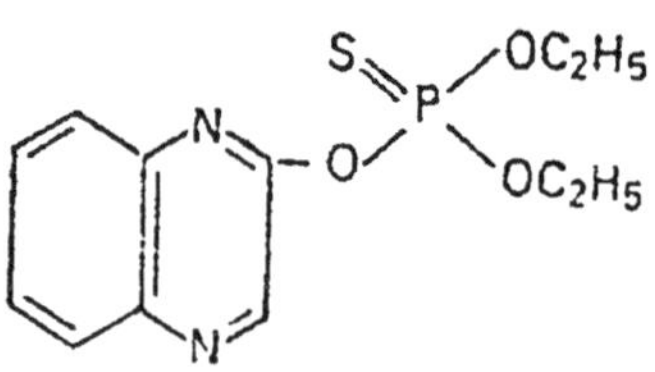

Qunialphos

Quinomethionate (oxythioquinox) $C_{10}H_6NOS_2$; acaricide and fungicide; pale yellow, odourless crystalline solid, in soluble in water, slightly soluble in cold organic solvents but soluble in hot aromatic hydrocarbons. Non-toxic to bees. Acute oral and dermal LD_{50} values for rat are 1100-1300 and, >1000 mg/kg respectively — Dithioquinox, Morestan® etc.

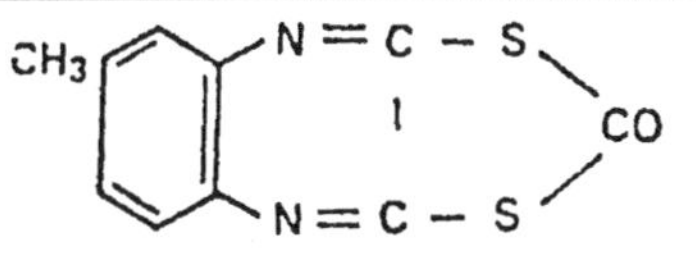

Quinomethionate

Quinones groups of chemical substance (cyclic diketones) involved in formation of some of the red and yellow pigment of insects and also in poisonous or repelælent secretion such as those emitted by cockroaches, earwigs and some bombardier beetles.

R

Radius (R) vein of insect's wing, it divides into an anterior branch R_1 and posterior branch R_s which again subdivides distally into 4 branches — R_2, R_3, R_4 and R_5.

Rami distal portion of a minute pair of appendanges (corpus) and free distally.

Rank level or position occupied by a category in taxonomic hierarchy — phylum, class, order, family, genus etc.

RAO, Rai Bahadur Y. Ramachandra — born September 11, 1885; passed M.A. zoology (1906) from Madras University and joined Madras Agricultural department. Since 1930, actively associated with locust research; retired in 1941 as Deputy Director, Plant Protection, Quarantine & Storage Directorate (Government of India), but continued to envince been interest in locust problem till death in 1972. His book, The **Desert Locust in India** (1960) is till today the bible for locustologist the World over.

Raphidiidæ Stephens 1936 major family of Raphidiodea (Megaloptera: Neuroptera)with character of superfamily, having snake-like neck and wings with pterostigma. Larvæ terrestrial with well developed spiracle — snakeflies; *Raphidia notata* is about 25 mm long, grayish-brown in colour and has large wings: *R. ratzebergi* Brauer is the European species and *R. hermandi* Navas occur in Japan.

Raphidiodea (Rhaphidiopea) Burmeister 1835 an order of snakeflies and serpentflies in an old system of classification; now regarded as a superfamily of Megaloptera (Neuroptera: Neuropteroidea); thorax exeptionally elongate setiform ovipositor. Found all over the except Australia and South Africa — Raphidiidæ, and Inocelliidæ.

Rephidiophoridæ family of apterous orthopteran insect (Gryllacridoidea: Ensifera); commonly called camel-crickets, primitive relatives of crickets and grasshopper, having long hind legs no tympanal organs; tarsi strongly compressed and usually without tarsal pulvilli (300 species) — *Daihinia brevipes, Tachycines asynamorus* (greenhouse camel-cricket) et al.

Raptorial adapted for catching the prey; forelegs of certain insect that are pincerlike with sharp claws and spines of bristles — preying mantid (Mantidae: Dictyoptera).

Rasp insect's stridulating apparatus.

Rat-tailed maggot apodous aquatic larva of various syrphid flies (Syrphoidea : Aschiza); having very long respiratory siphon, commonly called telescopic-tail; it can be extended upto 150 mm to reach water surface, at its free tip there are opening (spiracles) of two tracheal tubes surrounded by a fringe of floating hairs, for sucking fresh air — *Eristalis tenax* (Linnæus) et al.

Ratardidæ small family of Bombycoidea (Ditrysia: Lepidoptera); characterised by forewing having median vein preserved within the cell; Cu_2 absent from both wings and frenulum atrophied or vestigial. Found only in the Oriental region.

Receptaculum seminis alternate name for spermatheca of a female insect; a sac, where semina fluid can be received and stored and later released drop by drop, as the egg are laid. In this way a queen of social insect (ants, bee, termite etc.) can mate once in her lifetime and retain enough sperm cells in her body to fertilise all subsequently produced eggs.

Receptor sense organ or specialised structure of integument responsive to external stimuli.

Rectal pertaining to rectum.

Rectal sac enlarged anterior part of rectum.

Rectal gills series of gill lamellæ of thin-walled projection, well supplied with tracheæ and arranged in longitudinal rows on side of inner rectal wall of aquatic nymphs of Odonata; regular pulsation of hind-gut insures renewal of respiratory water.

Rectal pads (papillæ) six longitudinal bands of cellular layer found in insect's rectum; these play an important part in the maintenance of proper water balance through absorption of water from the fæces.

Rectum (pl. recta) posterior end of intestine that include rectal sac and opens at the anus.

Red cotton bugs *Dysdercus kœnigii* (Fabricius) and *D. singulatus* Fabricius (Pyrrhocoridæ: Heteroptera); medium-sizes, red bugs with black marking on the wings and white stripes ventrally on abdomen. Both nymphs and adults suck sap from leaves and green boils thereby impairing their vitality; affected bolls open badly and produce lint of poor quality; yellow excreta of bugs also stain the lint.

Red hairy caterpillar(s) *Amsacta moorei* (Butler) (Arctiidæ: Lepidoptera); polyphagous; moth having black antennæ; white forewing with costal margin red and red abdomen, black spotted above and white below. Caterpillar orange to dark brown in colour, densely hairy, feeds on leaves, specially of cotton (*Gossypium* spp.), maize and millets and move from field to field. *A. albistriga* Moore another *Polyphagous* pest is serious pest of groundnut (*Arachis hypogæa* Linnæus).

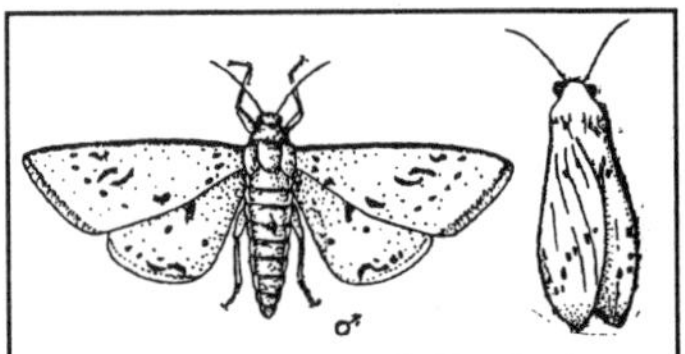

Amsacta moorei (Buller)

Red pumpkin beetle *Raphidopala* (*Aulacophora*) *foveicollis* (Lucas) (Galerucidæ: Coleoptera); small, 6 to 10 mm long, soft-bodied beetle are major pest of cucurbits specially pumpkin (*Cucurbita maxima* Duch.). Grub feed on roots and portion of stem below the soil as also borer into the fruits that are touching the soil; adults bite irregular holes in leaf lamina and also feed on flowers. Blue pumpkin beetle *R. intermedia* (Jacoby) and gray pumpkin beetle *R. cincta* (Fabricius) cause similar damage and are minor pest of cucurbits.

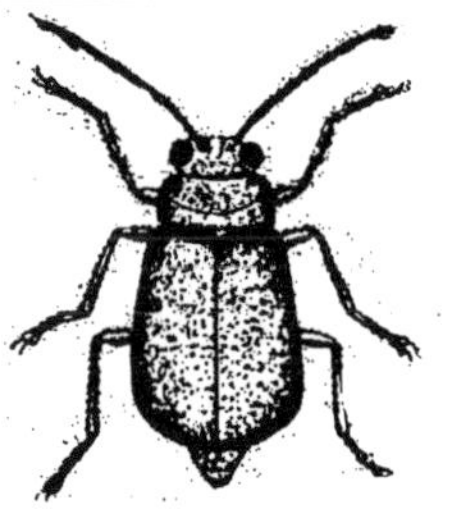
Raphidopala foveicollis (Lucas)

Reduviidæ Latreille 1807 large family of assassin bugs (Reduvioidea: Geocorisæ); having elongate-narrow head, antennæ often with intercalary segment (upto 40), rostrum 3-segmented, curved up and capable of friction against stridulatory grooves of prostrnum; forelegs not raptorial, tarsi 3-segmented; abdomen broad at the middle with margin of abdominal segment exposed. Nocturnal and mostly predaceous, a few blood-sucking, bite very painful (3000 species) — *Harpactor costalis* Rev. preys on *Dysdercus koengii* (Fabricius); *Reduvius personatus* (Linnæus) commonly called flying bug masked bedbug hunter is often found in the house under debris etc., feeding on small insect while *Apiomercus crassipes* (Fabricous) of North America feed upon bees which they capture in flowers.

Reduvioidea superfamily of Geocorisæ (Heteroptera: Hemipteroidea); predaceous bugs with long head and large eyes, ocelli present; antennæ usually geniculate after 1st segment; rostrum 3-segmented, powerful and curved away from ventral surface of head;

hemelytra with strong veins but no cuneus — Phymatidæ, Reduviidæ and Enicocephalidæ.

Regenerate produce a new: bring into released existence.

Regeneration replacement of regrowth of lost or injured part of an organism.

Regenerative cells cells that generate replacement cells of ventricular epithelium.

Registered pesticide a pesticide legally permitted to sale for the uses and purpose claimed.

Remigium remigial region: largest wing area, anterior to vennal fold, bearing costal, subcostal, radial, medial, cubital and postcubital veins.

Reniform kidney-shaped.

Repellant distasteful or disagreeable: substance that induce avoiding reaction.

Repellant, Chemical chemical that acts as stimulus, eliciting the insect to make direct move away from that area—diethyltoluamide (DEET) etc.

Replete(s) an additional special caste of ants, also called honey-pot ants, as these have crop capable of expanding greatly to store honey (honeydew) which they receive from worker ants and provide the same to the members of their colony in time of need.

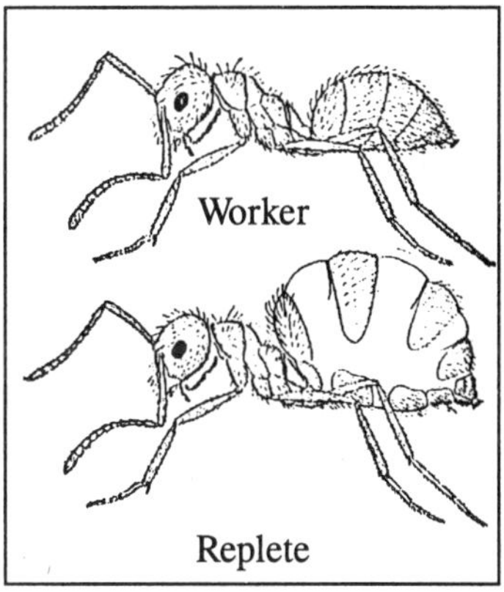

Reproduction giving rise to offspring. All living being (plants as well as animals) reproduce and like produces like. Non - living cannot reproduce.

Reproductive capacity fecundity: number of offsprings (eggs) produced by a female insect. On an average an insect species produce about 100 eggs but there are few exception, like queen bee lays 3000 eggs per day and termite queen of certain species may even give 6000 to 7000 eggs per day for several years.

Reptilia class of lowest vertebrates (Tetrapoda; Craniata); having dry rough non-granilar skin covered with cornified scales, terrestrial but now well adapted for running or walking, locomotion by creeping; evolved from primitive amphibians, more than 300 million years agro (Carboniferous period), these gave rise to aves and mammals. Common reptiles include, crocodiles, lizards, snake, tortoise, turtles etc. Whereas dinosaurs were abundant during Mesozoic era and are all extinct by now. The extant species are grouped into four subclasses — Anapsida, Diapsida, Parapsida and Synapsida.

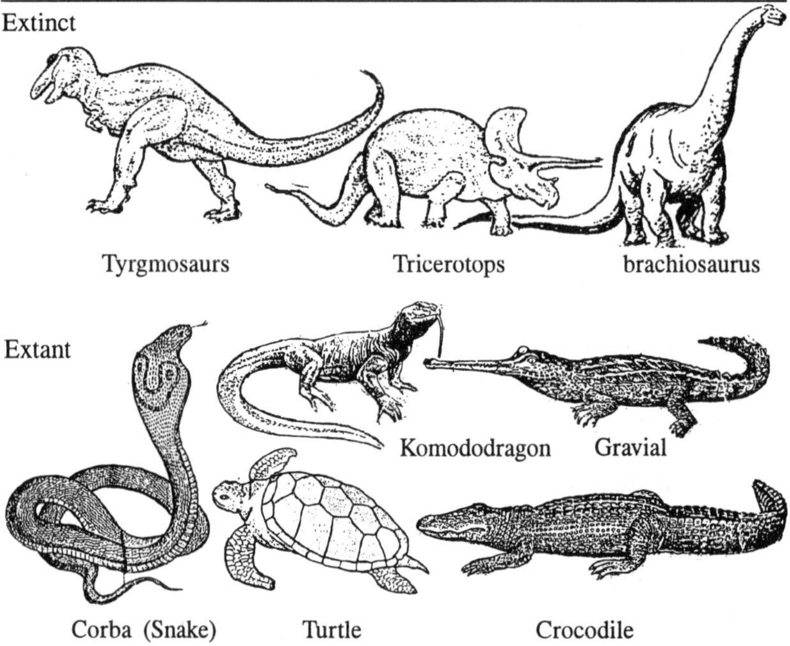

Residue reminant of pesticide and/or its degradation products and adjuvants found in/on plant tissue, usually expressed as part per million (ppm) or mg/kg on weight basis.

Residue tolerance maximum amount of pesticide residue expressed in parts per million that may lawfully be present in or on food, feed and other consumable product offered for sale.

Resilin elastic protein forming an important part of insect's cuticle.

Resistance ability to avoid or repel attack by pests (or parasites). Capacity to withstand toxic effect of a pesticide. Most of the plants that are highly tolerant to attack by certain insect pests are often called resistant. Certain mosquitoes, *Aedes aegypti* (Linnæus), *Culex pipiens* Linnæus etc. have developed resistance to DDT.

Respiration exchange of gases between the cell and its surrounding, as also between the organism and its environment. Series of physical and chemical processes accomplishing oxidation and removal of carbon dioxide; breakdown of food material within a cell, releasing energy for use of the cell. Insects usually breathe through spiracle leading into tracheæ and tracheoles, the latter penertate deeply into every part of body so that oxygen can diffuse directly into the cells; blood hardly plays any part in insect respiration. Insect respiration may be :

Holopneustic — most common respiratory system in insect, two thoracic and eight abdominal pairs of spiracles — all open — found in adults of most orders as also in larvae of many orders.

Hyperneutic — all spiracles open, but there are 3 to 4 pairs on thorax instead of usual two pairs.

Hemipneustic — having one or more pairs of spiracles non-functional. These may be.

Amphineustic — prothoracic and posterior abdominal spiracles open as in maggot of true flies (Diptera).

Metapneustic — only the last pair of spiracles open - larvae of most flies (Diptera) and aquatic beetle (Coleoptera).

Perineutic — only a few pair of spiracles closed along the side of the body - Neuroptera, Mecoptera, a few Diptera, Lepidoptera, Coleoptera and phytophagous Hymenoptera.

Polypneustic — having at least pairs of functional spiracles - most Diptera.

Propneustic — only the first pair of prothoracic spiracles open - pupae of Cuticidae (Culiocoidea: Nematocera).

Apenustic — spiracle wanting - Chaoboridæ (Cullicoidea: Nematocera) and first instar of certain parasitic Hymenoptera.

Branhiopneustic — spiracle replaced by gills - immature stages of Ephemeroptera, Plecoptera, Odonata, aquatic Neuroptera and a few aquatic Diptera.

Respiration, External exchange of gases with environment.

Resurgence significant increase in particular pest population after use of a broad-spectrum pesticide.

Retina receptive apparatus of eye, composed of optic sense cells.

Retinaculum hook-like structure beneath the abdomen of springtails (Collembola) which holds down the furcula. Hook-like process at the base of forewing of a moth (Lepidoptera) for holding frenulum of hind-wing, thus coupling the wings during flight.

Retinal rod rhabdom: transparent rod secreted by retinulæ beneath each crystalline cone in the compound eyes of an insect.

Retinene light sensitive pigment in retinal cells of insects, derived from vitamin A.

Retinulae (pl. retinula) group of cylindrical elongate light-sensitive cells in ommatidium of compound eye.

Rhabdomere one of the seven or eight segment that form a rhabdom or retinal rod of insect's compound eye.

Rhabdura one of the two suborder of Diplura (Diplurata: Apterygota); includes all campodids, characterised by mandible with prostheca; abdominal styli soft and setose; cerci many jointed — Projapygoidea and Campodoidea.

Rhadalidæ small family of beetles, often included in family Melyridæ (Cleroidea: Polyphaga).

Rhagionidæ Bezzi 1903 (Leptidæ Westwood 1840) family of fairly large predaceous snipeflies (Tabanoidea: Brachycera); bristleless flies having small head, short antennae, large circular eyes, long legs and tibia mostly with spurs. Maggot carnivorous, live in rotting wood and feed on other insect — *Rhagio sclopacea* (Linnæus) is 12 mm long having spotted wings and black and yellow abdomen.

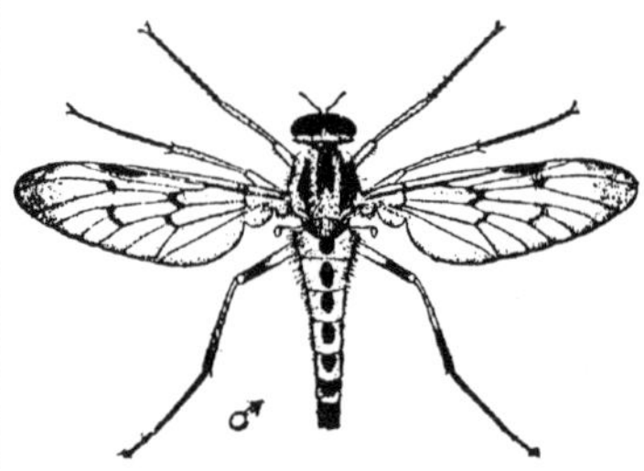

Rhagio scolopacea (linnæus)

Rhagophthalmidæ very small family of Cantharoidea (Polyphaga: Coleoptera); includes a few luminous species (glow-worms).

Rheo - current: flow (of water).

Rheotactic receptor(s) sensory cells, hairs or other structures that in case of aquatic insect, detect the changes in flow of water over the surface of insect's body.

Rheotaxis a response vide which an aquatic insect moves in a definite direction in relation to the flow of water.

Rheotropism orientation in relation to the direction of flow of water as in case of aquatic insect larvae.

Rhinomaceridæ old name for very small coleopteran family Nemonychidæ (Curculionoidea: Polyphaga).

Rhinophorinæ subfamily of Calliphoridæ (Calyptratea: Schizophora); often regarded as a independent family; include flies that are mostly parasitic on terrestrial isopods.

Rhinotermitidæ Light 1921 family of small subterranean termites

(Isoptera: Blattopteroidea); fontanelle and ocelli present; empodium absent; pronotum flat in worker; wings reticulate and without hairs; anterior wing scales long, hind-wing without anal lobe. About 150 species including a vast number of destructive pests — *Coptotermes ceylonicus* Holymgren (rubber and tea), *Nevtermes magniferæ* Roonwal and Sen-Sarma and *N.bosei* Synder (mango).

Rhinotoridae small family of small Neotropical and Australian dipteran flies (Anthomyzoidea: Schizophora) often regarded as a tribes of family Helomyzidæ (sunflies).

Rhipiceridæ (Sandalidæ) family of Rhipiceroidea (Polyphaga: Coleoptera), distinguished by membranous processes beneath first hour tarsal segment (100 species) — *Sandalus* spp. parasitise immature cicadas.

Rhipiceroidea superfamily of Polyphaga (Coleoptera: Coleopteroidea); antennæ strongly flabellate; labrum replaced by a nose-like projection; fore-coxa projecting with large exposed trochantin, hind coxæ with large femoral plate; empodium large and plurisetose. Only two families — Rhipiceridæ and Callirhipidæ.

Rhipidoptera Burmeister 1829 equivalent and obsolete term for insect order Strepsiptera Kirby 1813 (Coleopteroidea: Pterygota).

Rhipiphoridæ family of fan-beetles (Heteromera: Cucujoidea); having fan-shaped antennæ; pronotum posteriorly as vide as elytra at shoulder; elytra gaping apart with wings normally visible. Grubs parasitic or predaceous on wasps (unusual among beetle). About 400 species — *Metoecus paradoxus, Plecotoma* spp., *Rhipidius* spp., et al.

Rhipiptera Latreille 1817 old name obsolete term for insect order Strepsiptera Kirby 1813.

Rhizophagidæ small family of Clavicornia (Cucujoidea: Polyphaga); parallel-sided beetle, antennal club of one or two apical segments (X and XI); elytra truncate; fore-coxal cavities closed behind (200 species) — *Monotoma* spp., et al.

Rhopalidæ (Corizidæ Meyrs 1866) a small family of Coreidea (Pentatomomorpha: Heteroptera); often grouped with Coreidæ; body depressed, head over-lapping pronotum, antennæ inserted high above the upper side of head; rostrum very short; membrance with numerous indistint veins. Not be confused with Rhopaliidæ (Vespoidea: Hymenoptera) — *Corizus bohemaniii* Signoret, *C. hyalinus* Fabricious, *C. sidæ* (Fabricious), *Rhopalus* spp. et al.

Rhopalocera Dumeril 1806 Suborder of Lepidoptera in an old system of classification to include all butterflies and distinct from Heteroptera that covered all moths. Comprised of only two superfamilies — Hesperioidea and Papilionoidea.

Rhopalomeridæ family of Acalyptratæ (Schizophora: Cyclorrhapha); dipteran flies having eyes swollen, vertex developed between the two eyes; hind tibia compressed and curved; Cu_2 wanting or short, not reaching the wing margin.

Rhopalopsyllidæ family of fleas (Ceratophylloidea: Siphonaptera), characterised by a conspicuous tentorial arch in front of ocellus; no combs of long spines; both sexes with a single long antepygidial bristle; female with anal stylet.

Rhyacophilidæ Stephens 1836 family of primitive caddisflies (Rhyacophiloidea: Integripalpia); ocelli present; maxillary palpi with first two joint short and thick, the third joint much longer and thin. Larvae campodeiform having tracheal gills — *Rhyacophila fuscula* Walker, *Glossosoma americana* Banks, et al.

Rhyacophiloidea superfamily of intergripalpia (Tricoptera: Mecopteroidea); comprised of primitive caddis-flies resembling tineid-moths; fringed wings mostly folded over the dorsum, when at rest. — Hydroptilidae, Rhyacophilidae and Glossomatidae.

Rhynchitidæ small family of Curculionoidea (Polyphaga: Coleoptera); comprising of small weevils, 5 to 7 mm long that are often included in family of Curculionidae — rose snout beetle, *Rhychites bicolor* Fabricious and allied species.

Rhynachophora Bilberg 1820: suborder of Coleoptera (Coleopteroidea: Oligoneoptera) in an old system of classification. It included two superfamilies — Brentoidea Pierce 1916 and Curculionoidea with Brenthidæ (Brentidæ) as a families under it.

Rhynchophtherina (Phynchophthirini Ferris 1931) monofamilial suborder of Mallophaga (Psocopteroidea: Paraneoptera) having head prolonged anteriorly into rostrum with mandibles at the apex of rostrum; labium and maxillæ vestigial, thoracic segment fused — Hæmatomyzidæ.

Rhynchota Fabricious 1781 insect order proponed by Fabricius; that included bugs (Hemiptera Linnæus 1758), thrips (Thysanoptera Linnæus 1758) and fleas (Aphaniptera Kirby 1826 = Siphonaptera Latreille 1825); later the two were recognised as independent orders and the term Rhynchota used as synonym of Hemiptera (Homoptera and Heteroptera).

Rhyngota old and obsolete term suggested by Linnæus (1735) as an order for bugs and thrips; established by Fabricius (1781), as equivalent to Rhynchota.

Rhyphidæ Macquari 1838 dipteran family of Triassic period (Mesozoic era), true fossils of these flies found in Australia.

Rhysodidæ small family of wrinkled bark beetle (Rhysodoidea: Adephaga); brown, slender insect having head and prothorax with deep longitudinal grooves, hind-wing without oblongum; first abdominal sternite exposed between and hind coxæ and II, III fused; found in rotten wood — *Chinidium apterium, Rhysodes alerrimus, R. boysi, R. germari* Gangibaur et al.

Rhysodoidea Leng 1920 monofamilial superfamily of Adephaga, often included in Carabidea (Adephaga: Coleoptera); distinguished by stout and conspicuous moniliform antennæ and pronotum with three deep longitudinal grooves — Rhysodidæ.

Ribaga's organ Berlese's organ: sac-like organ for receiving spermatozoa and ventral surface of abdomen of female bedbug *Cimex* spp. (Cimicidæ: Cimicoidea: Geocorisæ).

Ricaniidæ tropical family of Fulgoroidea (Coleorrhyncha: Hommptera); moth-like is appearance with hind-trochanter ventrally and a short basal tarsal segment in hind-leg. Confined to Ethiopian and Indomalaysian region (350 species), lucerne (*Medicago sativa* Linnæus) and jasmine (*Jasminum* spp.) while *R. zebra* infest rice (*Oryza sativa* Linnæus) and grasses.

Rice caseworm *Nymphula depunctalis* (Guenee) (Pyraustidæ: Lepidoptera), small-sized moth having white wings with pale brown marking. Caterpillar first feed on leaf surface, then roll the leaves to from tubular case that are found floating on water or attached to plants with caterpillar inside.

Nymphula depunctalis (Guenee)

Rice fly *Hydrellia philippina* Ferino, *H. sasakii* (Ephydridæ: Dipter), small, black coloured, long-legged flies found in marshy or damp places. Maggots mine into leaves and inflict damages to the crops.

Rice gall midge *Orseolia* (*Pachydiplosis*) *oryzæ* (Wood-Mason) (Cecidomyiidæ: Diptera); minute fragile fly, male pinkish coloured and about 3 mm long; female reddish in colour and 4 mm long. Young maggot and burrow leaf sheaths near the base and reach the growing point, lacerate the tissues causing gall formation and producing long silver shoots.

Rice grasshoppter *Heirolyphus banian* (Fabricius) (Acrididæ: Orthoptera); hoppers yellowish-green in colour; both feed voraciously on the leaves, leaving behind only the midribs.

Heiroglyphus barnian (Fabricius)

Rice hispa *Dicladispa* (*Hispa*) *armigera* Olivier (Hispidæ: Coleoptera); small (4 to 5 mm long), shiny bluish-black beetle, fringed with short spines all over the body. Grub burrows into leaf tissues and feed on chlorophyll between the layers of epidermis, while the adults scrap away the green matter from leaf blades; affected leaves turn whitish, become membranous and die away.

Rice leaf roller *Cnaphalocrocis medinalis* (Guenee) (Pyraustidæ: Lepidoptera); small brownish-orange moths having light brown wings with two distinct dark wavy lines of forewing and one line of hind-wing, outer margin with a characteristic dark brown band. Pale yellowish-green caterpillar fastens together the edges of tender leaf and lives inside the rolled leaf, scrapping the green tissue.

Rice mealy bug *Heterococcus rehi* (Lindinger) (= *Ripersia oryzæ*) (Psedococcidæ: Homoptera); adults female apterous elongate oval to shape with distinct segmentation; antennæ present, legs well developed; body covered with mealy and waxy secretion; male winged.

Rice root weevil *Echinocnemus oryzæ* Marshall (Curculionidæ: Polyphaga); small gray weevil, grubs feed on fibrous roots of paddy crops, causing stunted growth and few tillers.

Rice skipper *Pelopidas mathias* (Fabricius) (Hesperiidæ: Hesperioidea: Ditrysia) medium sized brown butterfly with whitish specks on forewing. Caterpillar longitudinally folds the leaves making the edges stick together by means of silken threads and feed within, on the green matter.

Rice stem borer(s) *Tryporyza* (Walker) (Schoenobiinæ: Pyraustidæ: Lepidoptera); male light brown in colour with numerous small light brown spots of forewings; female yellow with distinct black spot on the centre of forewings; hind-wing of both

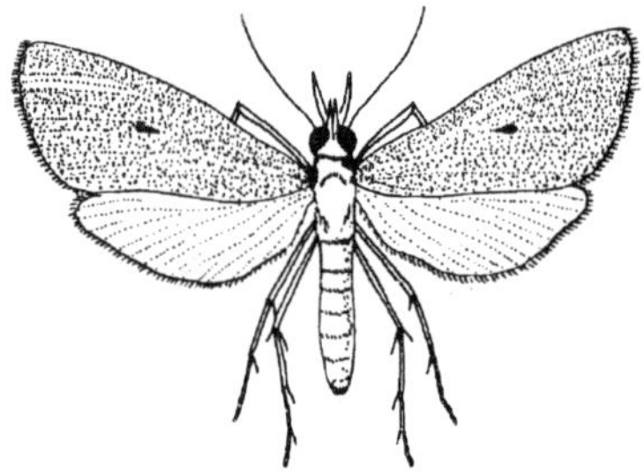

Tryporza incertulas (Walker)

sexes pale straw coloured; wing expanse of male and female 20-26 and 24-30 mm respectively. Caterpillar bore into the stem, killing the centre shoot and causing deadheart in early stage of the crop and white ears at later stages. Also known as yellow rice borer; another species *T. innotata* (Walker) called white rice borer, also cause similar damage.

Rice stem fly *Atherigona oryzæ* Marshall (Anthomylidæ: Muscoidea); small bluish - black fly whose maggots bore into the stems of seedlings.

Rice stripped borer, *Chilo suppressalis* (Walker) (Crambidæ: Lepidoptera); medium-sized moths (wing-span 26 mm), forewings light brown or straw coloured with black dots in the middle and hind-wing yellowish-white. Caterpillar dull yellow in colour with five distinct and two broken violet stripes; bore into the stalks causing deadhearts in the early stage and white-ears if the attack is at later stage.

Rice swarming caterpillar Armyworm: *Spodoptera (Prodenia) mauritia* (Biosduval) (Noctuidæ: Lepidoptera); serious pest of nurseries and newly transplanted crops. Stout medium-sized (wing-expanse 32 to 36 mm) dark brown moth, with a conspicuous spot on forewing, hind-wings whitish. Caterpillar smooth, cylindrical, dark green dorsally and greenish-yellow ventrally with three longitudinal stripe on either side.

Spoodoptera mauritia
(Boisduval)

Richardiidæ small family of Agromyzidæ (Schizophora: Cyclorrhypha); dipteran flies characterised by tibia having preapical bristles; costa interrupted well before R_1 and Sc complete; female with ovescapt (100 species).

Ricinidæ Enderiein 1927 (= Liotheidæ Burmeister) an important family of Amblycera (Mallophaga: Psocopteroidea); antennæ elongate 4-segmented; side of head straight or concave and not enlarged above the eyes; labial palpi wanting; first abdominal tergum fused in metanotum and spiracles on II and VII abdominal segment. Infest large variety of land and water birds - *Ricinus* spp. on song birds, *Trochilocoetes* spp. on humming birds.

Riodinidæ (Erycinidæ, Nemeobiidæ, Plebeiidæ) family of papilionoidea (Ditrysia: Lepidoptera); small butterflies having forelegs functional in females, reduced, brush-like and non-functional in males; costa of hind-wing thickened to meet the humeral angle, Sc with basal spur or humeral vein. About 1000 species grouped into two subfamilies — Libytheinæ (*Libythea celtis*) and Nemeobiinæ (*Dodonia* species, *Nemeobius lucina*) et al.

Roach(es) short name for cockroaches.

Rocking sprayer hand; pump and pressure chamber are clamped on a wooden board and operated manually with rocking movements of handle that helps in building up pressure in the pressure chamber. It can build-up pressure of 14 to 18 kg/cm^2 and given an output of 70 to 90 litres per hour.

Rodenticide chemical effective against rodents — cholecalciferol, warfarin, zinc phosphide etc.

ROONWAL, Dr. (Major) Mithan Lal - born September 18, 1908; M.Sc from Lucknow in 1930, served for a short period at Locust Research Laboratory, Lyallpur; joined Cambridge University, worked under A.D Imms and got Ph.D in 1935; the University also conferred upon him Sc.D. degree in 1962 in recognition of his published work on termites. Served (1936-39) as Entomologist-in-charge of Locust Field Research Station, Baluchistan and 1939 to 1949 at the Zoological Survey of India, Calcutta. During World war II, served as Major in 5th Punjab Regiment on Assam-Burma border. Joined as Forest Entomologist (1949-56) at Dehradun and as Director, ZSI, Calcutta in July 1956. Head of Zoology Department, Jodhpur University in 1965; Vice-chancellor 1966-69. On retirement, worked as Emeritus Scientist of C.S.I.R. for four years but continued his research at the Desert Regional Station of ZSI, Jodhpur, till his death, July 22, 1990.

Root-borer(s) insects that bore into roots of plants and feed therein. These may be grubs of flea-beetle and other coleopteran insects, maggot or some dipteran flies as well as certain caterpillar — apple root borer *Dorysthenes hugeli* Redtenbacher, rice root weevils *cyrtozemta* species, *Echinocnemus oryzae* Marshall, *Hydronomidius molitor* Faust; onion fly, *Delia* (*Hylemya*) *antiqua* (Meigen), sugarcane root-borer *Emmalocera depressella* Swinhoe et al.

Ropalidiinæ small subfamily of Vespidæ (Vespoidea: Apocrita); found in tropics of the Old World; construct naked combs and there is little or no distinction between queen and workers.

Ropalomeridæ family of Sciomyzoidea (Schizophora: Cyclorrhapha); dipteran flies having eyes swollen with vertex depressed between the two eyes; hind tibia compressed and curved. A few American species, mostly Neotropical, found in damp places.

Roproniidæ very small family of Proctotrupoidea (= Serphoidea) (Aporcrita: Hymenoptera) having antennæ 14-segmented; forewing with several enclosed cells, hind-wing without a closed cell; petiole much shorter, abdomen strongly compressed. Only three species reported from North America.

Rostrum beak: snout or proboscis of bugs.

Rote-organ special respiratory organ found in endoparasitic larvae of certain dipteran flies; comprising or symmetrical pairs of large red-coloured tracheal end cells and far bodies.

Rotenone $C_{23}H_{22}O_6$; insecticide and acaricide of plant origin, first used in 1848. Obtained from powdered routs of various papilionaceous plants (*Derris elliptica, D. malaccensis. Lonchocarpus utilis* and *L. uruca*) and used as stomach and contact poison. Not injurious to plants, harmless to mammal but highly toxic to fish. Acute oral LD_{50} value for rats 132. Used for control and eradication of fish population and control of ectoparasties of mammals. No more in usage — Protex, Rotacide.

Royal jelly brood-food: predigested substance that contain a high proportion of protein and is produced by the workers from a pair of large lateral pharyngeal and is fed to larval and adult queen and drones. Each gland is a long coiled of follicles packed away in antero-dorsal region of head; absent in drones and rudimentary in queen.

Ruby spot large damselfly, *Hetærina americana* Fabricius (Agriidæ: Agrioidea); copper-bronze and metallic green in colour with ruby-red spots covering the base of the pairs of wings in males; in female these spots are amber-yellow coloured; length about 45 mm and wing-span 60 mm; widely distributed in North America, along the river banks.

Rudimentary body part not fully developed: reduced and functionless part of body: vestigial.

Rugose wrinkled surface.

Run-off process of spray-shedding from a plant surface during and immediately after application, when the droplets coalesce to from a continuous film and surplus liquid drops down the sprayed surface.

Ruralidæ old and obsolete term for lepidopterous family Lycænidæ Leach 1815 (Papilionoidea: Ditrysia).

Rutelidæ Macleay 1819 family of medium-sized, brightly coloured, shining leaf-chafer (Scarabæoidea: Polyphaga); often considered as subfamily of Scarabæide. Differs from Melolonthidæ in having claws of hind tarsi unequal in length and movable; beside having labium chitinised, hind tibia with two apical spurs; pygidum exposed; abdominal spiracles in two rows each forming two lines (25000 species) — *Adoretus duvauceli* Blanchard, *A. versutus* Harold, *Anomala bengalensis* Blanchard, *A. varians* Olivier, *Mimela pectoralis* Blanchard *Popillia cyanea* Hope, et al.

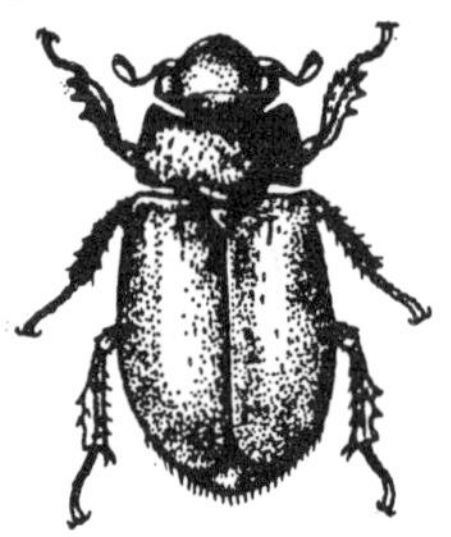

(*a*) *Adoretus duvauceli* Blanchard

(*b*) *Adoretus Versutus* Harold

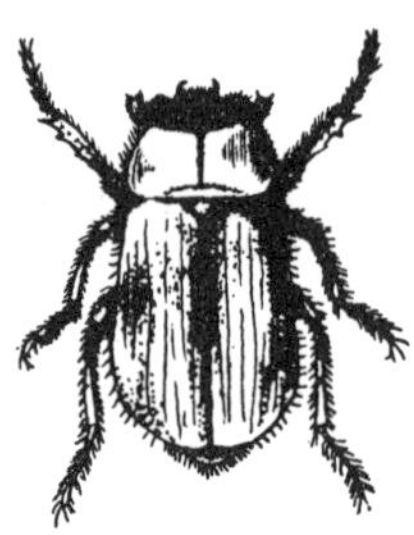

(*c*) *Anomala varians* Olivier

(*d*) *Mimela pectoralis* Blanchard

Ryania insecticide of plant origin; ryanodine extracted from root and stems of *Ryania speciosa* Vahl. (Flacourtiaceæ); a Tropical shrub of Trinidad; contact and stomach poison; less toxic to mammals, more stable and having longer residual action. Acute oral LD_{50} value for rats being 1200 mg/kg and dermal LD_{50} value for rabbit is 4000 mg/kg — Ryanicide, Ryanexcel.

S

Saccus a long narrow median process formed by invagination of vimculum in the genitalia of male lepidopterous insects.

Sagitta (pl. **sagittæ**) proximal lateral process of ædeagus in Apidæ (Hymenoptera).

Sagriinæ subfamily of Chrysomelidæ (Chrysomeloidea : Polyphaga); includes long characterised by having enormously elongated hind legs with greatly thickened femora found in Indonesia, Pacific Islands and Tropical America-*Sagria papuana* Jacoby and allied species.

Saldidæ (Amyot and Seville 1843) (Acanthiidæ Stephens 1829) family of Saloidea (Pentatomomorpha : Geocoisæ); small oval, delicate heteropteran bugs, 3 to 7 mm long, with large prominent eyes, tarsi 3-segmented; frequent marshy place (150 species)-*Sadula orthochila* (Fieber), *S. scotica* (Curtis), *Alda littoradis* (Linnæus) et al.

Saldoidea superfamily of Pentatmomorpha (Geocrisæ: Heteroptera); bugs with broad head and prominent eyes; antennæ and rostrum long; pronotum narrowed anteriorly; hemelytra having corium, clavus and membrance, the later with 4 to 5 cells but to longitudinal veins; divided into four families-Saldidæ, Leptopodidæ, Leotichiidæ and Omaniidæ.

Saliva colourless, tasteless watery secretion of neutral pH secreted by labial (salivary) glands, consists of enzymes (mostly amylases and invertases) that keep mouth (buccal) cavity moist, lubricates food during mastication and initiate digestion of food either outside the body or in foregut.

Salivarium small salivary pocket located between base of hypopharynx and labium wherein opens the duct of labial glands, often acts as salivary pump or syringe (buts) or as silk-press (caterpillars).

Salivary glands glands with mouth cavity that secrete saliva : in insects variable number located in head and thoracic regions which open by means of paired ducts (one to four) at the base of hypopharynx of adult and larva.

Salivate discharge (excess) saliva.

Saltatoria Latreile 1817 order of insects in some old system of classification, it includes jumping insects (*saltare* - to jump), now placed under order Orthoptera (Orthopteroidea : Polyneoptera).

San Jose scale armoured scale insect, *Quadraspidiotus perniciosus* (Comstock 1881) (Diaspididæ: Coccoidea); native to China but first reported as pest from San Jose (USA) in 1870 and form Kashmir (India) in 1923. Destructive to Temperate fruit trees all over the World. Female is ashy-gray in colour, oval in shape about 2 mm in diameter with dome-shaped or nipple-like covering (scale); ovoviviparous. Male winged and short-lived. Both nymphs and adult females suck sap from leaves, twigs and fruits, as a result, seedling way die away, young trees lose vitality and have poor growth, fruits show pink-coloured spots and have no market value. *Prospaltelal perniciosi* (Tower) has been reported as an effective parasite of this noxious pest.

Sandal tree issid *Sarina nigrochpeta* (Issidæ : Fulgoroidea); a common pest of sandal trees (*Santalum album* Linnæus) in South India; both adults and nymphs suck cell sap causing shedding of foliage, dying-back of twigs and ultimately death of the tree within tree months. Active all the year round with three overlapping generations.

Sapromyzidæ old name for Dipteran family Lauxaniidæ (Acalyptratæ: Schizophora).

Saprophagous organisms (insects) that feed on decaying or rotten organic matter-ceecidomyid files (Diptera) etc.

Sapygidæ (Laterile 1810) Leach 1819 small family of medium-sized hunting wasps (Scolioidea : Apocrita); 8 to 10 mm long; eyes deeply emarginate, clypeus with length and breadth subequal, legs short; both sexes winged with veins well developed, hind-wings having prominent anal lobe; abdomen not petiolate. Parasitic on bees and wasps; the adult frequent the nesting sites of wasps and bees and instead of making their own nests, stay as inquilines in those nests-*Sapya clavicornis* parasitises *Chelostoma florisomne.*

Sacrophagidæ (Macquart 1835) Brauer 1889 family of Muscoidea (Schizophora: Diptera); often regarded as subfamily of Calliphoridæ; characterised by arista being plumose upto middle and then base distally; uniformly coloured flies with a longitudinally striped thorax, and marbled abdomen-*Platyperas quinquistriata* Enderlein (Egyptian) is about 4 mm long while *Sarocophaga carnaria* (Linnæus) of Europe is 14 to 16 mm long.

Sarrothripinæ family of Noctuidæ (Noctuoidea: Ditrysia) in an old system of classification; small to medium-sized moths, palpi porrect and rostriform or upturned; legs with short and stout tarsal joints. Forewings with Costa and inner margin arched near the base, then of even width throughout. Caterpillars having four pairs of abdominal prolegs-*Bleniia donans* Walker, *Chlumetia transversa* Walker et al.

Saturniidæ (Harris 1841) Walker 1855 family of giant silkworms (Bombycoidea: Ditrysia); large crepuscular moths, antennæ bipectinated in both sexes; proboscis vestigial or wanting, legs short, hair and without spurs and a transparent eye-spot near center of each wing. Caterpillars smooth with spinferous tubercles and dorsal humps-*Attacus atlas* Linnæus, *A. royle* Moore, *Philosmia ricini* (Boisduval) et al.

Saturnioidea Dyar 1902 a superfamily of Frenatæ in an old system of classification of Lepidoptera that included large beautifully coloured soft-scaled and hairy moths with antennæ short and conspicuously bipectinate; proboscis rudimentary or wanting; tibial spurs short or absent and wings often having transparent eyespots. According to modern classification, it comes under Bombycoidea (Ditrysia: Lepidoptera).

Satyridæ (Boisduval 1832) Swainson 1840 (Agapetidæ) widely distributed family of shade-loving dusk butterflies (Papilionidea: Ditrysia); often regarded as subfamily of Nymphalidæ, it includes Meadow-browns, Heaths, Graylings and Marbled-whites; distinguished by strongly depressed palpi and veins at the base of forewings greatly swollen (except in *Melanitis* spp.). Caterpillars fusiform, green yellow or brown in colour with two born-like processes on head-rice butterfly, *Mycalesis gotama* Moore, rice horned-caterpillar, *Melanitis leda* (Linnæus) et al.

Sawfly not a true fly but a wasp (Tenthredinoidea: Symphyta), hymenopteran insects, female having well-developed, saw-like (toothed) ovipositor to cut plant tissues and insert its eggs therein. Larva phytophagous, caterpillar-like, having three pairs of thoracic and 6 to 8 pair of abdominal legs; pupate in elongate-oval silken cocoons or earthen cells-*Athalia colibori* Fabricius, *Sencclidia discoreæ* Rowher (Tenthredinidæ) et al.

Sawfly, Brich large sawflies *Cimbex femorata* and related species (Cimbicidæ: Tenthredinoidea); resembling bees in appearannce and colour. Larvæ feed solitarily on leaves of birch (*Betula* *alnoides* Buch-Ham), alder (*Alnus nitida* Enderlein), apple (*Malus pumila Miller*), cherry (*Prunus avium* Linnæus, *P. cerasus* Linnæus), elm (*Ulmus wallichiana* Planch.) etc., found in Europe and North America.

Sawfly, Cherry *Calivora cerasi* Linnæus is small shining balckfly (Tenthredinoidea: Hymenoptera) with four smoky transparent wings; reproduces parthenogenetically; native to Europe but found in almost all temperate regions; larvæ feed on leaves of several fruit-trees specially cherry (*Prunus cerasus* Linnæus), elm (*Ulmus wallichiana* Planch.) etc., found in Europe and North America.

Sawfly, Cherry *Calivora cerasi* Linnæus is small shining blackfly (Tenthredinoidea: Hymenoptera) with four smoky transparent wings; reproduces parthenogenetically; native to Europe but found in almost all Temperate regions; larvæ feed on leaves of several fruit-tress specially cherry (*Prunus cerasus* Linnæus), pear (*Pyrus communis* Linnæus) and plum (*Prunus domestica* Linnæus).

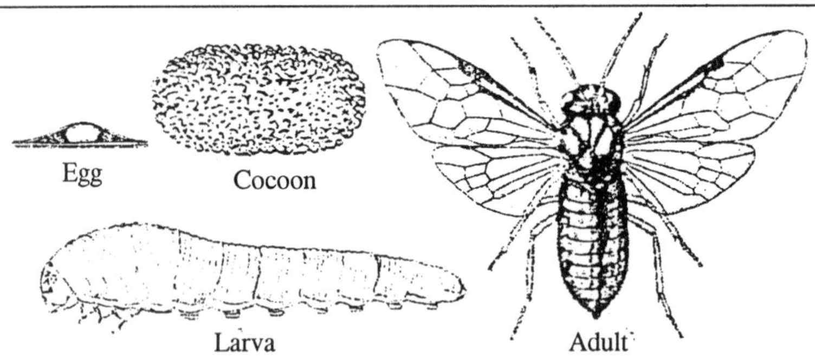

Calivora cerasi

Sawfly, Hawthorn largest sawflies *Trichiosoma lucorum, T. tibiale* and related species (Cimbicidæ : Tenthredinoidea) found feeding on hawthorn, willow (*Salix* spp., Salicaceæ) etc., in Europe and North America.

Athalia lugens proxima (Klug)

Sawfly, Leaforlling *Pamphilius persicus* MacGillivray (Pamphiliidæ: Tenthredinoidea).

Sawfly, Mustard *Athalia lugens proxima* (Klug) (Tenthredinidæ : Tenthredinoidea); shining black fly-like insect, 8 to 10 mm long with bright yellow thorax and femora and orange abdomen; cold-weather pest of various cruciferous crops specially mustard (*Brassica* spp., Brassicaceæ), radish (*Raphanus sativus* Linnæus) and turnip (*Brassica rapa* Linnæus).

Sawfly, Rose *Arge ochropus* (Argidæ Tenthredinoidea); well-known pest of roses distinguished by 3-segmented antennæ, apical segment of male being forked. Larvæ feed gregariously mostly on leaves of rose-plants (*Rose* spp.).

Sawfly, Webspinning *Neurotoma inconspicua* Norton (Pampiliidæ. Tenthredihoiea); major pest in orchards of North-east America.

Scale flat unicellular external process of body wall. Scales of insects are modified hair or setæ, lattened and of various shapes and overlap one another forming a protective covering for the wings and body.

Scale insect(s) common name of coccids (Coccoidea : Sternorrhyncha); minute inconspicuous and highly specilised insects. Female apterous and usually apodous; segmentation of body obscure; antennæ often atrophied; rostrum short. In most cases, there is waxy or powdery coating and the insects looks like a scale or gall. Male short-lived; have antennæ 10- to 25-segmented, mouthparts atrophied (do not feed); alate having only forewings, hind-wings reduced to halteres, rarely the males are apterous as in lac-insect, *Kerria lacca* (Kerremans) and sugaracane root mealy but *Succharicoccus sacchari* (Cockerell); abdomen usually terminates in stylet like process. Nymphs and adult females suck sap from various plant parts and many species are pests of economic crops-citrus fluted scale *lcerya purchasi* Maskell (Margarodidæ), rice mealy but *Heterocous rehi* (Lindinger) (Pseudococidæ), was scale *Ceroplastodes cajani* (Maskell) (Coccidæ), rose scale *Aulacspis rosæ* (Bouche) (Diapididæ) et al.

Scale, Chinese wax *Ericerus pe-la* Chavannes (Coccidæ Coccoidea); male secretes pure white wax which envelops its body as protection. In China and Japan these scales are commercially reared on evergreen *Rhus* spp. and *Hibiscus* spp., gathered in large number and the was used for making candles etc.

Scale Indian wax *Ceroplastes ceriferus* (Anderson) (Coccidæ: (Coccoidea) not only yields white was which is also of medicinal value in India, but is a destructive pest of Citrus (*Citrus* spp.), guava (*Psidium guajava* Linnæus), mango (*Mangifera indica* Linnæus), mulberry (*Morus* spp., Moraceæ) trees etc.

Scalpellum needle-like piercing stylet formed from one or more mouthparts-bugs, mosquitoes etc.

Scansorial legs clinging legs - Siphunculata (Anoplura).

Scape basal segment of insect's antenna.

Scaphidiidæ family of Staphylionidea (Polypahaga: Coleoptea); small, about 5 mm long, oval convex and very shining beetles, antennæ long, slender often filiform or slightly clavate; fore–coxæ hidden and hind–coxæ widely separated; elytra not entirely covering the abdomen. Mostly fungivorous (200 species - species of *Scaphidium, Scaphium, Scaphisoma* et al. *Scaphidium quadrimaculatum* is a glossy black coloured boat-beetle with two red marks on each elytron.

Scarabæidæ Latrielle 1802 large cosmopolitan family of nocturnal, lamellicorn beetles (Scarabæoidea: Polypahaga), commonly called dung beetles; antennæ variable but usually elbowed and capitate having clubs made of separate and movable plates; eyes large; prothorax large prothorax large and prominent; scutellum small or invisible; hind tibia with single terminal spur feed and breed in dung of mammals (18000 species)-*Heliocopris bucephalus, Polyphylla crinita* Le Conte, *Scarabæus sacer* Linnæus et al.

Scarabæoidea Bedel 1911 superfamily of Polyhaga (Coleoptera: Coleopteroidea); antennæ 10–segemented, 3 to 7 apical segments forming a lamellate–club; legs primarily fossorial, fore–tibia dentate with one apical spur; elytra covering most of abdomen; VIII abdominal tergite forming a exposed pygidium. Grub without tergal plates or cerci and with cribriform spiracles. Phytophagous or scavengers; nocternal and mostly having burrowing habit–Passalidæ, Trogidæ, Lucanidæ, Acanthoceridæ, Geotrupidæ and Scarabæidæ (Cetoniinæ, Dynastinæ, Glaphyrinæ, Melolonthinæ and Rutellinæ are also now regarded as separate families).

Scatophagidæ Robineau-Desvoidy 1830 a family of dung–flies (Calyptratæ; Schizophora), that cannot be easily separated form the family Muscidæ. Maggots of common saprophagous species live in dung and decaying vegetation while the adults are predaceous on other small flies.

Scatopsidæ a small dipteran family of minute, black scavenger–flies (Zygophthalma: Nematocera) distinguished by tibia without apical spurs and presence of scutellum and halteres.

Scavenger one that utilize or feeds on garbage, animal faeces or dead organisms which it has not killed–scarabid beetles (Coleoptera).

Scenopinidæ (= Omphralidæ) a family of Asiloidea (Brachycera: Diptera); small dark coloured narrow oblong flies, devoid of true pubescence or bristles; first two antennal segments short the third elongate and without any style or arista; empodium bristle-like, pulvilli small. About 200 species-*Senopinus* species et al.

Schizodactlidæ small, orthopteran family of large crickets (Gryllacridoidea: Ensifera); distinguished by tarsi with curious expansions that has very long wings with tips of tegmina coiled, to help in burrowing holes in sandy soil near rivers and streams where they hide during day. Carnivorous, preying at night. Only three genera, one alalte and two apterous - Indian species *Schizodactylus monstrosus* Durury, localy known as *bherw.*

Schizophora Becker 1882 suborder of Diptera; earlier considered as section of suborder Cyclorrhapha (Diptera; Mecopteroidea) that differs from suborder/section Aschiza in having distinct frontal suture and lunule; ptilinum always present; forewing with cell Cu short or vestigial (except in Conopidæ); further divided into 3 groups or series-Calyptratæ, Acalytratæ and Pupipara.

Schoenobiinæ subfamily of Pyraustidæ (Pyraliodoidea: Ditysia) often regarded as an independent family includes moths with proboscis usually absent or very small, palpi porrect, maxillary palpi well developed and dilated with scales at extremity. Hind wing with median nerve not pectinated on upper side and vein Sc + R_1 usually an astomosing with Rs - yellow rice borer *Tryporyza incertellus* (Walker), sugarcane top borer *T. nivella* (Fabricius), *Leucoides fusciostalis* Hampson et al.

Schradan phosphoramidate systemic insecticide; brown liquid, miscible in water, soluble in most organic solvents and only slightly soluble in petroleum hydrocarbons. Stable in alkali but gets hydrolysed rapidly in acids to form dimethylamine and phosphoric acid. Acute oral and dermal LD_{50} values for rat are 5 and 50 to 100 mg/kg respectively - Pestox-III®. Sytam®.

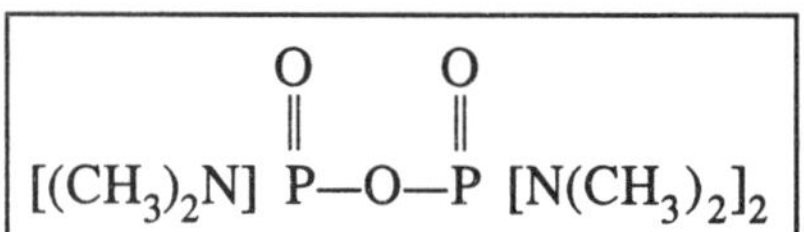

Schradan

Sciaridæ (Billberg 1820) (Lycoridæ) widely distributed family of dark-winged fungus-gnats (Nematocera: Diptera); living in damp forests of Asia, Northern Europe and North America. Maggots feed on fungi and often detroy mushrooms; migrate over the ground in large numbers.

Science tested knowledge: accumulation of facts derived from observations and experiments ceriticially tested, systematised and brought under general principles.

Scientific based on science : divided according to the rules of science.

Scientific name botanical or zoological name of an organism (plant or animal) based on binomial nomenclature.

Scientist man of science.

Sciomyzidæ (Tetanoceridæ) family of myodarian, cyclorrhaphous dipteran insect (Sciomyzoidea: Acalypterateæ Schizophora); small black flies, having big head and prominent antennæ, inhabit marshy places and edges of ponds and streams. Also called snail-flies as their maggots feed on snails (Mollusca).

Sclerogibbidæ monospecific family of Bethyloidea (Apocrita: Hymenoptera); antennæ 20- to 30-segmented, female apterous with short legs and fore-tarsus with row of pegs on posterior side; male alate; parasitic on embiids within whose colonies the females live.

Sclerite any sclorotised area of body wall.

Sclerotisation hardening of body wall in exocuticula.

Scolebythidæ small family of parasitic wasps (Bethyloidea: Apocrita); characterised by greatly reduced wing-venation, having few closed cells; abdomen pointed and ovipositor tubuler.

Scoliidæ (Latreile 1815) Westwood 1840 family of densely haired black wasps (Sclioidea: Apocrita); antennæ short, bowed or curled and not ending in a club; sternellum large and sharply defined; tibia usually flattened with bristles exteriorly. Wing-venation reduced, veins not extending to apical margin. Larvæ ectoparasites on Scarabæidæ and Curculionidæ-*Scolia porcer* Illiger (25 mm long and wing-span of 95 to 100 mm is probably the biggest species of this family), *Campsomeris annulata* Fabricius preys upon Japanese beetles *Popillia japonica* Newman; *S. manilæ* is used for controlling *Anomala orientalis* Waterhouse in Hawali.

Scolioidea superfamily of Apocrita (Hymenoptera: Hymenopteroidea); head more or less globular; antennæ 13-segmented in male (except Formicidæ), 12-segmented and often elbowed in female; pronotum extending back to tegula; legs short and stout. Many females apterous, alate forms with forewings having a pterostigma, hind-wings with atleast one enclosed cell and a distinct anal lobe-Scoliidæ, Tiphiidæ, Sapygidæ and Formicidæ.

Scolytidæ (Latreille 1907) Kirby 1836 (Ipidæ Ganglbaur 1903) large family of shot-hole borers (Curculionoidea: Polyphaga); minute to small, cylindrical beetles distinguished by rostrum rudimentary or wanting; antennæ clavate; prothorax large, wings well developed for flying elytra longer or shorter than abdomen. Grub apodous having curved, smooth or hairy body with well developed head. Bore into bark and wood-*Cryphalus javanus* (Eggers), *Xyleborus andrewesi* Blandford. *X. semigranousus* Blandford *X. dispar* are some destructive pests.

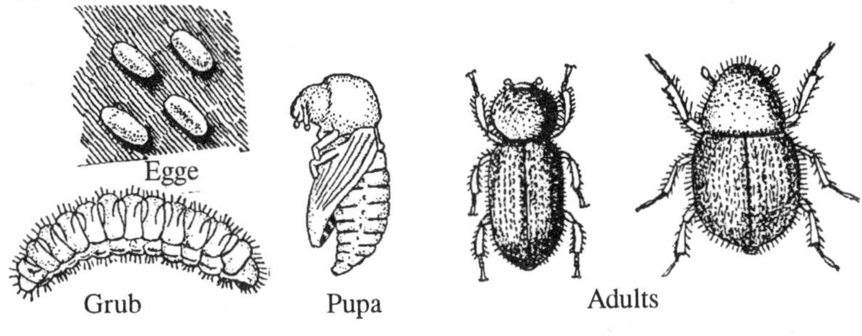

Xyleborus dispar

Scopa (pl. **scopæ**) brush-like structure formed of stiff bristles on inner side of flattened metatarsus of honeybees (Hymenoptera); used for removing pollen basket.

Scopulidæ old and equivalent term for family Acidaliidæ (Geometroidea: Ditrysia); that includes herbarium moth *Eois ptelearia* Riley, a destructive pest of preserved plant specimens.

Scopuridæ aberrant family of Filipalpain stoneflies (Plecoptera: Orthopteroidea); represented by a single wingless species, *Scopura longa* Ueno from Japan. Some Entomologists place this family under Euholognatha (Arctoperlaria: Plecoptera).

Scrobe(s) pair of grooves, one on each side of the snout of a weevil (Coleoptera) in which the antennæ can lie while the snout is busy boring.

Scutelleridæ (Leach 1815) widely distributed family of shield bugs (Pentatomidea: Pentatomomorpha); 8 to 10 mm long; scutellum very large extends upto apex of abdomen; membrane with veins; abdomen having 6 free sterna; mostly phytophagous-*Homæmus parvulus* (Germar) is 4 mm long; *Eurgastor alternatus* (Say) and *E. maurus* Linnæus are about 10 mm while *Calliphara imperialis* Fabricius *is* 25 mm long.

Scutellum dorsal thoracic sclerite, usually seen as triangular piece between bases of elytra.

Scutoscuteller suture external suture of V–shaped notal ridge of alinotum.

Scutum area of alinotum anterior to suture of V–shaped notal ridge.

Scydmænidæ (Leach 1815) family of small ant–like (Staphylinoidea: Polyphaga); maxillary papli have penultimate segment large and last one small; tarsi 5–segmented, elytra long and cover entire abdomen; occur mostly in moss or in ants nests (1300 species)– *Eumicrus* species, *Scydmænus tarsatus* et al.

Scythridæ (Scthrididæ, Butalidæ) a microlepidopteran family (Gelechiodea: Ditrysia); developed in South Europe and North Africa (not in India). moths having forewings with R_1 arising beyond middle of cell and about as long as R_2; Rs ends behind Costa; hind coxæ without whorl of bristles. Catepillar is found in web or silken tube among stems and flowers of various plants.

Scytinopteridæ Handlirsch fossil family of homopteran forms that lived during Permian period (Paleozoic era) in Australia (all extinct since ages).

Second auxillary pivotal sclerite of wing base that specifically manipulates radial vein.

Secondary not original or primary; auxillary; derivative.

Secondary parasite parasitoid which develops in or on primary parasite.

Secondary pest(s) that normally do not occur in economically important numbers, but which can exceed the economic–injury–level as a result of changes in cultural practices of crop varieties or the injudicious use of insecticides applied for key pests.

Secondary segment subdivision or segment of a segment: apparent segmentation produced by further subdivision of primary sclerites.

Seed borer(s) insects that bore into fresh developing seeds or dried and stored seeds for laying their eggs therein or feeding inside the same; accordingly boring is done either with ovipositor (chalcid wasps) or with snout (grain weevils).

Seed dressing coating of protective material (dry or wet) applied to seed just before plantings–addrin dust applied to sugarcane sets as safeguard agains termites; carbofuran and carbosulfan used on maize and sorghum seeds against borers and shoot–flies; application of organomercrial fungicide on cotton seeds as protection against angular leaf spot disease.

Segment subdivision of body or of appendage between areas of flexibility associated with muscle attachment-somite, podite metamere etc.

Segmental appendages paired ventrolateral segmental outgrowths of body wall serving primarily for locomotion.

Segmentation division of body or structure into similar parts.

Selandriinæ subfamily of Tenthredinidæ (tenthredinoidea: Symphyta); small, stout sawflies. Larvæ caterpillar-like, exclusively phytophagous; pests of ferns etc.-*Selandria* spp., *Strongylogaster* spp. et al.

Selidosematidæ old and obsolete term for lepidopteran family Geometridæ (Geometroidea: Ditrysia).

Sematuridæ small family of large Neotropical moths (Geometrφidea: Ditrysia resembling butterflies; characterieed by maxilary palpi vestivial or atrophied, no frenulum and without tympanal organs; hind-wings often with tails-*Sematua lunus* and allied species.

Semi-half.

Semicirular hemispherical : (in form of) half of circle or its circumference.

Semiochemical(s) signalling chemicals that convey messages between organism. Term coined by J.H. Law and F.E. Regnier for chemical substances secreted by insect into the environment; these chemicals profound influence and regulate insect behaviour and physiology. These are classified into two main groups-pheromones and allelochemicals.

Senescene to grow old : period between complete maturity and death of an organ or organism during which gradual deterioration occurs in its functioning and structure.

Sensation awareness of physical experience: power of sensing.

Sense faculty of receiving sensation.

Sense organ (sensillum) specialised innervated structure of body wall (set in most insects) receptive to external stimuli-feeling or touch (tactile), sight (vision), smell (olfactory) taste (gustatory) and hearing (auditary) etc.

Sensibility capacity of feeling.

Sensillum (p. **sensillæ**) sense-cell or group of sensory cells making up a receptor. These undergo several modifications depending upon the nature of the receptors-mechanoreceptors, chemoreceptors, phono-receptors, photoreceptors, temperature and humidity receptors etc.

Sensillum ampullacum sensory cells with external process sunken in a deep groove or pouch of body-wall.

Sensillum basiconium (pl. **basiconica**) and **styloconicum** (pl. **styloconica**) external process reduced to minute peg- or cone-like organs distinguished from trichoid sensilla by more thick-set appearance of projecting portion; found commonly on maxillary palpi of caterpillars where these respond to solutions of sugar, alcohol, amino acids etc., also widely distributed as antennal olfactory receptors.

Sensillum campaiformium (pl. **campaniformia**) external process not well developed that surrounds the distal process of sense cells or lacking, instead dome shaped thickening of cuticula is developed

Sensillum chæticum (pl. **chætica**) these are similar to trichoid sensillæ (sensory hairs), but shorter, stouter and spine-like.

Sensillum opticum (pl. **optica**) very distinct group of receptors, external cuticula forming transparent area that admits light to receptive cells of the eyes.

Sensillum placodeum (pl. **placodea**) thin, elongate, oval plate of cuticle that is flat or raised into a slight done-shaped structure; pores present all over the surface of the plate; found on antennæ of aphids (Aphidoidea) and bees (Apoidea).

Sensillum scolopoharum (pl. **scolopophara**) sensory cells with no external process but having two speciallised cells distal one ending in characteristics rod-like structure (scolopale).

Sensillum squamiformium (pl. **squamiformia**) sensory cells whose external part is scale-like process.

Sensillum trichoideum (trichoidea) thin-walled hair-like structures; most abundant and widespread chemoreceptors found on various parts of body-on labella and tarsi in Diptera (taste), antennæ of several species (olfactory), ovipositor of *Musca* spp. (gustatory) etc.

Sepsidæ family ant-like flies (Scoimyzoidea : Schizophora); small cylindrical, spiny-legged flies having spherical head; palpi reduce; scutellum usually convex and not elongate; tibia without preapical bristles (200 species).

Sericin water soluble gelatinous protein forming an outer layer surrounding an inner core of fibroin in the silk threads of insects, spiders etc.

Sericostomatidæ McLachlan 1876 large and widely distributed family of Limephiloidea (Intergripalapia: Trichoptera); caddisflies having maxillary palpi thickened and hairy; antennæ broadly developed at the base with large scape often longer than the head; ocelli wanting; dorsal part of tentorium reduced; hind-wings in female with R_2 not extending in R_1 nor forming a fork with R_3 Larva eruciform having head inclined at a distinct angle and with tracheal gills; found mostly in running water with portable case of sand or even small stone particles - *Gæra calcarata* Banks, *Philanisus plebejus* Walker, *Brachycentropus nigrosoma* (Banks).

Sericulture raising of silkworms (Lepidoptera) for production of raw silk.

Serosa Kieffer 1909 [Proctotrupidæ (Laterille 1802) Stephens 1829] small family of pointed-tailed wasps (Prototrupoidea = Serphoidea: Apocrita); small, 3 to 10 mm long, with 13-segmented antennæ inserted at the middle of face; forewings with stigma and one very small closed radial cell; abdomen with short petiole and segment II much longer than others; ovipositor straight and retractile. Cosmopolitan in distribution, it includes species of *Disogmus* Forster, *Exallonyx* Kieffer, *Serphus* Schrank etc.

Serphoidea Viereck 1916 [Proctotrupoidea (Laterille 1802) Stephens 1829] superfamily of Proctotrupoidea (Apocrita : Hymenoptera); includes minute to small, slender, dark or metallic coloured, primary and secondary parasitic wasps; antennæ elbowed or straight; venation greatly reduced; abdomen pointed; ovipositor tubular and issuing form extreme form extreme tip of the abdomen. Exibit hyperparasitism and polyembryony.

Serrate having saw-like notches.

Serration condition of being serrate.

Serritermitidæ Holmgren 1911 newly erected monogeneric family of termites (Isoptera: Blattopteroidea) that have unusual falcate mandibles in adult and worker; includes only one species *Serritermes serrifer* from Brazil.

Serrulate finely serrated.

Sesamum fly *Gitonides perspicax* Knab (Drosophilidæ : Schizophora); small yellowish fly with red eyes; more common in Gujarat; predaceous on sugarcane mealy but *Sccharicoceus sacchari* (Cockerell).

Seasmum gallfly *Asphondylia sesami* Felt (Cecidomyiidæ : Nematocera); a major pest of gingelly *(Sesamum indicum* Linnæus) in India. Larvæ feed on ovary of flower-buds causing malformation of pods which get shrivelled and do not set any seeds.

Sesamum lefhopper *Orosius albicinctus* Distant (Cicadellidæ Cicadomorpha); found all over India infesting gingelly (*Sesamum indicum* Linnæus) and sunnhemp (*Crotaloria juncea* Linnæus); also serves as vector of sesamum phyllody virus and is more predominant during warm wealther.

Sesiidæ Stephens 1829 (Æeriidæ Stephens 1829) family of clearwings (Yponomeutoidea: Ditrysia); brightly coloured moths, mostly with scaleless transparent wings and looking like wasps or bees. Forewings extremely narrow and hind-wings with Sc + R_1 concealed by fold of costa. Abdomen terminated by conspicuous fan-like tuff of scales. Caterpillars bore into veins, wood of trees and root-stock of plants-*Melittia eurytion* Westwood (pest of gourds in India), *Aegeria apiformis* (Clerck), *Bembicia hylæformis* (Laspeyres), *sphinx fusciformis* Linnæus et al.

Sessile incapable of moving; attached directly without a stem or petiode

Sessiliventres Haliday 1840 suborder of Hymenoptera; and old and obsolete term for Chalastogastra Konow 1897 or Symphyta Hehrstecker 1867, the last one being preferred is commonly used.

Seta (pl. **setæ**) slender, stiff, hair-like, unicellular external process of body-wall.

Setal membrane membranous floor of alveolus supporting seta.

Setipalpia suborder of Plecoptera (Orthopteroidea : Polyneoptera); primitive stoneflies found mostly in Eurasia and North America; having mandibles weakly developed, maxillary palpi subuliform; glossæ shorter than paragossæ; cross-veins reduced. Nymphs without segmentally arranged abdomen gills. Two superfamilies-Pteronarcoidea and Perloidea (= Subulipalpia).

Steomorphidæ family of Tinæoidea (Ditrysia: Lepidoptera); moths having mouthparts well developed, labial palpi scaled or short haired; forwings with hyaline aculeate forea in accessory cell distorting the venation; R_3, R_4 and R_5 seriately stalked.

Setulose bearing short blunt bristles.

Sex nature of being male or female; sum of characters that distinguish male and female individuals.

Sex-cell egg-cell or sperm.

Sex-chromosome chromosome that determines sex of an individual.

Sex-determination method or technique by which sex of species is determined.

Sex suppression suppressed sexuality as exhibited by certain individuals of various species of insects specially those leading social life–termites, ants, bees and wasps–in such insects, sexual organs through present are not sufficiently developed to permit reproduction and these individuals are known as neuters or workers.

Sexual pertaining to sex.

Sexual dimorphism diagnostic morphological difference between the two sexes or same species. Sexual variations are sometimes so marked that males and females of same species have frequently been relegated to different genera.

Shell outer covering (chorion) of an egg which affords protection to the inner contents. It is made up of two layers, a thick inner endochorion and a thin outer exochorion, composed of chemical substance scleroprotein or chorionin which is more or less same as chitin.

Shellac insects(s) lac–insect(s) (Lacciferidæ Homoptera) : insects that yield commercial lac or shellac.

Shiner nickname of small European cockroach or German roach, *Blattella germanica* (Linnæus) (Phyllodoromidæ : Blattaria); 10 to 12 mm long, yellowish–brown coloured with two dark brown longitudinal stripes on pronotum; common household pest in Europe and other Temperate regions of the World.

Shiny wings neuropteran insects (Stilbopterygidæ); large, high swift fliers; on wings resemble dragonflies except that they fly only straight, not darting and dodging like the latter, short knobbed antennæ, very large eyes; shining wings and abdomen found in Australia and South America; flying at dusk

Shoot–borer(s) common name for variety of insects that bore into stems of growing shoots of plants/trees; these may be beetles; weevils or caterpillars–*Diaphania cæsalis* (Walker), *Chilo infuscatellus* (Snellen), *Chlumetia alternans* Moore, et al.

Store earwig *Labidura riparia* (Pallas) (Labiduridæ Forficulina); large pale yellowish earwig, 30 to 35 mm long having black and brown wings; forceps of male thin and straight; lives at sea–shore on wet and among seaweeds; common in India and Hawaii.

Shot-hole borer(s) minute to small cylindrical, compact, bark and wood boring beetles (Curculioniodea: Polyphaga)- *Crypturgus pusillus* Gyllenhal and **C.** *atomus* Le Conte are about one mm long whereas *Dendroctonus micans* Kugelan and *D. valens* le Conte are 7 to 9 mm long. In India *Scolytoplyatypus raja* Blandford (2–3 mm long) is pest of apple trees (*Malus pumila* Miller).

Shouler enlarge basal portion of costal region of hind-wing over which hind margin of forewing rests. Part of body to which arm is attached.

Siagona Tropical ground beetle (Carabidæ Coleoptera), that stridulates by rubbing its anterior femora against a file on underside of the thorax.

Siallidæ Samouelle 1819 (Sialida Leach 1815) small family of carnivorous neuropteran insects (Siakoidea : Magaloptera); having long filiform antennæ, no ocellus or eye-sot, biting mouthparts; 4th tarsal segment dilated and conspicuously bilobed; wing span 40 to 80 mm. Larva with seven pairs of segmented lateral abdominal gill filaments without anal prolegs but with terminal anal filaments. Found in Temperate regions-*Sialis californicus* Banks, *S. infumata* Newman, *Syncordulia atrifroms* McLachlan et al.

Siebold's organ crista acoustica : the sensory organ comprising of a series of scolopophores attached to the walls of the tracheæ.

Sierohomophidæ very small family of Scolioidea (Apocrita: Hymenoptera) often regarded as subfamily of Tiphiidæ, but characterised by lack of an anal lobe in hind-wings. Only a few North American and one Hawaiian species.

Sigmoid S-shaped.

Signiphoridæ (Howard 1894) Vierick 1916 family of Scolioidea (Apocrita: Hymenoptera); minute 0.5 to 1.0 mm long, mostly yellow-coloured; primary, secondary and tertiary parasites usually on scale insects, whiteflies and parasitic Hymenoptera.

Silk glands the labial (salivary) glands in lepidopterous and trichopterous larvæ are transformed into a pair of very large cylindrical tubes occupying much of the body cavity of the caterpillars. These are meant for producing silk which is drawn through a single spinneret on the labium. The silk is produced in form of fluid which hardens when it comes in contact with air and is used for various purposes; such as forming a thick thread by which the caterpillar or pupa is suspended; stitching together of leaves to make a nest or constructing a sac or tent; silkworms utilise for formation of larval shelter or cocoon. In Embioptera and some dipteran flies these glands are on tarsi of forelegs and in certain Coleoptera the Malpighian tubules are modified to produce silk.

Silkworms caterpillars of silkmoths (Bombycidæ and Saturniidæ Bombycoidea). These insects secrete silken thread from their silk glands for constructing shelter (cocoon) around their body and pupate therein. The cocoons are commercially killed and spun into silk-thread-*Bombyx mori* Linnæus, *Attacus ricini* (Boisduval), *Antheræa assamensis* Helfer (= *assama* Westwood), *A. paphia* (Linnæus) et al. *Cf.* silkmoth.

Silphidæ Leach 1815 family of burrying or sexton beetles (Staphylinoidea: Polyphaga); having antennæ clavate, tarsi 4-segmented; anterior coxæ conical and contiguous, hind coxæ close together with plates wholly or partially covering femora; abdominal sternites distinct (not fused); feed and breed on carrion, hence also called carrion beetles-*Apathetica lebioides* Hope, *Catops vestita* Murray, *Diamesus oscculans* Vigelius et al.

Silvanidæ LeConte 1862 family of Clavicornia (Cucujoidea: Polyphaga); flat-grain beetles having elongated flat body, clubbed antennæ; third tarsal segment lobed beneath and fourth shorter than first (400 species)-saw toothed grain beetle, *Oryzæphilus surinamensis* (Linnæus) is a major pest of stored grains; *Silvanus unidentatus* Fabricuis is destructive to dried fruits and cereals.

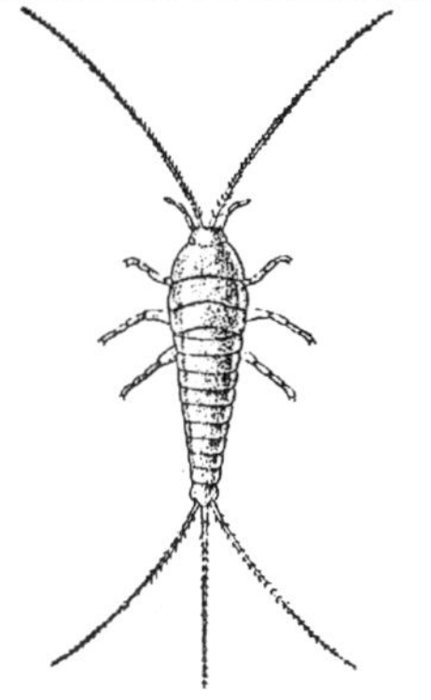

Lepsima saccharina Linnæus

Silverfish most primitive thysanopteran insect, *Lepisma saccharina* Linnæus (Lepismatidæ: Lepismatoidea); apterygote but very active runner, almost universal in distribution; having slender-body, 10 to 15 mm long, silver gray in colour, abdomen 10-or 11-segmented with a pair of cerci and a caudal filament at hind end. Household pest prefers dark and damp places and feeds of starchy material-starchy feed-stuff, papers, clothes etc.

Simple ye ocellus (pl. ocelli); simple photosensitive organ comprising of a single cuticular lens with a transparent corneagen laryer below it an a few retinulæ (sensitive cells) below this.

Simuliidæ Rondani 1856 family of minute, stoutly built, hump-backed, buffalognats (Culicoidea: Nematocera); having 11-segmented antennæ, elongated well developed

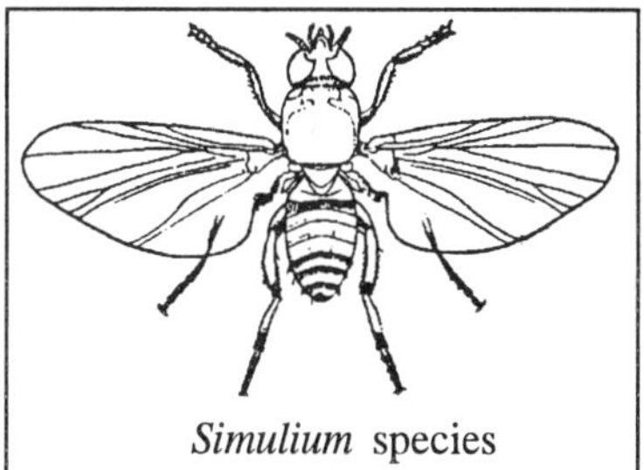

Simulium species

mandibles, no ocelli and legs short. Males holoptic, females voracious bitters and blood-suckers-*Simulium columbaczense* (Schiner), *S. bivittatum* Mallcoh, *S. meridionale* Riley, *Prosimulium hirtipes* (Fries) et al.

Sinentomidæ very small family of very primitive, tiny apteygote insects (Eosentomoidea: Protura); tracheal system and tow pair of spircles present; only the first pair or abdominal appendages two segmented (20 species)-*Sinentomon erythranum* found in China.

Sinodendridæ very small family of Scarabæoidea (Polyphaga: Coleoptera) in some old system of classification a few Holarctic species having completely cylindrical body short adn non-geniculate antennæ adn male with a cephalic horn; usually placed under family Lucanidæ.

Sinoxylon common and generic name of wood-boring bostrychid beetles (Bostrychoidea: Polyphaga); destructive pests of bamboo (*Bambusa* spp), felled trees and timber in India and Africa.

Siphlænigmastidæ monotypical newly errected family of Heplogenioidea (Ephemeroptera: Palæoptera) allied to Bætidæ and Siphlouridæ-*Siphlænigma janæ* discovered by Penniket (1962), from New Zealand.

Siphlonuridæ Klapalek 1909 large widely distributed family of Heptagenioidea (Ephemeroptera: Palæptera) mayflies having copious wing-venation, Cu_1 and Cu_2 of forewing close at the base but strongly divergent distally and Cu_2 of forewing close at the base but strongly divergent distally and Cu_1 with curved cross-veins joining it posteriorly to wing margin; only two caudal filaments-*Ameletus ornatus* Eaton and *A.percitus* Eaton are common in New Zealand while species of *Oniscigastor* MacLaclan are Indo-Australian.

Siphon breathing tube; usually with two spiracles and two tracheæ projecting from the hind end of an aquatic insect (larva) - dronefly, mosquito etc.

Siphonaptera Laterille 1825 (Aphaniptera Kirby 1826; Aphanipteroidea Jeannel 1945) order of Siphonpteroidea (Aphanipteroidea Jeannel 1945) (Oligoneoptera : Pterygota); small, one to 3 mm long, secondary wingless insects having hard smooth exoskeleton strongly compressed laterally and scletotides; antennæ short and stout; eyes wanting, two ocelli present; mouthparts piercing and sucking, maxillary and labial palpi present. Metamorphosis complete. Larva elongate, eruciform and apodous; pupa exarate and enclosed in cocoon. About 1500 species, free-living fleas, mostly ectoparasites of verterbrates; classified into six families-Hectopsyllidæ. Palicidæ, Dolichopsyllidæ, Ischnopsyllidæ, Hystrichopsyllidæ and Macropsyllidæ.

Siphonet honey-tube of aphids (Aphidoidea : Homoptera).

Siphunculata Laterille 1825 equivalent name for Paraneopteran order Anoplura (Pscocopteroidea Jeannel 1945); includes true (sucking) lice; about 400 species that are permanently ectoparasites of mammals.

Siricidæ Kirby 1837 family of wood-wasps or horn-tails (Siricoidea: Symphyta); giant sawflies (upto 50 mm in length) having antennæ filiform and many segmented; maxillary palpi one segmented; wings long and narrow with many veins and cells; abdomen long, cylindrical with anal spine, female also having a blunt ovipositor used boring a hole in the wood for laying an egg therein-*Urocerus gigas* (Linnæus), *U. imperialis* (Kirby), *Tremex magus* (Fabricius) et al.

Siricoidea superfamily of Symphyta (= Chalastogastra Konow 1897) (Hymenoptera: Hymonopteroidea); having antennæ inserted between the eyes and above base of clypeus; propodeum divided near the middle; fore tibia with one apical spur and ovipositor never coiled; embraces three families-Cephidæ, Siricidæ and Xiphydridæ.

Sisyridæ Handlirsch 1906 small family of Mantispodea (Planipennia: Neuroptera); small, brownish sponge-flies, found near water, first antennal segment quadrate, legs cursorial and forewings with costal veins simple. Larvæ aquatic with abdominal gills; live in association with fresh-water sponges, piercing the tissues of sponges with their mouthparts-*Sisyra fuscata* (Fabricius) Australia, *S. nikkoana* Navas (Japan), *S. vicaria* Walker (North America), *Climacia areolaris* (Hagen) (Nearctic) et al.

Size magnitude bigness or smallness. Insects are limited in size some are smaller than largest Protozoa while other are larger than smallest vertebrates. Giant dragonflies with wing span of 750 mm lived during Carboniferous era but are all extinct since ages. Present day insects are considerably

Attacus atlas (Linnæus)

smaller; females of Emperor moth *Coscinoscera hercules* Miskin of Australia and those of Atlas moths, *Attacus altas* (Linnæus) and *A. edwardsi* White have wing expanse of 260 to 280 mm while females of stick-insect *Pharnacia serratipes* (Phasmida) attain length of 320 mm. Smallest insects are fairy-flies (Mymaridæ Chalcidoidea), *Alaptus maganimus* Annandale is 0.2 mm long and *Melaptus torquatus* Malenotti is < 0.3 mm; the beetle *Nanosella fungi* (Ptiliidæ Staphylinoidea) is only 0.25 mm.

Skimmer common of larger and faster species of dragonflies *Libellula* spp. (Libellulidæ : Anisoptera) – back skimmer *L. luctuosa* Burmeister (wing-span 48 mm), red skimmer *L. saturata* Uhler (wing expanse 55 mm).

Skipjack click-beetles (Elateridæ : Elasteriodea), long slim, short-legged beetles having a peculiar habit–when at rest, they suddenly leap up into the air with a clicking sound. Larvæ known as wireworms.

Skipper(s) cosmopolitan butterflies belonging to family Hesperiidæ (Hesperioidea: Ditrysia) characterised by erratic darting (jerkily) flight and when at rest, wings flat or half open. Larvæ destructive pest of various economic crops–rice skipper *Pelopida* (*Parnra*) *mathias* (Fabricius), sugarcane skipper *Telicota augias* Linnæus, turmeric skipper *Udaspes folus* Cramer, et al.

Skipper, Chesse not a butterfly nor even a lepidopteran insect but maggot of dipteran fly, *Piophila casei* (Linnæus) (Piophilidæ : Acalytratæ); cosmopolitan; pale yellowish in colour, 8 to 10 mm long; saprophagous, fees on cheese, cured meats, carrion, fungi etc., and it jumps up when disturbed–hence the name cheese skipper. Adult fly is 3 to 4 mm long, blackish in colour with bluish reflections.

Skipper, Chequered *Carterocephalus palæmon* (Hesperiidæ), distinctive little skipper with all four wings chequered, inhabits grassy woodlands.

Skipper, Essex *Thymedicus lineola* (Hesperiidæ), having underside of antennæ with back tips and upper-wings golden-brown; usually found flying in pairs.

Skipper, Duskywing *Thorydes pylades* Scudder (Hesperiidæ) having wing-span of 35 to 40 mm, yellowish-brown in colour with white spots; found in America (South America and Canada).

Skipper, filed *Atalopedes campestris* (Boisduval) Hesperiidæ); found throughout North America, feeding on all types of grasses.

Skipper, Large *Oclalodes venata* larger in size and darker in colour than silver spotted skipper; underwings mottled greenish and without any silver–white spots.

Skipper, Regent *Euschemon raffesia* MacLeay (Hesperiidæ) of Australia: the males have both frenulum and retinaculum. Caterpillars gregarious, web the leaves of host (*Wilkiea* spp.) and feed within.

Skipper, Silver spotted small tawny brown butterfly, *Hesperia comma* (Hesperiidæ) having dark marking and pale yellowish spots on upper surface of wings, underside being greenish with number of large silver white spots. Caterpillars form shelter by weaving together grass blades with silken thread and feed within the same.

Skipper, Small *Thymelicus sylvestris* (Hesperiidæ) differs from Essex skipper in having orange tips to the underside of antennæ and orange–brown upper–wings with a more pronounced oblique black streak of scent scales on forewings of males.

Skipper, Turmeric *Udaspes folus* Cramer (Hesperiidæ), brownish–black butterfly, with eight white spots of varying sizes on each forewing and large with patch in the middle of hind–wing, wing expanse 35 to 45 mm. Major pest of turmeric (*Curcuma domestica* Val.) and arrowroot and minor pest of ginger (*Zingiber officinale* Rose) and cardamom (*Elettaria cardumomum* Maton). Green caterpillars feed in leaf folds and pupate therein a thick mass of white waxy substance.

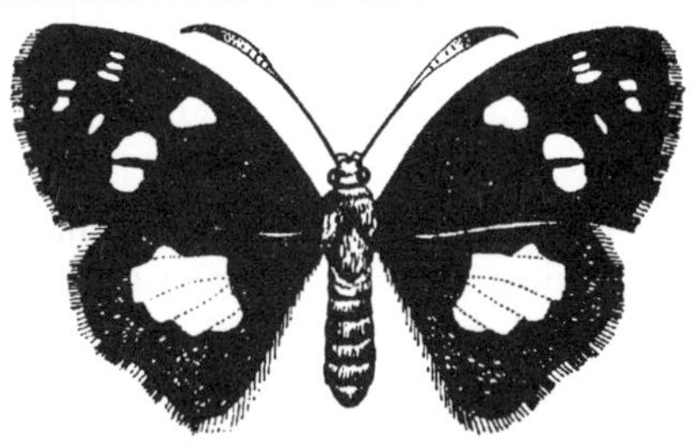

Udaspes folus Cramer

Slury semi–fluid (paste–like) mixture used for seed–coating.

Semerinthidæ old and obsolete name for lepidopteran family Sphingidæ (Sphingoidea : Ditrysia).

Sminthuridæ (Smythuridæ) Lubbock 1870 family of small globular spirngtails (Smyphyteona : Collembola); head vertical, antannæ as long as head; eyes present;

Sminthurus viridis *Sminthurides aquaticus*

thorax and so long as abdomen; abdomen bothriotrichia (250 species) – *Sminthurides aquaticus, Sminthurus viridis* Lubbock, S. *fuscuc et al.*

Smock aerial dispersal of minute solid particles of pesticide through the use of combustible mixture.

Smythuroidea Crampton 1920 superfamily of springtails, now regarded as synonym of suborder Symphyleona (Collembola).

Social pertaining to society or companionship.

Social insects highly organised insects that not only live in large colonies but also work together to build common nest; show marked division of larbour by having different castes; feed and protect their companions, have some means of communication with one another (sound, smell taste or touch) and above all exhibit distinct material care–termites (Isoptera), ants bees and wasps (Hymenoptera). Aphids and certain species of coccoids (Homoptera though live together in large colonies, are not social insects.

Socii lateral appendicular process of X abdominal segment in adult lepidoteran and trichopteran insects.

Sodium arsenate $NaAsO_2$ and Na_2HAsO_3; arsenical (inorganic) insecticide that is stomach poison and highly phytotoxic. It was used in baits (2%) against hoppers and locusts of as stock–dips against ticks etc. Oral LD_{50} value of rate is 10 to 50 mg/kg.

Sodium fluosilicate Na_3SiF_6; flourine (inorganic) compound used as insecticide since 1890 for about six decades; white crystalline powder; cheaper, safer and more efficacious than arsenicals; highly phytotoxic; used in baits and moth–proofing the fabrics. Acute oral LD_{50} value for rabbit being 150 to 200 mg/kg.

Soil injector plant protection equipment; a huge syringe with big needle (150 to 200 mm long) at one end, having several openings and pointed tip to pierce into the soil and a hand operated pump above, with which the liquid is pushed into the soil. Generally used for soil fumigation with a volatile liquid, such as carbondisulaphide, for control of nematodes and subterranean or soil–borne insects and diseases.

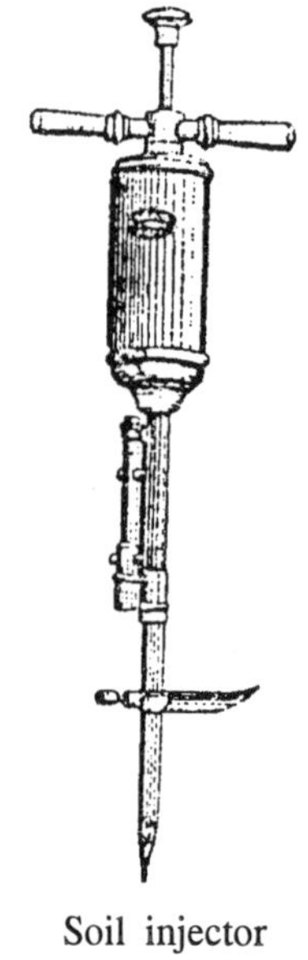

Soil injector

Soil sterilant the toxicant added to, or injected into soil of killing pathogens and pests inhabiting the soil–ethylene dibromide etc.

Solid cone jet with air–core reduced to give a cone of spray droplets.

Solute substance that is dissolved in solvent.

Solvent carrier solution in which a pesticide (Technical grade) is dissolved to form a concentrate; used for making emulsified concentrates (EC), oil–concentrates (OC) and also for coating granules-deodorised kerosine (Aromax Lomax), diesel oil, cyclohexane, methyl–isobutyl ketone, xylene etc.

Solvidæ (=Xylomyiidæ) small dipteran family of Tabanoidea (Brachycera), often regarded as subfamily of Stratiomyidæ (Tabanoidea : Brachycera); includes most aberrant species especially as regards their venation and presence of tibial spurs.

Soma body of animal (insect) as distinguished from germ cell.

Somatic cells body cells as distinct from germ cells.

Somite primary body segment.

Sorghum earhead bug *Calocoris angustatus* Lethierry (Miridæ: Heteroptea); full grown bug yellowish-green, measures about 6 mm in length; both nymphs and adults feed on sap of developing maize and millets/grains; sucked out grain shrink and turn black.

Sorghum shoot fly *Atherigona soccata* (Rondani) (Anthomyiidæ: Diptera); a small dark gray fly whose maggots bore into the shoots of young plants of maize and millets, kill central shoot, continue feeding on the decaying core of the shoot, ultimately causing dead heart.

Sorghum stem borer *Chilo* (*zonellus*) *partellus* Swinhoe (Crambidæ Pyraloidea); medium–sized (wing expanse 20–25 mm) moths, having forewing straw coloured with row of black dots along the outer margin an hind–wings smoky white. Caterpillar dirty white with four dull pinkish-brown longitudinal stripes, bore its way into the stem, kills the growing point, causing dead heart.

Chilo partellus (Swinhoe)

Spalangiidæ small family of Chalcidoidea (Apocrita: Hymenoptea) often placed under Pteromalidæ (Chalcidoidea: Apocirta); antennæ usually 12–segmented, abdomen distinctly petiolate; occipital line complete. Parasitic, usually on dipteran flies–*Spalangia muscidarum* is pupal parasite of houseflies (*Musca* spp.).

Spanioderidæ family of extinct order Protorthoptera comprising of Palæozoic forms having very long antennæ, strong mandibles and heteronomous wings that could be flexed backward; lived during Carboniferous period-*Gyrophlebia longicollis* and *Spaniodera ambulans* reported form Illinois (USA).

Sparaganothidæ old and obsolete term for lepidopteran family Tortricidæ (Trtrioidea: Ditrysia).

Spatha (pl. **spathæ**) dorsal lobe of ædeagus in male insects of Apidæ (Hymenoptera).

Species (sp., pl. **species,** spp.) taxonomic rank below genus : aggregation of organisms alike in structural and functional characteristics that normally breed only with each other; producing young ones which mate freely and produce fertile offsprings; reproductively isolated and genetically distinct individuals.

Speed velocity; quickness of flight. Speed of insects has been often exaggerated very highly [famous botfly (Oestridæ : Musocidea) used to be quoted @ 1300 km/hour]. In fact dragonflies (Anisoptera: Odonata) and horseflies (Tabanidæ: Tabanoidea) are considered to be the swiftest flying insects @ 50 km/hour; bees fly @ about 15 km/hour and most insects move slower than this Frequency of wing-beat is slower in bigger insects (moths and butterflies @ 50 to 70 per second) and faster in smaller insects (bee 200, mosquitoes 600 and midges 1000 or more per second).

Sperm male gamete: germ-cell.

Sperm cyst one of the cellular capsules in testes containing spermatocytes.

Sperm tube(s) testicular follicles that contain male germ-cells in successive stages of development.

Spermatheca seminal receptacle in female insects for receiving and storing sepromatozoa: receptaculum seminis.

Spermatid immature spermatozoon: cell that develops directly into spermatozoon.

Spermatocide (spermicide) toxin that can kill sperms.

Spermatocyte sperm mother-cell: male gametocyte that undergoes meiosis to form spermatid, which develops into spermatozoon.

Spermatogenesis process of formation of and development of sperms.

Spermatogeny sperm-formation.

Spermatogonium first stage in differentiation of sperm cell from primitive male germ-cell.

Spermatolysis destruction with dissolution of sperms.

Spermatophore gelatinous packet or cluster of sperms transferred by male to female of certain insects-cockroaches etc.

Spermatozoon (pl. **spermatozoa**) mature and functional sperm or sex-cell.

Sphæriidæ (Sphæritidæ) very small family of rare, minute, hemispherical beetles (Sphærioitidæ : Myxophaga); earlier Sphæriidea (Polyphaga); antennæ -segmented club; fore-tibia with external spines but not dentate; fore-codæ transverse and trochantin exposed. Wings fringed with long hairs. Occurs in damp places-*Sphærites glabratus, Sphærius acaroides* et al.

Sphærioidea only superfamily of Myxophaga (Coleoptera: Coleopteroidea); maxilla without a distinct galea; sulcus usually distinct in prothorax; wings with oblongum distinct, often rolled apically, when at rest. Four families-Sphæriidæ (Sphæriidæ), Hydroscaphidæ, Torridinoclidæ and Lepiceridæ.

Sphecidæ (Leach 1815) Comstock and Comstock 1895 large cosmopolitan family of thread-waisted, mud-daubers and digger wasps (Sphecoidea: Apocrita); characterised by long narrow waist (petiole) between thorax and abdomen; head large and transverse, antennæ near middle of face, clypeus long and slender; forelegs fossorial middle tarsi with two apical spurs. Wings slender with three subical cells; propodium long with three subical cells; propodium long with spiracles distant from the base-*Anmophila humbertiana, Sceliphron bilineatum* and *Sphex lobatus* etc. are some Indian species, *S. destillatorium* Illiger is European while blue mud wasp *Chalybion cæruleum* (Linnæus) and steel-blue mud-dauber *Sphex luctuosa* Smith are North American wasps.

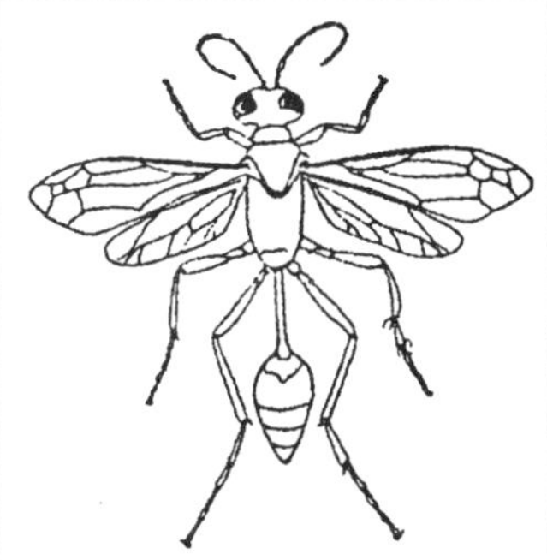

Scliphron bilineatum

Spheciformia large group of aculeate hymenopteran insects that includes bees (Apodea) and digging wasps (Spheoidea having pronotum not extending beyond tegulæ, near the wing bases.

Spheciidæ monogeneric family of Sphecoidea (Apocrita: Hymenoptera) that includes one of the largest North Amercian wasp, giant cicada killer *Sphecius specious* (Drury).

Sphecoidea Ashmead 1899 superfamily of Apocrita (Hymenoptea : Hymenopteroidea); solitary or colonical predaceous wasps distinguished by pronotum terminating laterally in a round lobe that does not reach the tagula and abdomen with more than three dorsal segments visible. Mostly fossorial and occasionally construct free mud cells; the nests are provisioned with various orthopteran insects, bugs, caterpillars and spiders (Arachnida) that are stung and paralysed, to be stored and fed to their larvæ–Sphecidæ, with about 20 subfamilies that are often regarded as independent families.

Sphegidæ old and obsolete name for hymenopteran family Sphecidæ (Sphecoidea: Apocrita).

Sphindidæ a small family of Clavicornia (Cucujoidea: Polyphaga); characterised by tarsi 5-5-4 in males and 5-5-5 in females, first segment being the smallest; epodium small. About 30 species mostly feed on Mycetozoa.

Sphingidæ Leach 1819 large and widespread family of diurnal or crepuscular hawk moths (Sphingoidea: Ditrysia); powerful flies that feed on the wings; antennæ gradually thickened but again pointed at the tip and nearly always hooked; proboscis and frenulum well developed, trympanal organs absent; forewings elongated and narrow, hind-wings small; legs strong with well developed spurs and spined tarsi. Caterpillars smooth with a horn on XI somite (1000 species) - *Acherontia styx* (Westwood), *Deilephila nerii* Linnæus, *Hippotion celerio* (Linnæus) et al.

Sphinginæ subfamily of Sphingidæ (Sphingoidea: Ditrysia); distinguished by very long proboscis, abdomen conical of both sexes and male without lateral expansions to abdomen. Caterpillar with anterior segments slightly smaller than posterior, head bent downwards in repose; horn when present, long and curved. Pupa generally with external sheath for proboscis-*Agrius convolvuli* (Linnæus), *Pseudosphinx discistriga* Walker, *Nepaphele hespera* Fabricius et al.

Sphingoidea Dyar 1902 monofamilal superfamily of Ditrysia (Lepidoptera: Mecopteroidea); antennæ apically pointed and hooked; frenulum strongly developed; Cu_1 absent from both wings; hind-wing with Sc + R_1 connected with cell by a cross-vein (r_1) and approximated to Rs beyond the cell–Sphingidæ.

Sphinx common name of somc hawk-moths (Sphingidæ : Sphingoidea)–Morning glory sphinx *Agrius convolvuli* (Linnæus), Tomato sphinx

Protoparce quinquemaculata (Haworth), White-lined sphinx *Deilephila lineata* (Fabricius) et al.

Spina median apodemal process of spinasternum.

Spinastermum one of the spina-bearing intersegmental sclerites of thoracic venter united with the sternum preceding.

Spine movable multicellular external process of body wall.

Spinipedes Walker 1872 old and obsolete term for heteropteran family Phymatidæ (Laporte 1833).

Spinner nickame of brightly coloured, shining adult mayfly (Ephermeroptera).

Spinneret(s) small tubular appendages from which silk threads are exuded by spiders and larvæ of several species of insects. In spiders these open at the hind-end of abdomen, but in insects it is quite different. In caterpillars, it is a large cylindrical tube coming from a pair of modified salivary glands and opening beneath the labium.

Spinose beset with spines.

Spiracle(s) paired, segmentally arranged lateral tiny openings on thorax and abdomen of insects, for intake of oxygen (from air) needed for respiration. Spiracles lead into a complex system of trachleæ and tracheoles through which the oxygen reaches all the tissues.

Spiracular gill(s) filamentous outgrowths of ectoderm without any connection to the body cavity, as found in some aquatic insects-larvæ of several dipteran flies and coleopteran beetles. The cuticular wall in such case becomes excavated to form air spaces (temporary air store), which become connected to tracheal system.

Spiracular sound organ a series of leaf-like folds inside the tracheal system of certain insects that vibrate and produce a height pitched sound; though not common, it is present in some beetles (Coleoptera) and flies (Dipetra).

Spotted bollworm(s) bud and shoot bores: *Earias insulana (Boisduval)* and *E. vittella* (Fabricius (Noctuidæ: Lepidoptera); medium-sized moths, *E. insulana* with wing expanse of 20–25 mm has forewings uniformly green and hind-wings creamy-white top silvery while *E. vittella* has wings expanse of 30–35 mm, forewings with broad white stripes along costal and anal margins, leaving only a wedge-shaped green patch in the middle. Caterpillars attack young apical shoots, flower buds and bolls; attacked shoots wither and idea away and the growing point is killed; flower buds and young bolls drop down whereas in case of mature bolls, there is bad opening with poor quality of raw-cotton.

Spray sprinkle with shower of fine droplets may be perfume on body or clothes; paint on any surface; pesticides/fertilisers on crops etc.

Spray boom(s) spray bar carrying more than two nozzles; normally coupled to power sparyer(s) mounted on tractors or attached to aircraftrs. Somes short booms (3 to 4 nozzles) may also be used with manually operated sprayers; it may be a tail boom or rig.

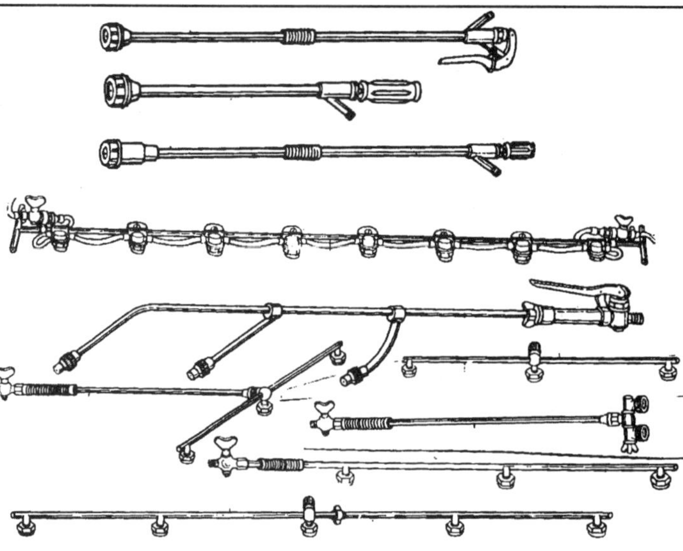

Spray booms

Spray gun(s) impeller gun; a compact integral unit consisting of cut-off device, extension rod and the nozzle; generally used for heavy duty, attached to power sprayer(s).

Spray lance(s) extention rod(s) usually of brass and about a metre long; fitted with stop-cock at one end which is attached to the hose pipe of the sprayer and nozzle (or set of nozzles) at the other end for discharging the spraying materal. There are various types to suit different requirements including bamboo-lance that is about two metres long, used for spraying trees.

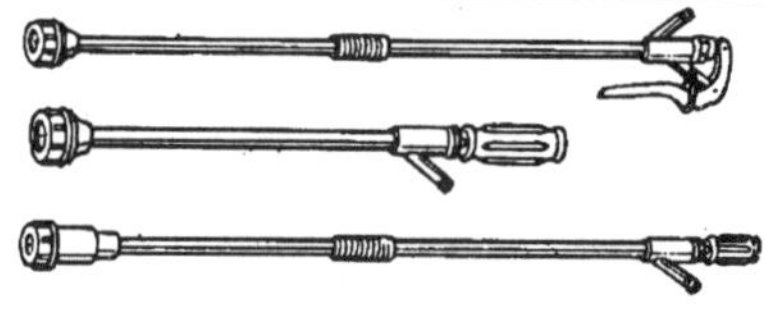

Spray lances (with stop-cock and nozzle)

Spray rig(s) a sort of spray–boom, generally comprising of three nozzles attached to a lance wit cut-off device; the nozzles are so designed or arranged that both upper and lower surface or the leaves are sprayed simultaneously, resulting in through spraying of the crop.

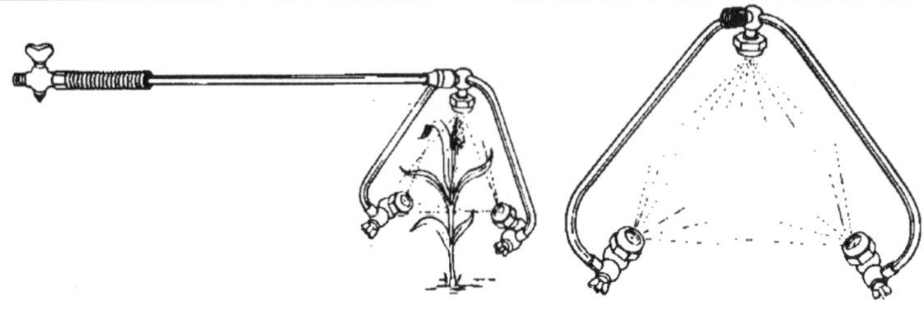

Spray rigs

Sprayer(s) appliances or equipment (manual or power–driven) for applying spray–solution. Its main function is to atomise or break the solution (pesticide) into fine droplets to ensure even distribution of these on (crop) surface.

Sprayer, Tractamount power sprayer mounted on a tractor or carried in a tractor–trolley. It is designed for economic and effective application of pesticides/fertilisers; saves time and labour (one hectare of low vegetation can easily be sprayed in one hour); suitable only for large spraying operations in orchards and on farm crops–spray booms, gums or lances will have to be properly selected to suit individual spraying requirements.

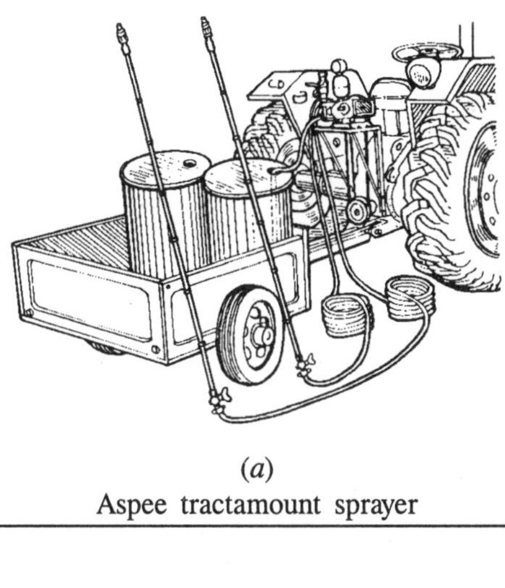

(*a*)
Aspee tractamount sprayer

(*b*)
Aspee tractamount sprayer

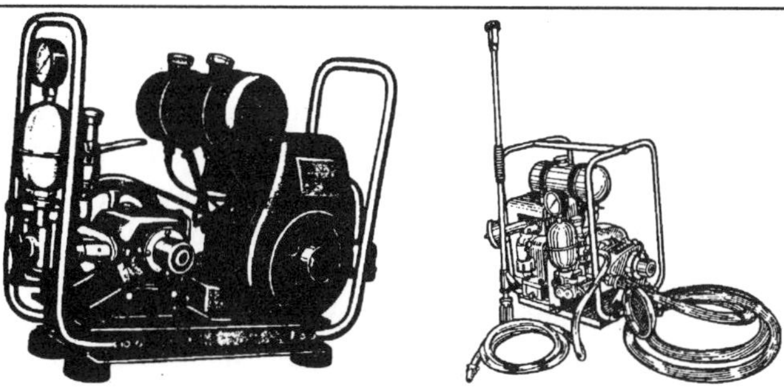

Courtesy : Amercan Spring & Pressing Works Pvt. Ltd., Mumbai

Springtail(s) very small, inconspicuous and primitively apterygote insects (Collembola: Entotropha); less than 6 mm long, biting moutparts; short antennæ and legs without tarsi; abdomen 6-segmented with a well developed collophore under first segment and a muscular springing organ (furcula) attached beneath near the hind end; no styli or cerci. Prefers damp soil, feeds on organic matter and fungi (including mushrooms); greenhouse pest-garden springtail *Bourletiella hortensis* (Fitch), yellow springtail *B. lute* (Lubbock) et al.

Spur(s) movable thick cuticular hairs or spines on legs of certain insects.

Spurious vein vena supria: false or spurious vein lying between veins R and M, being free at both ends; it is a distinctive characteristic of dipteran family Syrphoidæ: Aschiza).

SQH status que hormone or juvennile hormone secreted by corpora allata of certain insect larvæ; it inhibits premature metamorphosis.

Squama (pl. **squamæ**) one or two additional small dorsal lobes on wings of dipteran flies, close to alula; when two squamæ are present, the one nearest to alula is called alar squama or antisquama and the other thoracic squama.

SRIVASTAVA, Dr. Brij Kumar born July 15, 1927; M.Sc. (Agri.) Agra University 1948; Research fellow (1948–50); Cane Protection Inspector (1950–53). Ph.D. (Entomology) Kansas State University, Manhattan (USA) 1955. Lecturer Government Agricultural College Kanpur, 1955; Seed-testing Entomologist IARI New Delhi, 1956; Professor of Entomology, Rajasthan college of Agriculture, Udaipur, 1958; Entomologist Government of Rajasthan 1960; Director Agricultural Experimental Station, Udaipur, 1964, died with his boots on, September 14, 1975.

Stability ability of pesticide formulation to resist chemical degration over a period of time.

Stadium (pl. **stadia**) interval or time gap between two ecdyses. Enclosed athletic or sport ground with tiers of seat for spectators.

Stainer(s) brightly coloured pyrrhocorid bugs (Pyrrhocoroidea: Pentatomomorpha); having red and black markings; elongate-oval in shape; differ from lygæid bugs by want of ocelli and having more branched veins and cells in hemelytra; legs posses rotary coxæ, 3-segmented tarsi and pulvilli-cotton stainers *Dysdercus kœnigii* (Fabricius) and *D. cingulatus* Fabricius pierce the cotton bolls, making the same vulnerable to fungal attack resulting in stained lint.

Staphylinid type of wing-venation found in hind (metathoracic) wing of coleopteran insect characterised by disappearance of all cross-veins and atrophy of proximal portion of vein M, remainder portion being isolated in the apical part of the wing.

Staphylinidæ (Leach 1815) 1817 family of rove beetles (Staphylinoidea: Polyphaga); elongated, parallel-sides, flattened or cylindrical beetles; labial palpi usually 3-segmented; elytra short; tarsi having two equal claws and abdomen dorsoventrally flexible with seven or eight visible somites-*Myremedonia* spp. prey upon dead or diaabled ants, *Dinarda* spp. are synoekete in the nests of *Formica* spp. while species of *Atemeles* and *Lomechusa* are symphiles that are assiduously tended by ants. In India *Termitodiscus heimi* are found in termitaria; *Aleochara bilineata* and *A. algarum* are pupal parasites of dipteran flies. *Creophilus erythrocephalus* Fabricius is known as Australian devil's coachhorse while *Staphylinus olens* Muller is devil's coach-horse of Europe.

Staphylinoidea Ganglbauer 1895 superfamily of Polyphaga (Coleoptera: Coleoopteryoidea); elytra truncate, leaving more than two abdominal tergites exposed; veins M and Cu not connected. Mostly mould-eaters or predatory, rarely phytophagous.

Stegomyia old nickname of mosquitoes *Aedes* spp., specially A. *ægypti* (Linnæus) (Culicidæ: Nematocera) that transmits yellow fever and dengue.

Stelididæ (Schenk 1859) Schmied 1882 (Stelidæ) family of Apoidea (Aculeata: Apocrita); characterised by having no subtriangular area on apical abdominal tergite and abdomen of the female being without a ventral pollen brush; arolia present between tarsal claws; these are parasites or inquilines in the hives/nests of other bees.

Stem-borer(s) any insect (adult or larvæ) that bores into the stems of plant/tree borer *Sphenoptera gossypii* Cotes (Buprestidæ: Coleoptera), thistle stem-borer *Euribia cardui* (Linnæus) (Trypetidæ: Diptera), brinjal stem-borter *Euzophera perticella* Ragonot (Phycitidæ: Lepidoptera), sorghum stem-borer [*Chilo* (*zonellus*) *partellus* (Swinhoe) (Chrambidæ Lepidoptera)] et al.

Stemma (pl. **stemmata**) adaptive or later ocelli (ocelanæ or simple eyes) as present in insect larvæ: may occur singly or in groups on either side of the head. These vary in number, from one in Tenthredinidæ (Hymenoptera) to maximum of six or seven on either side in Neuroptera, Lepidoptera and Coleoptera.

Stenogamy ability to mate in confined space; rare in insects, as most species mate while flying (in the open), some even after extended courtship (nuptial) flights.

Stenogasteridæ small family of Vespoidea (Aculeata : Apocrita) that includes hunting wasps, mostly insectivorous or araneidivorous.

Stenomidæ (Stenomatidæ) family of Gelechioidea (Ditrysia : Lepidoptera); head smooth scaled, ocelli present, maxillary palpi 4-segmented; hind tibia with long scale-hairs; forewings having R_4 and R_5 separate to apex; hind-wing broader than forewing.

Stenoneuridæ extinct family of Carboniferous period belonging to order Protorhoptera, perhaps the original stock from which the Blattaria (Dictyoptera) of present day took their origin.

Stenopelmatidæ Burmeister 1838 family of cave cricket and sand cricket and wetas (Tettigoinioidea : Ensifera); robust-bodies antennæ very long and filiform; usually wingless or wings partially developed; tarsi 4-segmented; ovipositor inconspicuous, cerci short but conspicuous. Mostly carnivorous and nocturnal (350 species)-gaint Australian king cricket, *Anostostoma australasœ* Greay (Australasia); giant weta, *Dienacioa neteracantha* White (New Zealand), green house cricket *Distrammena japonica* Blatchley (Japan and USA) et al.

Stenophagy ability of an organism (insect) to feed and subsist on a narrow range of closely related food plants.

Stenoplastic an organism which exhibits a limited capacity for modification or adapation to a new environment.

Stenopsychidæ Martynov 1924 family of Hydropsychopida (Annulipalpia: Trichoptera); most beautiful caddisflies separated form family Philopotmmidæ McLaclan 1876 by having mandibles robust, second segment of maxillary palpi short and anal angle of hind-wing much wider than the forewing Larvæ inhabit swift mountain streams-species of *Stenopsyche* (Indo-Malaysian), *Stenopsychodes* (Australian), *Pseudostenopsyche* (Neotropical) et al.

Stenopteridæ Beach 1896 old and obsolete name for thysanopteran family Thripidæ Stephens 1829, Uzel 1895 (Thripoidea: Tereorantia).

Stenoviciidæ highly specillised extinct Permo-triassic family that is believed to have given rise to present day Cicadelloidea (Auchenorrhyncha: Homoptera).

Stenurothripidæ Bagnall 1930 a family of fossil thrips (Thysanoptera: Thysanopteroidea) unassigned to any suborder or superfamily.

Stephanidæ (Leach 1816) Haliday 1840 family of Ichneumonoidea (Apocrita: Hymenoptera); antennæ very slender having 30 or more segments; hind femora with teeth; forewing having costal cell broad; sternites of gaster evenly sclerotised ovipositor long. Mostly parasitic on larvæ of wood boring beetles (100 species) - *Stephanus serrator* of Europe parasitise *Xyloterchus* species while *Schletererius cincitipes* of USA is parasitic on *Sirex noctilio. Staphanus crassicauda* Morley of Queensland, probably the largest known species, has antennæ 14 mm long, body 36 mm and ovipositor 40 mm.

Stephanocircidæ minor family of fleas (Ceratophylloidea: Siphonaptera); characterised by head with one or two combs; occiput with a subdorsal internal tubercle on each side; coxal ridge and hind tibial tooth present. Confined to Australian and Neoptropical regions.

Stereopkinesis thigmotaxis phenomenon by which reflex movements and responses of an insect to light or other stimuli are inhibited by action of tactile stimuli.

Sterile barren; unable to produce (offspring or fruit).

Sterile male technique scientific method of pest control, first suggested by Knipling in 1938; males of unwanted insect species are sterlised in the laboratory and released in filed; these mate with females which lay sterile or unfertilised eggs and thus there is no increases in population of that species; sterility may be included either by irradiation (5000 to 10000 rads) or by using chemosterilants.

Sterilisation act or process of making sterile.

Sterilise render sterile by chemical treatment, radiation or surgery.

Sterility inability to produce offspring; barrenness.

Sternacosta transverse internal ridge of sternal suture through bases of sternal apophyses.

Sternacostal suture external suture of sternacosta, separating basisternum form sternellum.

Sternal apophyses lateral apodermal arms of eustermum.

Sternal plate sternite.

Sternal spatula breast-bone-an elongate sclerite either toothed, pointed or bilobed anteriorly -found in most Cecidomyid larvæ (Mycetophiloidea: Nematocera) situated mid-ventrally on the thorax.

Sternellum area of esusternum posterior to bases of sternal aprophyses of sternacostal suture.

Sternite(s) sclerites that compose the sternum.

Sternopleurite sternopleura : infracoxal sclerotisation of thoracic pleuron as found in dipteran insects.

Sternorrhyncha (= Phytophthires Burmeister 1835; Hymenelytra Latreille 1825) suborder of Homoptera (Hemipteroidea : Paraneoptera); antennæ usually well developed with terminal arista, rarely atrophied origin of rostrum more thoracic, arising between the fore-coxæ. Many species with immature stages and adult females incapable of locomotion. Grouped into four superfamilies-Psylloidea, Aleyrodoidea, Aphidoidea and Coccoidea.

Sternum ventral region of typical segment of most adult insects and their larvæ; differentiated into two regions-segmental (eusternum, comprising of three sclerites-presternum, basisternum and intersegmental (spinasternum or postsernellum). Vertebrate breast-bone.

Sthenaropodidæ family of Protorthoptera-an extinct order of Carboniferous period-that gave rise to Orthopteroid insects of today-*Stenaropoda fischeri,* fossil recorded form France.

Stichotermatidæ Hofeneder 1910 very small family of Strepsiptera (Coleopteroidea: Oligonoeoptera); known only from females of *Stichotrema* species, characterised by cephalothorax with look like projection behind the spiracles. Parasitic on tettigonids, gryllids and mantids; confined to Dutch new Guinea.

Stick-insect common name of a large number of elongate, cylindrical, often wingless insects (Phasmida: Orthopteraoidea) resembling twigs in appearance; head prognathous, compound eyes present; antennæ and legs exceptionally long and slender; meso-and metathorax also elongated and long; no auditory and stridulatory organs; foul smelling fluid emitted from thoracic glands as means of defence-*Caraussius morosus* Brunner is 8 to 10 mm long while

Eurycantha horrida Boisduval and *Pharnacia serratipe* are 300 to 350 mm long (excluding antennæ) with 6 to 8 mm diameter.

Stictococcidæ monogeneric family of Cocooidea (Sternorrhyncha; Homoptera); adult female flat and circular in shape having wax impregnated integument-*Stictococcus* spp.

Stictopterinæ subfamily of Noctuidæ (Noctuoidea: Ditrysia); palpi slender; forewing generally long and narrow increasing in width to outer margin, hind-wing usually hyaline at the base with vein 5 fully developed and from the lower angle of the cell; abdomen hardly longer than hind-wing. Caterpillar with four pairs of abdominal prolegs *Odontodes aleuca* Guenee, *Stictoptera cucullioidea* Guenee et al.

Stigma pterostigma : conspicuous thickened opaque spot, along the costal margin of wings of many insects, present specially in psocopteran, most hymeno-pteran and in both pairs of wings of wings of odonotan insects. Alternative name for spiracle or breathing pore of an insect.

Stigmellidæ (= Nepticulidæ Tutt 1899, Spuler 1910) family of smallest lepidopteran insects (Stigmelloidea = Nepticuloidea : Monotrysia); wings narrowly lanceolate and clothed with aculei wing span 3 to 10 mm, forewing with branched veins and usually a small cell. Wing spread of *Stigmella (Nepticula) microtheriella* (Stainton) is only 3 mm. Caterpillars minute, mostly leaf-miners.- *Stigmella hoplometallus* and *S. elachistarcha* are leaf-miners of Bengal kino (*Butea monosperma* Lamarck) and berries (*Zizyphus rugosa* Lamarack and allied species) respectively.

Stigmelloidea (Nepticuloidea Forbes 1923) superfamily of Monotrysia (Lepidoptera: Mecopteroidea); wing venation reduced specially in hind-wings; male having frenulum and female with fleshy ovipositor and short cloacal duct-Stigmellidæ, Opostegidæ and Tischeriidæ.

Stiboterygidæ Tillyard 1926 family of Myrmeleontoidea (Planipennia : Neuroptera); large, delicate fragile insects, swift fliers, fly at dusk and only striaght ahead, have short knobbed antennæ, very large eyes, and shining wings and abdomen. Larvæ large, black and spiny, live in vegetable refuse and prey upon small living insects; confinend to Australia and South America-*Albardia* spp., *Stilbopteryx* spp. et al.

Stimulus (pl. **stimuli**) any change in the environment or internal condition of an organisms (insect) that effects response by the organism.

Sting modified ovipositor; composed of stylets or lancets fitting together to form a hollow piercing terebra and stylet–sheath with posion sac and basal plates. Found in females of ants, bees and wasps (Apocrita : Hymenoptera) as organ of self defence.

Stinger sting : sharp piercing organ of female bee or wasp, usually connected with poison gland.

Stipes distal subdivision of maxillæ, at its ventral extremity; it bears two endite lobes-lancina and galea.

Stizidæ (= Arpactidæ) family of Sphecoidea (Apocrita: Hymenoptera) now placed under family Sphecidæ; included small number of species that nest singly or in colonies and store therein grasshoppers, crickets etc. for feeding their young ones hence considered beneficial-*Sizus* species prey mostly on orthopteran insects, while *Stizoides unicinctus* parasitises *Sphex atratus* and allied species.

Stomach mid-gut, mid intestine or ventriculus; middle region of alimentary canal, lined by a thin peritrophic membrane which surrounds the food: complex in many insects.

Stomach poison toxicant that after being eaten acts by absorption, through intestine and kills the animal (insect); used as insecticide—HCH (BHC), DDT, lead-arsenate etc.

Stomodæum foregut or fore-intestine; the anterior ectodermal part of alimentary canal of an insect.

Stomoxyidæ Meigen 1824 small family of dipteran flies (Muscoidea: Calyptratæ); hæmatophagous having proboscis long, rigid and harp with small labella for piercing and sucking blood (of man and mammals) – stablefly *Stomoxys calcitrans* (Linnæus) et al.

Stratiomyidæ Latreille 1802 family of soldier flies (Tabanoidea : Brachycera); flat, bristleless flies having head short and wider than prothorax, eyes large, ocelli present, third antennal segment annulate, proboscis short; scutellum conspicuous; tibia suprless, pulvilli and empodia pad–like. Larvæ terrestrial or aquatic and carnivorous or saprophagous (1500 species) – *Clitellaria heminopla, Cyphomyia indica, Hermetia illucens* (Linnæus) *Sargus metallicus* Febicius, *Stratiomy badia* Walker, *S. potamida* et al.

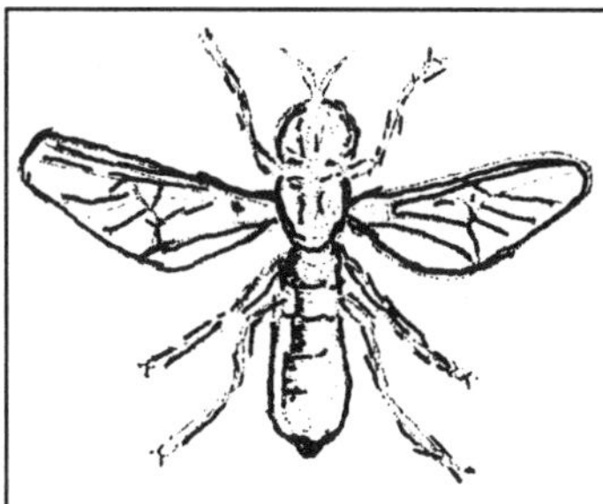

Sargus metallicus Fabricius

Streblidæ family of batflies (Pupipara : Cyclorrhapha); having head not flexed on dorsum or thorax; eyes when present small, ocelli absent; antennæ in pits and 2-segmented; palpi leaf-like projecting in front of the head; hind coxæ enlarged and claws not distinctly toothed. Ectoparasities of bats (50 species-species of *Nycterobosca, Raymondia, Strebla* et al.

Strepsimanidæ small family of Yponomeutoidea (Ditrysia : Lepidoptera); that includes Oriental moths, having maxillary palpi vestigial; forewings elongate and triangular with costa slightly curved near middle; M_1 absent, M_2 and M_3 ending in costal margin.

Strepsiptera Kirby 1813 order of small oligoneopteran insects (Coleopteroidea); 2 to 10 mm long; antennæ conspicuous and flabellate; mouthparts degenerated, biting and chewing type; metathorax exceptionally large; male winged and free-living, forewings modified into flattened club shaped appendages, hind-wings large and fan-like; trochanters absent. Holometabolous; females endoparasitic on bees, leaf hopper, wasps etc., remain inside the host in a puparium. About 500 species, arranged into five superfamilies-Mengenilloidea, Corioxenoidea, Halictophagoidea, Bohartilloidea and Stylopoidea.

Stretch receptor(s) proprioceptors or sensory receptors comprising of stands of connective tissues associated with nerve cells; these aid an (insect to be conscious of its body movements.

Stria (pl. **striæ**) longitudinal depressed line or furrow frequently punctured, extending form base to apex of elytra.

Striate (surface) having grooves or depressed lines.

Striated border striated surface of epithelial cells lining the mid-intestine and Malpighian tubules of insects-honeycomb order and brush type border.

Stridulation production of sound by certain insects (also by a few other invertebrates) by rubbing one part of their body (rasp or scraper) against another (file). Among the males of short-horned grasshoppers (Acrididæ), sound is produced by rubbing a row of pegs on inner side of hind femur against outer surface of tegmen; while males of long-horned grasshoppers (Tettigoniidæ) produce sound by rubbing together two tegmina. Mole crickets (Grylloptaphidæ) do not possess stridulatory organs yet the males make sound by vibration of their wings. Similarly many other insects such as *Drosophila* spp. (Drosophilidæ) make buzzing or humming sound during flight by vibration of their wings. Some beetles make sound by tapping some part of their body against an external object. Death-watch beetle *Xestobium rufovillosum* and allied species (Anobiidæ) tap their head against the walls of burrows; some plecopteran and psocopteran insects rap the end of their abdomen against the subtrate.

Stridulatory sound producing.

Stridulatory organs (aparatus) organs that when fubbed with each other, produce sound, may be rasping, chriping, whistling to shrill shriekeing. Thse modified organs my be present on throax, wingns, abdomen or even on legs and are of two types–plate type and comb type the former is found in many coleopteran insect (Cerambycidæ etc.) while the latter in orthopteran insect (Acrididæ, Tettigoniidæ etc.) Male cicadas (Cicadidæ) are the noisiest insects having most complex stridulatory organs (plate type) situated on ventral side of basal abdominal segment.

Strigil comb–like structure on the leg of honeybee (Apoidea : Apocrita) used for cleaning antennæ. Similar structure on fore–tibia of certain lepidoperan insects.

Strophandria one of the two groups of suborder Symphata (Hymenoptera) distinguished by the male genitalia which are inverted with the ventral surface uppermost. Comprises of only one superfamily–Tenthredinoidea. *Cf.* Orthandria.

Structural colours brilliant iridescent colours produced by refraction, reflection, diffraction or interference of light of different wave-lengths and not by the presence of pigments in insect body. These colours are usually found in bodies of dragonflies (Odonata), wings of some butterflies (Lepidoptera) as also elytra of certain beetles (Coleoptera).

Stygiaridæ very small family of Cossoidea (Ditrysia : Lepidoptera) often include in Cossidæ; areole present is forewings and vein Cu_2 in both wings. Moths nocturnal; caterpillars wood–borers.

Stylate antenna(e) of certain dipteran insects having bristle–like process at distal end.

Style slender, bristle–like process at the apex of antenna.

Stylet median dorsal process in shaft of ovipositor (sting).

Stylopidæ Kirby 1813 family of strepsipterans (Coleopteroidea : Oligoneoptera); male has 4–to 6–setmented antennæ with third segment only laterally flabellate; cephalothroax in female having only one pair of spiracles and abdomen with five genital tubes. Parasitic on aculeate hymenopteran insect–*Stylops pacifica* Bohart,

S. saliciflorís Pierce, Polistes *fuscatus* Fabricius, *Xenos vesparum* Rossi, et al.

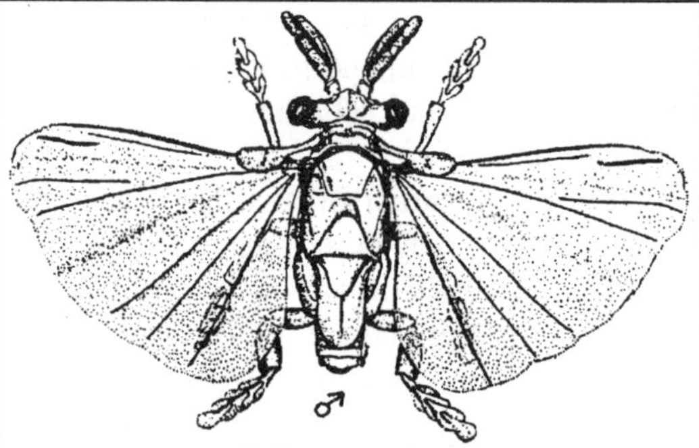

Stylops species

Stylopoidea superfamily of Strepsiptera (Coleopteroidea : Oligoneoptera), stylopids distinguished by very long mandibles and base of maxilla not swollen. All are parasites and grouped into three families–Elenchidæ (parastiic on delphacids), Myrmeocolacidæ (male attacks ants and female orthopteroids).

Stylotermitidæ Holmgren and Holmgren very small family of damp-wood termites (Isoptera : Blattopteroidea); subterranean insects; chatacterised by flat pronotum and wings when present without any hairs–*Stylotermes fletcheri* (Holmgren) found only in India.

Stylus (pl. **styli**) short, slender finger like–process.

Sub–under less than.

Subalare epidermal epiplurite giving insertion to posterior pleural muscle of insect's wing.

Subclass below the class : a taxonomic group that may be further divided into superorders or directly into orders – Apterygota and Pterygota are subclass of class Insecta (Hexapoda).

Subcosta (Sc) second vein of insect's wing situated next to and behind vein costa; typically forked distally into two short branches (Sc_1 and Sc_2).

Subcoxa (pl. **subcoxæ**) segment preceeding the coxa on legs of certain primitive insects.

Subcuticle endocuticle : flexible inner layer of cuticle of an insect or other arthropod; composed of chitin and protein. During moulting this layer is digested away by the enzymes released form the epidermal glands.

Subdominant a species which may appear more abundant at particular periods of the year than its true dominance in a climax.

Subfamily a taxonomic category immediately below the level of family; comprising of one to several related tribes or genra: their names always end in inæ-Eumenin, Polistinæ and Vespinæ are the subfamilies of Vespidæ (Apocrita : Hymenoptera).

Subgalea lobe of maxillary stipes bearing galea.

Subimago dun : a stage before adult stage as fund in some hemimetabolous (prometabola) insects; the premature adult that emerge for the aquatic larva (naiad) has fringed wings with legs and cerci not fully developed and a thin dull opaque skin covering the entire body including wings. After sometime, it moults and become a full-fledged, brilliantly coloured adult mayfly (Ephemeroptea).

Submentum basal segment of labium of an insect : proximal plate of postlabium.

Suboesophageal ganglion (pl. **gangila**) pair of ganglia located beneath the oesophagus; it connect the ventral nerve cord with nerve-collar.

Suborder taxonomic subdivision of an order comprising of superfamilies/ families-Apocrita and Symphytra are suborder of order Hymenoptera

Subphylum below the phylum; further classified or divided into classes-Crustacea, Chelicerata and Uniramia are subphyla of phylum Arthropoda.

Subspecies (race) geographically separated population of an species (animal/plant) capable of interbreeding with each other. Trinomial nomenclature is used for naming the subspecies; first and second being generic and species names respectively-*Pyrilla aberrans aberrans, P. a. plagati, P. a. consors, P. a. comes* and *P. a. achates* are subspecies of *Pyrilla aberrans* Kirby (Lophopinæ : Fulgoridæ).

Subulipalpia old and equivalent name for Perloidea, a superfamily of stoneflies (Setipalpia : Plecoptera); no more in usage.

Sucking insects all such insects that suck nectar, plant-sap, blood or any other liquid food. They all have piercing and sucking type of mouthparts, variously modified to suit their purpose-bees, bedbugs, plant bugs, lice mosquitoes, fruit sucking moths etc.

Suctoria suborder or section of old order Aptera (Linnæus 1758) to separate the fleas, as suggested by the Geer (1778). Retzius (1783) placed Suctoria under order Saltatoria (jumping insects) and finally Latreille (1825) recognised it as an independent order and named it Siphonaptera.

Sugarcane green bore *Raphimetopus* (*Anerastia*) *ablutella* (Zeller) (Pyralididæ: Lepidoptera); medium-sized moth (wing-span 22-25

mm), having ochreous forewings and white hind–wings; female has filamentous antennæ whereas those of male are comb–shaped. Specific pest of sugarcane *Saccharum officinarum* Linnæus. Copper–green, thick–set, cylindrical caterpillar cause similar damage as any other stem bore.

Sugracane internode bore *Chilo sacchariphagus indicus* Kapoor (Crampbidæ : Lepidoptera); pale brown moth having white hind–wings. Caterpillar white with dark spots on the body and down head, bores near the nodes and feeds on inner contents; plugs the entry holes with excreta; migrates and attacks number of nodes.

Sugarcane purple bore *Phragmatœcia purpurens* Fletcher (Cossidæ: Lepidoptera); rare pest of *Saccharum* spp., including sugacane. Caterpillar faint violet in colour having in stripes; one it enters the host, it tunnels the entire stem as it has to go down into the roots or stubble for hibernating.

Sugarcane root borer *Emmalocera depressella* Swinhoe (Pyraldidæ Lepidoptera); specific pest of sugarcane confined to Indian subcontinent. Caterpillar uniformly creamy–white in colour without any stripes; feeds exlusively on underground portion of the stem (not roots) causing deadheart and side shoots (tillering) that produce very thin and often non–millable canes.

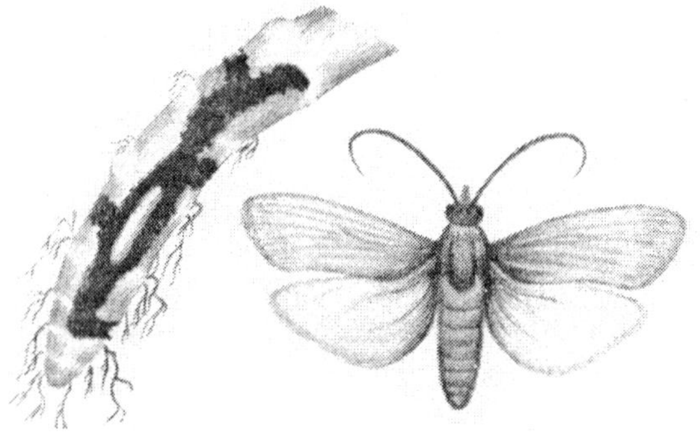

Emmalocera depressella Swinhoe

Sugarcane shoot bore *Chilo infuscatellus* (Snellen) (Crambidæ: Lepidoptera); small pale grayish–brown moths, forewings with darker markings specially along the outer edge and hind–wings whitish. Caterpillar dirty white with five light violet longitudinal stripes, bores into central shoot and tunnels downwards causing deadheart; migrate and attack number of shoots.

Sugarcane top borer *Tryporyza nivella* (Fabricius) (Pyraustidæ Lepidoptera); milky white moths; female with buff coloured tuft of hairs at anal end of abdomen. Pale-creamy-yellow caterpillar tunnels into midrib of the leaf upto its base, enters the central leaf spindle, kills the growing point resulting into the bunchy top appearance; a post-monsoon pest that flourishes under humid conditions.

Sulfex wettable sulphur; an off-white powder based on sulphur as the active ingredient; highly effective fungicide and acaricide on a wide range of economic crops (including fruits and vegetables) and ornamentals.

Super-above over.

Super-development hypermetamorphosis.

Superfamily taxonomic subdivision of an order of animals/plants further classified into families having broadly more or less same characteristic (the terms suggested always end in -oidea) - Bacterioidea and Phasmoidea are superfamilies of order Phasmida.

Superlinguæ pair of small ventral lobes, developed form mandibular somites found attached to the sides of hypopharynx in certain adult insects.

Superorder major taxonomic group of a class of subclass that is further divided into various order (names of superorder of Pterygota, end in-oidea)-Odonatopteroidea, Orthopteroidea, Hemipteroidea, Hymenopteroidea et al.

Supraoesophageal ganglion brain of an insect, situated in the head region, just above oesophagus, it innervates antennæ, eyes and labrum.

Surfactan spreader. wetter - Tergitol (sodium sulphate derivative).

Susceptible (host plant) that is unable to suppress or retard.

Suspension fluid containing solid particles larger than 100 μ distributed throughout, but will ultimately settle down.

Suture any external groove or narrow space of body wall that separates scleortic areas of cuticula.

Swallowtail(s) big, beautiful, brightly coloured butterflies, *Papilo* spp. (Papilionidæ Papilionoidea); having a conspicuous tail-like prolongation on its hind-wing. Most species have only one tail on each hind-wing *P. daunus* Boisduval the largest Western species has two tails on each hind-wing while a rare species. *P. pilumnus* Boisduval has three tails - African swallowtail *P. dardanus* Brown, Parsnip swallowtail *P. ajax* Linnæus, Spicebush swallowtail *P.*

troiles Linnæus, Yellow swallowtail *P. glacucus* Linnæus, Zebra swallowtail *P. marcellus* Cramer, et al.

Swallowtail moth *Ouraoterys sambucaria* (Selidoematdæ : Geometroidea); pale sulphur-yellow with two thin greenish-brown transverse lines on forewings and one each on hind-wing; tail on each hind-wing short, pointed, edged and spotted with reddish-brown. The yellow pigment of the moth is brilliantly fluorescent in ultraviolet light.

Swarm throng of insects or other small animals: to go off in a swarm may be in search of food, mate or migration. Swarms of dragonflies, specially *Libellula* spp. (Anisoptera; Odonata) are often seen travelling long distances out to sea. Adults of locusts (migratoria forms) move in dense swarms over great distances and settle where environmental conditions are conductive for breeding and feeding. Mating swarms of mayflies (Ephemeroptera) consists of males; females enter the swarm and leave with a male and mating takes place in flight. Emission of swarms form ants nest (formicaries), bee-hive or termitarium is the typical method of founding new colones by the ants, bees and termites respectively.

Sweet potato weevil *Cylas formicarius* Fabricius (Curculionidæ : Coleoptera); small (6–8 mm long), slender bodied having bluish-brown head, with elongate snout, bright red thorax and legs and brownish-red abdomen. Specific pest of sweet potato (*Impomoea batatas* Linnæus); grubs bore into stems tunnel inside and feed on inner tissues; both grubs and adults bore into tubers in fields as well in godowns.

Swift(s) extremely rapid flying moths (Hepialidæ: Hepialoidea); found flying during dusk; mouthparts vestigial, no proboscis; antennæ very short; wing coupling Jugate type, jugal lobe elongate and resting upon the hind-wing-common swift *Hepialus lupulina* (Linnæus) orange swift *H. sylvina* et al. Caterpillars of *H. hecata* (Linnæus) of Europe and *H. gracilis* Grote on North America attack roots of ferns.

Symbiont organism living in symbiosis with another organism.

Symbiosis harmonic association of different organisms : living together of two species in more or less intimate association, both benefitting from each other.

Sympherobiidæ Brues and Melander 1932 very small, newly erected neuropteran family (Hemerobioidea Planipennia) akin to Hemeropidæ but distinguished from the same by having wing–

venation of forewings with two distinct radial sections, R_{2+3} and R_{4+5} - brown lacewings, *Sympherobius angustatus* (Banks), *S. elegans* Stephens, *Psectra diptera* Burmeister, et al.

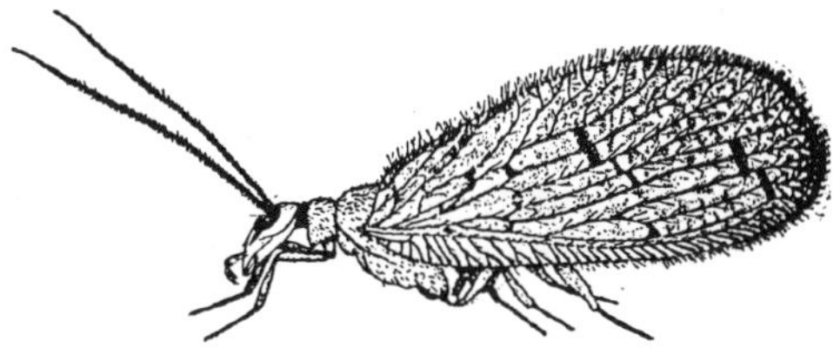

Sympherobius angustatus (Banks)

Symphile another species kept as guest in the nest of ants or termites.

Symphypleona Borner 1901 (Symnthuroidea Crampson 1920) one of the two suborders of Collembola; very small insects, body subglobose; head hypognathous; thorax and first four abdominal segments fused together, head tracheæ often present, scales small, furca large–Neelidæ, Sminthuridæ and Dicyrtomidæ. *Cf.* Arthropleona.

Symphyta Gehrstecker 1867 (= Chalastogastra Konow 1897) one of the two suborders of Hymenoptera (Hymenopterpoidea : Oligoneoptera); primitive species distinguished by abdomen broadly attached to thorax, no marked constriction between I and II abdominal segments; ovipositor adapted for sawing and boring. Larvæ mostly with thoracic and abdominal legs, except in *Orussus* spp. (Orussidæ), parasitism wanting. *Cf.* Apocrita.

Syn– together.

Synapse site where the axonic end of a neurone comes in close approximation (not direct contact) with the dendritic end of an adjacent neuron, synaptic gap is about 100 Å (0.1 mμ or 0.000001 mm).

Synecphalon secondary composite head, formed of prostomium and one or more succeeding somites.

Syndrome occurrence of symptoms that collectively indicate abnormal conditions of the organism.

Synechthry myrmecophile beetles (Staphylinidæ: Polyphaga); that are considered by ants as their enemies and hunted by them, but escape notice because of the strong body armature–*Myrmedoma* spp., *Xantholinus* spp. etc.

Synecology scientific study of inter-relationship of population of different species (of insects) inhabiting a given area. *Cf.* antecology.

Synergism increased pesticidal activity of a mixture of pesticides.

Synegry combined or coordinated action.

Synexenic rearing of one or more individuals of a single species in association with one or more known species of organisms-monoxenic, dixenic, trixenic or polyxenic depending whether the number of associated species is one, two three or more respectively of axenic.

Synistata old order of insects as proposed by Fabricius (1775); it included Apterygotes, Ephemera, Neuroptera, Phryganes, Panorpes, Perlids, Hymenoptera, Termites and certain Crustacea; subsequently splitted into several separate orders and groups.

Synlestidæ Tillyard 1926 small family of primitive damelflies (Coenagrioidea : Zygoptera) distinguished by vain Cu_2 arching strongly forward on leaving discodial cell; discodial cell very narrow and pointed apically; two (nearly three) antenodal cross-veins - species of *Megalestes* (Indian), *Perilestes* (Neotropical), *Chloroleste* (Ethiopian), *Synlestes* and *Chorismagrion* (Australasia).

Synockete individual of another species (of insect), tolerated as guest, with indifference in a termitarium or ants' nest.

Synomone substance (allelochemical) released by an organism, which benefits both the emitter and the receiver.

Synonym(s) two or more names given to same taxon, according to Laws of Priority, the oldest one is taken as proper name and is called senior synonym while the rest are all junior synonyms.

Synonym, Junior more recently published, of the two or more available synonyms of the same taxon.

Synonym, Objective (nomenclatural, obligate, homotypic) synoymies based upon same type these are always absolute and indicated by sign of congruence (≡).

Synonym, Senior earliest published, of the two or more available synonyms, of same taxon.

Synonym, Subjective (heterotypic, taxonomic) synonymies based upon different types belonging to same taxon; these are not absolute and are indicated by sign of equality (=).

Synonymity quality of being synonymous.

Synonymous having same or nearly same meaning or description.

Syntelliidæ very small family of Histeroidea (Polyphaga : Coleoptera); head with gular sutures separate; antennæ geniculate with

3-segmented club; fore–tibia dentate, coxæ all approximated and front ones projecting; embiodum distinct bisetose. Found in Mexico and the Orient sucking sap exuding for wounded trees–*Syntelia indica* and allied species.

Syntexidæ family of Siricoidea (Symphyta: Hymenoptera); distinguished by abdomen not constricted, cenchri present; fore–tibial spur (only one) pectinate; includes ingle North American species having wood–boring larvæ.

Synthetic artificially prepared form component clements; not obtained directly form natural sources. Most of present day pesticides are synthetic (organic) compounds as against inorganic (arsenicals, fluorine, sulphur etc.) and botanicals (of plant origin) used earlier.

Syntomidæ (Syntomididæ) old and obsolete name of a small Tropical lepidopteran family Ctenuchidæ (Sphingoidea : Ditrysia).

Syntonopteridæ family of fossil insect order Palæodictyoptera that flourished during Palæozoic era, these insects had intercalary veins as a distinctive feature–species of *Syntonoptera* and *Lithoneura.*

Syntype each of the several specimen cited in describing a new species, before selecting finally a type speciment.

Syringogastridæ Pardo 1969 monogeneric family of Diopsoidea (Schizophora: Cyclorrhapha); flies having eyes not stalked, ocellar bristle present hind femora thickened with spines; alula and spuamæ greatly reduced, ædeagus long and coiled. Eight Neotropical species of *Syringogaster.*

Syrphidæ (Leach 1815, 1819) one of the largest dipteran family (Syrphoidea: Aschiza); hover–flies or flower flies, having brightly coloured markings, almost always bristleless; head large, antennæ 3–segmented, eyes dichoptic in females and holoptic in males, ocelli present, wings with *vena spuria* between Rs and M, cell Rs distally closed. Mostly predaceous, a few phytophagous, inhabiting decaying vegetation and rotting wood (3200 species) - *Sphærophoria javana, Xanthogramma scutellare* and *Paragus serratus* etc. Prey upon aphids; *Baccha sapphirina* is predaceous on psyllids; *Syritta pipiens* is found in rotting paddy and gourds while *Chrysotoxum* species bred in decaying vegetation.

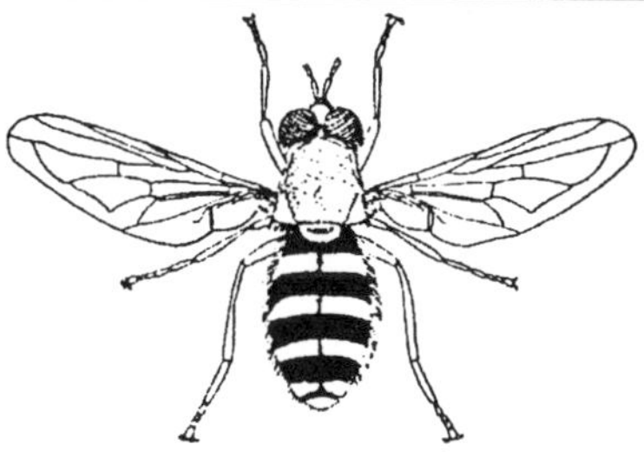
Chrysotoxum species

Syrphoidea superfamily of Aschiza (Cyclorrhapha: Diptera); frontal suture lacking, ptilinum wanting; wings with cell Cu elongated and extending more than halfway to wing-margin; marked hovering habit. Two main families-Pipunculidæ and Syrphidæ.

Syssphingidæ old name of a family of royal moths (Saturnioidea : Ditrysia), now regarded as Cithernoiidæ (Bombycoidea: Ditrysia).

System association of organs of an organism for performing some general activity of body, such as digestion, reproduction etc.

systematics taxonomy : science of classification its method and principles of procedure; often includes identification and nomenclature.

Systemic insecticide pesticide which when applied to roots or aerial parts of plant/tree, is rapidly absorbed by plant surfaces and translocated through plant vascular system to the various part of the plant, in amount lethal to the organisms (insects) feeding thereon–carbonfuran, phorate, phosphamidon, etc.

Systole rhythmical contraction of (insect) heart.

T

Tabanidæ Leach 1819 family of bristleless, blood-sucking flies (Tabanoidea: Brachycera); comprising of clags, gadflies and horse-flies; stout build, third antennal segment annulate but devoid of style; eyes large and laterally extended; proboscis projecting, adapted for piercing in female; squamæ large, pulvilli and arolium pad-like. Larvæ soil-dwellers and predaceous (25000 species) – *Erephopsis guttata* Donovan of Australia, *Tabanus intermedius* Egger of Europe and *T. americanus* Foster are 25 to 30 mm long while *Chrysops coloration* Bigot, C. *dimidiata* C. *nigra* Macquart, C. *vittata* Wiedmann of North America are only 7 to 9 mm.

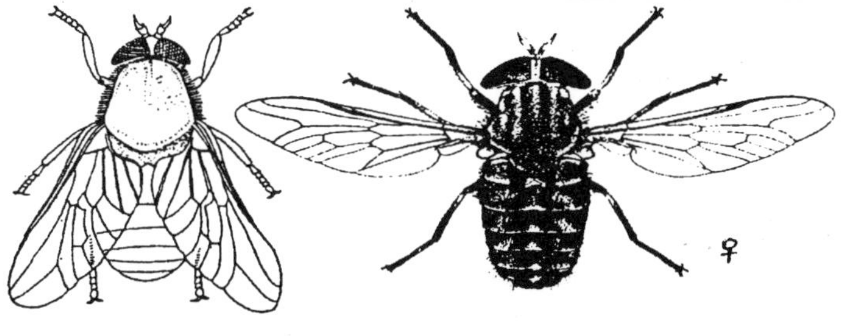

Tabanus Species

Tabanoidea superfamily of Brachycera (Diptera : Mecopteroidea); that includes five families – Rhagionidæ, Xylophagidæ, Tabanidæ, Pelecorhynchidæ and Stratiomyidæ.

Trchardidæ Green 1896 old and equivalent name for homopteran family Lacciferidæ Chamberlin 1925 (Cocoidea : Sternorrhyncha) comprising of lacinsects.

Tachinidæ Bezzi and Stein 1907 large dipteran family of Calyptratea (Schizophora : Cyclorrhapha); stout flies with grey or striped body, head large and free; metanotum prominent; wings large rarely mottled with basal and anal cells closed and squamæ large; abdomen with conspicuous apical, marginal and dorsal bristle. Oviparous or larviparous; fecundity 50 to 5000 eggs or larvæ per female. Larvæ (maggots) muscoid, cylindrical tapering anteriorly and distinctly segmented; all parasitic on other insects, spiders, centipedes etc. and important in biological control of insect pests specially in plantation crops and forest trees (6000 species)- *Exorista quadrimaculata* Baranov and *Sturmiopsis inferens* Townsend are larval parasites of sugarcane borers (Lepidoptera).

Tachiniscidæ large, stout, bristly flies, like bumblebees with only two monotypical genera (one South American and other South African); often placed under family Otitidæ (Agromyzoidea : Schizophora).

Tachinoidea Enderlein 1936 superfamily of dipteran flies (Calyptratea: Schizophora) suggested by some taxonomists, to include 7 families - Hypodermatidæ, Œstridæ, Tachinidæ, Cuterebridæ, Calliphoridæ Sarcophagidæ and Dexiidæ.

Tachygenesis shortened development involving omission of two or more larval instars.

Tactile senses of touch.

Tactile receptor(s) insect's organs of touch in their simplest form, comprising of hairs, bristles or spines with sensory cells and nerve ending at their base.

Tænidium spiral thickening of chitin strengthening walls of an insect's tracheæ.

Tæniopterygidæ family of filipalpian stoneflies (Nemouroidea: Plecoptera); wing-venation with enormous number of veins and large anal lobe in hind-wing; cerci 5-or 6-segmented; confined to Northern Hemisphere (70 species) - *Rhabdiopteryx lunata* (occurs in Himalayas at 5000 metres height), *Tæniopteryx fasciata* Burmeister, *T. occidentalis* Banks, *T. nivalis* Fitsch, *T. pacifica* Banks, et al.

Tagmosis union of body segments into functional groups.

Taint unwanted flavour in fresh, stored or processed food (from pesticides applied in field, orchards or godowns).

TALGERI, Professor Ganesh Mangesh - born December 6, 1919; B.Sc. (Agri) in 1942 and M.Sc. (Agri) in 1949 from Bombay University. Joined government service in 1942 and worked in various capacities as Entomological Research Officer till 1950 (Lecturer in entomology 1745-47); Crop Protection Officers (1950-56), Assistant Entomologist (1956-60), Deputy Director of Agriculture Extension (1960-64); Chief Plant Protection Officer (1964-67); State Entomologist and Professor of Entomology (1964-67); State Entomologist and Professor of Entomology (1967-72) and finally Deputy Director of Agriculture, Pune/Nasik till 1976; After retirement, served as Consultant and Technical Advisor to Pest Control (India) Private Limited till 1983. Died January 1990.

Tamarind fruit borer *Phycita orthoclina* (Phycitidæ: Pyralidoidea); a serious pest of tamarind fruits (*Tamarindus indica* Linnæus) and dry fruits as also seeds of red silk-cotton (*Bombax ceiba* Linnæus); pink, cylindrical caterpillars bore inside and feed on pulp of tamarind fruits wherein their excrement, cast skins and webbings render the fruits unfit for any culinary purpose.

Tanaostigmatidæ small family of encytid-like wasps (Chalcidoidea : Apocrita); characterised by mesocutum convex in both sexes with fin but complete notauli and marginal vein in forewing usually long. Parasitic on coleopterous insects specially weevils.

Tango receptors sense organ responsive to touch.

Tanyderidæ family of small-sized, most primitive dipteran flies (Psychodoidea : Nematocera); no ocelli, mandibles present but reduced; veins Rs and M 4-branched, R reaching the margin; intermedian cell present; female with horny ovipositor. Larva aquatic (30 species)-*Eutanyderus oreonympha, Protoplasa fitchii,* et al.

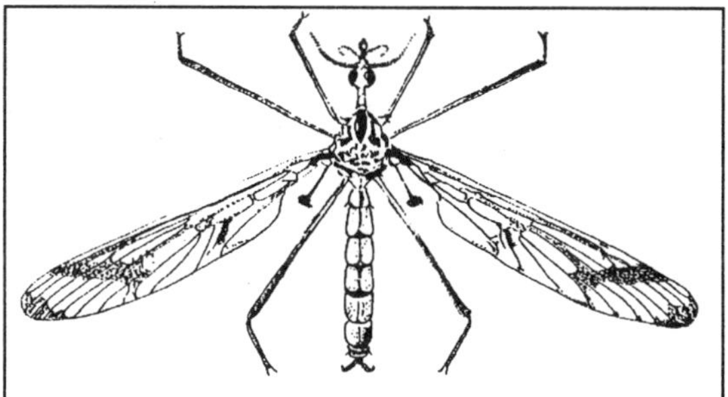

Eutanyderus oreonympha

Tanypezidæ a dipteran family of Tanypezoidea (Acalyptratæ : Schizophora); flies having propleura completely hairy, oral vibrissæ absent; legs long, slender and still-like; mid and hind tarsus larger than their tibiæ. Male holoptic-*Strongylophthalmomyia* species, *Tanypeza* spp., et al.

Tanypezoidea superfamily of Schizophora (Diptera : Macopteroidea); distinguished by antennæ with second segment not cleft; no prestomal teeth; costa broken at end of Sc with runs to wing margin, cross-vein *im* present; abdomen with spiracles 2 to 5 in membrane - Tanpezidæ, Heteromyzidæ.

Tapetum mirror-like reflecting surface in an insect's eyes; formed either of pigments (ocelli) or densely massed tracheæ surrounding the retinal rods, as in compound eyes of nocturnal moths. These reflect the light back into the retinal cells resulting into greater sensitivity in dim light.

Tarantula hawk(s) very large (upto 40 mm in length), Tropical wasps metallic green in colour and having infuscated, violaceous, orange or fiery-red wings-*Pepsis* spp. (Psmmocharidæ: Apocrita)-*P. femoratus, P. formosa* (Say), *P. obliquerugosa* Lucas, et al hunt tarantula and other large spiders.

Tardigardia newly erected phylum of Triploblastica (Non-chordata), regarded by some biologist as a class of uncertain Arthropida; microscopic (< one mm) aquatic animals mostly inhabit fresh-water, a few marine and some terresttial also; cylindrical unsegmented soft-bodied fat invertebrates; mouth a sucking proboscis with stylets; four pairs of stumpy unjointed legs; no circulatory or respiratory organs, sexes separate; undergo periodic moults (300 species) - water-bears (*Echiniscus* spp.) and bear-animacules (*Macrobiotus* spp.)

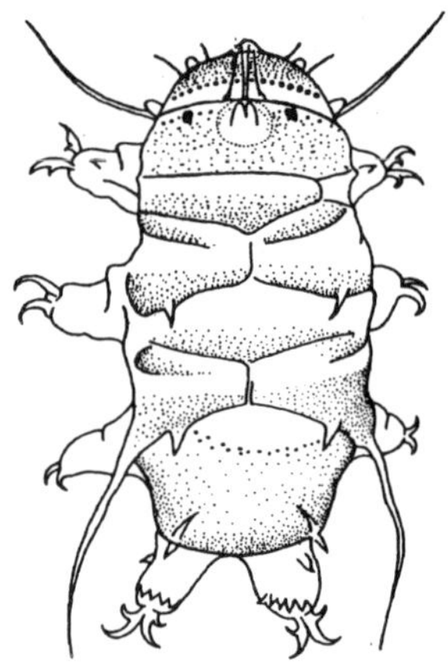

Echiniscus granulatus (larva)

Tarsal pertaining to tarsus.

Tarsal taste organs chemosensory organs situated on legs of certain butterflies and possibly some other insect. On painted lady, *Vanessa cardui* (Linnæus) (Nymphalidæ: Lepidoptera), these are estimated to be 200 times as sensitive to sugar as human tongue.

Tarsomere(s) subsegment of tarsus, not provided with any muscle: tarsite.

Tarsus fifth and last segment of insect's leg usually comprising of two to five small, short, tarsal segments (Tarsomers). Ankle bones in vertebrates.

Tascinidæ (= Neocastniidæ) very small family of beautifully coloured, day-flying moths (Castnioidea : Ditrysia); and closely related to butterflies; often placed under family Castniidæ; characterised by having prodoscis vestigial, tympanal organs wanting; all wings with first A vein reduced and discal cells open; found mainly in Australian region; *Tascinia* species et al.

Teste organs taste receptors in insects may be simple hairs to placoid and basiconic types; situated on the antennæ, palpi or other mouth-parts and often on legs as well. In insect it is generally difficult to distinguish between the senses of smell and taste, as both are usually of same form and type, found at the same place.

Tautonym name of species or subspecies in which second or even third component of name is same as generic name-*Atals atals* (Saturnidæ), *Apus apus apus*. This is permissible only under Zological code of nomenclature and not the Botanical code.

Taxis movement of organism in definite direction in response to a particular stimulus; the reaction may be positive (towards the stimulus) or negative (away from the stimulus)-chemotaxis (chemical), phototaxis (light), thigmotaxis (touch) etc. Arrangement or order.

Taxon (pl. **taxa**) taxonomic unit that is sufficiently distinct to be ranked in a definite category.

Taxonomic pertaining to taxonomy.

Taxonomist person having sound knowledge of taxonomy.

Taxonomy branch of biology concerning study of principles and practices of classification.

Tegma (pl. **tegmata**) group of segments, forming a unit in the body of an arthropod. In insect there are three tegmata - head, thorax and abdomen; while in mites and spiders there are only two tegmata-cephalothorax (head and thorax fused together) and abdomen.

Tegmen (pl. **tegmia**) thickened leathery forewing of an insect-Dictyoptera, Orthoptera and Phasmida.

Tegula (pl. **tegulæ**) small, arched, scale-like lobe or sclerite found at the base of forewing in some insects; it covers the place where wing articulates with the body-Lepidoptera, Hymenoptera (except Formicoidea).

Tegumen dorsal hood-like process that covers the male genitalia of certain insects: enlargement of IX abdominal segment.

Telegeusidæ very small family of Cantharoidea (Polyphaga: Coleoptera); only three species of beetles resembling phengodids, found in North America.

Telephoridæ Leach 1815 old and equivalent name for coleopteran family Cantharidæ (Cantharoidea: Polyphaga); that includes leather-winged beetles and soldier-beetles.

Telolecithal egg in which yolk lies at one end of the egg.

Telotaxis reaction vide which an organism (insect) orientates itself and moves in a definite direction towards a source of stimulus at a distance–moths flying towards light.

Telson XII abdominal segment present in embryos of certain insects and in adult of primitive insect order Protura (Entrotrophi)

Temperature intensity of heat or cold; usually measured in degrees by a thermometer.

Temperature resistance temperature plays vital role in the growth and activity of all organisms including insects. Power of resistance in insects is comparatively higher–aquatic larvæ of gnats and other insects can live in hot springs where the temperature goes as high as 50°C while certain caterpillars hibernate for several weeks at temperatures below–150°C. To tide over the unfavourable or unbearable temperatures the insects often go into a state of dormancy or diapause (æstivation or hiberanation).

Temperature sense like any other organism, insects too are sensitive to change in temperature of their surroundings. Generally rise in temperature causes increased activity and a fall brings about inactivity. Extremes of heat or cold eventually cause death. Exception of this generality is blood–sucking insects; most of these have fine hairs on their antennæ that are extremely sensitive to temperature changes and help the insect to locate a warm–blooded animal and perhaps the warmest part of the victim where the skin is thin and the blood flows near the surface.

Tenacity property of pesticide deposit or residue to resist removal by weathering.

Tenacity index ratio of quantity of residue per unit area at the end of given period of weathering to that present at the beginning.

Tenaculum minute organ situated medially on concave ventral surface of III abdominal segment of collembolan insects (springtails), having conical base and two laterally divergent distal prongs; it serves to hold furcula in place.

Tendipedidæ old and equivalent name for Chironomidæ Rondani 1841 (Culicoidea: Nematocera); includes small, delicate, gnat–like flies (midges) – *Chironomus socius, C. vicarius, C. zelandicus* Hudson, et al.

Tendipedidæ (Leach 1915, 1917) one of the large coleopteran family (Tenebrionidea : Polyphaga); dark brown to black, elongated, flattened, nocturnal beetles; antennæ short and 11–segmented; eyes prominent; legs stout often long and smooth with claws simple. Wings mostly vestigial (non-functional) or absent and wing cases firmly fused together. Maggots large cylindrical yellow-coloured, chitinous or leathery and distinctly segmented. Largely phytophagous and scavengers, also feed on cured seeds and stored cereal product and some even fungi (16000 species) - *Alphitobius lævigatus* (Fabricius) fee don stored seeds and dry tamarind fruits; *Dasus hoffmannseggi* Steven on grapevine, *Tribolium castaneum* (Herbest) on Temperate fruits; *Tenebrio* spp. (mealworms) are cosmopolitan, found in flour and stored products.

Tenebrionoidea Seminov 1902 superfamily of Polyphaga (Coleopptera: Coleopteroidea); nocturnal beetles, also called darkling or ground beetles, having slow awkward gait, strong offensive odour. Tenebrionidæ and other families of this are often placed under section Heteromera of Cucujoidea (Polyphaga : Coleoptera).

Tenent hairs cluster of minute tubular hair that secrete adhesive substance; present on the pulvilli or pads of certain adult insects; help in walking on smooth surface-flies etc.

Teneral incompletely hardened : said of adult insects that have just emerged form the nymph or pupa, whose cuticle has not yet hardened fully or whose colours are not yet wholly apparent.

Tenthredinidæ Leach 1819 a family of Tenthredinoidea (Symphyta: Hymenoptera) : stout, wasp-like sawflies; head wide, antennæ setiform; eyes large, ocelli three, arranged in a triangle on vertex; wings large with complete venation; abdomen distinctly eight segmented and broadly attached to thorax without abdominal pedicel. Larvæ caterpillar-like; cylindrical, naked, wrinkled, spiny or hairy and having 6 to 8 pairs of prolegs, without crochets. Phytophagous, adults feed on foliage, larvæ mine the leaves, produce galls and feed within. Over 3000 species-*Caliroa limacina* Retzius, *Pristiphora erichsoni* (Hartig), *Senoclia dioscoreæ* Rowher et al.

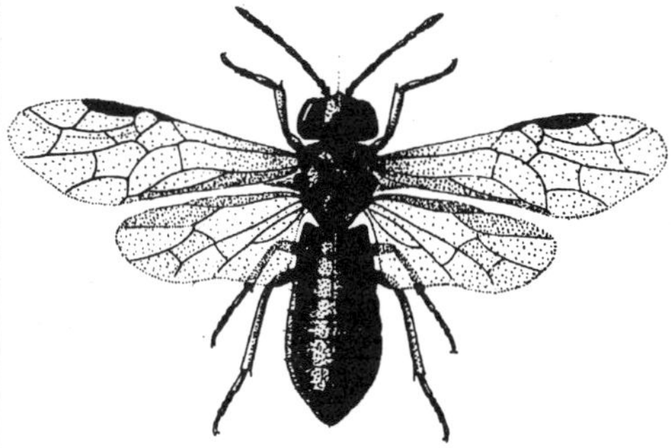

Senocli(di)a dioscoreæ Rowher

Tentorial pits four external depressions in cranial wall where arms of tentorium are invaginated.

Tentorium (pl. **tentoria**) endoskeletal structure found by two pairs of cuticular invaginations that unite within the head to compose a framework arching over the ventral nerve cord.

Tephritidæ Newman 1845 (= Trypetidæ Loew 1862) dipteran family of Agromyzoidea (Schizophora: Cyclorrhapha); head large, broad and with very small neck, antennæ decumbent, short; arista dorso-basal; lower frontal-orbital bristles present and close to eyes; vibrissæ absent and eyes large, frequently iridescent green. Wings conspicuously marbled and usually spotted or banded, subcosta turned up sharply near apex, basal and anal cells present. Male with long, flexible, coiled ædeagus and female having horny flattened oviscapt. Cosmopolitan in distribution (1600 species)–fruit–flies, *Carpomyia vesuviana* Costa *Bactrocera ciliatus* (Loew) *B. dorsalis* (Hendel), *Anstrepha ludens* (Loew) *Ceratitus capitata* (Weidemann) et al.

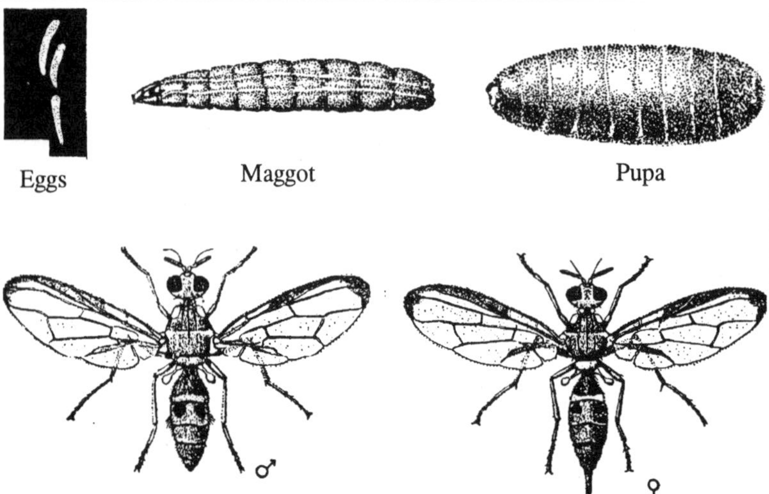

Adult flies

Bactrocera ciliatus (Loew)

TEPP (tetra–ethyl, pyro–phosphate) $C_8H_{20}O_7P_2$; contact poison, derivative of phosphoric acid; colourless, odourless, hygroscopic liquid, readily

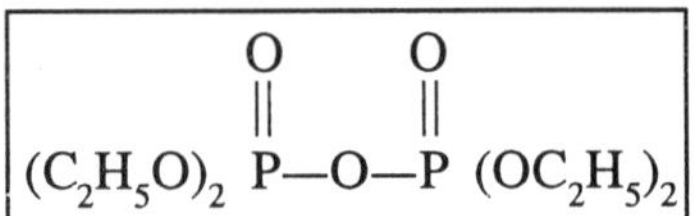

TEPP

miscible with water and most organic solvents; highly toxic to man and mammals; acute oral and dermal values for rate being 0.5 and 20 mg/kg respectively-Nifos, Tetron®, Vapotone®, Neotox®, etc.

Teragiidæ old and equivalent name for lepidopteran family, Metarbelidæ (Zygænoidea : Ditrysia), no more in usage.

Teratembiidæ Krauss (= Oligembiidæ Davis) large family of Euembioptera (Embioptera: Orthopteroidea); often considered as two separate families; embiids or oligotomids distinguished by wings having radial sector three branched with R_{2+3} bifid and R_{4+5} simple; includes small or diminutive insects mainly form Africa and the New World -species of *Diradius* Friederichs, *Oligembia* Davis, *Teratembia* Krauss et al.

Teratomyziadæ very small dipteran family of Nothyboidea (Aschiza: Cyclorrhapha); comprising of a few species, confined to Austalasia.

Terebrant (terebrate) adapted for boring: insects having an organ for boring, may be a modified ovipositor or rarely a modified proboscis.

Terebrantia Haliday 1836 one of the two suborders of Thysanoptera (Thysanopteraoidea: Paraneoptera); includes thrips having ovipositor (saw-like) and characterised by intermediate antennal segments annulate, clothed with microtrichia, forewing with atleast one longitudinal vein reaching to apex; X abdominal segement conical in female and bluntly rounded in male-Æolothripoidea, Merothripoidea and Thripoidea.

Tergite a constituent sclerite of tergum.

Tergum dorsal sclerotisation of body segment : notum.

Terminal filament slender thread-like apical prolongation of peritoneal layer of an ovariole. Filaments of ovarioles of one vary combine to form a common thread that unites with that form the ovary of opposite side to form a median ligament which is attached to body wall and helps to maintain ovaries in position.

Terminology nomenclature : set of terms used in any art of science.

Termitaphididæ family of heteropteran bugs (Aradoidea : Pentatomomorpha); apterous, flat, broadly oval body, without eyes (blind); antennæ 4-segmented, concealed below the head; rostrum 4-setmented; resemble wood-lice and are found in termite galleries-about ten tropicopolitan families.

Termitarium (pl. **termitaria**) termitary : nest, mount or abode of termite colony, may be underground or above the ground (mounds) in latter case it is completely covered with earth or mud as termites shun light. Mounds of *Nasutitermes exitiosus* (Hill) of Australia are as high as six metres and elaborately constructed with several million individuals inside the same.

Termite(s) whiteants (Isoptera : Blattopteroidea); highly social insects that live in colonies (termitaria), comprises of several castes; reproductive forms (kings and queens) have four long narrow, semi-transparent wings, superposed flat over the back when at rest and capable of being shed; soldiers and workers are apterous and sterile majority subterranean, a few mound-building; all shun light, exhibit remarkable division of labour and feed on almost anything containing cellulose, strach or sugar-obnoxious pests that destroy buildings, wooden furniture, books field crops etc. are *Microtermes* spp. and *Odontotermes* spp. commonly found on mango (*Mangifera indica* Linnæus), sugarcane (*Saccharum officinarum* Linnæus), tea [*Camellia sinensis* (Linnæus)] vegetables etc., while *Kalotermes* spp. bore into dry wood (buildings, wooden furniture, poles railway slippers etc.)

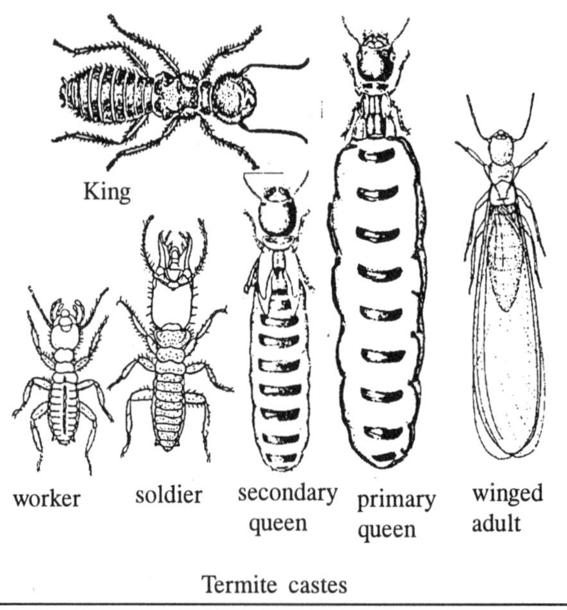

Termite castes

Termitidæ (Westwood 1840) Light 1921 largest family of subteranean termites (Isoptera: Blattopteroidea); characterised by fontanelle and ocelli present; pronotum of workers and soldiers saddle-shaped but narrow with raised median anterior lobe; wings slightly reticulate, hind-wings without anal-lobes; venation reduced anterior wing-scales short-*Microtermes obesi* Holmgren, *Odontotermes obesus* (Rambur) *Trinervitermes biformis* (Wasmann) are some polyphagous pests.

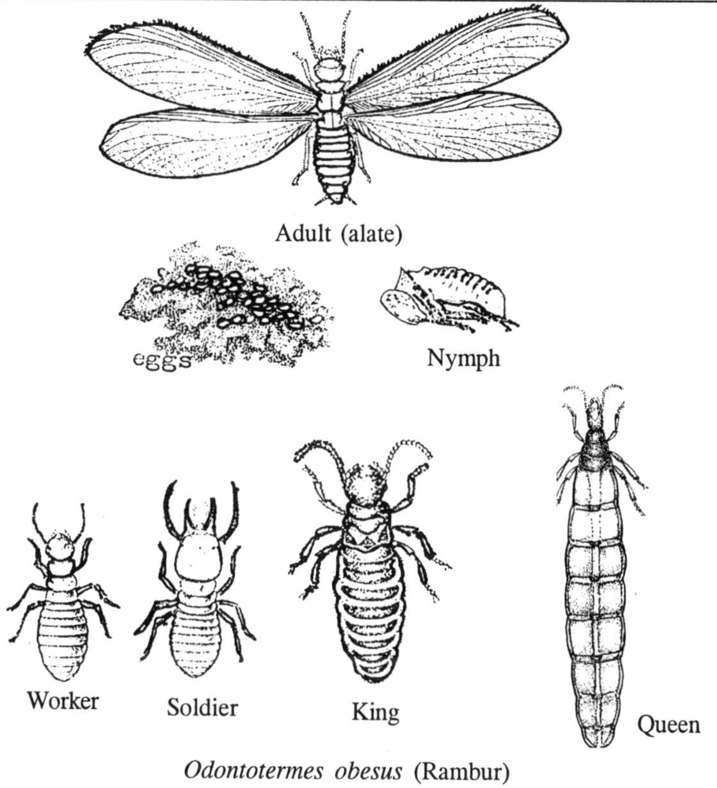

Odontotermes obesus (Rambur)

Termitomastidæ Silvestri 1901 dipteran family of Phoroidea (Aschiza: Cyclorhapha); very small nearly wingless flies, having long antennæ (20-segmented), fused eyes ocelli and reduced wing. Live as guests or inquiline in termitaria in South America-*Termitomastus* spp.

Termitophiles common name for various insects that live as inquilines in the termitaria, scavenging or sharing the food of their hosts.

Termitoxeniidæ Wasmann 1901 dipteran family of Phyoroidea (Aschiza: Cyclorrhappa); highly specialised physogastric flies having arista plumose, mouthparts free, eyes small and circular;

legs long; wings vestigial; abdomen swollen and deflexed forward over the thorax. Inhabit the termitoria and excrete fluids that are eagerly sought by termites-*Termitoxenia* spp.

Termopsidæ Homgren 1911 family of damp-wood termites (Isoptera: Polyneoptera); most primitive; ocelli lacking, fontanelle absent; pronotum flat, narrower than head; tarsi 4-segmented; hind-wings without anal lobe; anterior wing-scales large; no true workers but pseudogates-*Archotermopists wroughtoni* Desneux, *Termopsis swinhoei* Cockerell et al.

Tertiary parasite parasitoid that feeds and lives on secondary parasite.

Tessaratomidæ very small family of brilliantly coloured heteropteran bugs (Pentatomoidea: Pentatomomopha); often regarded as subfamily of Pentatomidæ. Stribulatory (sound-producing) organs in both sexes, these comprise of a file at the base of abdomen contacting comb on venter of wing base-*Tessarathoma jawanica* Thumberg (25 mm long).

Test critical trial; means of so examining; experiment for comparating or confirming the qualities or efficacy of a substance or organism: to determine the chemical nature of substance by means of reagents.

Test insect insect species against which a test is being evaluated.

Test tube long, cylindrical glass container (heat resistant), of varying length and diameter, rounded (closed) at one end and open at the other; used in labouratories for carrying out various tests or growing culture media or tissue culture.

Testis (pl. **testes**) male reproductive organs comprising of a group of a short sperm tubes that contain male germ cells.

Telanoceratidæ (= Telanoceridæ, Sciomyzidæ) a family of cyclorrhaplous dipteran marsh-flies (Acalypteratæ: Schizophora); head big, antennæ prominent; inhabit marshy places and edges of ponds, streams etc.; larvæ feed on snails. Also known as snail-flies.

Tethinidæ dipteran family of small gray-flies (Agromyzoidea: Schizophora); often regarded as synonym of Opomyzidæ; having postvertical bristles convergent; labella elongated, Sc close to R and evanescent. Male with VI, VII abdominal sternites and VII, VIII tergites lost. Female having two speromathecæ. Found near sea-shores.

Tetrabrachinæ subfamily of Coccinellidæ (Cucujoidea: Polyphaga); distinguished by mandibles being carnivorous type and tarsi evidently 4-segmented.

Tetrachlorvinphos $C_{10}H_9O_4Cl_4P$; contact insecticide with slight systemic action; white crystalline solid, slightly soluble in water, acetone and xylene but readily soluble in chloroform and methylchloride. Acute oral LD_{50} value for rat is 4000 to 5000 mg/kg and dermal for rabbit is > 25000 mg/kg - Gardona® etc.

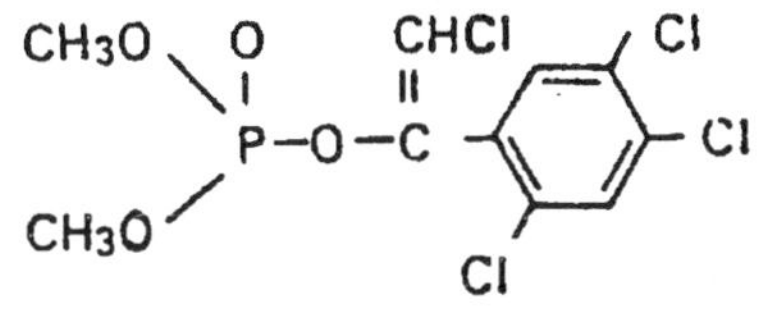

Tetrachlorvinphos

Tetradifon $C_{12}H_{16}Cl_4O_2S$; organic sulphur compound, used as acaricide; insoluble in water but stable in acid and alkali media, light and temperature; having very long residual effect and non-toxic to bees. Acute oral LD_{50} value fore rat 5000 to 14700 mg/kg and dermal for rabbit 10000 mg/kg–Tedion® etc.

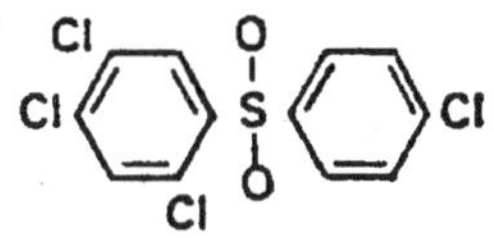

Tetradifon

Tetrapterous having four wings.

Tetratomidæ family of Cucujoidea (Polyphaga : Coleoptera), often regarded as subfamily of Melandryidæ (Heteromera : Cucujoidea) and differentiated by tibial spurs being simple.

Tetrigidæ (Tettigidæ Walker 1870) family of grouse–locusts or pigmy–locusts (Acridoidea: Orthoptera); characterised by backwardly directed process of pronotum which covers the abdomen and conceals hind–wings; tegmina reduced to small scales; fore–and mid tarsi 2–segmented and no arolia. About 100 species < 15mm long, extremely active, feeding on algæ, fungi, lichens, mosses, grasses and decaying vegetable matter–*Acrydium dpressum* (Brisout), *A. arnatum* Say, *Telmatettix aztecus* (Saurssure), *Paratettix meridionalis* (Rambur), *Tettigidea lateralis* (Say), *Tetric undulata* (Swrb.) et al.

Tettigractidæ family of pilose cicadas (Cicadoidea : Auchemorryhyncha); having complete terminal ventation; tymbal present in both sexes; tympanal auditory organs absent; empodia present–*Tetigarcta tomentosa* White from Tasmania and *T. crinita* Distant from Victoria (South–east Australia) are only living species. Also represented by fossils of Mesozoic era and Tertiary period form Northern hemisphere.

Tettigometridæ family of Fulgoroidea (Auchenorrhyncha : Homoptera); differs from other fulgorids in having segmented antennal flagellum and the ocelli lying inside the lateral carinæ of face; forewings opaque with reduced venation; ovipositor very short or absent (120 species).

Tettigonidæ Uhler 1876 old and obsolete term for homopteran family Cicadellidæ (Latreille1802).

Tettigoniidæ Karny 1903 (Locustidæ Burmeister 1838 nec. Kirby 1910) family of Tetigonioidea (Ensifera : Orthoptera); long-horned or angular-winged grasshoppers and katydids; body somewhat compressed, tarsi 4-segmented, auditory organs on fore-tibia; winged and apterous forms; ovipositor conspicuous and sickle-or sword-shaped; mostly herbivorous, a few predaceous as well (7500 species) - *Holochlora indica* Kirby, *Lea floridensis* (Beutenmuller), *Scudderia furcata* Brunner, *Sathroyhyllia* spp., et al.

Tettigoniodea Karny 1903 : suborder of Orthoptera in old system of classification. The families of this suborder are now placed under various superfamilies of suborder Ensifera (Orthoptera, Orthopteraoidea), while this term is no more in usage.

Tettigonioidea superfamily of Ensifera (Orthoptera : Orthopteroidea); having tarsi 4-segmented; foretibia with tympanel organ and forewings with stridulatory apparatus. Ovipositor well developed with three pairs of valves-Prophalangopsidæ (Haglidæ) and Tettigoniidæ.

Thamyrididæ old and equivalent name for lepidopteran family Hesperiidæ Leach 1815, 1819 that includes skippers, no more in usage.

Thanatosis feigning death : going into a rigid, motionless position when disturbed; though common among beetles, it also occurs in some earwigs, water-bugs and many other insects.

Thaumaleidæ small family of Culicoidea (Nematocera : Diptera); small sluggish, stream-midges flies having antennæ apparently 2-segmented, terminated by 10 or 11-segmented style-like appendage; palpi longer than antennæ; eyes holoptic in both sexes; wings with strong veins without network of folds or creases and venation not reduced. Larvæ aquatic and amhipneustic-*Thaumalea testacea* et al.

Thaumastocoridæ (= Thaumastotheriidæ) small heteropteran family of Lygæoidea (Pentatomorpha : Geocorisæ); small phytophagous bugs with broad head, ocelli present, rostrum short 3-segmented; thoracic scent-glands lacking; coxæ rotatory, widely separated, tibia with membranous apical appendage-Royal palm bug, *Xylatodoris luteolus* Butler et al.

Thaumastotheriidæ old and equivalent name for heteropteran family Thaumastocoridæ (Lygæoidea: Pentatomomorpha).

Thaumatoxenidæ Breddin and Borner 1904 small dipteran family of Phoroidea (Aschiza : Cyclorrhapha); oval, dorsally depressed wingless insects, having little resemblance to flies; head large extending back over thorax; antennæ resting in grooves and with apical arista; eyes small and circular legs long; abdomen not visibly segmented. Found in termitaria in Mediterranean basin–*Thaumatoxenia* species.

Thaumetopœidæ (Eupterotidæ) very small palæarctic family of Noctuoidea (Ditrysia : Lepidoptera); often treated as subfamily of Notodontidæ (Noctuoidea: Ditrysia), large dull-coloured moths; proboscis aborted; frenulum present, cubitus of forewing 2–or 3-branched. Caterpillars tufted with long hair and numerous secondary satæ pupate in silken cocoons–*Thaumetopœa processionea* Linnæus, *T. pinivora* Tr. and allied species.

Thelaxidæ small family of Aphidoidea (Strenorrhyncha: Homoptera); includes species of *Anoecia, Hormaphis, Mindarus* et al.

Thelytoky mode of reproduction wherein unmated females produce (parthenogenetically) only female progeny.

Therevidæ Wetswood 1840 family of stilleto flies (Asiloidea : Brachycera); pilose or bristly having antennæ 3–segmetned, apical segment much enlarged and terminating in a short, slender, simple or segmented style; proboscis large with fleshy labells and 2-segmented palpi; eyes large, not protruding ocelli present wings with posterior cells the fourth often closed, costa continuing around the wing and the fourth vein ending behind apex of the wing (500 species–*Tabula* spp., *Thereva* spp., are common Europe and North America.

Thigmotaxis a locomotory response : an inhibition of movement, caused by stimulus of touch. Due to this many insects as also other small organisms cling to the surfaces with which they come in contact.

Thiodemeton (disulfoton) $C_8H_{19}O_2PS_3$; systemic insecticide and acaricide; colourless oily liquid with characteristic odour, slightly soluble in water but readily

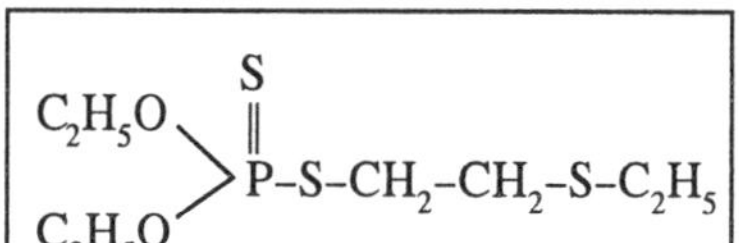

Thiodemton

soulble in most organic solvents (acetone, alcohol, either etc.) Acute oral and dermal LD_{50} values for rate being 4 and 50 gm/kg respectively-Disyston®, Solvirex®, etc.

Thiometon $C_6H_{15}O_2PS_3$; systemic insecticide and acaricide; effective against sucking insects and mites. Acute oral and dermal values for rate being 100 and 200 mg/kg respectively-Ekatin® etc.

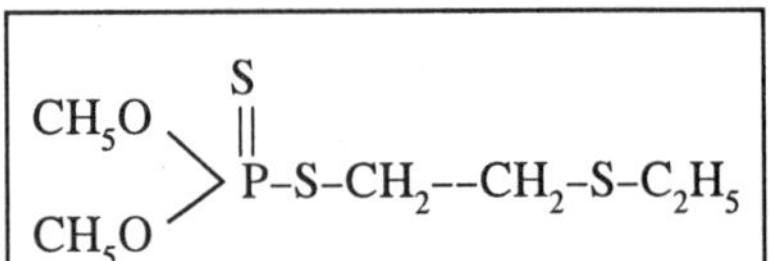

Thioemton

Third axillary posterior hing plate of wing base on which flexor muscle is inserted.

Thoracic pertaining to thorax.

Thoracic glands glands situated in the thoracic region of an insect which when stimulated by other hormones from the brain secrete the moulting hormone.

Thorax second region of insect body between head and abdomen comprising of three segments-prothorax, mesothorax and metathorax-each bearing a pair of legs (without exception) while wings, when present, are borne on meso-and metathorax.

Thorictidæ small family of Dermestoidea (Polyphaga: Coleoptera) often considered as tribe or subfamily of Dermestidæ; head usually reflexible on prostermum; fore-coxæ projecting and hind-coxæ ovate and widely separated, front margin of pronotum as wide as hind margin. Mostly myrmecophiles (80 species)-*Thorictodes* spp., et al.

Thorn insect horned tree-hopper; small brown plant bug (Memberacidæ: Auchenorrhyncha); having thorn-like projection on its thorax; found feeding mainly on thorn bushes which also given them certain degree of protective resemblance-*Centrotus cornutus* (Linnæus) et al.

Thripidæ Stephens 1829 largest and most injurious family of thrips (Thripoidea : Terebrantia); antennæ 6 to 9-segmented, fourth being largest; style 1 to 3-segmented; alate and apterous forms; forewing more pointed, veins distinct with seatæ on veins; last abdominal segment of female conical and poorly sclerotised; ovipositor well developed and curved downwards. Sap-feeding, causing serious crop losses. Over 200 species.

Thripinæ subfamily of Thripidæ (Terebrantia : Thyanoptera); includes thrips distinguished by antennæ 6-segmented, segment 5 and 6 being short and thick, dorsum of body not polygonally reticulate, almost with transverse striæ–species of *Sciothrips, Scirthothrips* Shull, *Thrips* Linnæus etc.

Thripoidea Hood 1915 superfamily of Terebrantia (Thysanoptera : Thysanopteoidea); having antennæ 6 to 10–segmented, the last one to three forming a thin style; wings narrow and pointed; ovipositor curved downwards–Heterothripidæ, Thripidæ, Uzelothripidæ etc.

Thripoides Burmeister 1829 an earlier and outdated name for insect order Thysanoptera Haliday 1836; no more in usage.

Thrips (erroneously **thrip**; pl. **thrips**) al insects belonging to order Thysanoptera; head hypognathous produced below into a mouth-cone or rostrum; three ocelli in winged species one or two in brachyperous and nil in apterous ones; prothorax well developed and distinctly separate from head and distinctly separate from head and mesothorax; wings, when present, have well–developed fringes and reduced venation; XI abdominal segment rudimentary; ovipositor present (Terebrantia) or absent (Tubulifera). Mostly sap-feeding, a few predaceous, some act as vector of certain bacterial, fungal and viral diseases of plants :–

1. *Retithrips*

2. *Selenothrips rubrocintus*

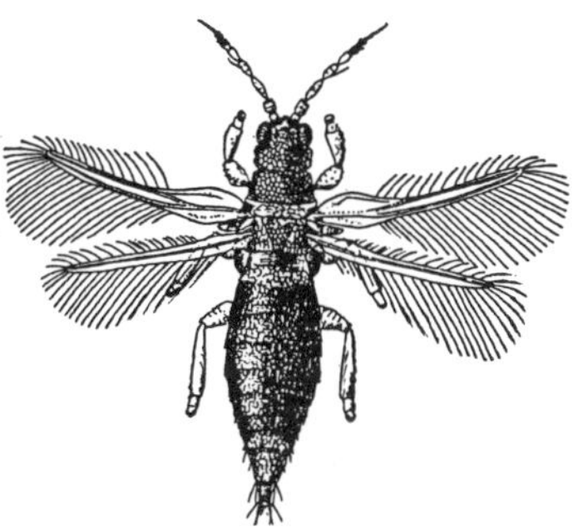

4. *Rhipiphorothrips cruentatus*

5. *Leeuwania ramarkrishnæ*

5. *Leeuwania ramarkrishnæ*

6. *Neoheegeria mangiferæ*

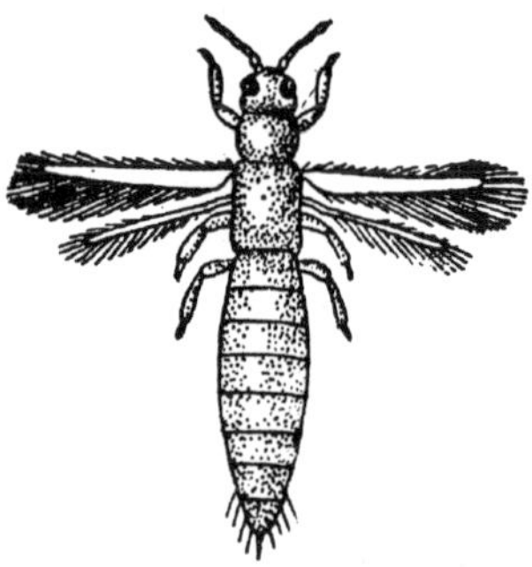

7. *Thrips tabaci*

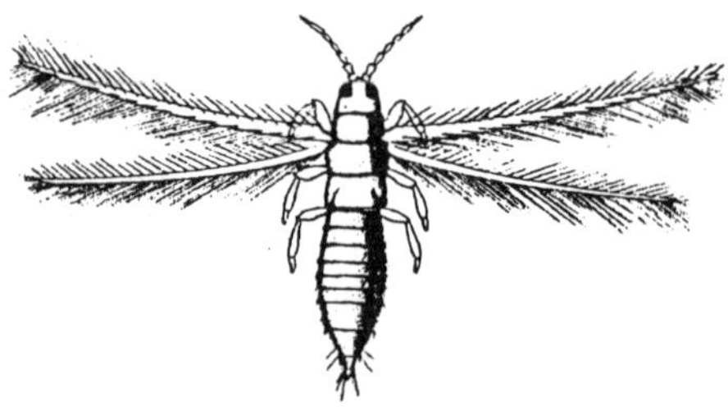

8. *Chloethrips oryzae*

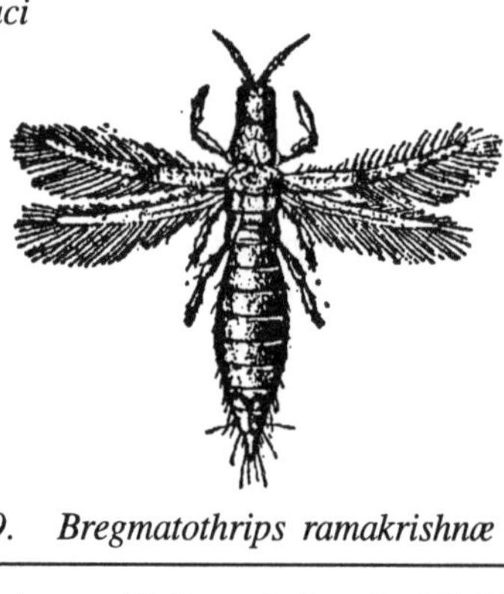

9. *Bregmatothrips ramakrishnæ*

Banana leaf thrips - *Helionothrips kaddiphilus* (Ramakrishna & Margabanhu)
Banana rust thrips - *Chætanaphotrips signipennis* (Bagnall)
Cardamom thrips - *Sciothrips cardamomi* (Ramakrishna)
Castor thrips - *Retithrips syriacus* (Mayet)
Chilli thrips - *Scirtothrips dorsalis* (Hood)
Citrus thrips - *Scirtothrips citri* (Moulton)

Cocoa thrips - *Selenothrips rubrocinctus* (Giard)
Coffee thrips - *Scirtothrips bispinsous* (Bagnall)
Composite thrips - *Microcephalothrips abdominalis* (Crawford)
Corn thrips - *Limothrips cerealium* Haliday
Croton (greenhouse) thrips - *Heliothrips hæmorrhoicdalis* (Bouche)
European thrips - *Thrips physapus* Linnæus
Flower thrips - *Frankliniella damfi* Priesner
Thrips hawaiiensis Morgan
Gladiolus thrips - *Tæniothrips simplex* (Morison) (= *gladioli* Moulton)
Grapevine thrips - *Rhipiphorothrips cruentatus* Hood
Groundnut thrips - *Caliothrips indicus* (Bagnall)
Jamun thrips - *Leeuwenia ramakrishnæ* Ananthalkrishnan
Jasmine thrips - *Isothrips orientalis* (Bagnall)
Lily thrips - *Liothrips vaneeckei* (Prisener)
Lithchi thrips - *Dolichothrips indicus* (Hood)
Mango thrips - *Neoheegeria mangiferæ* (Prisener)
Mimusops gall thrips *Arrhenothrips ramkrishnæ* (Hood)
Olive thrips - *Liothrips oleæ* (Costa)
Onion (tobacco) thrips-*Thrips tabaci* Lindemans
Paddy (rice) thrips - *Chloethrips oryzæ* Williams
Pear thrips - *Tænithrips inconseqwuens* (Uzel)
Propper gall thrips - *Sorghothrips jonnaphilus* (Bagnall)
Sorghum thrips - *Sorghothrips jonnaphilus* (Ramakrishna)
South American thrips - *Uzelothrips scabrosis*
Sugarcane thrips - *Bregmatothrips ramarkrishnæ* Bagnall
Tamarind thrips - *Haplothrips ceylonicus* Schmutz
Turmeric thrips - *Pauchætothrips indicus* Bagnall
Wheat thrips - *Anaphothrips sudaensis* Trybon

Thripsida Leach 1815 family under insect order Omoptera, tribe Aphides in a very old system of classification; it included all the thrips, now placed under order Thysanoptera; no more in usage.

Thripsites Newman 1835 old and obsolete term for insect order Thysanoptera Haliday 1936 that included all thrips.

Throscide old and equivalent name for Coelopteran family Trixagidæ (Elateroidea: Polyphaga) that included pseudo-click beetles.

Thryididæ small tropicopolitan family of small moths (Pyralidoidea: Ditrysia); probably ancestral group from which the butterflies have been derived. White or yellowish translucent areas on wings; vein Cu_2 in hind-wings rudimentary or absent. Caterpillars burrow in twings and stems causing gall-like swellings (120 species) - *Rhodoneura* spp. *Thyris* spp. et al.

Thynnidæ Erickson 1842 very small family of Australian flower wasps (Vespoidea: Aprocrita); social insects, mostly black in colour but often metallic and spotted with yellow orange and red; females have well developed and effective sting; predaceous on grubs etc., nests underground-*Diamma bicolor* Westwood (blue ant), *Thynnoturneria cerceroides* Smith et al.

Thyreocoridæ small family of negro bugs. often included in family Cydnidæ (Pentatomoidea: Pentatomomorpha); small black plant-bugs found in large numbers on flowers during Spring season in North America-*Thyrecocris* species.

Thyerophoridae very small dipteran family (Agromyzoidea: Schizophora); rare flies, distinguished by antennæ lying in deep grooves; found in dry carcases.

Thyreopterinæ subfamily of Carabidæ (Caraboidea : Adephaga) that includes arboreal beetles distinguished by lateral borders of elytra produced into broad leaf-like expansions-*Mormolyce* spp. of Malaysia.

Thyridium a semi-transparent whitish spot, devoid of hairs, found on both pairs of wings of caddisflies (Trichoptera) and wanting in Lepidoptea.

Thyrsophoridæ Enderlein 1903 very family of Psocomorpha (Psocoptera: Psocopteroidea); booklice characterised by having 3rd and 4th antennal segments larger and more thickly pubescent than others; tarsi 2-segmented-About 20 Neotropical species.

Thysanidæ very small family of Chacidoidea (Apocrita : Hymenoptera); very curious small insects having bristly tibiæ; parasitic on Cocoidea - *Thysanus* (*Signiphora*) species et al.

Thysanoptera Haliday 1936 (Physpida Leach 1817; Physapoda Burmeister 1838; Physopoda Comstock 1888) order of Thysanopteroidea (Paraneoptea: Pterygota); very small 0.5 to 1.0 mm long. slender-bodied, fragile insects having 6- to 10-segmented antennæ, asymmetrical resping and lapping mouthparts with maxillary and palpi. Alate and apterous forms, wings very narrow, heavily fringed and with reduced venation; cerci absent. Metamorphosis incomplete but accompanied by one or two inactive pupa-like instars. Mostly sap-feeders, a few predaceous. (About 600 species placed under two suborders - Terebrantia and Tubulifera.

Thysanura Latreille 1796 order of Ectoropha (Apterygota); primitive, small three-pronged bristletails, having elongated flattened body, with or without scales, mouthparts ectognathous, adapted for biting and chewing; long many jointed antennæ; compound eyes well developed; vestigial or absent; ocelli present or absent; coxæ small; tarsi 2 or 3-setmented with two or three claws; cerci and medium caudal filament long and many-segmented. Metamorphosis slight with post-adult ecydses. About 800 species grouped under two suborders-Microcorphia and Zygentoma.

Tibia (pl. **tibiæ**) fourth segment of insect's leg, usually long and slender and often provided with downward projecting spines.

Tineidæ (Stephensn 1829) Wocke 1891 a world-wide family of Trineoidea (Ditrysia: Lepidoptera); characterised by head being rough-haired, proboscis short or absent; maxillary palpi often long 5- or 6-segmented and usually folded, labial palpi porrected; posterior tibia hairy; wings narrow, veins free. Caterpillars mostly saprophagous, feed upon dried animal and vegetable matter and fungi. Over 2500 species-Species of *Setomorpha* and *Demobrotis* attack tobacco; *Tinea vastella* feeds on dried fruits and horns of antelops whereas *T. pellionella* (Linnæus), *Tineola bisselliella* Hummel and *Trichophaga tapettiella* Linnæus are household pests attacking clothes, carpets, furs etc.

Tineodidæ a very small family of Pyral(id)oidea (Ditrysia : Lepidoptera); related to Thyrididæ; includes a few species of Australian and Indian moths characterised by presence of maxillary palpi and having relatively long legs and narrow wings; forewings with one anal vein but hind wing with two anal veins and Cu_2 vestigial or absent.

Tineodidæ small Indo-Australian family of Pyrlidoidea (Ditrysia : Lepidoptera)

Tineoidea Dyar 1902 a superfamily of Ditrysia (Lepidoptera), moths having head usually with rough scales, proboscis scaled with present; haustellum not scales; maxillary palpi folded, labial palpi either bristled or with first segment enlarged; legs moderately long; wings entire, having soft scales and usually fringed, vein M often present in the discal cell. Caterpillars phytophagous and coprophagous-Tineidæ, Psychidæ, Gracillariidæ, Phyllocnistidæ etc.

Tingidæ (Laporte 1832) the only family of Tingoidea (Cimicomorpha: Geocorisæ); it includes lacewing bugs having body strongly flattened; antennæ 4-segmented, ocelli absent; hemelytron with reticulations enclosing small membranous area; pronotum with an angular process extending over scutellum and often with an anterior hood covering the head; tarsi 2-setmented and arolia present. About 800 species-*Leptostyla oblonga* (Say), *Stephanitis pyri* Fabricius, *S. Pyrioides* Scott, *S.typicus, Teleonemia lantanæ* Distant, et al.

Tiphiidæ (Leach 1815) Thunberg 1870 hymenopteran family of Vespoidea (Aculeata : Apocrita); small (7 to 12 mm long), slender wasps clothed with dark or pale hairs; head roundish, as wide as thorax with a slender neck; eyes circular, ocelli very small; antennæ short, straight in males and curled in females; wings narrow, transparent or infuscated; abdomen long with short petiole. Mostly parasitic especially on white-grubs through the adults of some species frequent flowers-*Methoca ichneumonides* Latreille is ectoparasite on grubs of tiger beetles while *Tiphia popilliavora* Rohwer parasitises grubs of Japanese beetles.

Tipulidæ Leach 1819 family of crane-flies (Nematocera : Diptera); black brown, orange or gray in colour with spotted or clouded wings, a few apterous also; head prominent, ocelli absent; antennæ pectinate or serrate (not plumose); mesonotum with distal or obscure V suture; tip of abdomen enlarged to form hypopygium bearing coxites, styles and ædeagus; female with fleshy or sclerotised ovipositor-*Conosia irrorata* Wiedmann (India), *Longurio dux* Hudson (Australia), *Tipula simplex* Doane (USA) et al.

Tissue groups of similar cells organised to carry out one or more particular functions.

Titillator terminal process of insect's ædeagus.

Tolerance ability of plant to withstand infestation and to support insect population that would severely damage susceptible plants.

Tolerance limit maximum amount of toxicant permitted in food stuffs for human consumption; expressed as parts per million (ppm).

Tongue term loosely used for the proboscis of an insect particularly when it is lengthened for sucking nectar or blood; actually it should refer only to glossæ and paraglossæ.

Tongueworm(s) acorn-worms; free living solitary animalcules (Enteropneusta: Hemichordata); having no exoskeleton; usually burrow in sand or mud.

Tonofibrillæ non-contractile fine cuticular fibrils that connect muscle fibres with inner surface of cuticula.

Tonus condition of the muscle that remains continuously in a state of partial contraction to enable the insect to maintain particular posture for hours together.

Topochemical sense perception of scents along a path or tract enabling ants and other insects to follow that path perviously used by them and to return by the sam.

Topotaxis any reaction vide which an insect orientaes iteself and moves in a definte direction in response to a stimulus.

Tormogen epidermal cell that forms setal membrane or socket of a setta or bristle on the body of an insect.

Tornus anal angle of insect's wing; distinct notch on the hind margin of the wing, dividing it into two parts.

Tortricidæ Stephens 1829 (including Olethreutidæ Walsingham 1900 or Eucosmidæ Durrant 1918) family of leaf roller (Trotricoidea : Ditrysla); forewings with R_3 and R_4 separate, rarely stalked, hindwing with M_1 and M_2 absent. (4000 species) — *Cydia pomonella* (Linnæus), *Homona coffearia* Nietner.

Tortricoidea McDunnough 1939 superfamily of Ditrysia (Lepidoptera: Oligoneoptera), maxillary palpi vestigial or absent, labial palpi with 2nd segment more or less rough-scaled, Cu_2 present in both pairs of wings, often vestigial in forewings — Tortricidæ, Olethreutidæ (Eucosmidæ), Cochylidæ (Phaloniidæ).

Torus thickened part or pedical on which an organs or appendages is borne — base of antennæ.

Torymidæ Walker 1833 (= Callimomidæ Viereck 1916) a large family of Chalcidoidea (Parasitica: Apocrita); minute (2 to 4 mm), brilliantly metallic-blue or -green coloured and seed-infesting chalcids; mandibles stout with 3- 4 teeth; thorax not gibbous, nataulices present. hind coxæ 5-6 times larger than the frontal one; tarsi 5-segmented; ovipositor long and exserted Mostly parasities of gallicolous insect, a few seed-infesting (phytophagous) — *Syntomaspis druparum* Boheman

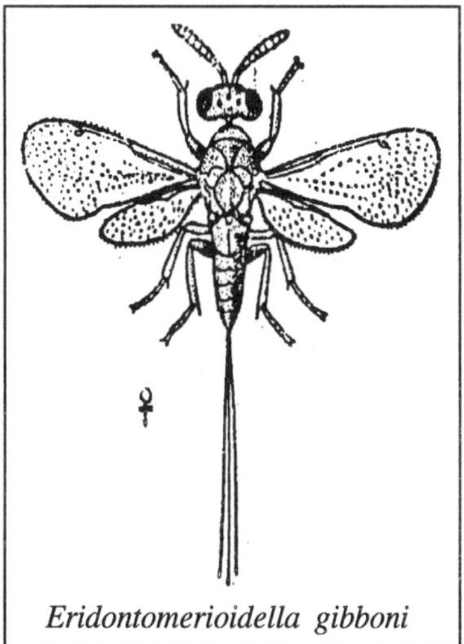

Eridontomerioidella gibboni

infests apple seeds in North America and Europe; species of *Megastigmus* are phytophagous as well as parasitic on eggs of preying mantids as also the dipteran and hymenopteran gall-makers. In India, *Eridontomerioidella gibboni* parasitises grubs of *Bruchophagus mellipes G.* (seed-wasp).

Touch sense of feeling: tactile sense.

Toxaphene $C_{10}H_{10}Cl_8$; camphechlor or chlorinated camphene containing 68 to 70 percent chlorine; amber coloured waxy solid; stomach poison, introduced by M. LeRoy Parker and J.R Beacher in 1948. Insoluble in water but soluble readily in organic solvents (benzene, ether, xylene etc.), unstable in presence of alkali; non-toxic to bees but toxic cucurbitaceous plants. Acute oral and dermal values for rat are 283 and 1000 mg/kg — Drexel®, Toxaphene® etc.

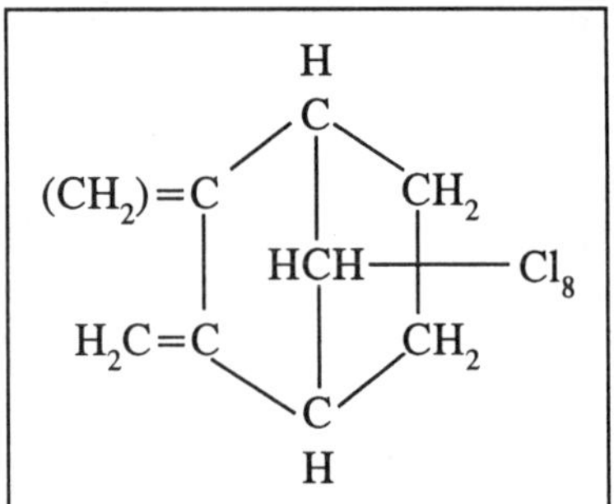

Toxaphene

Toxic poisonous: potentially fatal.

Toxicant poison: chemical exhibiting toxicity.

Toxicity ability of substance (toxin) to cause damage to living tissues, when inhaled, absorbed or indigested.

Toxicity, Dermal measure of toxicity of a pesticide through the skin; usually expressed as milligrams per kilogram (mg/kg) of body weight.

Toxicity, Inhalation poisonous effect of a pesticide breathed (into the lungs) while working.

Toxicity, Oral toxicity of chemical (pesticide) when ingested through the mouth; expressed as milligrams of pesticide per kilograms of body weight (mg/kg).

Tracheal pertaining to trachea or windpipe.

Toxicologist person having sound knowledge of toxicology.

Toxicology branch of medical science dealing with nature and effect of poisons; qualitative and quantitative study of injurious effects of chemical.

Toxin specific poisonous substance of animals or animals or plants origin: metabolic product of an individual that is poisonous to another individual.

Trachea (pl. tracheæ) windpipe; tube/tubule conveying air into the body for respiration; found in arthropods (including insects) and terrestrial vertebrates.

Tracheal pertaining to trachea or windpipe.

Tracheal gills an exclusive network of fine tracheal beneath the integument as found in the immature stages of many aquatic insect specially those of Ephemeroptera, Trichoptera and Plecoptera.

Tracheal system part of breathing or respiratory system comprising of tracheæ and tracheoles.

Tracheoles microscopically minute, capillary-like tubelets of uniform dimensions which arise independently of and anastomose with tips of tracheæ, to aid in transmission of oxygen to all organs of the body.

Translaminar pesticide that passes through from one leaf lamina to other through leaf tissues.

Translocation uptake of pesticide into a part of plant and its dispersal to other parts of the plants.

Transverse broader than long: set crosswise.

Trap-crop crop or plant grown in between the rows or periphery of another crop with a view to attracting the insect pests and when infested, either sprayed profusedly (heavily) or destroyed along with the insects thereon.

Triatominæ subfamily of kissing bugs (Reduviidæ: Geocorisæ); carnivorous insects characterised by long slender rostrum; suck blood of birds and mammals — *Triatoma infestans* Klug, *T. megista* (Burmeister), *T. protracta T. rubrofasciata* de Geer, *T. rubrovaria* Blanchard, *T. uhleri* Nieva et al.

Tribe taxonomic rank which is further division of subfamily comprising of group of related genera (name end in -ini).

Trichiidæ family of brush beetles (Scarabæoidea: Polyphaga); often considered as a subfamily (Trichiinæ); chaffer beetle having a coat of fine hairs on elytra and long shaggy hairs on rest of the body — *Trichius fasiatus* (banded brush beetle) is pest of wild roses in Scotland.

Trichlorphon $C_4H_8Cl_3O_4P$; contact and stomach poison with slight fumigant action; white crystalline powder, soluble in water and low alcohol, ketone, aromatic hydrocarbons, chlorinated hydrocarbons etc. Acute oral and dermal LD_{50} value for rat are 650 and 2800 mg/kg respectively — Diptrex®, Neguvon®, Tugon® etc.

Trichoceridæ (Trichoceratidæ) a family of winter gnats (Tipuloidea: Nematocera); small crane-flies, akin to those of Tipulidæ but having ocelli and legs not deciduous. Adults flies commonly dance in swarms during cold winter. Found feeding on humus.

Trichogen epidermal cell that forms seta.

Trichogrammatidæ (Forster 1856) Walker 1871 family of Chalcidoidea (Apocrita: Hymenoptera); smallest insects 0.3 mm and 1.0 mm long, having tarsi 3-segmented wings broad and fringed, pubescence arranged linearly, marginal and stigmal veins united; ovipositor short and ventro-apical. About 250 species, mostly egg parasite — *Trichogramma evanescen* Westwood is universal egg-parasite extensively used for biological control of sugarcane borers and codling moth. *Prestwichia aquatica* Lubbock swims under water and parasitises eggs of aquatic beetle and bugs in Europe, while *Austromicron zygopterorum* Tillayard dives for eggs of damselflies in Australia.

Trichogramma evanescens Westwood

Trichopore cylindrical internal cavity of cuticular opening beneath the base of seta.

Trichoptera Kirby 1813 order of small to medium-sized moth-like caddisflies (Mecopteroidea: Oligoneoptera); soft-bodied with setaceous multi-articulate antennæ; mouthparts rudimentary (licking type); mandibles vestigial or absent maxillæ single lobed with elongate palpi, wigs membranous, densely hairy and held roof-like over dorsum when at rest, tarsi pentamerous. Metamorphosis complete; larvæ aquatic with tracheate gills on sternites; found usually in vicinity of water (6000 species), two suborder — Annulipalpia and Integripalpia.

Trichopterygidæ family of world's smallest beetle (Coleoptera), 0.25 to 1.00 mm long, fringed wings; free-living and fungus-feeders.

Trichoscelidæ dipteran family of Acalyptratæ (Schizophora: Cyclorrhapha).

Tricorythidæ family of Leptophlebioidea (Ephemeroptera: Ephemeroidea), mayflies having fore-wings with very few cross-veins, hind-wings absent. Nymphs often with concealed gills.

Trictenotomidæ family of large prionid-like beetles (Cucujoidea: Polyphaga); antennæ long, last three segments differentiated; mid-coxal cavities closed by the sterna, fore-coxæ strongly transverse. About 12 species found in forests of the Oriental region.

Tridactylidæ Brunner 1882 family of small orthopteran insect (Tridacyloidea: Cælifera); that live near water; burrowing into sandy ground; antennæ short; hind-legs saltatorial with very large femora and tibia bearing a pair of long apical appendages — sand cricket (*Tridactylus apicalis* Say), pigmy mole-crickets (*Ellipes minula* Scudder).

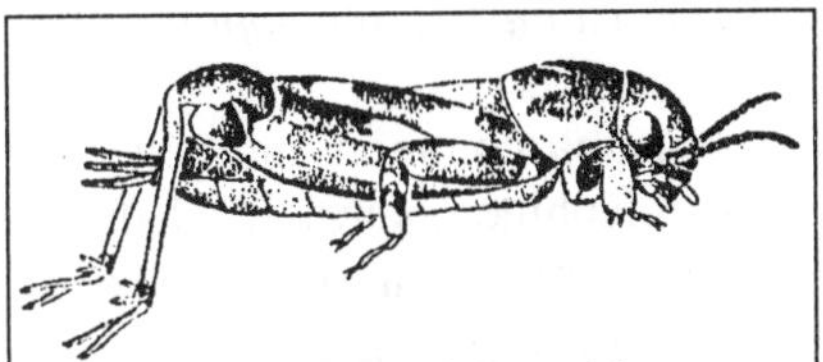
Ellipes minutà Socudder

Tridactyloidea superfamily of Cælifera (Orthoptera: Orthoteoidea); adapted for burrowing and leading subterranean life, antennæ shorter than body with 12 or less segments; tarsi 1- or 2-segmented; forelegs normal, hind-legs saltatorial; arolium wanting and no ovipositor — Cylindrachetidæ, Tridactylidæ.

Tridymidæ hymenopteran family of Chalcidoidea (Parasitica: Apocrita); comprising of small wasps having antennæ 12-segmented with funicle 5-segmented; pronotum scarsely visible in the abdomen almost sessile.

Poecilogonalos kerala

Trigonalidæ small family of rare but widely distributed parasitic wasps (Trigonaloidea: Apocrita); forewings with two or three closed sub-marginal cells; costal cell present; venter sclerotished; gaster attached to propodeum, not petiolate, ovipositor vestigial. Prey mostly upon immature stages of other insect — *Ischnogonatos dubia, Lycogaster rufiventris, Poecilogonalos kerala, Pseudogonatos narmadni* et al.

Trigonaloidea monofamilial superfamily of Apocrita (Hymenoptera: Hymenopteroidea); head large, mandible with four strong teeth, antennæ having more than 20 segments; hind femur with trochantellus; forewings with pterositgma, Rs two branched, hind-wing with enclosed cells — Trigonallidæ.

Trilobita fossil (extinct) marine arthropods that flourished throughout Paleozoic ear and were most abundant during Cambian and Ordovician periods. Body oval and flattened, comprising of head, thorax and pygidium; lengthwise apparent triple division into an elevated median area (rhachis) and two depressed lateral areas (Pleuræ).

Trimenoponidæ Harrison 1917: small family of Amblycera (Mallophaga: Psocopteroidea); lice having antennæ clavate and 4-segmented; prothorax and mesothorax similar in size and shape but inverted; tarsi with two claws — species of *Cummingsia, Harrisonia, Trimenopon* infest rodents in Central and South America.

Triphlosoba common and generic name of fossil insect, *T. putchella* (Protephmeroptera) lived during Carboniferous period in France and looked intermediate between mayfly (Empheroptera) and dragonfly (Odonata).

Triquetra Brazilian treehopper (Membracidæ: Cicadoidea) with dorsal thoracic shield covering the entire body and having three horn-like processes on its anterior side.

Tritocerebrum hind brain: third part of insect's brain formed of ganglia of postoral second antennal somite.

Tritogenaphis bright red aphid (Aphioidea) from which a vermilion anthocyanin pigment can be extracted.

Triungulin (triunguloid) primary larva: minute, active, spiny, first instar larva of certain beetle (Meloidæ: Coleoptera) and stylops (Sterpsiptera).

Trixagidæ (= Throscidæ) Coleopteran family of minute black or brown beetle (Elateroidea: Polyphaga); differ from Elateridæ in not having two tibial spurs. Prothorax firmly attached to mesothorax and not movable; frontal coxal cavities closed behind by mesosternum, presternal process prolonged but not movable; all abdominal segment free and movable. Grubs live in decaying vegetable matter. (200 species) — pseudo-click beetles.

Trochanter small second segment of insect's leg between coxa and femur; it may not be visible from all sides of the leg.

Trochantin small basal articular sclerite on trochanter of an insect's leg; any small sclerite interposed between two other sclerites.

Troctidæ Emderiein 1903 old and equivalent name for psocopteran family Liposcelidæ Enderiein 1911 (Troctomomorpha: Psocoptera).

Troctomorpha suborder of Psocopter (book lice) hemipteroid insects having antennæ 11- to 17-segmented; flagellum secondary annulated, labial palpi 1- to 2-segmented; winged forms with pterostigma not thickened; tarsi 3-segmented divided into four small families — Amphientomidæ, Lipsoscelidæ, Pachytropctidæ and Sphæropsocidæ.

Trogidæ Macleay 1819 small family of oblong, rugose, dusky-looking, scaly or setose, brownish beetle (Scarabæoidea: Polyphaga); body roughly sculptured, antennal club with 3 closely co-adapted lamellate; empodium small or absent; elytra often tuberculate. Found in dry place, feeding on carrion or dung (200 species — skin beetles, *Cryptogenius* spp., *Glaresis* spp., *Trox hispidus* Pontoppidon and allied species).

Trogiidæ Enderlein 1911 (= Atropidæ Kolbe 1884) very small family of Trogiomorpha (Psocoptera: Psocopteroidea); often referred as Atropidæ Kolbe; head and transverse; inner side of maxillary palpi with peg-like sense organs; claws not toothed; apterous or brachypterous (20 species) — *Lepinotus inquilines* Hyden, *L. reticulatus* Enderlein, *Ptrroxanium squamosum* Enderlein, *Trogium pulsatorium* (Linnæus) are pest of museums, libraries and house.

Trogiomorpha suborder of Psocoptera (Psocopteroidea: Paraneoptera); booklice having antennæ will more than 20 segments, never secondarily annulated labial palpi 2-segmented; tarsi 3-segmented; alate species with pterostigma not thickened. Comprises of five families — Lepidopsocidæ, Trogiidæ, Psoquillidæ, Psyllipsocidæ and Prionoglaridæ.

Trogos(s)itidæ (= Ostomatidæ: Ganglbauer 1899) family of grain and bark-gnawing beetle (Cucujoidea: Polyphaga); head out pronotum large, antennæ clubbed, fore-coxæ transverse, first tarsal segment much shorter than second; elytra not truncate. Inhabit decaying trees, stored products (flour, grain etc.), a few prey upon larvæ of other lignicolous insect (600 species) — cosmopolitan Çadelle beetle *Tenebrioides mauritanicus* (Linnæus), *Temnochila cærulea* Olivier, *T. virescens* (Fabricius) et al.

Trophallaxis mutual exchange of nutrients in symbiosis.

Trophocyte(s) nutritive or nuerse cells of ovary or testis.

Truncate ending abruptly as it cut straight across.

Trupaneidæ old family name (Agromyzoidea: Schizophora) to include all fruit-flies, antedated by Tephritidæ Newman 1845.

Trypoxylidæ (= Trypoxylonidæ) small family of Sphecoidea (Apocrita: Hymenoptera) often placed as subfamily of Sphecidæ — species of *Pison, Trypoxylon* et al.

Tubulifera Haliday 1836 suborder of Thysanoptera (Thysanopteroidea: Paraneoptera); that include, alate and apterous thrips, the former without microtrichia and forewings veinless or with a single abbreviated median vein, further distinguished by last abdominal segment being tubular in both sexes and female without ovipositor. Mostly predaceous on insects and mites, some feed upon exudation of plants and decaying vegetable, a few phytophagous found on economic crops — Phlæothripoidea, Urothripoidea.

Tumbler(s) pupæ of mosquitoes (Culicidæ: Culicoidea) distinguished by large cephalothorax (head and thorax fused together) with a pair of conspicuous respiratory horns or trumpets and small, slender, 9-segmented tail-like abdomen which has a pair of large chitinous anal peddles of fins. Very active and move by succession of quick peddle-like motions.

TVO tractor vapourising oil - kerosine oil.

Tylidæ old and obsolete term for dipteran family Micropezidæ (Micropezoidea: Schizophora).

Tymbals pairs of membranes used for producing communicative sound in cicadas (Cicadidæ: Homoptera).

Type zoological speciment on which the original published description is based; once designated the type cannot be changed — primary types of proterotypes (allotpye, chirotype, holotype, paratype), Supplementary types (neotype, heautotype, plesiotype) and lcotypes (homotype, ideotype, metatype, topotype).

Tyrosine white crystalline amino-acid obtained from most proteins, important in all living organisms: in insect it is basis for formation of melanin in the cuticle.

U

Ugly nest nest of shelter made by caterpillar of certain tortricid moths (Torticoidea: Ditrysia); a common site in cherry orchard. Leaves are tied together with silken threads produced by developing caterpillar. The caterpillar live in colonies and enlarge their nest to cover the foliage being used for food; ultimately the outer leaves also turn brown and die away, giving the nest an ugly appearance — *Archips cerasivorana*. et al.

Ulidiidæ family of Agromyzoidea (Schizophora: Cyclorrhapha); these dipteran flies were placed under family Otitidæ. Maggots mostly saprophagous (*Physiphora* spp. and *Seioptera* spp.) or less often phytophagous (*Euxesta* spp.).

Ultra low volume (ULV) spraying technique or method of applying pesticides in concentrated or undiluted from. The spraying is done either by aircraft (aeroplanes and helicopters) or by motorised mist blowers and quantity of pesticide may very from 1 to 4 litres per hectare. This is more effective and cheaper than conventional aerial spraying.

Ultraviolet light spectrum (electromagnetic radiation), beyond the violet band; it is invisible to men but visible and highly attractant to nocturnal insects.

Ultraviolet vision most insects possess ultraviolet vision. They can perceive wide range of light-frequencies including all colours that are visible to human being as also infrared and ultraviolet which are beyond the range of human eye, many moths are attracted to ultraviolet light while ants avoid it.

Umbrella ant parasol ant or sauba: common name for species of tropical leaf-cutting ants; that carry to their underground nests, pieces of leaves, holding these like an umbrella over their body. In the nest these leaf pieces are mixed with saliva and used as medium for rearing fungi — *Atta* spp. (Myrmicinæ: Formicidæ).

Uncus a secondary median process, associated with male genitalia; usually hook-like or bifid, attached to hind margin of IX tergum of many male lepidopterous insect; often regarded as tergum of X segment.

Terminal Portion of springtail's tarsus

Underwing(s) colloquial name for moths of various different families all having cryptically coloured forewings and brightly coloured hind-wings, which give good camouflage — copper underwing (*Amphipyra pyramidea*). red underwing (*Catocala nupto*), straw underwing (*Catocala nupto*), straw underwing (*Thalpophila matura*), yellow underwing (*Noctus pronuba*) etc.

Unguis (pl. ungues) one or two curved, simple, toothed or bifid claws, used mainly for clinging to plants, animal hosts and other object; more common in springtails (Collembola). The least segment of aphid anetenna is also often called unguis.

Uniramia subphylum of Arthopoda (often considered as an independent phylum); includes bilaterally symmetrical metamerically segmented invertebrates having uniramous jointed appendages and chitinous exoskeleton that is moulted periodically is permit growth of the animal. About one million described species grouped into two classes — Myriapoda and Insecta (Hexapoda).

Unisexual having only one sex — male or female; almost all insects are unisexual; hermaphroditism being very rare.

Univoltine (species) having only one generation in a year.

Unstable easily decomposible.

Uraniidæ family of large, slender-bodied, most beautiful moths (Geometroidea: Ditrysia); widely distributed in Tropics of the Old and New World; having tympanal organs ventral in female and at sides of 2nd tergite in males; it is tridescent blue-green, black and white; hind-wings swallow-tall type, average wing span about 70 mm — *Chrysisdia medagascariensis* is velvety black with band of tridescent green, shading to orange-red; hind-wings having white-edge tails. Other common species are *C. ripheus. Urania leilus, Nyctalæmom* app.

Uranioidea Forces 1923 a lepidopteran superfamily in the old classification system that included Uraniidæ, Mimallonidæ and Epiolemidæ now placed under superfamily Geometroidea (Ditrysia).

Urbicolidæ Tutt 1906 old and obsolete from for a lepidopteran family (Rhopalocera) that include skipper butterflies, now placed under family Hesperiidæ (Hesperioidea: Ditrysia).

Urgomphi fixed and non-segmented appendage of the anal end of abdomen of some grubs (Coleoptera).

Urostylidæ (= Urolabidæ) a small insignificant family of brilliantly coloured bugs, often included as subfamily of Pentatomidæ (Pentatomoidea: Pentatomomorpha).

Urothripidæ Bagnall 1909 family of Urothripoidea: Thysanoptera); earlier considered as subfamily of Phæothripidæ; have distance between hind-coxæ greater than between fore- and mid-coxæ and anal setæ exceptionally long — *Bænothrips asper, Bournieria indica, Stephanothrips occidentalis* et al.

Ursala butterfly a pale blue green and black nymphalid butterfly of North America — *Basilarchia* spp.

Uterus misnomer and absolutely term when applied to insect; denoting the enlargement of vagina where the two oviduct meet. In certain larviparous and pupiparous insect like forest-flies and fleshflies (Muscoidea: Schizophora), the eggs hatch here and the larvæ remain in the female's body. These are deposited either in their instar (Sarcophagidæ) of are nourished by the product of special nutritive glands and deposited only when nearly full brown and ready to pupate (Glossinidæ).

Uzelothripidæ family of Thripoidea (Terebrantia: Thysanoptera); comprising of only of one species of thrips, *Uzelothirps scabrosis* having 7th (terminal) antennal segment unusually long.

Uzuchidæ old name obsolete name for a Neotropical family Xyloryctidæ (Tinæcoidea: Ditrysia).

V

V-shaped notal ridge V-shaped endoskeletal ridge of mesonotum or metanotum; ita arms divergent posteriorly, marked externally by scutocutellar suture.

Vacuole clear within cytoplasm, usually filled with air or fluid and bounded by membrane; may function in digestion, storage, secretion of excretion.

Vagility ability of organism (insect) to disseminate.

Vagina (pl. vaginæ) a medium duct formed by joining two oviducts in female insect; it opens into the genital chamber at the egg-guids and is associated with copulation. In some insects (Coleoptera, Lepidoptera, Trichoptera etc.) the vagina continues posteriorly and opens on IX sternum, serving both for copulation and discharge of eggs.

Valid describing a taxon classified on the bases of distinctive characters of accepted importance.

Valvate having valve-like parts.

Valve flat of tissue, as found in veins etc., it permits movement of fluid in one direction only.

Valvifer(s) basal plates of ovipositor.

Valvula (pl. valvulæ) small valve: small process forming a sheath for the ovipositor in certain insects.

Valvulate having valvules.

Vamidothion $C_8G_8O_4NS_2P$; organo-phosphate, systemic insecticide and acaricide of high persistence, used for control of sap-sucking insect specially aphids as also mites. Colourless waxy solid, readily soluble in water and most organic solvent but insoluble in light petroleum oils and cyclohexane; compatible with most pesticides. Acute oral LD_{50} value for rat 100 mg/kg Kilval®, Vamidoate® etc.

CH_3O
CH_3O $>$ $PO\text{-}S\text{-}(CH_2)\text{--}SCHCH_3.CONHCH_3$

Vamidothion

Vanessid(s) big and brightly coloured butterflies (Vanessidæ: Papilionoidea); strong fliers; frequent flowers in the garden — painted lady [*Vanessa cardni* (Linnæus) is cosmopolitan; *V. indica* Herbest is Indian species].

Vanessidæ family of nymphid butterflies (Papilionoidea: Ditrysia) often included in family of Nymphalidæ, comprise of vanesside — red admiral [*Vanessa atanlanta* (Linnæus), painted beauty *V. virginiensis* Drury etc.].

Vanhorniidæ monogeneric and monospecific family of Proctotrupoidea or Serphoidea (Apocrita: Hymenoptera); very small pointed-tailed wasps, having short mandible with three large outwardly directed teeth; ovipositor long and directed forward; parasitic on grubs of euenmid beetles, reported only from North America.

Vannal region (vannus) posterior wing area usually triangular in shape, separated from remigium by vannal fold (*Plica vannalis*) often expanded into fan-like area and bearing vannal (anal) veins.

Vannal veins (A_1 to A_5) veins arising from third axillary sclerite and occupying vannal region of wing — anal veins except postcubitus.

Vannus anal or posterior lobe of insect wing: fan-like expansion separated from the rest of wing by a furrow.

Variety ambiguous term of classical taxonomy for a heterogenous group of taxon having non-generic variation of phenotype; term proposed by Linnæus (Carolus A.) that later because controvertial and subsequently deleted from zoological nomenclature though it is still being used in botanical nomenclature.

Vas deferens (pl. vasa deferentia) lateral duct leading from testis and ending in vasicula seminalis in male reproductive system.

Vas efferens (pl. vasa efferentia) one of many small ducts connecting sperm tube of testis with vas deferens.

Vector agents or carrier that transmits bacterial, fungal or vital pathogens from one host to another.

Vedalia alternative name for ladybird beetle (Cocoinellidæ: Coleoptera).

Vein thread-like- thickening in membranous wings of insects arising from wing base and branching distally, serve as support for the wing; hæmolymph flows through these at their formative stages. Tubelet(s) in body, for carrying blood back to heart.

Vein, False spurious vein or vena spuria thought to be vestigial, M.A lying between R and M and at both ends, distinctive characteristic of syrphid flies (diptera).

Veliidæ Douglas and Scoot 1865 (Microveliidæ Reuter 1910) family of heteropteran bugs (Gerroidea: Gymnocerata); includes small water-striders having vertex with longitudinal groove between the eyes; hemelytra homogenous; legs short and claws ante-apical (250 species) — *Velia* spp. occurs in streams, *Rhagovelia* spp. swim against the current of swift stream, while *Halovelia* spp. marine found in Indian and Pacific oceans.

Velocipedidæ Tropical family of heteropteran bugs (Cimicoidea : Gymnocerata); often regarded as subfamily of Nabidæ, having rostrum 3-segmented, eyes large and projecting, ocelli present; hemelytra macropterous; caneus present — *Scotomedes* and *Velocipeda* species are found in the Orient.

Velocity speed at which insect fly, maximum and speed has been achieved by some large dragonflies (Anisoptera: Odanata); *Austrophiebia* species can fly at nearly 100 km per hours, the average for dragonflies being 45 to 50 km per hour, followed closely by some flies (Diptera) that are said to fly upto 40 kilometers per hours; most of the insect fly @10 to 15 km/hour (figure given by some of the worker are often greatly exaggerated).

Vena arcuata first jugal vein in insect wing; arcuate vein.

Vena cardinalis second jugal vein in insect wing.

Vena dividens secondary jugal in wings of some Orthoptera.

Vena spuria false vein.

Venation pattern or arrangement of veins or nervures in an insect wing; often a diagnostic character. Forewing typically has 12 veins — internal vein with one of the tree branches (1a, 1b, 1c), median vein with its three branches (2, 3, 4), lower radial (5), upper radial (6), subcostal with five branches (7, 8, 9, 10, 11) and costal vein (12); but almost any of these may be absent. Hind-wing typically has only eight veins, four of the subcostal being absent. In Hepialidæ (Monotrysia: Lepidoptera) and Micropterygidæ (Zeugiloptera: Lepidoptera) however, twelve veins are present as in the forewing.

Venom position.

Venter lower surface of insect abdomen. Entire under surface of an animal.

Ventral undersurface of an organism or organ.

Ventral diaphragm delicate membrane, present in some insects, stretched between lateral margins of abdominal sterna, often extending into thorax, separating ventral sinus from perivisceral sinus.

Ventral gland gland located in head of some insects and prothorax of other; secretes hormone that is found in the blood and induces moulting.

Ventral nerve cord chain of connected ventral gangila; often applied to thoracic and abdominal gangila only.

Ventriculus endodermal stomach of insect.

Vermes plural of worms. In old animal classification suggested by Carolus A. Linnæus, *a* subkingdom that included all invertebrate (flatworms, roundworms, threadworms etc.) except arthropods; term abandoned since long.

Vermicide (vermicidal) toxin that can kill worms.

Vermicular worm-like: caused by worm(s).

Vermiculate worm-eaten: having zigzag marking indicating worm-tracts.

Vermicule very small worm.

Vermiform having form or shape of a worm.

Vermiform larva headless (acephalic), legless (apodous), worm-like larva, typical of some true flies (Diptera).

Vermipsyllidæ Wagnen 1889 (= Dolichopsyllidæ) large family of blood-sucking fleas (Pulcoidea: Siphonaptera); having no eyes, occipital region without a transverse dorsal suture and each abdominal tergite with atleast two rows of setæ — American squirrel flea *Diamanus montanous* (Baker), Candian chicken flea (*Ceratophyllus gibsoni* Fox), European rat flea *Nosopsyllus fasciatus* (Bose) at al.

Vermivorous worm eating: feeding on worms.

Verruca (pl. verrucæ) scoli wrat or compound spines found covering the body of various caterpillars.

Versicoloured variously or diversely coloured: changeably coloured.

Versonian cell(s) specialised large apical cells at the tip of testis in most Lepidoptera and some other insects, believed to assist in nutrition of germ cells.

Verson's glands moulting glands of caterpillars: epidermal glands that secrete a fluid which helps in loosening epicuticle, prior to moulting.

Vertex dorsal surface of two parientals forming top of head, between and behind compound eyes.

Vertical resistance gene-for-gene resistance: resistance exhibited against some biotpyes of insect species which is governed by one or more genes in host plant each gene corresponding to a matching gene for parasitic ability in insect species.

Vesica (pl. vesicæ) terminal membranous plate of ædeagus. Bladder or sac specially urinary bladder.

Vesicula seminalis dilation of vas deferens, in which spermatozoa are retained.

Vespiary wasps' nest.

Vespidæ (Leach 1815) Stephens 1829 widely distributed family of hornets, paper-wasps and yellow-jackets (Vespoidea: apocrita); yellow and red with black or brown marking or bands. Wings long and narrow, hind pair without anal lobes and veins not extenting to apical margins; claws simple; three forms — queen of fertile females, worker or sterile female and drones or males.

Vespiformia group of hymenopterous insects, comprising of both ants and wasps.

Vespinæ subfamily of Vespidæ (Vespoidea: Apocrita); includes common parasitic social wasps that makes nests of papery material made out of chewed vegetable fibres — *Vespa basalis* Smith, *V. crabo* Linnæus, *V. orientalis* Linnæus, *V. tropica* Linnæus, et al.

Vespa orientalis Linnæus

Vespine (vespoid) wasp-like.

Vespoidea Ashmead 1899 superfamily of Apocrita (Hymenoptera: Oligoneoptera); that includes true wasps — mostly social, a few solitary; wings folded longitudinally when at rest; fore tibia with a comb for cleaning antennæ; petiole long and slender; abdomen sessile and conical with well developed sting — more than 20 families, including Eumenidæ (solitary) and Vesidæ (social wasps).

Vespophile(s) common name for insect that lives as guest or inquilnes in wasps' nest — rove beetle (Staphylinidæ: Coleoptera).

Vestibulum external cavity formed above VII abdominal sternum when the latter extends beyond VIII sternum.

Vestige degenerate: useless reminant of an organ that was once functional in earlier stage of embryological or evolutionary development.

Vestigial poorly developed: degenerated: often non-functional.

Vestiture covering; such as setæ hair, spines and even tubercles that cover the insect larvæ or scales on adults.

Vianaidæ very small family of South American bugs (Tingoidea: Cimicomorpha: Heteroptera).

Vibrissæ stout cephalic bristles placed closed to the sides of epistoma at the lower extremities of facial ridges on the head in dipteran flies.

Vicarint(s) two or more closely related taxa, presumably derived from one another or from a common immediate ancestor that inhabit geographically different areas.

Villose covered with villi.

Villus (pl. villi) long soft hairs: hairs-like projection from surface of a membrane.

Vinculum entire coxosternal plate of IX abdominal segment in Lepidoptera.

Vine phylloxera notorious pest of vineyard in Europe, *Phylloxera vitifoliæ* Fetch (Phylloxeridæ: Aphiodoidea) is native of America. Its complicated life-cycle varies according to the variety of grape; leaf-gall form is more prevalent in Eastern USA while root-gall from is more common on Pacific coast.

Vipionidæ Viereck 1918 small family of very small parasitic wasps (Apocrita: Hymenoptera) often included in Braconidæ; distinguished by 14- to 18-segmented antennæ and undivided mesothorax — *Apanteles glomeratus* (Linnæus), *Microbracon brevicornis* Westmael, *Opis fletcheri* Silvestri, *O. watersi* Fullaway, *Vipio* (Stenobracon) *deesæ* (Cameron) et al.

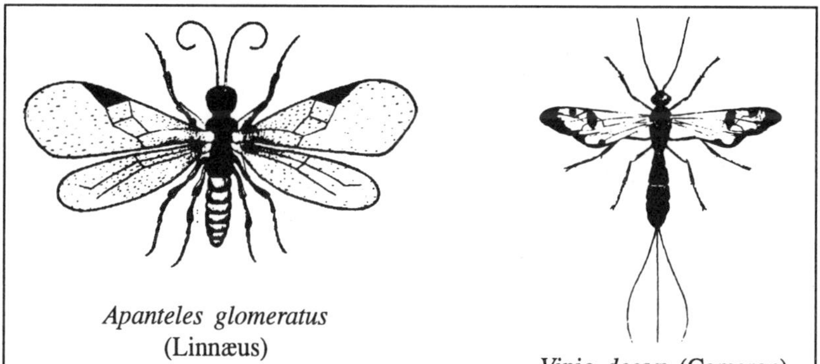

Apanteles glomeratus (Linnæus) *Vipio deesæ* (Cameron)

Virga terminal phallic spine, arising from endophallus.

Virus non-cellular, ultra-microscopic, micro-filterable, infectious organism composed of protein shell and core of nucleic acid; having power growth and multiplication. Discovered by Iwanowsky in 1892 as bacterium.

Visera collective term for all internal organs of an animals contained in ceolom.

Vision faculty of sight: sense that perceives the form, colour, distance and movement of an object; for this purpose, most adult insects have simple eyes (ocelli) and compound eyes whereas larvæ usually have only ocelli.

Vitamin(s) complex organic substances, present in minute quantity in nutritive foods or made synthetically; though do not yield any energy, their presence is essential for normal metabolism of an organism, hence supplied in their diet, absence or shortage causes various deficiency diseases.

Vitellarium major portion of an ovariole, it contain developing eggs and nurse-cells (trophocytes); epithelial layer of wall grows inwards enclosing each oocyte in definite egg-chamber of follicle; cells of follicle secrete the chorion of egg and often serve to nourish the oocytes.

Vitelline membrane wall of mature egg-cell.

Vitellophage(s) primary yolk cells; cleavage cells that remain scattered in the yolk; their main function is to liquefy the yolk and bring about assimilation.

Vitrellæ cells that secrete crystalline cone in each ommatidium.

Vitreous bodies crystalline cone in the compound eye of an insect.

Vivacious long-lived.

Vivi- alive: living.

Vivicum all living things.

Vivid full of life.

Viviparism viviparous reproduction.

Viviparity ability of animals (including insects) to give birth to living young ones.

Viviparity, Adenotrophic phenomenon in certain insect whereby the larvæ are retained in the parent's body after hatching, these develop by feeding on uterine secretions, moult twice and are finally deposited in a mature stage ready for pupation — Tsetse flies (*Glossina* species): species of *Pupipata* (Schizophora: Diptera) produce puparia.

Viviparous bringing fourth or producing live young ones that have reached advanced stage of development — Aphidoidea, most mammals etc. *Cf.* oviparous.

Viviparousness quality of being viviparous.

Volatile capable of rapid evaporation at ordinary temperature.

Volsella (pl. volsellæ) ventral process or lobs on each side of phallobase in Hymenoptera.

Vulva external opening of female genital chamber of vagina meant for mating and discharge of eggs.

W

WADHI, Dr. Sukhdev Raj — born April 11, 1932; B.Sc. (1950) from Agra University, M.Sc. (1952) from Banaras Hindu University, Associate IARI (1958) and Ph.D. (1972) from IARI, New Delhi. Joined Division of Entomology, IARI in 1953 as Senior Research Assistant, later (1956) shifted to Plant Introduction and Exploration section, which in 1976 became an independent organisation, National Bureau of Plant Genetics Resources (ICAR) and he took over as Head of Plant Quarantine Division. Beside bruchids and coccids, he also worked on control of mango stone weevil, *Sternochaetus mangifeae* (Fabricius) and Oriental fruitfly, *Bactrocera dorsalis* Hendel. Developed X-radiography technique and quarantine disinfection schedule. Died February 12, 1986, with his boots on. Catalog of Bruchidæ, compiled by him and Mrs. Sujaya Udayagiri (nee Rao) was published after his death (1989).

Walking-stick nickname given to various stick-insect (Phasmida: Orthopteroidea).

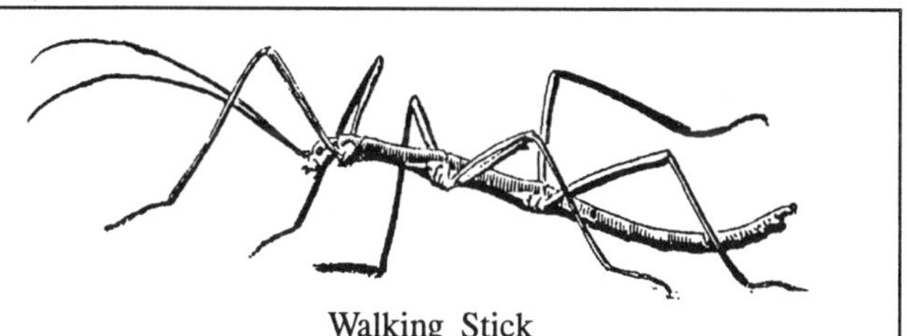

Walking Stick

Warfarin $C_{19}H_{16}O_4$; colourless crystalline substance, anti-blood coagulant, slow poison, boiling point 161°C; used for killing mice, rats and rodents in house, godowns and fields.

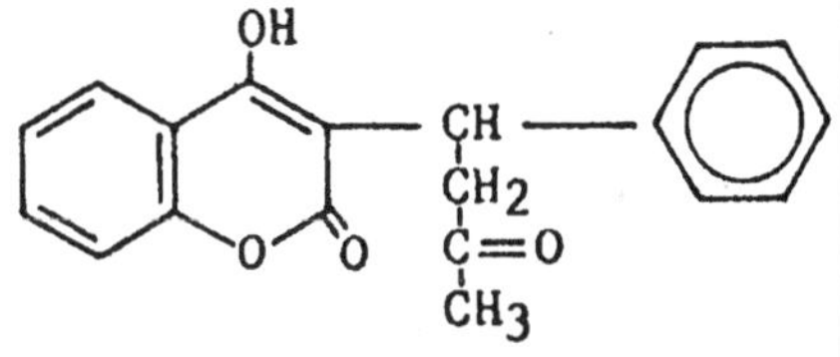

Warfarin

Warning colouration bright colours or marking on body or wings of insect (or any other animal) that often serves as protection against parasites and predators.

Wart biter nickname of stout, handsome, long-horned grasshopper, *Decticus verrucivorous* (Tettigoniidæ: Ensifera); about 40 mm long, mottled with brown and green having short elytra, long abdomen and very long hind-legs; females have long, curved ovipositor; found on grassy hillsides, where males can often be heard stridulating. Bite painful.

Wasp common name for a large number of hymenopteran insect — social and solitary forms — belonging to over 66 families of Apocrita and differ from ants, bees and sawflies; slender body, thorax and abdomen joined by narrow stalk or petiole; ovipositor modified to form a sting; males have no sting. Adults feed on nectar and fruit juices, larvæ parasitic or predaceous on other insects. (slang) petulant and spiteful person.

Wasp, Aphid small, 6 to 10 mm long, pemphredon wasps (Sphecoidea: Apocrita), black with or without red markings, abdomen cylindrical with short petiole; nest in pithy plant stems or dry grass stems and provision the cells with aphids, psyllids etc. — *Psenulus atratus* (Fabricius), *P. fascipennis* Dahlblm, *P. pallipes* Panzer, et al.

Wasp, Common wasps belonging to family of Vespidæ (Vespoidea: Apocrita); social insect; after hibernation during Winter, the queen starts building the nest, having completed single layer of cells, she lays an egg in each cell, these develop into worker (imperfect females) and take over of further expansion of the nest, feeding of young ones etc., the nest generally does not last for more than one season, a generation of perfect females and males is produced; these leave the nest, mate in open and white male dies away, fertilised female (queen) hibernate and starts a new colony after hibernation.

Wasp, Cuckoo gold-wasp (Chrysididæ: Bethlyloidea); less than 12 mm in length, bright metallic blue or green in colour, body covered with fine indentation, lays its egg in nests of other wasps and bees; on hatching the larvæ parasitise the host or consume the food stored therein, thereby starving the intended inmates — *Parnopes edwardsii* Cresson, *Chrysis viridula* Linnæus, *Omalus punctulatus* Dahlborn, *Stilbum splendidum* Fabricius et al.

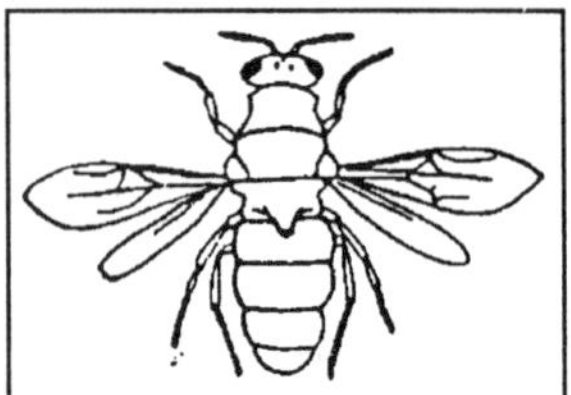

Pamopes edwardsii Cresson

Wasp, Digger group of fossorial solidtary wasps (Sphecidæ: Apocrita), that burrows into (sandy) ground to build nests, like many other wasps, these also provision their nests with paralysed but alive insect (caterpillar etc.) that serve as food for their developing larvæ — *Sphex hirsula* Scopoli, *S. sabulosa* Linnæus, *S. punctipennis* (Walsh) et al.

Wasp, Flower remarkable and interesting wasps (Thynnidæ: Vespoidea) of Australia; males winged and usually longer than females which are often apterous. For mating the male grasps female with its long mandibles and carries her up into the air on a nuptial flight; soon the couple drops down, male dead and gravid female to start her progeny - *Thynnoturneria cerceroides* Smith (♂ 8 mm, ♀ 3 mm), *Diamma bicolor* Westwood (blue ant) et al.

Wasp, Gall wasps that produce galls or live within the galls (Cynipidæ: Synipoidea); very small, 1 to 6 mm long, dark coloured insect; eggs are laid on various plant parts and galls or swellings begin to develop thereon, soon after the hatching of the eggs, each species produces one or more chracteristic kind of galls; oaks (*Quercus* app., Fagaceæ) roses (*Rosa* spp.) and plants of Asterceæ family are preferred hosts — *Biorrhiza pallida* Olivier, *Andricus californicus* (Basett) el al.

Wasp, Mason mud- or potter-wasp (Eumenidæ: Vespoidea); solitary wasp that builds no nest but only flask-shaped hard cells of clay or mud and small pebbles glued together with their saliva and attached to twigs or stem of various plants. Eggs are laid in these nest and paralysed caterpillar are also stored therein to serve as food for newly hatched larvæ — *Abispa ephippium* (Fabricius), *Eumenes coarctatus* (Linnæus), *E. fraterna* Say, *Odynerus birenimaculatus* Saussure, *Zethus* spp., el al.

Wasp, Paper hornets or yellow-jackets: most organised wasps (Vespidæ: Vespoidea) widely distributed; yellow and red coloured with black or brown bands or marking; 8 to 10 mm long. Predaceous, mostly on caterpillar which are paralysed and dragged for storing in the cells of their nests. The nests are generally large and made of papery material made out of chewed vegetable fibre - *Vespa basalis* Smith, *V. magnifica* Smith, *V. orientalis* Linnæus el al.

Wasp, Polistes social wasp (Polistidæ: Vespoidea); 10 to 25 mm long characterised by short and very slender petiole; adults hunt forage and food upon caterpillar, fruit-juice and honeydew; the wasps pinch, squeeze and crush the victim with their mandible and carry away that dead and macerated caterpillar to their nest for feeding the young ones — *Polistes annularis* Linnæus, *P. aurifer* Saussure, *P. hebracus* Fabricius, *P. tepidus* Fabricius, *P. pallipes* Lepeletier and allied species.

Wasp, Sand large (15 to 20 mm long) robust, semisocial wasps (Bembicidæ: Sphecoidea); cosmopolitan in distribution build their nests in sunny sandy places, keep busy throughout the day in opening the burrows, feeding their young ones and again closing the entrances, as these wasps do not provision their cells like other wasps — *Bembix beutenmulleri* Fox, *B. rostrata* (Linnæus), *B. spinolæ* Lepeletier, *Stictiella* spp., at al.

Wasp, Spider ground nesting wasps (Pomphilidæ or Pammocharidæ: Pomplioidea); that provision its nest with spiders and occasionally with cricket and cockroaches as well — *Psammochraes ilus* Bingham and *P. mirandus* (Saussure) are Indian species; *P. plants* (Fox) is trapdoor spider-wasp of North America.

Wasp, True all species belonging to superfamily Vespoidea (Apocrita: Hymenoptera) including hornets, yellow-jackets, paper wasps etc.

Wasp, Wood horn-tail: large hymenopterous insect (Siricidæ: Symphyta); having at anal end of abdomen a horn-like process (stout spine), ovipositor blunt and stiff constiting of two piece enclosed in a sheath: phytophagous pests of forest trees; maggots make large burrow in sapwood and heartwood of conifers and broad-leaved deciduous tree — *Sirex juvencus* (Linnæus), *S. gigas* Linnæus, *Tremex columba* (Linnæus) *T. magus* (Fabricius), et al.

Water H_2O; colourless, tasteless and transparent fluid that freezes (ice) at 0°C and boils (steam) at 100°C substance of rain, lakes, rivers, sea etc., Constitutes about 80 per cent of most of insect tissues as also other animals and plants. An important requirement in insect nutrition, even grain weevils and other stored-product pest cannot survive if moisture content in their diet is below 8 per cent.

Water boatman group of aquatic bugs (Corixidæ: Hydrocorisæ); found at bottom of ponds feeding on plants debris. Breathe from bubble of air carried inside the water, trapped between wings and abdomen; periodically rise to surface of water to renew air bubble — *Micronecta striata* Fieber, *Sigara* (*Corisa*) *hieroglyphica* Dufour, *Corixa geoffroyi* Leach (European) Is the largest species (15 to 18 mm long).

Water creeper(s) needle-bugs; small, flattened, leathery aquatic bugs (Naucoridæ: Notonectoidea); inhabit running and quite freshwater — brachypterous and macropterous — predaceous, devouring all form of aquatic insect — *Ambryans signoreti* Stal, *Polocoris femoratus* (Polisot-Beauvois), *Aphelocheirus* spp., et al.

Water cricket stoutly built predatory aquatic plant bugs (Veliidæ: Gerroidea) found in Neotropical and the Oriental regions — fresh water and marine form — *Halovelia capitata* is confined to Central America and *Velia* spp., are common in Europe.

Water scorpion flattened aquatic bug (Nepidæ: Notonectoidea) found on bottom of Shallow ponds; forelegs thickened and strongly hinged for grasping prey; anal end of abdomen drawn out of into a long respiratory tube that is pushed out into a long respiratory tube that is pushed up to water surface for breathing purpose — *Laccotrophes rubra* Linnæus, *L. robustus* Stal, *Ranatra brevicollis* Montandon, *R. filiformis* Fabricius, et al.

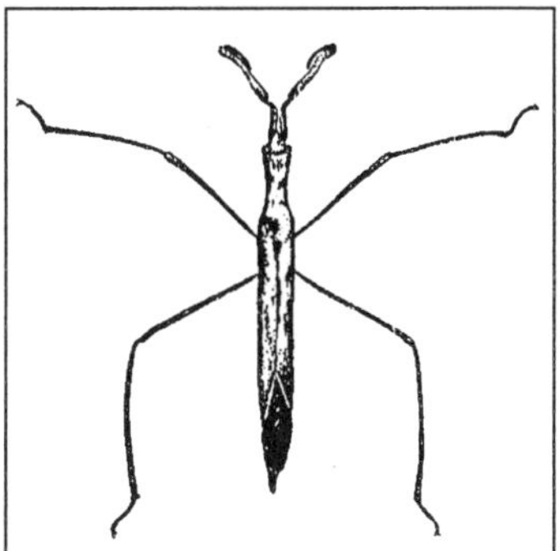

Ranatra brevicollis Montandon

Water stress drought.

Water strider pond skates: medium-sized subaquatic bugs (Gerridæ: Amphibicorisæ); that walk, run, glide of skate on surface of water or over water plants, apterous, alate or brachypterous; body slender venter clothed with silvery pubescence which is waterproof and enable the bugs to cruise about; mostly predaceous upon smaller insect and also feed on dead insect falling in the water — *Gerris lacurstris* Linnæus, *G. paludum* Fabricius, *G. najas* de Geer, *G. thoracicus* Schumacher, *Halobates micans* Escherich, *H. sericeus* Escherich, et al.

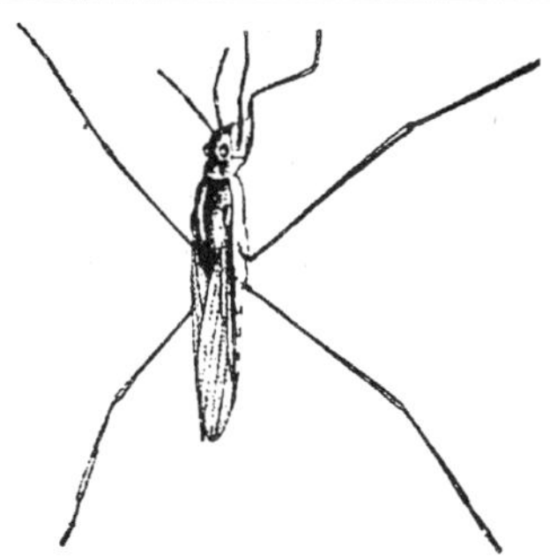

Water Strider

Water tiger(s) grubs of some carnivorous beetles (Dytiscidæ: Adephaga); voracious feeder on various aquatic insects and other water animals including small fish.

Waterbug(s) any bug (Cryptocera/Hydrocorisæ: Heteroptera) that lives an aquatic habitat during all phase of its life-cycle — back swimmer, water boatman, water scorpions etc.

Wax type of lipid, found in epicuticle of many insects, secreted mainly for protection. It is non-greasy insoluble solid substance that melts at fairly low temperature; commercially used for sealing and making candles, models etc.

Wax, Bees $C_{30}H_{61}COC_{15}H_{31}$; soft yellowish non-greasy substance, insoluble in water; produced by honeybees for building their cells in the beehive.

Wax, Parafin Petroleum product; substance more or less similar to beeswax, except that beeswax is natural and this synthetic.

Wax, glands glands found in many homopteran bugs and honeybees (Apoidea: Hymenoptera).

Wax, insect common name for various plant bugs and scale-insect that secrete wax-like substance — Chinese wax-scale-insect *Ericerus pela* Chavannes (Coccidæ: Coccidea) is bred commercially for obtaining wax, used in candle making.

Wax pick a spine or spur on second leg of honeybee; used by the bee for removing wax and pollen from ventral part of the body.

Waxworm caterpillar of wax moth, *Galleria mellonela* Linnæus (Galleriidæ: Pyraloidea); feed on wax in the beehives.

Weather atmospheric condition: sum of abiotic factor like sunlight (duration, intensity), temperature, cloudiness, wetness (humidity, rainfall, snow), wind (direction, velocity) and pressure. It exerts very great influence on activity of all insects.

WEBER, Dr. Herman — born November 27, 1899; Director Zoological Institute, Eberhard-Karis University at Tubingen (Germany); well known Entomologist for his original work on insect morphology; authored a large number of publications including some monographs — **Biologie der Hemipteren** (1930), **Lehrbuch der Entomologie** (1933), and **Grundriss der insektenkunde** (1938). Died November 18, 1956.

Weevil beetle (Curcolionoidea, except Scolytidæ) with head characteristically drawn out into long snout of rostrum; antennæ arise from sides of snout and are usually bent sharply near the middle and generally also clubbed; mouthparts mandibulate, situated at the tip of the snout; thorax rounded and without a lateral margin; body usually heavily armoured and often clothed with scales. Feeds on leaves, buds, fruits, grains, nuts and even tree trunk as also stored-grains (over 60,000 species) — species of *Brachythinus* Latreille (= *Otiorhynchus* Germar), *Pissodes* Germar, *Sitophilus* Schænherr, *Tychius* Germar, et al.

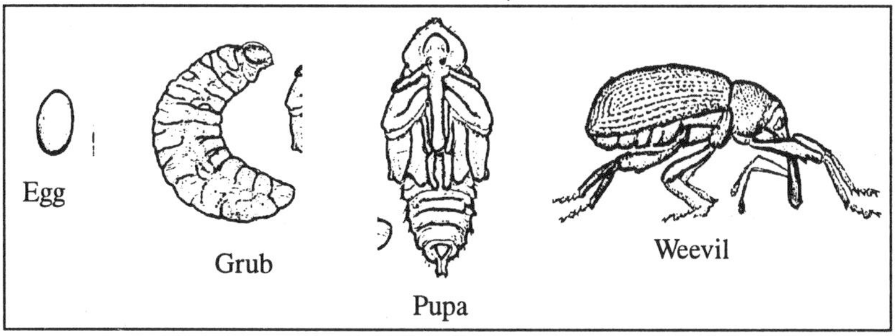

Weevil, Acorn over 50 species of *Curculio* (*Balaninus* Germar) (Curculionidæ: Curculionoidea); long-legged, brown of gray weevils having long, slender and curved snout used for tunnelling into acorns, hazelnuts (*Corylus avellana* Linnæus) etc., to oviposit therein; grubs develop inside the nuts and eat their way out to pupate; found in North America and Indo-Malayan region.

Weevil, Almond *Myllocerus lætivirens* Marshall; polyphagous pest, prefers almond (*Prunus amygdalus* Batsch) and pear (*Pyrus communis* Linnæus); adults nibble irregular holes and gradually eat away the entire leaf lamina.

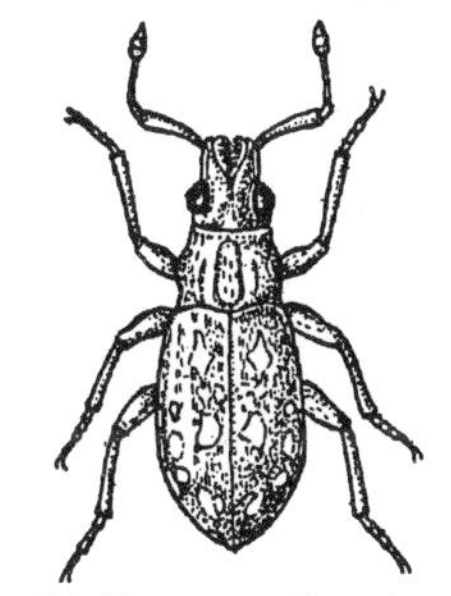
Myllocerus discolor Boheman

Weevil, Apple *Myllocerus discolor* Boheman; polyphagous pest; grubs feed in the soil on root of different plants and adults come out to feed of foliage. *Dereodus pollinosus* Redtenbacher is known as lesser apple weevil and *Anthonomus pomorum* (Linnæus) is apple blossom weevil of Europe.

Weevil, Apricot *Emperorrhinus defoliator* Marshall polyphagous pest of temperate fruit trees; active in Summer when swarms of these weevils appear and defoliate the trees, preferred host being apricot (*Prunus armeniaca* Linnæus).

Weevil, Aquatic *Bagous trapæ* Parshad, small black weevil that feed on submerged soft stem of water nut (*Trapa bispinosa* Roxbergh) crop, near the internode affecting adversely the branching of plants, resulting in low yield.

Weevil, Banana *Cosmopolites scordidus* (Germar) specific and most destructive pest of banana (*Musa* spp.); adults feed during night on pseudostems and bore into suckers; the tunnels made by these weevils are occupied by fungi and bacteria, accelerating the process of rotting.

Cosmopolites sordidus (Germar)

Weevil, Ber *Xanthochelus superciliosis* Gyllenhal, relatively large weevil, about 15 mm long; grubs feed on roots and adults on leaves and flower buds; jujube (*Ziziphus* spp., Rhamnaceæ) are the preferred hosts.

Weevil, Bollworm *Anthonomus grandis* Boheman, a notorious pest of cotton in USA.

Weevil, Cardamom *Prodioctes hæmaticus* Chevrolat; adults about 12 mm long, brownish with stripes and spots on elytra; grubs tunnel into rhizomes and shoots of cardamom (*Elettaria cardamom* Maton), often causing serious damage.

Weevil, Cherry *Myllocerus lefroyi* Marshall is small weevil, 2 to 3 mm long, black in colour with light green scalling and elytra having irregular subdenuded darker marking.

Weevil, Clover leaf (alfalfa, lucerne) *Hyper punctata* (Fabricius), *H. variabilis* Herbest and *H. postica* Gyllenhal as the name suggests, are serious pests of alfalfa or lucerne (*Medicago sativa* Linnæus).

Weevil, Coffee bean *Aræcerus fasciculatus* de Geer infest coffee berries in fields and dry beans in storage.

Weevil, Cotton stem *Pempherulus affinis* Faust, small white grubs bore into cotton stem near the base and cause gall-like swellings; attached plants lodge with strong winds.

Weevil, Granary (grain) *Sitophilus* (*Calandra*) *granarius* (Linnæus); the cereal-infesting weevil about 4 mm long, shiny brown in colour, incapable of flying and causes great financial losses every year around the world.

Weevil, Gray *Myllocerus maculosus* Desbrochers, polyphagous pest; grubs feed on root, adults on leaf margin, buds and even flowers; serious pest of cotton in India.

Weevil, *Gujhia Tanymecus indicus* Faust; adults earthern-gray coloured, about 5 mm long hide during day and night under loose soil (crepuscular in habit). A serious pest of wheat (*Triticum æstivum* Linnæus); grubs feed on roots while adults cut and feed on the plumules of young seedlings.

Weevil, Jack bud *Ochyromera artocarpi* Marshall small grayish-brown weevil found damaging jack-fruit trees (*Artocarpus heterophyllus* Lamarck); grubs bore into tender buds and fruits, which rot and fall prematurely.

Weevil, Jute stem *Apion corchori* Marshall (Apionidæ: Curculoinoidea); adults easily recognised by its non-geniculate antennæ and long trochanters. Serious pest of jute (*Corchorus capsularis* Linnæus); grubs attack nodal joint below leaf base; often internode, petiole and buds are also attacked.

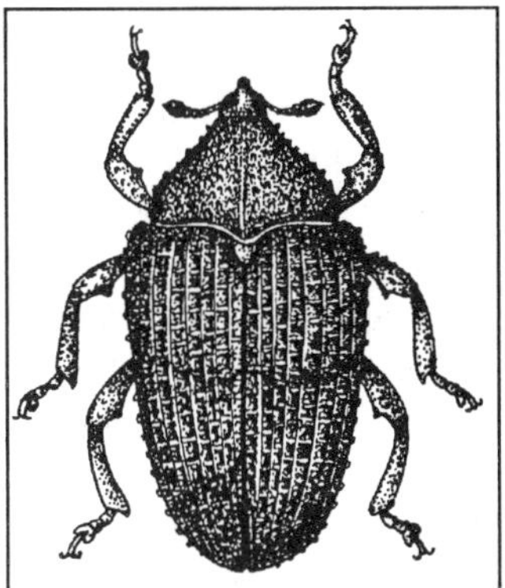

Sternochaetus mangiferae (Fabricius)

Weevil, Mango (nut) stone *Sternochaetus mangiferæ* (Fabricius), specific pest of mango (*Mangifera indica* Linnæus). Eggs laid on ripening fruits; grubs bore through the pulp and feed on seed coat and later damage the cotyledons; pupate inside the seed and adults eat their way out of ripe fruit.

Weevil (Red) Palm *Rhynchophorus ferrugineus* Olivier (*Signaticollis* Chevrolet); major pest of coconut and datepalm confined to the Old World. Grubs are voracious feeders, once inside the tree trunk, they seldom come out; bore into head woody portion, tunnel through the tissue in all directions and feed on soft succulent tissue; the adults are not harmful.

Weevil, Pepper *Eugnathus curvus* Faust is minor pest of black pepper (*Piper nigrum* Linnæus) in India while *Anthonomus eugenii* Cano is found America.

Weevil, Plum *Amblyrrhinus poricollis* Boheman; adults feed on upper epidermis of fresh and tender leaves, skletenosing the same, such leaves dry up and fall.

Weevil, Red-legged large wingless weevil, *Otior(r)hynchus clavipes* (Otiorrhynchidæ: Adelognatha: Curculionidæ); adults have scattered patches of pale hairs on wing-cases; found in Temperate region in moss and roots of hedges, shrubs and fruit trees; female lay fertile egg without having mated; males being very rare.

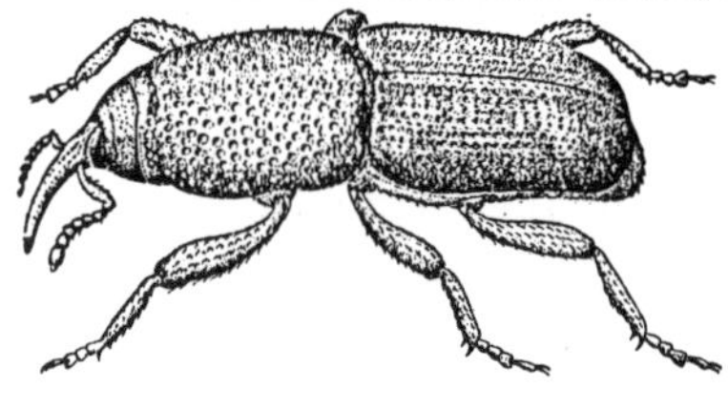
Sitophilus oryzæ (Linnæus)

Weevil, Rice *Sitophilus oryzæ* (Linnæus); notwithstanding its name, this weevil is serious pest of all stored-grains. Adults about 3 mm long dark brown in colour with four reddish spots on elytra, capable of flying; infest ripening grains in the field, female makes a small hole in the grain, deposits and egg into the kernel and seals the opening; larva feed inside while the grains are in godown.

Weevil, Rose *Pantamorus godmani* (Crotch); brown weevil, 8 to 10 mm long, widely distributed in northern hemisphere, as a minor pest of roses (*Rosa* spp.).

Weevil, Strawberry root *Brachyrhinus ovatus* (Linnæus); adults (parthenogenetic female) are about 6 mm long, black, stout and having elytra, with parallel longitudinal grooves, growing together, that makes the weevil incapable of flight. Common in Europe and USA.

Cylas formicarius (Fabricius)

Weevil, Sweet potato *Cylas formicarius* (Fabricius) (Apionidæ: Curculionoidea); specific pest of sweet potato *Ipomoea batatas* (L.) Lamk; Pantropical in distribution; grubs and adults bore into the rubbers both in and in godowns; grubs also bore into the stems and feed on soft tissues. Adults are slender ant-like, having black head, red thorax and legs and bluish-brown elytra; prothorax is conspicuously narrowed.

Weevil, Walnut *Alcidodes porrectirostris* Marshall, specific pest of walnut (*Juglans* spp.). Eggs laid on fruits; grubs bore inside and feed on kernels, reducing the same into useless black mass; pupation takes places within the fruits and adults bites a circular hole and come out. Also causes premature fruit drop.

Weevil, White pine *Pissodes strobi* (Peck); an important pest of white pines in North America. Adults about 6 mm long, reddish-brown with a white spot on each elytron; these overwinter in ground litter around March-April, start feeding on main shoots (leaders) of pines and lay eggs; grubs feed tunnelling downward till August and pupate therein; on emergence, adults escape out through round holes.

Weevil, White spotted *Alcidodes bubo* Fabricius is reddish-brown; 8 mm long weevil with longitudinal white marking on elytra; often found feeding and breeding in tender shoots of indigo (*Indigofera tinctoria* Linnæus).

Weevil, white fringed *Pantomorus* (*Graphonathus*) *leucoloma* Boheman of North America. Adults 12 mm long, dark gray with paler longitudinal lines and rudimentary wings under elytra; females parthenogenetic and males not known.

Wettable powder (WP) dry dust formulation of pesticide that contain 15 to 95 percent toxicant blended with a carrier and a dispersing agent — HCH, carbaryl, DDT etc.

White(s) butterflies of family Pleridæ (Papilionoidea: Ditrysia); distinguished by proportionately larger hind-wing that is without tail and white in colour and their peculiar migratory instict and irresistible impulse to congregate in swarms is most stricking. Clouds of these butterflies are often seen migrating from place to place usually against the wind, passing by for hours at a time during the heat of the day and for several days in succession — *Appias* spp., *Catopsilia* spp., *Pieris* spp. (*P. brassicæ*, *P. canidia, P. napi, P. rapae, P. naganum, P. chumbiensis, P. devta* are found in India, the last three species being rare, have been reported from Naga hills, Sikkim and Kashmir respectively).

Whiteant common name (misnomer) for termites (Isoptera: Blattopteroidea).

Whitefly not true fly; adult a tiny aleurodid (Aleyrodoidea: Homoptera); 1 to 2 mm long; having four rounded wings; body and covered with white powdery wax, highly reduced venation; wing-expanse 2 to 4 mm, most species polyphagous and pest of various economic crops, some even transmit plant diseases — *Aleurocanthus spiniferus* (Quaintance), *Dialeurodes citrifoli* (Morgon) etc.

Whitefly, Anar *Siponinus phillyreæ* (Haliday) (= *finitiaus* Silvestri), major pest of pomegranate (*Punica granatum* Linnaeus), cosmopolitan in distribution.

Whitefly, Bean *Zaphanera publicus* (Singh) pest of beans and pulses, confined to India.

Whitefly, Ber *Alerocanthus zizyphi* Priesner and Hosny, pest of *Zizyphus* spp., in UAR (Egypt) and Sudan.

Whitefly, Castor *Traileurodes ricini* (Misra), Polyphagous pest, widely distributed, major pest of castor (*Ricinus communis* Linnæus) in India.

Whitefly, Citrus *Dialeurodes citri* (Ashmead) and citrus blackfly, *Aleurocanthus woglumi* Ashby; highly polyphagous and most destructive pests; cosmopolitan in distribution.

Whitefly, Cotton (tobacco) *Bemisia tabaci* Gennadius, major pest of cotton and tobacco is more destructive as vector of yellow mosaic and leaf curl virus disease of various crops.

Whitefly, Greenhouse *Trialeurodes vaporariorum* (Westwood) found all over the World on maximum number of host plants; serious pest in greenhouse as condition therein are very conducive for its rapid multiplication.

Whitefly, Jamun *Singhiella bicolor* (Singh) specific pest of jamun (*Eugenia jambolana* Lamarck and *E. jambos* Linnæus) in India.

Whitefly, Litchi *Dialeurolonga elongata* (Dozier), a destructive pest of litchi (*Litchi chinensis* Sonnerat) in India and Pakistan is also found on *Citrus* spp. and some ornamental plants (*Ixora* spp.).

Whitefly, Sugarcane *Aleurolobus barodensis* (Maskell) major pest of sugarcane (*Saccharum offcinarum* Linnæus and allied spp.) in the Indian subcontinent and South-east Asia. *Neomaskellia andropgonis* (Corbett) (*bergii* Signoret) is another species widely distributed, feeding on sugarcane and other Poaceæ plants.

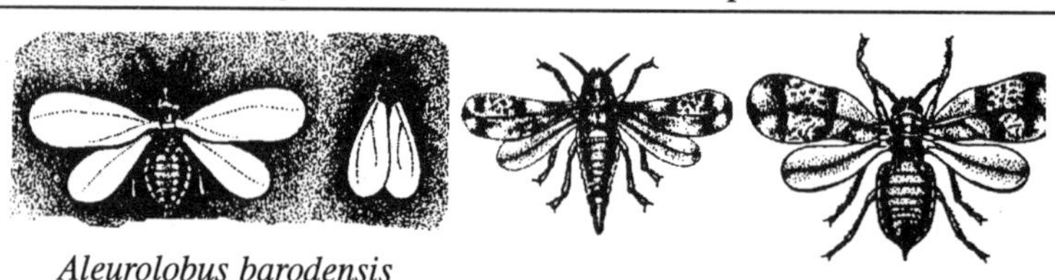

Aleurolobus barodensis (Maskell)

Neomaskellia andropgonis (Corbett)

Widow, The dragonfly, *Libellula luetuosa* Brumeister (Libellulidæ: Libelluloidæ), a blackish species, 45 to 50 mm long with broad dark basal wing bands; ranges over the country east of Rocky mountains (North America); Red skimmer, *L. saturata* Uhier (55 to 60 mm long) occurs generally West of the Rocky mountains.

Wing(s) structure of flight in bats (Mammalia), birds(Aves) and insect (Insecta). In insects it is lateral chitinous outgrowth of body-wall, arising as flattened, double-layered evaginations protruding from the dorsoventral margins; the hollow space between the two layer is filled with fluid, nerves or veins and tracheæ that give support and hold it stiff. Present only in adult stage and used locomotory organs. Lack of wings in apterygote insect is primitive used as locomotory organs. Lack of wings in apterygote insect is primitive characteristic and constant feature: Pterygote insect usually have two pairs of wings each arising from meso- and metathorax. First pair of wings in Strepistera and hind pair in Diptera and male Coccids are reduced to short club-like processes (balancers or halters). Wingless condition of Mallophaga, Siphunculata, Siphonaptera, sterile caste of ants and termites, female of coccids, stylops and embiids is a secondary acquired character while loss of wings in case of bedbug is transitory. The wing of an insect in general, is more or less triangular with humeral, apical and anal angles, embracing costal (upper), apical (outer) and anal (inner) margin and having network of venaton comprising of longitudinal veins and cross-veins.

Longitudinal veins		Cross veins	
Costa	(C)	humeral	(h)
Subcosta	(Sc)	radial	(r)
Radius	(R)	sectorial	(s)
Radial sector	(Rs)	radio-medial	(r-m)
Media	(M)	medial	(m)
Cubitus	(Cu)	medio-cubital	(m-cu)
Anal	(A)	cubito-anal	(cu-a)

The various modification of insect wings are :-

Elytron (pl. elytra) — thickened, horny forewing of coleopterous insect (beetles and weevils) that cover and protect the membranous hind-wings.

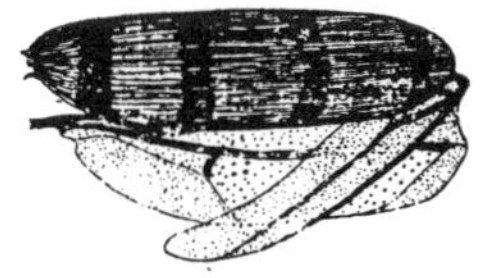
Elytron (Coleoptera)

Fringed — having well developed fringes or long hair as in thrips (Thysanoptera).

Fringed (thrips)

Halteres (balancers) — hindwings of dipteran flies; act as gyroscopic stebilisers.

Haltere (Diptera hind-wing)

Hemelytron (pl. hemelytra) — major basal portion of forewing thickened and well sclerotised like elytron and dorsal portion membranous, as in case of most true bugs (Heteroptera).

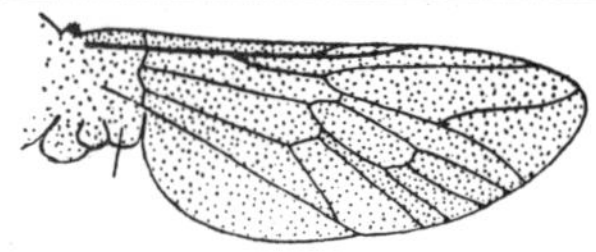

Membranous (Diptera forewing)

Membranous — thin firm more or less transparent; found in Odonata, Diptera, Ephemeroptera, Mecoptera, Neuroptera, Trichoptera and Hymenoptera.

Scally — covered with scales or setæ as in Lepidoptera (butterflies and moths).

Tegmen (pl. tegmina) — thickened leathery forewings of insects belonging to orders Dictyoptera, Orthoptera and Phasmida.

Tegmen (Orthoptera)½

Tegmen base proximal part of wing between bases of veins and the body, containing humeral and auxillary sclerites.

Tegmen buds wing-pads: hollow, flattened outgrowths of body wall that appear usually in second instar nymphs along the lateral margin of dorsum in meso- and metathorax.

Tegmen region differentiates of alar surface of wing into principal areas often marked by distinct lines of folding: axillary, jugal, remigial and vannal.

Tegmen veins narrow sclerotised tubes between upper and lower epidermis of insect's wing. Circulation of hæmolymph (blood) takes place along these and in some cases, theses also mark the position of tracheal tubes.

Winter-gnat common name for large number of non-blood sucking dipteran insects, resembling small craneflies, seen in swarms dancing on warm winter days — *Trichocera* spp. (Trichoceridæ: Tipuloidea).

Wireworm long, cylindrical larva, yellow or reddish-brown in colour, having tough, elastic, smooth and shining body with slender, well sclerotised thoracic legs but no prolegs, as those of Elateridæ (Elateriodea: Polyphaga); feed on roots of cereals, pasture and root-crops — *Agriotes lineatus* (Linnæus), *A. mancus* (Say), *Athous* spp., *Limonius* spp., et al. Type of millipede, *Julus* species (Diplopoda: Arthropoda); having 4-segmented thorax, each of the last three with a pair of walking legs; beside each abdominal segment also having two pairs of legs; herbivorous, feeds on roots of living plants.

Wohlfahrtia dipteran fly whose maggots were formerly used by physicians for cleaning up decaying tissue and bacteria in wounds.

Wood-borers wood boring insect, these may be lepidopterous, coleopterous or hymenopterous, beside termites (Isoptera). Caterpillar, beetle and weevils bore by means of jaws while sawflies (Hymenoptera) use their ovipositor. Very few insects (wood eaters) possess the enzyme (cellulase) necessary for digesting the cellulose of cell-walls.

Wood-eaters insect that can digest wood completely : termites and some wood-boring beetle that feed on fresh wood, have a dilated hind-gut containing vast assortment of micro-organism which live and breed there as symbionts and provide the necessary enzymes to break down the wood. All wood-boring insect are not necessarily wood-eaters.

Wood-nymph nickname of some dark brown butterflies (Satyrinæ: Nymphalidæ); having one or two white-centered black spot on each forewing — *Satyrus alope* et al.

Woodland grayling smoky-brown coloured butterfly, *Hipparchia fagi* (Satyrinæ: Nymphalidæ) of Europe having a white along each forewing and one smaller one on each hind-wing; found in beachwoods. Caterpillar feed on grasses.

Wooly clothed with wood-like substance such as hairs, filament etc.

Wooly aphid *Eriosoma lanigerum* (Hausmann) (Eriosomatidæ: Aphidioidea); nymphs secrete waxy filament to cover their body; pest of Temperate fruit tree.

Wooly bear densely hairy, brown caterpillar of various moths — tiger moths (Arctiidæ: Ditrysia).

Worker(s) sterile female caste in social insects (Hymenoptera — ants, bees and wasps); they build and expand the nest, store food and feed other members of the colony, specially immature individuals of either caste.

Worm(s) very small elongated soft-bodied invertebrate creature, term more commonly used for annelids, often for caterpillars also.

Worm, Canker caterpillars of *Paleacrita* spp. (Geometroidea: Ditrysia) and *Alsophila* spp. (Œnochromatidæ: Geometroiea: Ditrysia) that are pests of fruit and shade trees, found all over the world — *P. vernata* (Peck), *A. pometaria* (Harris).

Worm, Railroad nickname of a rare glow-worm, *Phryxothrix* species of South America that emits red light from near its head and greenish-yellow from either side of segmental abdomen.

Worm'-eye view (slang) view from below or from humble position.

Worm-eaten full of worm-holes.

Wormly full of worms.

X

Xanth(o)- denoting yellow.

Xanthophyll $C_{40}H_{56}O_2$; yellow pigment of leaves that is absorbed by certain insect feeding of these leaves — caterpillar of cabbage butterfly (*Pieris brassicæ* Linnæus).

Xanthopterin yellow pigment derived from pteridine — basic chemical constituent of pteroic acid — found in wings of certain butterflies (Lepidoptera) and bodies of some wasps (Hymenoptera).

Xenophagous feeding on non-specific host plants.

Xiphyriidæ (Leach 1817, 1819) Stephens 1829 family of Siricoidea (Symphyta: Hymenoptera); includes wood-wasps; antennæ inserved above the eyes; prothrax conical and pronotum cylindrical. Larvæ bore into deciduous trees (alder, oaks, poplars etc.) - *Konowia* spp. (North America), *Xiphydria* spp. (Europe, Australia and New Zealand).

Xyelidæ Haliday 1840 small only family of Xyeloides (Symphyta: Hymenoptera); includes xyelid sawflies. Larvæ feed on foliage of forest trees, mostly deciduous, sometimes conifers as well — *Xyela julii, Macroxyela* spp., et al.

Xyeloidea monofamilian superfamily of Symphyta (Hymenoptera: Oligoneoptera); earlier placed under Tenthredinoidea; antennæ inserted between the eyes well above clypeus, always having more than 3 segments, the third segment being very long ovipositor is also comparatively very long — Xyelidæ.

Xylocopidæ (Lepeletier, 1841) Shuckard 1859 Tropical family of large robust, solitary bees (Apoidea: Hymenoptera) with dense scopa; tongue long; thorax thickly pubescent; legs stout and hairy; forewing with three submarginal cells, first one shorter than second; radial cell very long and narrow — black carpenter bees [*Xylocopa latipes* (Drury)], yellow carpenter bee [*X. æstuans* (Linnæus)], Californian carpenter bee (*X. californica* Cresson), great carpenter bee. *X. virginica* (Drury) et al.

Xlomyiidæ (= Solvinæ) family of orthorrhapous flies (Brachycera: Diptera); most aberrant species annectent between Stratiomyidæ and Xylophagidæ.

Xlophaga (xylophage; xylophagous) larva of insect or molliusc eats or bores into wood.

Xlophagidæ family of Brachycera (Diptera: Oligoneoptera); formerly a subfamily of Rhagionidæ; sombre coloured flies, having third antennal segment short, not annulated with an arista on style and calypteres or squama small — *Xylophagus* spp., the maggots bore into and live in decaying wood.

Xlophilidæ old and obsolete name for coleopteran family Aderidæ (Polyphaga), characterised by deflexed head that is constricted immediately behind the eyes which are comparatively large.

Xlorictidæ (Cryptophasidæ, Uzuchidæ, Syntomidæ) family of Tinæoidea (Ditrysia: Lepidoptera); includes large-sized (wing expanse upto 75 mm), beautifully coloured moths; characterised by blunt forewings with long anterior vein, R_5 long, stalked and extending to outer margin; Cu_1 and Cu_2 widely separated and M_3 terminating beyond wing-tips. Caterpillar are borers and feeders. In. India, *Procometis* (*Hyostola*) *trochala* Meyrick feeds on dry and fallen leaves of sugarcane. (*Saccharum officinarum* Linnæus).

Xlotomous (insects) that are wood-cutting or wood-boring.

Y

Yeast type of fungi (Saccharomyetacea: Endomycetales); unicellular micro-organisms that reproduce sexually by ascus formation and asexually by budding or cell division; produce zymase that converts sugar into alcohol and carbon dioxide; used in baking and brewing.

Yellow jacket(s) hornets: common name of wasps belonging to subfamily of Vespinæ (Vespidæ: Vespoidea: Apocrita); large yellow wasps having malar-ocular spaces well defined. Their nests commonly found above ground in hollow trees, stumps, caves etc., comprise of a number of circular combs attached to one over the other and each containing numerous hexagonal paper cells. Common Indian species include, *Vespa basalis, V. cincta, V. ducalis, V. orientalis* et al.

Yellow legged clearing clearwing moths, *Sesia vespiformis* (Sesiidæ: Yponnomeutoidea) resembling a wasp; thorax with black and yellow stripe of each side, legs black and yellow; and abdomen also banded with black and yellow, male having black tuft or hair and female yellow tuft at their anal end.

Yellow meadow ant most common European species of ant, *Lasius flavis* de Geer (Formicidæ: Aculeata: Apocrita); about 4 mm long, light yellowish-brown body with pale yellow legs; builds grass-covered mounds as its nest, wherein it farms green-aphids for the sake of honeydew secreted by the aphids and relished by the ants.

Yellow-necked termite a Southern European species of termite, *Calotermes flavicollis* (Kalotermitidæ: Isoptera); characterised by the soldier having greatly enlarged head that is nearly half the length of their body.

Yellow springtail *Bourletiella lutea* (Lubbock) (Smynthuridæ: Symphypleona) is found in Northern Europe and Eastern USA feeding upon weeds, flowers, vegetable and occasionally on field crops causing considerable damage.

Yellow scale *Aonidiella citrina* (Coquillett) (Diaspididæ: Coccoidea: Sternorrhyncha); a pest of *Citrus* spp. in India, Japan and USA.

Yellow underwing(s) colloquial name for several noctuid moths (Noctuoidea: Ditrysia) having hind-wing bright yellow or orange with black border. These include — Large yellow underwing

(*Noctua pronuba,* having wing span of 60 mm); Lesser yellow underwing (*Euchesis comes*); Lunar yellow underwing (*E. orbona* having prominent black spot at the tip of each forewing); Least yellow underwing (*Lampra interjecta*), Broad-bordered yellow underwing (*L. fimbriata*), Beautiful yellow underwing (*Anaris myrtilli*) et al.

Yolk stored food material in ovum.

Yolk cells nucleated cytoplamic masses derived from egg nucleus that did not pertake in formation of blastoderm.

Yolk cleavage division of yolk masses containing one to several cleavage nuclei.

Yponomeutidæ (Hyponomeutidæ) Stephens 1829 family of brightly coloured Erimne moths (Yponomeutoidea: Ditrysia); maxillary palpi well developed and projecting; ocelli small or wanting (900 species) — *Prays citri* Milliere and *P. endocorpa* Meyrick are widely distributed pests of citrus; diamonds back moth, *Plutella xylostella* (*maculipennis*) (Linnæus) (Plutellinæ) a notorious pest of cruciferous crops is cosmopolitan while *Argyresthia* spp. are common in Europe and America.

Prays citri Milliere

Plutella xylostella (Linnæus)

Yponomeutoidea Mosher 1916 superfamily of Ditrysia (Lepidoptera: Oligneuroptera); charterised by forewing having no jugum or fibula and hind-wing without frenulum and reduced venation — Sesiidæ, Glyphipterygidæ, Heliodinidæ, Yponomeutidæ and very small family Eperimeniidæ.

Z

Zealure synthetic insect attractant.

Zebra caterpillar nickname of black and yellow striped caterpillar of a North American noctuid moths, *Ceramica picta,* (Noctuoidea: Ditrysia); common garden pest in USA.

Zebra swallowtail *Papilio marcellus* Cramer (Papilionidæ: Papilionidea); the large butterfly (wing-span 80 mm) conspicuous by broad vertical bands of dark chocolate colour and transucent white across both the wings; Summer from having long tails on the hind-wings. Common in Eastern North America, catepillar feeding on leaves of papaya (*Carica papaya* Linnæus); trees.

Zectran phenylcarbamate pesticide; a derivative of carbamic acid, 4-dimethylamino 3, 5-xylyi N-methylcarbamate; its acute oral and dermal LD_{50} to rats are 15 to 65 and 1500 to 2500 mg/kg respectively. Effective against various insect pests of ornamentals and sails (gastropod mollusc).

Zero cipher: nothing: starting point of any scale of measurement; its importance was pointed out by the Indian astronomers and mathematician Aryabhatta (476-520) Bhaskaracharya (1114-1185).

Zero gravity weightlessness.

Zethidæ a small family of Vespoidea (Aculata: Apocrita); mostly insectivorous wasps: female having well-developed and effective sting; often considered as subfamily (Zethinæ) in a family Vespidæ (Vespoidea).

Zeugloptera Dyar 1902 one of the four sub-order of Lepidoptera (Oligoneoptera: Pterypota); comprising of primitive tiny diurnal moths with biting mouthparts; mandible present; maxillary palpi vestgial or absent; proboscis not well developed and tibial present. Often regarded as not being true Lepidoptera and considered as an independent order — Micropterygoidea (Micropterygidæ).

Zeuzeridæ (Biosduval 1829) Butler 1886 family of medium-sized moths (Cossoidea: Ditrysia); characterised by presence of areole in forewing and absence of haustellum — red borer, *Zeuzera coffeæ* Nietner is major pest of coffee. *Coffea arabica* Linnæus; leopard moths, *Z. pyrina* (Linnæus) is serious pest of fruit trees, while acacia borer, *Z. eucalypti* (Boisduval) attacks acacia and fruit trees.

Zinc arsenate $Zn_3(AsO_4)$; arsenical compound (inorganic insecticide); used during earlier days before the inception of synthetic organic insecticide, for controlling the insect pests; it acted as stomach poison and was effective against mandibulate insect pests.

Zinc arsenite $Zn(AsO_2)_2$; Inorganic insecticide that was used earlier in poison-baits and not on plants, as it is phytotoxic, causing burning injury to the plants; no more in usage.

Zinc flurosilicate $ZnSiF_6.6H_20$; water soluble salt of zinc (Zn); used as fungicide and wood preservative.

Zinc phosphide Zn_3P_2; a dark, free-flowing, heavy powder having garlic-like odour; insoluble in alcohol but soluble in benzene and carbon disulphide (CS_2); a quick-acting and effective rodenticide; used to baits to control field rats, mice and other harmful rodents in fields, orchards, godowns and warehouse — Ratol®.

Zoe life.

Zoetic pertaining to life: vital.

Zoic denoting life: of rocks containing evidence of life in form of fossils.

Zoism doctrine that life originate from specific vital principle.

Zoo animal: collection of animals: gardens or park where wild animals (lives) are kept mainly for public exhibition.

Zoobiotic parasitic on another animals: living in association with another animals.

Zooblast animals cell.

Zoocecidium plant gall caused by insect.

Zoochemistry chemistry of animals (insect) body.

Zoocide toxic substance (chemical) that can kill animals specially the mammals such as rabbits, rodents, (wild) dogs, jackal etc.

Zoodynamics animals physiology.

Zoogamy sexual production in animals.

Zoogeny doctrine of theory of origination of living organisms.

Zoogeographical realms region of the Earth's surface; originally six, now nine.

Palæarctic — the Old World: Europe, Asia (North of Himalayas) and Africa (North of Sahara).

Ethiopian — Africa and Arabia (South of Tropic of Cancer) and Malagasy. Of late Malagasy is being regarded as separate division Madagascan.

Oriental — South Asia east of Persian gulf, including the Indian subcontinent, Malaysia, Combodia, Vietnam, Indonesia, Philippines and Taiwan (Formosa).

Australian — Australia, New Guinea, Tasmania, New Zealand and Pacific oceanic island and Hawali. Now pacific islands of Volcanic origin are considered as Polynesian while New Zealand and neighbouring islands from New Zealand.

Nearctic — the New World: the North America (North of Tropic of Cancer) including Greenland and other islands of Canada.

Neotropical — Tropical New World: Central America (South of Mexican platea), South America, West Indies, Cuba and Falkland islands.

Zoogeography biological science dealing with distribution of animals (including insects) in space and time.

Zoogonous viviparous.

Zoogony viviparity.

Zoography descriptive zoology.

Zooid free moving cell: animals-like.

Zoolite fossil animals (including insect).

Zoologist person having specialised in zoology.

Zoology one of the main branches of biology — the science that deal with natural history of animals. Some of its specilised branches are Spongology (Sponges — Porifera), Helminthology (parasitic worms), Nematology (nematodes), Malcology (molluscs), Conchology (shells of molluscs), Arthropodology (arthropods), Carcinology (crustaceans), Entomology (insects), Myriapodology (myriapods), Acarology (mites); lcthylology (Pisces of fishes), Herpetalogy (amphibians and reptiles), Orinthology (Aves or birds), Mammalogy (mammals) etc.

Zoometry comparative measurement of part of animals (insects).

Zoon (pl. zoa) total product of a fertilised ovum: individual.

Zoonomist person well-versed in zoonomy.

Zoonomy branch of Zoology that deals with function of various part of animals body: animals physiology.

Zoopathologist person having sound knowledge of zoopathology.

Zoopathology (zoopathy) biological science that deals with diseases of animals (insects).

Zoophagous feeding on animals (insects).

Zoophyte plant-like-animals- coral, sponge etc., also animals-like plants.

Zooplankton microscopic insect and other animals that move passively in aquatic ecosystems. *Cf.* phytoplankton.

Zootaxy systematic zoology.

Zootomist animals (insect) anatomist.

Zootomy dissection of an animals (insect), to study its comparative anatomy.

Zootoxin toxin produced by an animals (insect).

Zootype an animals (insect) serving as type.

Zopheridæ Boving and Craighead 1931 small coleopteran family, separated from Tenebrienidæ (Tenebrionoidea: Polyphaga); distinguished by having movable abdominal stermites.

Zorabiid(s) any species of insect belonging to genus *Zorotypus* (Zoropidæ: Zoraptera) — *Z. ceylonicus* Silvestri, *Z. snyderi* et al.

Zoraptera Silvestri 1913 primitive hemipteroid order of little known minute (<3 mm long) insect; head distinct with 9-segmented moniliform antennæ; mouthparts mandibulated, adapted for mastication; thorax as long as abdomen, prothorax well developed; mostly apterous, wings when present, membranous having 2 to 3 veins and capable of being shed; cerci very short; ovipositor wanting Metamorphosis slight or wanting. Found in small colonies in decaying wood, feeding on mites, small insect and dead arthropods — one family (Zorotypidæ), one genus (*Zorotypus*) with 22 described species — Zorabiids.

Zorotypidæ Silvestri 1913 only family of Zoraptera (Palæcoptera: Pterygota); comprising of single genus *Zorotypus*; minute insects, about 2.7 mm long; compound eyes and ocelli wanting, wingless of alate (wing-spread maximum 5 mm); thorax as long as abdomen, prothorax well developed; abdomen 11-segmented having 10 pairs of spiracles: feeding on dacaying organic matter (22 species)—*Z. guineensis* Silvestri (apterous), *Z. hubbardi* Caudell (alate) et al.

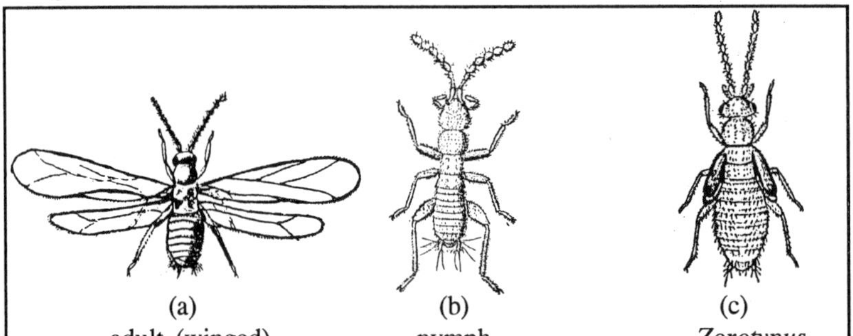

(a) adult (winged) (b) nymph *Zorotypus hubbardi* Caudell (c) *Zorotypus guineensis* Silvestri

Zygentoma suborder of Thysanura (Ectotropha: Apterygota); primitive wingless insect having elongated flattened body, ectognathous mouthparts adapted for chewing; cerci and median caudal filament long and many joined; single superfamily — Lepismatoidea (fish brats, fire moths etc.).

Zygo- yolk.

Zygoentomata group of primitive wingless insects, characterised by presence of abdominal styli and absence of copulatory organs, sperms transfer being by externally deposited spermatophores, moulting continues even after attaining sexual maturity — Thysanura.

Zygopleural bilaterally symmetrical.

Zygoptera Selys 1854 suborder of Odonata (Odonatoptera: Pterygota); includes damselflies; weak filter; eyes small, well separated and protrudine laterally; wings petiolated basally, fore- and hind-wing almost subequal in form and venation four superfamilies —Coenagrioidea, Hemiphlebioidea, Lestinoidea and Agrioidea.

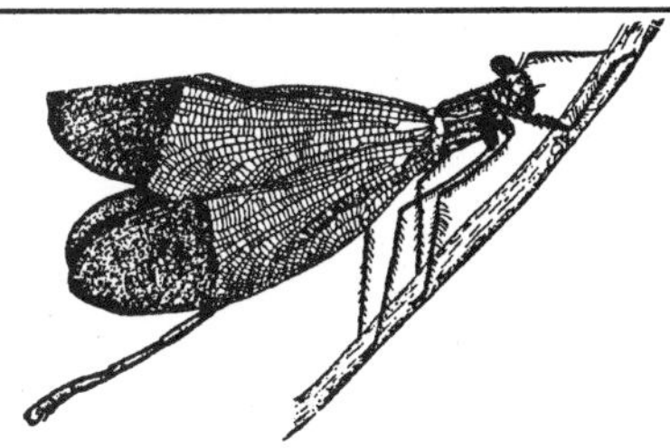

Agrion æquabile (Sa)
(Agriodea)

Zygote fertilised egg: product of union of male and female gametes.

Zygotic number diploid chromosome number.

Zymase complex of enzymes that catalyse glycosis.

Zymogen proenzyme: inactive precursor of enzyme; liberates an active enzyme on reaction with appropriate kinose.

Zymogenesis transformation of zymogen into active enzyme.

Zymophore active portion of an enzyme.

Zythiaceæ family of Sphæropsidales (Coleomycetes: *Fungi imperfecta*) that contain many plants and insect pathogens.

Zunzier common name of some pyralid moths (Ditrysia: Lepidoptera); used especially in Germany for the stored-grain pests — Indian meal moth *Plodia interpunctella* Hubner, Mediterranean flour moth, *Ephestia kuhniella* Zeller, chocolate moth, *E. elutella* (Hübner), raisin moth *E. figuleilla* Gregson et al.

Zygænidæ (Anthroceridæ, Chalcosiidæ, Pyomorphidæ) family of Zygænoidea (Ditrysia: Lepidoptera) that includes forester and burnet moths having slender body and broad, large, brilliantly coloured wings resembling butterflies. Bright colour warn their predators (birds etc.) that they are distasteful (batesian mimicry).

Caterpillar with tufted hairs and pupæ enclosed in tough elongate membranous cocoons — *Cyclosia imitans* Butler closely resembles Nymphalid butterfly *Danais aglea* (Cramer); other Indian species includes *Eterusia œdea* Linnæus, *Lophasoma quadricolor* Walker and *Trypanophora semihyalina* Kollar.

Zygæninæ a subfamily of Zygænidæ (Zygænoidea: Ditrysia) includes Burnets and foresters; confined of Palæarctic region; diurnal in habit and characterised by antennæ being distally enlarged (100 species) — six spotted Burnet *Zygæna filipendulæ;* bloodwort Burnet *Z. læta*; metallic green Foresters *Ino* spp., et al.

Zyænoidea Gravenhorst 1843 superfamily of diurnal moths (Ditrysia: Lepidoptera); characterised by maxillary palpi vestigial or absent; posterior tibial spurs reduced and middle often absent; wings large,

Appendix

1. CLASSIFICATION

ARISTOTLE (384-320 B.C) — the father of zoology — was first to suggest the grouping of plants and animals; but is was Carolus A. Linnæus (1707-78), rightly called the father of classification, who placed all the living organism into two major groups (kingdoms), namely Animalia and Plantæ. He also suggested binomial nomenclature; described and named a large number of species of plants and animals, including 3000 of insects. His classification of insects was bases only on wing-venation (alary) and he established (1758) eight classic (orders) — Eleutherata (Coleoptera), Ulonata (Orthoptera, Blattes, Dermaptera), Synistata (Ephemera, Phryganes, Apterygotes, Perlids, Neuroptera, Psocids, Panorpes, Hymenoptera, Termites and some Crustacea), Unogata (Libellula, Myriapoda and Arachnida), Glossata (Lepidoptera), Rhyngota (Hemiptera, Aphaniptera and Thysanoptera) and Antiliata (Diptera, Anoplura, Mallophaga, certain Arachnids and some Crustacea), beside Agonata (Scorpion and certain Crustacea). His student J.C Fabricius (1745-1808) — father of entomology — improved upon this system by studying structure and function of mouthparts and added one more order Odonata (1793). This classification remained in vogue till about end of 19th century during which period, the insect were gradually separated from crustaceans , myriapods and arachnids.

Foundation of modern system of classification was laid by Brauer (1885) who recognised the fundamental division of insects into two subclasses — Apterygota (Primitive wingless) and Pterygota (winged or secondary wingless). This was modified from time by Borner (1904), Handlirsch (1908), Martynov (1925, 1938) , Imms (1934) Brues and Melander (1945), and other.

The classification followed in this dictionary is mainly based on the one suggested by Imms and revised by Richards and Davies (1977) with slight modification — elevating the suborder of Hemiptera (Heteroptera and Homoptera) as independent orders as suggested by

Martynov (1938). Thus there are 30 orders, grouped into 18 superorder and divided into more than 1200 families, comprising of several hundred genera and approximately 937000 described living species, besides 13000 fossil forms.

Borner, C, 1904: Zur Systematik der Hexapoda. *Zoologisher Anzeiger,* 27: 511-533, Leipzig. Brauer, F, 1885: Systematisch-zoologische studien. *Sitzungsberichte der Kais Akademic der Wissenchaften in Wien,* 91: 237-413, Wien (Austria).

Brues, C.T. and A.L Melander, 1945 : *Classification of Insect.* Cambridge Mass. Fabricius, Johann Christian, 1781: *Species insectorum,* 1: 552 pp., Hamburgi.Fabricius Johann Christian, 1793: *Entomologia Systematica,* 2, Copenhagen, Hafniæ. Handirsch, A. 1906. *Die fossilen inseckten und die Phylogenie der* rezenten Formen, 1430 pp., Englemann, Leipzig.

Imms, A.D., 1934: *A General Textbook of Entomology, 727* pp. Methuen & Co., London. Linnæus, C.A., 1758: *Systema Naturea, Editio decima,* Reformato, Holmiæ.

Martynov, A.V 1925: Ueber zwei Grundtypen der Flugel bei den Insekten and ihre Evolution. *Zeitschrift fuer Morphologie und Oekologie Tiere,* 4: 465-501 Berlin.

Martynov, A.V 1938: Etudes sur 1' historie geologique et de phylogenie des orders des insect (Pterygota) — 1 partie: Palæeoptera et Neoptera: Polyneoptera. *Travaux de'I' Institul palezoologique de I' Academic des Science de I'URSS,* 7(4): 150 PP.

Richard, O.W and R.G. Davies, 1977: *Imms' Genral Textbook of Entomology,* 2 volumes, 1354 pp., Chapman and Hall Ltd., London.

1.1 LIVING ORGANISM (2200000)

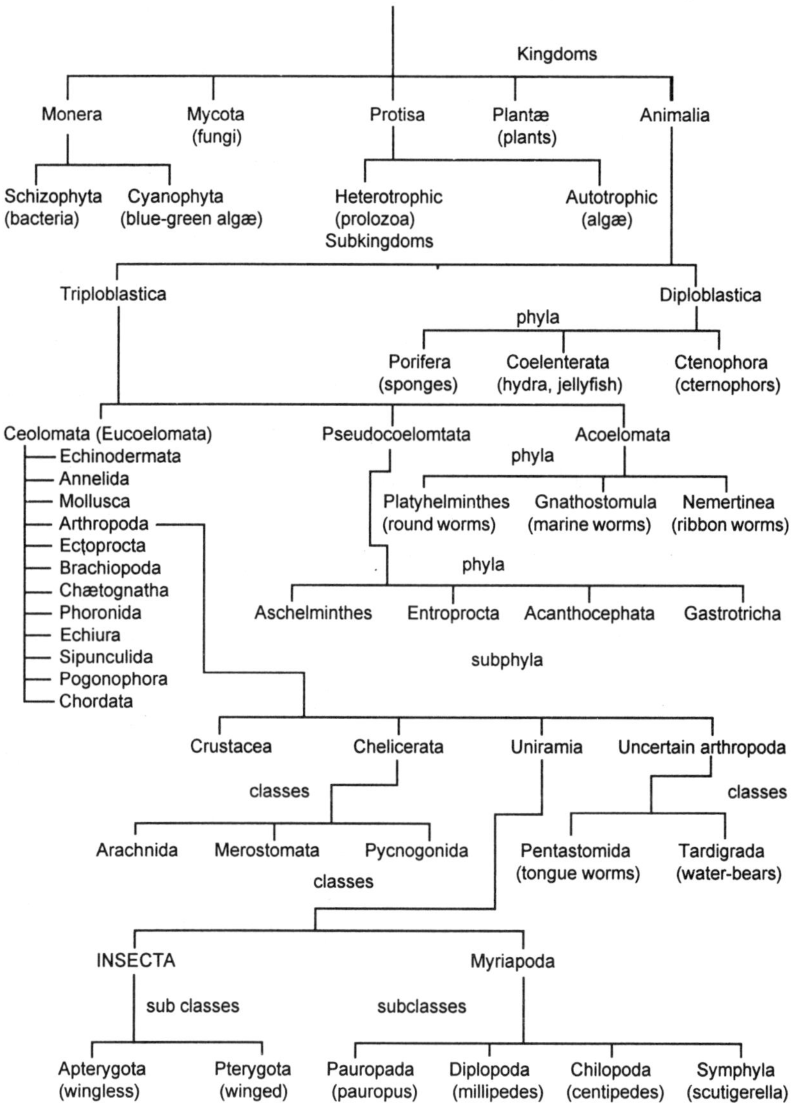

1.2 INSECTA Linnæus 1758 (HEXAPODA Latreille 1825)

sub-classes	infra-classes	Super-orders	Orders -suborders	Described species	Example
APTERYGOTA BRAUER 1885	Ectotropha Grassi 1889	Zygoentomata	Thysanura Latreille 1796 Microcoryphia Zygenotoma	800	 bristletails silverfish
		Diplurata	Diplurata Bomer 1904 Dicellurata Khabdura	700	 japigids campodide
	Entotropha Grassi 1889	Myrientomata	Protura Silvestri 1907 Eosentomoidea Acerentomoidea	300	telson tails
		Oligoentomata	Collembola Lubstock 1870 Arthopleona Symphypleona	2200	snowflies
PTERYGOTA BRAUER 1885	Palæoptera Martyov 1923	Ephemeroidea Martynov 1938	Ephemeroptera Hæakel 1896	2500	mayflies
		Odonatoptera	Odonata Fabricius 1793 Zygoptera Anisozygoptera Anisoptera	6500	 damselflies dragonflies
	Polyneoptera Martyov 1923	Orthopteroidea Handlirsch 1903	Plecoptera Burmeister 1839 Archiperlaria Filipalpia Setipalpia	1800	stoneflies
			Grylloblattodea Brues & Melander 1932	16	grylloblatids
			Orthoptera Olivier 1789 Ensifera Cælifera	28000	 crickets locusts
			Phasmida Leach 1815 Phasmoidea Bacteroidea	4000	 leaf-insects stick-insects
			Embioptera Shipley 1904 Protembioptera Euembioptera	400	embiids
		Dermapterioidea Martoyov 1925	Dermaptera Leach Hemimerina Arixentina Forficulina	1500	earwigs
		Blattopteroidea	Dicyoptera Leach 1818 Blattaria Mantodea	8000	 cockroaches mantids
			Isoptera Comstock & Comstock 1895	2800	termites
			Zoraptera Silvestri 1913	22	zorabiids

1.3 OTHER SYSTEMS OF CLASSIFICATION

A. **Based on wings and wing-venation** (*pteron* = wing)

I. APTERYGOTA (without wings): Ametabolous — Thysanura, Diptura, Protura, Collembola

II. PTERYGOTA (having wings): Metabolous —

1. **Exopterygota (Heterometabolous):**

(*i*) Palæoptera (Hemimetabolous) = Ephemeroptera, Odonata.

(*ii*) Polyneoptera (Paurometabolous) — Plecoptera, Grylloblattodea, Orthoptera, Phasmida, Embioptera, Dermaptera, Dictyoptera, Isoptera, Zoraptera.

(*iii*) Paraneoptera (Paurometabolous) — Psocoptera, Mallophaga, Siphunculata, Homoptera, Heteroptera Thysanoptera.

2. **Endopterygota (Holometabolous)**

Oligoneoptera — Neuroptera, Coleoptera, Sterpsiptera, Mecoptera, Trichoptera, Lepidoptera, Diptera, Siphonaptera, Hymenoptera.

B. **Based on metamorphosis**

I. AMETABOLA (no metamorphosis): Apterygotes — Thysanura, Diptura, Protura, Collembola.

II. METABOLA (having metamorphosis): Pterygotes —

1. **Heterometabola (incompelete or direct metamorphosis)** : Exopterygotes —

(*i*) Hemimetabola (nymphs and adults not alike) — Ephemeroptera, Odonata.

(*ii*) Paurometabola (nymphs and adults alike) — Plecoptera, Grylloblattodea, Orthoptera, Phasmida, Embioptera, Dermaptera, Dictyoptera, Dictyoptera, Isoptera, Zoraptera, Phasmida, Embioptera, Dermaptera, Dictyoptera, Isoptera, Zoraptera, Psocoptera, Mallophaga, Siphunculata, Homoptera, Heteroptera, Thysanoptera.

2. **Endopterygota (Holometabolous)**

Neuroptera, Coleoptera, Strepsiptera, Mecoptera, Trichoptera, Lepidoptera, Diptera, Siponaptera, Hymenoptera.

1.4 KEYS TO SUBORDERS/SUPERFAMILIES

THYSANURA Latreille 1796
(suborder)

Body strongly convex; eyes large with numerous facets and contiguous above; maxillary palpi long, 7-segmented; abdomen with paired stylets on II to IX segments .... Microcoryphia

Body flattened; eyes absent or small, with few facets and never contiguous above; maxillary palpi 5- or 6-segmented; abdomen with paired styletes only in VII, VIII and IX segments .... Zygentoma

DIPLURA Borner 1904
(suborder)

Cerci unsegmented, heavily sclerotised and forcep-like .... Dicellurata

Cerci segmented (short, stout and few segmented or long and strong segmented) .... Rhabdura

PROTURA Silvestri 1907
(superfamilies)

Spiracles (2 pairs) and tracheæ present; abdominal appendages vestiges blarticulate ; VIII abdominal segment without pectines .... Eosentomoidea

Spiracles and tracheæ absent; appendages on II and III abdominal segments unlarticulate; VIII abdominal segment with a pair of pectines .... Acerentomoidea

COLLEMBOLA Lubbock 1870
(suborder)

Body more or less elongate and segmentation distinct; last 2-3 segments partly fused together .... Arhtropleona

Body subspherical or globose with segmentation obscure, obliterated in the abdomen .... Symphyleona

EPHEMEROPTERA Hæckel 1896
(subfamilies)

1. Hind tarsi four free segments; fifth, it present, fused with tibia 2
   Hind tarsi having five free segments 3

2. M_{1+2} and Cu_1 of forewing diverging strongly at base ....Ephemeroidea

   M_{1+2} and Cu_1 of forewing parallel at base ....Baetoidea

3. M_{1+2} of forewing parallel at base to Cu_1 ....Ecdyuroidea

---

For keys to families please consult:

Mani, M.S., 1973: **General Entomology**, Oxford and IBH Publishing Co., New Delhi.

Purthi, Hem Singh, 1969: **Textbook on Agriculture Entomology**, ICAR, New Delhi.

ODONATA Fabricius (suborder)

1. Both pairs of wings equal (similar), more or less stalked at the base — damselflies .... Zygoptera

   Hind wing much wider at the base with venation different from forewings — dragonflies 2

2. Wing slightly unequal; discoidal cell a simple quadrilateral, different in fore-and hind-wings .... Anisozygoptera

   Wings unequal; discoil cell divided into a triangle and supratriangle — true dragonflies ... Anisoptera

PLECOPTERA Burmeister 1839 (suborder)

1. Primitive forms, wings having entire margin without any notch and anal lobe of hind-wings having archedictyon .... Archiperlaria

   Hind-wing with anal lobe separated by a marginal notch and anal lobe without archedictyon 2

2. Mandibles well developed; clypeus and labrum not concealed under the frons .... Fillipalpia

   Mandibles weakly developed or vestigial; clyperous and labrum concealed under frontal shelf; third tarsal segment longer than other two combined .... Setipalpia

GRYLLOBLATTODA Brues and Melander 1932 (NOTOPTERA Crampton 1915)

No suborder or superfamilies. Only one family Grylloblattidæ, comprising of three genera and 16 species — *Gryllobatta* (9), *Galloisiana* (6) and *Grylloblattina djakonovi.*

ORTHOPTERA Olivier 1789 (suborder)

Antennæ longer than the body and having more than 30 segments; auditory organs, if present, on foretibia; tarsi usually 4-segmented; ovipositor long and needle - of sword-sharped .... Ensifera

Antennæ shorter than the body and having less than 30 segments; auditory organs, it present, near the base of abdomen; tarsi usually triarticulate; ovipositor short. .... Cælifera

PHASMIDA Leach 1815
(superfamilies)

Winged or wingless with leaf-like expansion on forefemora — leaf-insect .... Phasmoidea

Wingless, thin elongated body with long antennæ and legs, without any expansionon forefemora: I abdominal segment free from thorax — stick-insect .... Bacterioidea

EMBIOPTERA Shipley 1904

Two suborder —

Protembioptera (Permian period); only one family Protembidæ often under fossil order Protothoptera Euembioptera; seven families — Clothodidæ, Oligotomidæ, Teratembiidæ, Oligembiidæ, Anisembiidæ, Notoligotomidæ and Embiidæ.

DERMAPTERA (de Geer 1773) Leach 1815
(suborder)

1. Compound eyes atrophied or wanting; instead of forceps a pairs of long unsegmented cerci at the anal end of abdomen; ectoparasitic or rodents in South Africa .... Hemimerina

   Compound eyes present; body terminating in a pair of movable forceps 2

2. Mandibles densely fringed with hairs on the inner margin and toothed only at the tip; compound eyes much reduced; forceps weak and setose. Ectoparastic on Old World bats on occurring in caves of bats .... Arixenina

   Mandible dentate both at the tip and base on the inner side; compound eyes normal; forceps strongly sclerotised, smooth and opposables; free living forms; cosmopolitan .... Forficulina

DICTYOPTERA Leach 1818
(suborder)

Body oval, much flattened; two ocelli represented by whitish fenestræ; pronotum large projection over the head and concealing it from above; legs cursorial, large and similar — cockroaches .... Blattaria

Body elongated; three dilate ocelli; prothorax greatly enlarged but not concealing the head; forelegs raptorial and usually heavily spined beneath — mantids .... Mantodea

ISOPTERA (Bruile 1832) Comstock and Comstock 1895

No suborder or superfamilies. Nine families — Hodotermitidæ, Termopsidæ, Kalotermitidæ, Rhinotermitidæ, Termitidæ, Mesotermitidæ, Stylotermitidæ, Indotermitidæ and Serritermitidæ.

ZORAPTERA Silvestri 1913

No suborder of superfamilies. Only one family Zorotypidæ, comprising of only one genus *Zorotypus* including 22 described species.

PSOCOPTERA Shipley 1904
(suborder)

1. Antennæ usually 13-segmented; labial palpi single segmented and tarsi 2- or 3-segmented .... Psocomorpha

   Labial palpi 2-segmented and tarsi 3-segmented 2

2. Antennal segment without annulations and with more than 20 segment .... Trogimorpha

   Antennal segment annulated and having 12 to 17 segments .... Troctomorpha

MALLOPHAGA Nitzsch 1818*
(suborder)

Antennæ 4-segmented and capitate; mandibles more horizontal maxiilary palpi 4-segmented; meso- and matathorax usually separate .... Amblycera

Antennæ 3- to 5-segmented and filiform; mandibles vertical and without maxillary palpi; meso- and metathorax generally fused .... Ischnocera

Antennæ 3- to 5-segmented; head prolonged anteriorly into a rostrum, mandible at apex of rostrum working outwards; labium and maxillæ vestigial .... Rhynchophthrina*

SIPHUNCULATA Latrielle 1825*
(ANOPLURA Lucas 1840)

No suborder, no superfamilies. Three (five) families — Echinopthiriidæ, Pediculidæ (including Pthiridæ), Hæmatopinidæ (including Hæmatopinoididæ).

HOMOPTERA Latreille 1817 (suborder)

Antennæ with 2 basal large segments and a proximal arista; rostrum more cephalic arising from ventral base of head; tarsi usually triarticulate; active forms .... Auchenorrhyncha

Antennæ well developed but without arista; rostrum more apparently from between the forelegs; tarsi uni- or blarticulate; inactive forms, incapable of locomotion .... Sternorrhyncha

Antennæ short, 3-segmented without terminal arista and concealed beneath the head .... Coleorrhyncha (not in India)

HETEROPTERA Latreille 1810 (suborder)

Antennæ longer than head, usually free, concealed in the groove; if short, eyes and ocelli absent; arolia present or absent .... Gymnocerata

Antennæ shorter than head, usually hidden in a cavity beneath the eyes; arolia absent; mostly aquatic .... Cryptocerata

---

* Mani (1973) consider Mallophaga, Rhynchophthirina and Anoplura (Siphunculata) as suborder or order Phthiraptera.

THYSANOPTERA Haliday 1836 (suborder)

Intermediate antennal segments annulate and clothed with microtrichia; forewings usually longer than hind-wings; X abdominal segments nor tubiform and rarely sclerotised; female with saw-like ovipositor .... Terebrantia

Intermediate antennal segments smooth; forewings similar to hind-wings; X abdominal segment tubular or conical and strongly sclerotised; female without ovipositor .... Tubulifera

NEUROPTERA Linnæus 1758

There appears to be no agreement among faxonomist as to the proper grouping of this order. Some entomologists divide Neuropteroidea into three orders — Megaloptera Latreille 1802 (alderflies, dobsenflies and fishflies), Raphidioptera Burmeister 1835 (snakeflies) and Neuroptera or Planipennia (antlions and lacewings); majority consider only Neuroptera as order with Megaloptera and Planipennia as suborder and Raphidioptera as superfamily (Raphidoidea) of Megaloptera.

COLEOPTERA Linnæus (suborder)

1. Hind coxæ immovably fixed to metasternum, completely dividing first visibleabdominal sternite; hind-wings having 2m-cu cross-veins .... Adephaga

   Hind coxæ not immovably fixed to metasternum and not completely dividing the first visible abdominal sternite 2

2. Prothorax without notoplural sutures; hind-wings without cross-veins ....Polyphaga (including Rhynchophora*)

   Prothorax with notoplural suture, wings with distal part spirally coiled in repose .... Archostemata (not in India)

   Prothorax with notopleural sutures; head sunk into prothorax; antennæ short; wings well developed, more or less distinct oblongum but Rs reduced .... Myxophaga (Crowson, 1955)

STREPSIPTERA Kirby 1813

No suborder. seven families — Mengeidæ, Halictophacidæ, Elenchidæ, Myrmecolacidæ, Stylopidæ, Callipharixenidæ and Stichotrenmatidæ.

---

* Some entomologists place Rhynchophora as separate suborder that include beetle having head extented to form a rostrum or snout in front of the eyes; comprising of only one superfamilies — Curculionoidea (Anthirbidæ, Apionidæ, Brenthidæ, Curculionidæ and Scolytidæ).

MECOPTERA Comstock and Comstock 1895 (suborder)

Body generally depressed, wings and short and broad; male genitialia not swollen .... Protomecoptera

Body generally cylindrical; wings narrow and slender; male genitialia swollen .... Eumecoptera

TRICHOPTERA Kirby 1817 (suborder)

Last segment of maxillary palpi annulated, flexible, generally much longer than the others; dorsal part of tentorium not developed. Larvæ making shares of nets .... Annulipalpia

Last segment of maxiilary palpi not annulated generally less elongated. Larval IX abdominal tergite with a sclerotised plate. .... Integripalpia

LEPIDOPTERA Linnæus 1758
(a) old and artificial

Moths — generally nocturnal, having comparatively large body and small wings; antennæ variable but not clubbed .... Heterocera

Butterflies (including **skippers**) — generally diurnal, slender body with broad colourful wings and clubbed antennæ .... Rhopalocera

Antennæ inserted wide apart; antennal club mostly recurved apically; vein R of forewing with five branches, all arising separately from the cell .... Hesperioidea (skippers)

Antennæ inserted close together; antennal club rounded apically; vein R of forewing with 5 branches, but R_4 and R_5 stalked .... Papilionidea (butterflies)

(b) Old — based on wing coupling

Wings interlocked by jugum; venation of fore- and hind-wings similar .... Jugatæ

Wings interlocked by frenulum, ventation of fore- and hind-wings not similar .... Frenetæ

(c) Old — based obsolete — based of mouthparts and venation

Mouthparts mandibulate or rudimentary, wings venation almost identical in both fore- and hind-wings .... Homoneura

Mouthparts suctorial; wing venation not identical in fore- and hind-wings .... Heteroneura

(d) Current/Recent
(suborder)

1. Mandibles functional, maxilla lacinia well developed, galea not haustellate.
Wings with aculet, venation of fore- and hind-wing identical; no frenulum. Caterpillar with 8 pairs of abdominal legs, each terminating in a single claw. Pupa dicticuous, exarate .... Zeugloptera

Mandible non-functional or vestigial, maxilla without lacinia, galea mostly haustellate unless mouthparts are much reduced. Caterpillar with not more than 7 pairs of abdominal legs 2

2. Wings not a rarely a little aculeate, venation of fore- and hind-wings not similar, forewing without jugum or fibula, hind-wing with frenulum or with amplexiform coupling. Caterpillar with not more than 5 pairs of crochet-bearing abdominal legs. Pupa obtect without functional mandibles .... Ditrysia
Wings distinctly aculeate (except Hellozelidæ) 3

3. Wing venation reduced and different in fore- and hind-wing. Male with rudimentary fibula on forewing and strong frenulum on hind-wing; female with strong fibula and no frenulum. Pupa obtect with rudimentary mandibles .... Monotrysia (part) (Neticuloidea, Incurvarioidea)
Wing-ventation of both Paris identical, frenulum not developed ... 4

4. Forewing with fibula, hind-wings with some costal spines. Caterpillar apodous. Pupa exarate with hypertrophied mandible .... Dacononypha

Forewing with strong humeral veinlet and a long jugum, hind-wing without costal spines, aculet not numerous. Caterpillar subterranean with 5 pairs of crochet-bearing abdominal legs. Pupa obtect with rudimentary mandible .... Monotrysia (part) (Hepialoidea)

DIPTERA Linnæus 1758
(a) old and not in usage

Straight-seamed flies; pupæ split longitudinally along the middorsal line; Nematocera, Brachycera .... Orthorrhapha

Circular-seamed flies; adults have short antennæ and emerge from pupæ by a circular slit — Aschiza, Schizophora .... Cyclorrhapha*

*Some Entomologist still consider Cyclorrhapha as suborder with two sections — Aschiza and Schizophora.

(b) Current/Recent (suborder)

1. Abdomen longer than head and thorax; many-segmented; discal cell usually absent, cubital cell, if present, open. Larvæ eucephalic; pupæ free .... Nematocera

   Antennæ shorter than head and thorax; with fewer segments; discal cell usually present; cubital cell closed. Larvæ acephalic or hemicephalic 2

2. Arista when present terminal. Larvæ with incomplete retractile head; pupæ free .... Brachycera

3. Ptilium absent; frontal suture obscure; lunula absent or indistict; cell Cu elongate, extending to more than half the length of wing .... Aschiza

   Ptilium present (rarely absent); frontal suture and lunula distinct; cell Cu short or vestigial, wings sometimes absent or reduced .... Schizophora

SIPHONAPTERA Latreille 1825* (AHANIPTERA Kirby 1826)

No suborder. Nine families — Hectopsyllidæ, Pulicidæ (including Archæopsyllidæ, Dolichophsyllidæ, Ischnopsyllidæ, Hystrichopsyllidæ, Macropsyllidæ, Ctenophthalmidæ, Tungidæ and Uropsyllidæ.)

HYMENOPTERA Linnæus 1758 (suborder)

Hind-wings with 3 basal cells; trochanters always biarticulates; abdomen broadly sessile .... Symphyta

Hind-wings with two or one or no basal cells; trichanters uni- or biarticulate; abdomen petiolated, base sharply constricted, never broadly sessile .... Aprocrita

*Oudemans (1909) errected two suborder — Fracticipita and Integricipita, but the grouping was of little systematic value; no more in usage.

1.5 FOSSIL INSECT ORDERS AND FAMILIES (Palæozoic era)

Superorder PAL.ÆDICTYOPTERA Goldenberg 1854

Order EUPALÆODICTYPTERA Jeannel 1945 — Breyaridæ, Calvertiellidæ, Dictyoneuridæ, Fouqueidæ, Lamprotilidæ, Paoliidæ, Polycreagridæ, Protagrionidæ, Thesoneuridæ, Syntonopteridæ, Spilapteridæ at al.

Order PROTHOTEMIPTERA Handlirsch 1903* — Eugereonidæ, Mischopteridæ, Brodiidæ, Aspidothoracidæ, Bariohymenidæ, Foririidæ, Protohymenidæ, Scytohymenidæ, Sphecopteridæ, Ischnoptilidæ, Elmoidæ, Diaphanopteridæ, Marynoviidæ, Asthenohymenidæ and Biarmohymenidæ.

Superorder EPHEMEROPTERA Martynov 1935

Order PROTEPHÆMEROIDEA Handlirach 1908 — monofamilial, monogeneric and monospecific — Triplosobidæ (*Triplosoba pulchella* recorded from France).

Superorder ODONATOPTERA Martynov 1938.

Order MEGANISOPTERA Martynov 1932 PROTODONATA* — Meganeuridæ, Erasipteridæ and Paralogidæ.

Superorder BLATTOPTEROIDEA Martynov 1938

Order PROTOBLATTOIDEA Handlirsch 1908 — Stenoneuridæ, Protophasmidæ, Eoblattidæ, Asyneritidæ, Epideigmatidæ, Ecuænidæ, Adiphlebiidæ, Anthracothremmidæ et al.

Superorder DERMAPTEROIDEA Martyov 1925

Order PROTELYTROPTERA Tillyard 1931 — Elytroneuridæ, Megelytridæ, Blattelytridæ, Protelyridæ and Archelytridæ.

Superorder COLEOPTEROIDEA Handlirsch 1908

Order PROTOCOLEOPTERA Tillyard 1924 — monofamilial and monospecific — *Protocoleus* and Archelytridæ.

Superorder COLEOPTEROIDEA Handirsch 1908

Order PROTOCOLEOPTERA Tillyard 1924 — monofamilial and monospecific — *Protocoleus mitchelli* from Upper Permian of Belmont, New South Wales.

Superorder ORTHOPTEROIDEA Handirsch 1903.

Order PROTORTHOPTERA Handlirsch 1908* — Ischnoneuridæ, Geraridæ, Cenmidolestidæ, Oedischiidæ, Sthenaropodidæ, Spanioderidæ, Homeodictytidæ, Blattinopsidæ, Euryptilinidæ, Liomopteridæ, Dermopteridæ, Stereopteridæ, Protokollariidæ, Probnisidæ, Adeloneuridæ, Protembiidæ, Chelopteridæ et al.

Order CALONEURODEA — Caloneuridæ, Synomaloptilidæ, Euthygrammidæ, Permobiellidæ, Paleuthygrammidæ, Plesiogrammidæ and Anomalogrammidæ.

In addition to these orders, Mani (1973) mentions Protoperlaria, Hadentomoidea, Sypharoteroidea, Mixotermitoidea, Reculoidea, Hapalopteropidea, Synarmogoidea, Palæohemiptera, Paramecoptera, Protodiptera and Paratrichoptera*.

---

* These insects continued to live till close of Triassic period (Mesozoic era).

2. ETYMOLOGY

THE NAMES of insect orders are derived from Greek language *pteron* = wings (*apteron* = without wings, *opteron* = with wings); *oura* = tail.

Palae*optera*	palaeo = ancient	ancient-winged
Megal*optera*	mega = big	big-winged
Ne*optera*	neo = new	new (modern)-winged
Polyne*optera*	poly = many	many-veined, modern-winged
Thysan*ura*	thysanos = fringe	having fringed-tail
Dipl*ura*	diplos = double	having double tail
Prot*ura*	protos = first/simple	simple-tailed
Ephemer*optera*	ephemeros=living for a day	winged insects that live for less one day.
Odonata	odos = tooth	conspicuous toothed maxiillæ
Plec*optera*	plecos = fold	folded (platted)-winged
Orth*optera*	orthos = straight	straight-winged
Phasmida	phasma = apparition	like dead-leaf or stick
Embi*optera*	embios = living	lively-winged
Derm*aptera*	derma = skin	skin-winged
Dicty*optera*	dictyon = net	net-winged
Is*optera*	isos = equal	having equal wings
Zor*aptera*	zoros = completely	completely wingless (first species described was wingless)
Psoc*optera*	psocus = to grid	gnawing insects (book-lice)
Mallophaga	mallos = wool;	wool-feeding (bird-lice) phagein= to eat
Hom*optera*	homos = homogenous	homogenous-winged
Heter*optera*	hetero = different	fore- and hind-wings different
Thysan*optera*	thysanos = fringed	fringed-winged
Neur*optera*	neuron = nerve (lace)	lace-winged
Cole*optera*	coleos = sheath	sheath-winged (elytra)
Strepsi*ptera*	strepsis = twisted	twisted-winged
Mec*optera*	mecos = long	long-winged
Trich*optera*	trichos = half	hairy-winged
Lepid*optera*	lepis = scale	scaly-winged
Di*ptera*	di = two	two-winged
Siphon*aptera*	siphon = tube	sucking and wingless
Hymen*optera*	hymen = membrane	membranous-winged

3. ABBREVIATIONS AND ACRONYMS

Latin

ad. lib	*ad libitum* — freely
aff.	*affinis* — akin to
al.	*alorum* — others
ap.	*apud* — with; in the work of
auct. non.	*auctorum non* — not (of), to indicate misapplied name
ca.	*circa* — about
cf.	*confer* — compare (with)
cit.	*citatus* — cited
eg. (ex. gr.)	*exempli gratia* — for example
emen.	*emendavit* — he amended
emend.	*emendatus* — amended
e.p.	*ex parte* — in part
etc. (& c.)	*et cetera* — and the rest, and so on
fide	— on authority of
h.c.	*honoria de causa* — as an honour
ib.	*ibidem* — in the same (journal or place)
i.e.	*id est* — that is
ined.	*ineditus (-a, —um)* — unpublished
in litt.	*in litteris* — in correspondence
l.c	*loco citalo* — in the place cited
m.	*mihi* — to me Writers'
n.	*nobis* — to us responsibility for its publication
nec.	— not
no.	*numero* — number
nom. nud.	*nomen mudum* — naked name
nom. oblit.	*nomen oblitum* — forgotten name
nom. rejie.	*nomen rejicinedum* — rejected name
op. cit.	*opera citato* — in work cited
pp.	*pro parte* — in part
q.v.	*quae vide (pl. qq.v)* — which (please) see
sec.	*secundum* — according to
sic.	— so, thus

***Only those commonly used in Entomology have been listed.**

e.g.	*verbi gratia* — for example
vide.	— see
viz.	*videlicet* — namely
vs.	*versus* — against

Measurements

$\mu\mu$	micro micron = 10^{-12} m
μ	micron = 10^{-3} m
mm	millimetre = 10^{-3} m
cm	centimetre = 10^{-2} m
m	meter = 10^{6} m
km	kilometer = 10^{3} m
sq m	metre carre = 10^{0} sq m
a	are = 10^{2} sq m
ha	hectare = 10^{2} a (10000 sq m)
sq km	square kilometer = 10^{4} a

Weight

μg	microgram = 10^{-6} g
mg	milligram = 10^{-3} g
g	gram = 10^{0} g
kg	kilogram = 10^{3} g
Q	quintal = 10^{5} g
mT	metric ton = 10^{6} g
ml	mililitre = 10^{-3} l
l	litre = 10^{0} l
kl	kilolitre = 10^{3} l

General

ADI	acceptable daily intake
AOAC	methods of analysis
(methods)	suggested by Association of Official Analytical Chemist
Aq.	aqueous (solution)
b.p.	boiling point
CGR	crop growth rate
d	density (@ 20° C)

DAG	days after germination
DAS	days after sowing
DAT	days after treatment
DI	days' interval
dw	dead weight
f.p	freezing point
GLC	gas liquid chromatography
HPLC	high performance liquid chromatography
IPM	integrated pest of management
IRS	infrared spectroscopy
NA	not applicable (available)
RH	relative humidity
sp (spp.)	species (plural)
ULV	ultra low volume
UVS	ultraviolet spectroscopy
VP	vapour pressure

Statistics

CC	correlation coefficient
CD	critical difference
CV	critical variance
NS	non-significant
SD	standard deviation
SE	standard error
SEM	standard error of mean

Pesticides (formulations)

AF	aqua flowable
a.i.	actual ingredients
D	dust
DP	dustable powder
EC	emulsifiable concentrate
FI	flowable
G	granules

IGR	insect growth regular
LC_{50}	lethal concentration to cause 50% mortality
LD_{50}	lethal dose to cause 50% mortality
MAC	maximum allowed concentration
m.l.	miscible liquid
MRL	maximum residual limit
OC	oil concentrate; organochlorine(s)
OP	organo-phosphate(s)
ppm	parts per millions
ppb	parts per billion
SL	suspension liquid
WDG	water dispersible granules
WDP	water dispersible powder
WP	wettable powder
WS	water soluble
WSC	water soluble concentrate
WSG	water soluble granules
WSP	water soluble powder

Virus groups

CPV	cytoplasmic polyhedrosis virus
EV	entomopox virus
GV	granulosis virus
NPV	nuclear polyhedrosis virus

Insecticides Registered under section 9 (3) of the Insecticide Act, 1968 as on 28/12/2006

S. No.	Name of Pesticide
1.	2, 4-ichloro-phenoxy Acetic Acid
2.	Acephate
3.	Acetamiprid
4.	Alachlor
5.	Allethrin
6.	Alphacypermethrin
7.	Alphanaphthyl Acetic Acid
8.	Aluminium Phosphide
9.	Anilophos
10.	Atrazine
11.	Aureofungin
12.	Azadirachtin (*Neem* Products)
13.	Azoxystrobin
14.	*Bacillus thuringiensis* (B. t.)
15.	*Bacillus thuringiensis* (B. S.)
16.	Barium Carbonate
17.	*Beauveria bassiana*
18.	Bendiocarb
19.	Benfuracarb
20.	Benomyl
21.	Beta Cyfluthrin
22.	Bifenazate
23.	Bifenthrin
24.	Bitertanol
25.	Bromadiolone
26.	Buprofezin
27.	Butachlor
28.	Captan
29.	Carbaryl
30.	Carbendazim
31.	Carbofuran
32.	Carbosulfan
33.	Carboxin
34.	Carfentazone Ethyl
35.	Carpropamid
36.	Cartap Hydrochloride
37.	Chlorofenvinphos
38.	Chlorfenapyr
39.	Chlorimuron ethyl
40.	Chlormequat Chloride (CCC)
41.	Chlorothalonil
42.	Chlorpyriphos
43.	Chlorpyriphos Methyl
44.	Cinmethylene
45.	Clodinafop-propargyl (Pyroxofop-propargyl)
46.	Clomazone
47.	Chlothiandin
48.	Copper Hydroxide
49.	Copper Oxychloride
50.	Copper Sulphate
51.	Coumachlor
52.	Coumatetralyl
53.	Cuprous Oxide
54.	Cyfluthrin
55.	Cyhalofop-butyl
56.	Cymoxanil
57.	Cypermethrin
58.	Cyphenothrin
59.	Dazomet
60.	Deltamethrin (Decamethrin)
61.	Diazinon
62.	Dichloro Diphenyl Trichloroethane (DDT)

63. Dichloropropene and Dichloropropane mixture 9DD mixture)
64. Diclorvos (DDVP)
65. Diclofop-Methyl
66. Dicofol
67. Difenocenazole
68. Difenthiuron
69. Diflubenzuron
70. Dimethoate
71. Dimethomorph
72. Dinocap
73. Dithianon
74. Diuron
75. Dodine
76. D-trans Allethrin
77. Edifenphos
78. Endosulfan
79. Ethephon
80. Ethion
81. Ethofenprox (Etofenprox)
82. Ethoxysulfuron
83. Ethylene Dibromide and Carbon Tetrachloride mixture (EDCT mix 3:1)
84. Fenarimol
85. Fenazaquin
86. Fenitrothion
87. Fenobucarb (BPMC)
88. Fenoxaprop-p-Ethyl
89. Fenpropathrin
90. Fenthion
91. Fenvalerate
92. Fipronil
93. Fluchloralin
94. Flufenacet
95. Flufenoxyuron
96. Flufenzine
97. Flusilazole
98. Fluvalinate
99. Fosetyl-Al
100. Gibberellic Acid
101. Glufosinate Ammonium
102. Glyphosate
103. Hexaconazole
104. Hydrogen Cyanamid
105. Imazethapyr
106. Imidacloprid
107. Imiprothrin
108. Indoxacarb
109. Iprobenfos (Kitazin)
110. Iprodione
111. Isoprothiolane
112. Isoproturon
113. Kasugamycin
114. Lambdacyhalothrin
115. Lime Sulphur
116. Lindane
117. Linuron
118. Lufenuron
119. Magnesium Phosphide Plates
120. Mancozeb
121. Malathion
122. Mepiquate Chloride
123. Metasulfuron Methyl + Iodosulfuron methyl sodium
124. Metalaxyl
125. Metalaxyl-M
126. Metaldehyde
127. Methabenzthiazuron
128. Methomyl
129. Methoxy Ethyl Mercury Chloride (MEMC)
130. Methyl Bromide
131. Methyl Chlorophenoxyl Acetic Acid (MCPA)
132. Methyl Parathion

133. Metiram
134. Metolachlor
135. Metoxuron
136. Metribuzin
137. Metsulfuron Methyl
138. Milbemectin
139. Monocrotophos
140. Myclobutanil
141. Novaluron
142. Nuclear polyhyderosis virus of *Helicoverpa armigera*
143. Oxadiargyl
144. Oxadiazon
145. Oxycarboxin
146. Oxydemeton-Methyl
147. Oxyfluofen
148. Paclobutrazole
149. Paraquat dichloride
150. Penconazole
151. Pencycuron
152. Pendimethalin
153. Permethrin
154. Phenthoate
155. Phorate
156. Phosalone
157. Phosphamidon
158. Primiphos-methyl
159. Prallethrin
160. Pretilachlor
161. Profenophos
162. Propanil
163. Propergite
164. Propetamphos
165. Propiconazole
166. Propineb
167. Propoxur
168. Pyrethrins (pyrethrum)
169. Pyridalyl
170. Pyriproxyfen
171. Pyrithiobac sodium
172. Quinalphos
173. Quizalofop ethyl
174. S-bioallethrin
175. Sirmate
176. Sodium Cyanide
177. Spinosad
178. Streptomycin + Tetracycline
179. Sulfosulfuron
180. Sulphur
181. Tebuconazole
182. Temephos
183. Thiacloprid
184. Thiomethoxain
185. Thifluzamide
186. Thiobencarb (Benthiocarb)
187. Thiodicarb
188. Thiometon
189. Thiophanate-Methyl
190. Thiram
191. Transfluthrin
192. Tricontanol
193. Triadimefon
194. Triallate
195. Triazophos
196. Trichlorofon
197. *Trichoderma viride*
198. Tricyclazole
199. Tridemorph
200. Trifluralin
201. Validamycin
202. Zinc Phosphide
203. Zineb
204. Ziram

List of Pesticides/Pesticides formulations banned in India

A. Pesticides Banned for manufacture, import and use (25 Nos)

1. Aldrin
2. Benzene Hexachloride
3. Calcium Cyanide
4. Chlordane
5. Copper Acetoarsenite
6. Clbromochloropropane
7. Endrin
8. Ethyl Mercury Chloride
9. Ethyl Parathion
10. Heptachlor
11. Menazone
12. Nitrofen
13. Paraquat Dimethyl Sulphate
14. Pentachloro Nitrobenzene
15. Pentachlorophenol
16. Phenyl Mercury Acetate
17. Sodium Methane Arsonate
18. Tetradifon
19. Toxafen
20. Aldicarb
21. Chlorobenzilate
22. Dieldrine
23. Maleic Hydrazide
24. Ethylene Dibromide
25. TCA (Trichloro acetic acid)

B. Pesticide/Pesticide formulations banned for use but their manufacture is allowed for export (2 Nos.)

26. Nicotin Sulfate
27. Captafol 80% Powder

C. Pesticide formulations banned for import, manufacture and use (4 Nos)

1. Methomyl 24% L
2. Methomyl 12.5% L
3. Phosphamidon 85% SL
4. Carbofuron 50% SP

D. Pesticide Withdrawn (7 Nos)

1. Dalapon
2. Ferbam
3. Formothion
4. Nickel Chloride
5. Paradichlorobenzene (PDCB)
6. Simazine
7. Warfarin

List of pesticides refused registration

1. Calcium Arsenate
2. EPM
3. Azinphos Methyl
4. Lead Arsonate
5. Mevinphos (Phosdrin)
6. 2, 4, 5-T
7. Carbophenothion
8. Vamidothion
9. Mephosfolan
10. Azinphos Ethyl
11. Binapacryl
12. Dicrotophos
13. Thiodemeton /Disulfoton
14. Fentin Acetate
15. Fentin Hydroxide
16. Chinomethionate (Morestan)
17. Ammonium Sulphamate
18. Leptophos (Phosvel)

Pesticides restricted for use in India

1. Aluminium Phosphide
2. DDT
3. Lindane
4. Methyl Bromide
5. Methyl Parathion
6. Sodium Cyanide
7. Methoxyl Ethyl Mercury Chloride (MEMC)
8. Monocrotophos (ban for use on vegetables)

sub-classes	infra-classes	Super-orders	Orders -suborders	Descirbed species	Example
PTERYGOTA BRAUER 1885	Parneoptera Martyov 1923	Psocopteroidea (Jeannel 1945)	Psocoptera Shipley	3000	book-lice
			Psocomorpha		
			Trogiomorpha		
			Troctomorpha		
			Mallophaga Nitzsch 1818 Amblycera	4000	bird-lice
			Ischocera		
			Rhynchophthrina		
			Siphunculata Laterille 1825	400	sucking lice
		Hemipteroidea Geoffroy 1763 (Hemiptera Linnæus 1758)	Homoptera Latreille	58000	aphids, phyrilla, scales
			Auchenorrhyncha		
			Stemorrhyncha		
			Coleorrhyncha		
			Heteroptera Latreille 1810	48000	true bugs
			Gymnocerata Cryptocerata		
		Thysanopteroidea	Thysanoptera Haliday 1836	6000	thrips
			Terebrantia		
		(Jeannel 1945)	Tubulifera		
		Neuropteroidea Handlirsch	Neuroptera Linnæus 1758	6000	antilions, snakeflies

sub-classes	infra-classes	Super-orders	Orders -suborders	Descirbed species	Example
	1923	Coleopteroidea Handlirsch 1908	Coleoptera Linnæus 1758	350000	beetles, weevils
			Adephaga		
			Polyphaga		
			Archostemata		
			Strepsiptera Kirby 1813	500	stlyopids
	Martynov	Mecopteroidea (Petanoptera) Brauer 1885	Mecoptera Comstock & Comstock 1895	600	scorpionflies
			Protomecoptera		
			Eumecoptera		
			Trichoptera Kirby 1813	6000	caddisflies
			Annulipalpia		
			Integripalpia		
			Lepidoptera Linnæus 1758	160000	butterflies, moths
			Zeugloptera		
			Ditrysia		
	Oligoneoptera		Monotrysia		
			Dacnonypha		
			Diptera Linnæus 1758	98000	ture flies (gall-midges, mosquitoes)
			Nematocera		
			Brachycera		
			Aschiza		
			Schizophora		
		Siphonapteroidea	Siphonaptera Latreille 1825	98000	true flies
		Hymenopteroidea Handlirsch 1908	Hymenoptera Linnæus 1758	135000	
			Symphyta		sawflies
			Apocrita		ants, bees, wasps

INSECTS & MITE

List of the agriculture, pests infesting some econmic important crops, Rice, *Oryza sativa*, Linnaeus Rice, *Chavaal, Dhan.*

Common Name (1)	Insect pests (2)
Stem borer	
Gold fringed rice-borer	*Chilo auricilius* Dudgn
Maize of sorghum stem borer	*Chilo partellus* (Swinhoe)
Dark-headed striped borer	*Chilo polychrysus* (Meyr.)
Pale-headed striped rice borer	*Chilo suppressalis* (Walker)
Yellow rice borer	*Tryporyza incertulas* (Walker)
Pink borer	*Sesamia inferens* (Walker)
Leaf Feeders	
Small rice grasshoppers	*Oxya bidentata* Willernse
Large rice grasshoppers	*Hieroglyphus banina* (Fabricius)
Green rice leafhoppers	*Nephotettix nigropictus* (Stâl).
	N. virenscens (Distant.)
Zig-zagged leafhoppers	*Recilia dorsalis* (Motschulsky)
White-backed planthopper	*Sogatella furcifera* (Horv.)
Brown planthopper	*Nitaparvata lugens* (Bredd.)
Slender rice bug	*Cletus trigonus* (Thnb.)
Black paddy bug	*Scotinophara coorctata* (Thnb.)
Green stink bug	*Nezara virdula* Linnaeus
Thrip	*Baliothrip biformis* (Bagnall)
Slug caterpillar	*Latoia bicolor* (Walker)
Rice leaf folder	*Cnaphalocrocis medinalis* (Guenèe)
Rice skipper	*Parnara guttata* (Bremer & Grey)
	Telicota ougias (Linnaeus)
Army worms	*Spodoptera exigua* (Hbst.)
	S. mauritia (Biosaduval)
	S. litura (Fabricius)
	Mythimna separata (Walker)
Rice hispa	*Dicladispa armigera* (Olivier)
Flea beetle	*Chetocnema concinnipennis* Baly
Mite	*Oligonychus oryzae* (Hirst)

(1)	(2)
Mole criket	*Gryllotapla africana* P.de.B
Rice root aphid	*Rhopalosiphum ruffabdominalis* Sasaki
Rice root aphid	*R padi* L.
Rice mealy bug	*Heterococcus rehi* (Ldgr)
Rice gall midge	*Orseolia oryzae* (Wood Mason)
Shootfly	*Atherigona* oryzae Mall.
Rice root weevil	*Echinecnemus oryzae* Mshl.

Wheat, *Triticum aestivam* L. emend, Flori & Paol, *Gehun*
***T. duram* Desi, Duram wheat,** *Gehun*
T. dicoccum Schubi, Emmer, *Gehun*

(1)	(2)
Termite	*Odontotermes obesus* (Rambur)
	Microtermes obesi Holmgren
Stem borer	*Sesamia inferens* Walker
Gujhia weevil	*Tanymecus indicus* Faust
Cutworm	*Agrotis ipsilon* (Hufnagel)
	A. flommatra Schiffer Mueller
Armyworm	*Mythimna separata* (Walker)
Thrip	*Analphothrips flavicinctus* Karny
Aphids	*Schizaphis grominum* Rondani
	Rhopalosiphum maidis Fitch
	Sitobion anenae Fabricius
Surface grasshopper	*Chrotogonus trachyterus* Blanchard
Shootfly	*Artherigona naqvii* Steyskal
Brown wheat mite	*Petrobia latens* (Muller)

Sorghum, *Sorghum bicolor* (L.) Moench *Jowar*
S. vulgare* Pers, var, *sacchartum Koerh, Sugar sorghum Deodhan

(1)	(2)
Soil pests	
White grubs	*Holotrichia serrate* Faricius
	Lachnosterna consanguinea (Blanchard)
Termite	*Odonatatermes* sp.
	Microtermes sp.
Cutworm	*Agrotis ipsilon Hufngel*
Seedling pest	
Shootfly	*Atherigona soccata* (Rondani)
Flea beetle	*Chaetocnema indica* Wse.
	Longitaises sp.
	Phyllotreata chotanica Duviver
Cotton grey weevil	*Myllocerus undecipunctulus maculosus* Desbroachers
Plants Pests	
Stems borers	*Chilo partellus* (Swinhoe)
Pink borer	*Sesamia inferens* Walker
Green stirped borer	*Maliarpha separatella* Ragonot
Defoliators	
Oriental armyworms	*Mythimna partellus* (Swinhoe)
Red hairy caterpillar	*Amsacta moorei* (Butter)
Sap-sucking Pests	
Shoot bug	*Peregrinus maidis* (Ashmead)
Mites	*Oligonychus indicus* (Hirst)
Aphids	*Rhopalosiphum maidis* (Fitch)
	Melanaphis sacchari (Zehntner)
Sugarcane leafthopper	*Pyrilla perpusilla* Walker
Green sting bug	*Nezara viridula* Linnaeus
	Lygaeus bandurus (Scop)
	Taeniothrips traegardhi (Trybon)
	Xylaplothrips pellucidus (Ananthakrishnan)

Earhead Pests

(1)	(2)
Sorghum midge	*Stenodiplosis sorrghicola* Coquillett
Earhead caterpillar	*Autoba silicula* (Swinhoe)
	Cryptoblabes gnidiella Milliere
	Dichocrocus punctiferalis Guenèe
	Euproctis fraternata (Moore)
	Helicoverpa armigera (Hübbner)
	Mythimn separata (Walker)
	Plodia interpunctella (Hübbner)
	Prothesia xanthorrhoea Kollar
Earhead bugs	*Calocoris angustatus* (Leth.)

Pests of pearlmillet, *Pennisetum glaucum* (L.) R. Br. *Bajra*

(1)	(2)
Blister beetle	*Lytta tenuicollis* Pallas
	Zonabris pustulata Thunberg
Leaf roller	*Marasima trapezalis* Guenèe
Shootfly	*Atherigona opproximata* Malloch
Black hairy caterpillar	*Estiginena lactinea* Cramer
Red hairy caterpillar	*Amsacta moorei* Butler
Bihar hairy caterpillar	*Spilarctia obliqua* (Walker)
Midge	*Geromyta penniseti* Harris
Grasshoppers	*Hieroglypus nigrorepletus* Boliver
	Chrotogonus trachypterus (Blanchard)
	Colemania sphenaroides Boliver
Stem borers	*Chilo partellus* (Swinhoe)
	Sesamia inferens Walker
White grub	*Holotrichia serrata* Fabricius
	Lachnosterna consanguinea Blanch
Sugarcane pyrilla	*Pyrilla perpusilla* Walker

Maize, *Zea mays* Linnaeus Corn, *Makai, Bhutta*

(1)	(2)
Shootfly	*Atherigona* spp.
	A. naqvi Steyskal
Stem borer	*Chilo partellus* (Swinhoe)
	Sesamia inferens Walker
Phadka grasshopper	*Hicroglyphus nigrorepletus* Boliver
Hairy caterpillars	*Amsacta moorei* Butler
	A. albistriga Walker
Aphid	*Rhopalosiphum maidis* Fitch
Termites	*Odontotermis obesus* Rambur
Grey weevils	*Myllocerus spp.*
Army worms	*Mythimna separata* (Walker)
Pink ear-head worm	*Pyoderces simpler* Walsingham
Thrip	*Anaphothrips sudensis* Trybom
Mite	*Oligonychus indicus* Hirst.

Chickpea (*Cicer arittinum* Linnaeus) *Gram Chana*

(1)	(2)
Cutworm	*Agrotis ipsilon* (Hufnagel)
Gram pod-borer	*Helicoverpa armigera* (Hübner)
White grub	*Lachnosterna consanguinca* (Blanchar)
Gram semilooper	*Autographa nigrising* (Walker)
Gujhia weevil	*Tanymecus indicus* Faust
Armyworm	*Spodoptera exigua* Hübner
	S. litura (Fabricius)
Leaf miner	*Litromyza cicerina* (Rondani)
Semilooper	*Plusia signata* (Fabricius)
Aphid	*Aphis crassivora* (Koch.)
Termite	*Odontotermes* sp.
Surface grasshopper	*Chortogonus trachypterus* (Blanchrad)
Pest of stored seeds	
Storage weevils	*Callosobruchus analis* (Fabricius)
	C. chinnesis (Linnaeus)
	C. maculates (Fabricius)
Tobacco beetle	*Laisoderma serricorne* (Fabricius)
Red red flour beetle	*Tribolium castaneum* (Herbst)
Rice moth	*Coreyra cephalonica* (Stainton)
Dried currant moth	*Ephestia coutella* (Walker)

Pigeonpea, *Cajanus cajan* (Linnaeus) Millsp. Redgram, *Archar, Tur*
Pod borer complex infesting pigeonpea

(1)	(2)
American Cotton bollworm	*Helicoverpa armigera* Hübner
Cowpea spotted bollworm	*Maruca testulalis* Geyer
Pigeonpea leaf webber	*Exelastis atomosa* Wlsm
Slug caterpillar	*Lampides boeticus* Linnaeus
	Catochrysops strabo Fabricius
Pod borer	*Etiells zincknella* Treitschke
Podfly	*Melanogromyza obtusa* Malloch
Pod weevil	*A. benignun* Faust:
	Apion clavipes Gerst
Pod bugs	*Clavigralla gibbosa* Spinola
	Riportus pedestris Fabricius
Green stink bug	*Nerara viridula* Linnaeus
Weevil under storage	*Callosobruchus maculatus* Fabricius
	C. analis Fabricius
	C. chinensis Linnaeus
	C. theobrome (Linnaeus)
Pod damaging bugs	*Anoploenemis curvipes* Fabricius
	Clavigralla gilbosa Spin
	C. orientalis orientalis Dolling
Root beetle	*Adoretus caliginosubicolor* Br.
	A. stolickae Ohs.
Stem weevil	*Alcidoes collans* Pasc,.
	A. fabricii Fabricius
Flower damage	*Oxycetonis versicolor* (Fabricius)
Scarabaeids	*Sphenoptera laplunei* Kerr
Laef feeding weevil	*Tanymecus indicus* Fst.
Cotton bug	*Dysdercus koenigii* Fabricius
Melay bug	*Ferrisia virgata* Cockerell
	Laccifer longispina Misra
	Nipaecocus vastator M.
Pod sucking bug	*Lygaeus panduras* Scop.
Bud weevil	*Indozocladius aspenilus* Faust
Blister beetle	*Cantharis setacea* Cast.
Leaf hoppers	*Empoasca kerri* Pruthi
	E. binotata Pruthi

(1)	(2)
Whitefly	*Bemisia tabaci* (Gennadius)
Leaf feeding weevil	*Cyrtozemis cognata* Marshall
Ground nut leafminer	*Aproaerems modicella* Deventer
Seed damaging grub	*Goneaerems depressum* Fabricius
Whitegrub	*Lachnosterna* (*Holotrichia*) *consanguinea* (Blanchard)
Gelerucid beetle	*Madurasia obscurella* Jacoby *Monolepta signata* Ol,
Blister beetle	*Mylabris palerate* Pallas *M. pustulata* Thunb
Leaf eating weevils	*Myllocerus,dicolor* sp. *variegatus* Boh.
Aphids	*Aphis craccivora* Koch., *A.gossypii* Glover., *A. medicaginus* Kle.
Mite	*Teranychus cinnabarinus* Boisd. *Aceria cajani* Channabasavanna

(1) Blackgram, *Vigna mungo* (Linnaeus) Hopper, *urd*
(2) Greengram, V. *radiata* (Linnaeus) Wilkezek, Walp. *Mung*
(3) Cowpea, *V. unguiculata* (Linnaeus) Walp. *Lobia*

(1)	(2)
Defollator Grasshopper	*Gastrimagus marmoratus* (Thunb.)
Grasshoper	*Oxya velox* Fabricius
Leaf feeding Grassopper	*Chrotogonus trachypterus* Blanchard
	Colemania sphenarioides Boliver
Aphids	*Aphis gossypii* Glover
	A. craccivora Koch
	A. glycines Mais
	Myzus persicae Sulz.
Cutworm	*Agrotus ipsilon* Hufagel
Leafhopper	*Empoasa kerri* Pruthi
	E. moti Pruthi
	E. terminals Distant
Mealy bug	*Geococcus coffeae* Green
Pod feeding bug	*Anoplocenemis phasiana* Fabricius
Pod feeding bug	*Clavigralla gibbosa* (Spinola)
Galenrucid beetle	*Modurasia obsurcella* Jacoby
Pod bugs	*Riptortus fuscus* (Fabricius)
	R. linearis Fabricius
Sugarcane Pyrilla leafthopper	*Pyrilla perpusilla* (Walker)
Hadda beetles	*Epilachna dodecastigma* (Wied)
	E. varivertis (Mulsant)
Leaves, flower, pod bug	*Chaulips fallex* Scott
Green slender bug	*Cerontides pallidifer* Walker
Leaf eating weevils	*Myllocerus undecimpustulatus maculosus* Desbrocher
Green painted bugs	*Bagrada hilaris* Burm
Blister beetle	*Mylabris pustular* Thunberg
Green stink bug	*Nezara viridula* Linnaeus
Whitegrub	*Lachnosterna consanguinea* Blanchard
Podflies	*Melanogromyza sojae* Zehnter
	M. obtusa Malloch
Thrips	*Caliothrips indicus* Bagnali
	Thrips tabaci Lindermann
	Megalurothrips distalis Bagnali

(1)	(2)
Storage beetle	*Callosobruchus ananlis* Fabricius
	C. chinensis Linnaeus
	C. maculatus Fabricius
Stemfly	*Ophimyia phaseoli* Tryon
Red hairy caterpillar	*Amsacta albistriga* Walker
Red hairy caterpillar	*A. moorei* Butter
Bihar hairy caterpillar	*Spilarctia obliqua* Walker
Leafminer	*Acrocercops caerulea* Meyrick
Sap sucking bugs	*Ragmus importunitas* Distant
Slug caterpillar	*Euchyrsops cnejus* Fabricius
	Lampides boeticus Linnaeus
Lima bean pod borer	*Etiella zinckenella* Treitschka
Defoliator caterpillar	*Autohrapha nigrisigna* (Walker)
Cowpea spotted podborer	*Maruca testulalis* (Geyer)
American cotton bollworm	*Helicoverpa armigera* Hübner
Pigeonpea leaf webber	*Exelastis atomosa* Wlsm.
Armyworm	*Spodoptera litura* (Fabricius)
Sesmum Hawk moth	*Acherontia styx* Westw.
Mite	*Palyphagotarsonemus latus* (Banks)
Whiteflies	*Bemisia tabaci* (Gennadius)
Mites	*Tetranychus cinnabarinus* (Biosduval)
	T. telarious Linnaeus;
	T. urticae Koch.

Garden Pea, ***Pisum sativum arvense*** **Linnaeus,** ***Matar***

(1)	(2)
Cow bugs	*Leptocentrus obliquis* Walker
	Tricentrus bicolor Distant
Whitefly	*Bemisia tabaci* (Gennadius)
Pea Aphids	*Acyrthosiphan pisum* Harris
	Cerontiades stramineus (Walker)
Thrips	*Callothrips indicus* (Bagnali)
Pod borer	*Etiella zoncknella* (Treitschke)
Egg plant borer	*Leucinodes orbonalis* Linnaeus
Bihar hairy caterpillar	*Spilarctia obliqua* Walker
	Plusia errosoma Doubleday
	Mythimna (Pseudoletia) seprata Walker
Armyworm	*Spodoptera exigua* Hübner
	S. litura Fabricius
American cotton bollworm	*Helicoverpa armigera* (Hübner)
Pod - fly	*Melanagromyza phaseoli* Coquillett
Pea leafminer	*Chromatomyia* (*Phytomyza*) *horticola* (Goureau)
Leaf eating weevil	*Tanymecus indicus* Fst.
Pod borers	*Adisura atkinsoni* Moore
	Exelastis atomosa (Walsingham)
Cowpea spotted podborer	*Maruca testulis* (Geyer)
Pea stem borer	*Leguminivora tricentra* (Meyrick)
	Santhrobrota simplen (Walsingham)
	Paraspistes plapigera (Walsingham)
Pod borers	*Sphenarches caffer* (Zeller)
	Anarsia Spp.
Pigeonpea pod borer	*Euchrysops cnejus* (Fabricius)
Deccan Grass-hopper	*Colemania sphenarioides* Boliver
Leafthoppers	*Enpoasca kerri motti* Pruthi
	E. binotata Pruthi
Pod sucking bugs	*Ragmus importunita* Distant
	Anoplocenmis phasiana (Fabricius)
Pea leaf roller	*Hedylepta indicata* (Fabricius)
Leaf eating caterpillar	*Diachrysia orinchalcea* (Fabricius)

Rape and Mustard

1. ***Brassica alba*** Linnaeus white mustard.
2. ***B. chinensis*** Jusien *Kali sarson*, chinese cabbage (non-heading)
3. ***B. napus*** L. var. **glauca** Roxb Schulz. Yellow sarson.
4. ***B. napus* L. var *napus*** Indian rape *toria*, *laha* or *achara rai*
5. ***B. juncea*** (L.) Czern & Coss sub sp. ***juncea*** Indian mustard. *Rai.* ***b. juncea*** L. Czern & Coss. subsp. ***intergrifolia*** (West) Thell. Leaf mustard *Rai*, ***B. napus*** L. var ***napus*** *cv. Kali Sarson.*
6. ***B. nigara*** (L.) Kich Black mustard *Kali rai* or *Banarasi rai*
7. ***Eruca sativa*** Milli *Taramira*
8. ***B. pepinesis*** (Lous) Rupr Chinese cabbage (heading)
9. ***B. juncea*** var, ***rugosa*** (Roxb) Pahadi rai
10. ***B. juncea*** (L.) Czern, Duhue and Fuller, brown sarson
11. ***B.tourneforti*** Gouan, *Punjabi rai* or *jangli rai*
12. ***B. napobrassica*** (Linn Mill cross between cabbage ***B. oleracea*** Linn and turnip, ***B. rapa*** Linn Emend, Metzger, Rutabga)
13. ***B. deracea*** var ***intalica*** Broccoli
14. ***B. deracea*** var. ***gemmifera*** D C Brussels sprotes *Chotee gobee.*

(1)	(2)
Surface grasshopper	*Chrotogonus trachypeterus* Blanchard
Mustard sawfly	*Athalia lugens proxima* (Klug)
Painted bug	*Bagrada cruciferarum* Kirkaldy
	Modates limbabta (Fabr.)
Fica beetle	*Phylltreta cruciferae* Goeze
Mustard aphid	*Lipaphis erysimi* Kaltenbach
Cabbage aphid	*Brevicoryne brassicae* (Linnaeus)
Green peach aphid	*Myzus persicae* (Sulzer)
Whitefly	*Bemista tabaci* (Gennadius)
Diamondback moth	*Plutella xylostella* (L.) Curtis
Bihar hairy caterpillar	*Spilarctia (Spilosoma) obliqua* (Walker)
Cabbage butterfly	*Pieris brassicae* Linn
Cabbage leaf webber	*Crocidolomia binotalis* Zeller
Cabbage borer	*Hellula undalis* Fabricius
Semilooper	*Plusia orichalcea* Fabricius
Armyworm	*Mythimna loreyi* (Duponchel)
Cutworm	*Agrotis ipsilon* Hufnagel
Cutworm	*Ochropleura flammatra* Schiffermüller
Gujhia weevil	*Tanymecus indicus* Faust
Leaf miner	*Chromatomyia* (*Phytomyza*) *horticola* (Goureau)
	Lirromyza brassicae Riley

Groundnut, ***Arachis hypogenea*** Linnaeus Peanut, *Mungphali*

(1)	(2)
Termite	*Odontotermes obesus* (Rambur)
pod spot	*Trinervitermes bigormis* (Wasm.)
Root feeding insects	
Cutworm	*Agrotis ipsilon* (Hfn.)
Foliage and stem feeding insects	
Aphid	*Aphis craccivora* Koch.
Green stink bug	*Nezrara viridula* (Linnaeus)
Thrip	*Frankliniella fusca* (Hinds)
Armyworm	*Spodoptera litura* (F.)
Gram podborer	*Helicoverpa armigera* Hübner
Red hairy caterpillar	*Amsacta moorei* Butter
Leafminer	*Stomopteryx subsecivella* Zell.
Blister beetle	*Mylabris pustulata* Thunberg
Mites	*Tetranychys cinnabarinus* (Biosduval)
	T. telariust (Linnaeus)

Sunflower, *Helianthus annuus* Linnaeus *Surajmukhi*

Sucking pests

(1)	(2)
Jassids	*Amrasca biguttula biguttula* (Ishida)
Whitefly	*Bemisia tabaci* (Gennadius)
Foliage pests	
Tubacco caterpillar	*Spodoptera litura* (Fabricius)
Bihar hairy caterpillar	*Spilarctia obliqua* (Walker)
Green semilopper	*Plusia orichalcea* Fabricius
	Trichoplusia ni Hübner
	Plusia signata Fabricius
Capitulum - borer	*Helicoverpa armigera* (Hübner)
Banded sunflower moth	*Cochylis hospes* Walsingham
Seeding pests Cutworm	*Agrotis ipsilon* (Hufnagel)
Grasshopper	*Attractomorpha crenulata* Fabricius

Seasamum, *Sesmum indicum* Linnaeus Gingelly, *Til*

(1)	(2)
Surface grasshopper	*Chrotogonus* sp.
Thrips	*Thrips palmi* Karny
	Trypactothrips rutherfordi Bagnall
Bugs (Sucking)	*Eysarcoris ventralis* Westwood
	Nezara viridula Linnaeus
	Lygaeus pandurus (Scopoli)
Pod sucking bug	*Elasnolomus sordidus* Fabricius
	Nysius inconapicuus Distant
Green peach aphid	*Myzus persicae* (Sulzer)
Jassid	*Orosius albicinctus* Distant
Shoot and leaf - webber	*Antigastra catalaunalis* (Duponchel)
Sesamum sphins	*Acherontia styx* (Westwood)
	Hippotion echeclus (Boisduval)
	Euproctis scintilans (Walker)
Red hairy caterpillar	*Amsacta moorei* (Butter)
	A. lactinea (Cramer)
Bihar hairy caterpillar	*Spilarctia obliqua* (Walker)
Linseed caterpillar	*Spodoptera exigua* (Hübner)
Tobacco caterpillar	*S. litura* Fabricius
Leave and pods damaging beetle	*Pachephorus bistriatus* Muls
	Episomus laceria Fabricius
	Serica insanbilis Brenske
	Oxycetonia albopunctata Fabricius
Dipteran pests	*Dasineura sesami* Grover & Prasad
Seaamum galfly	*Ashondylia sesami* Felt.
Pea leafminar	*Chromatomyia* (*Phytomyza*) *horticola* (Goureau)
Sesamum fly	*Gitonides perspicas* Knab.

SAFFLOWER, ***Carthamus tinctorius*** **Linnaeus,** ***Kusum***

(1)	(2)
Safflower aphids	*Dactynotus compositae* Theobald *D. carthami* (HIlla Ris Lambes) *D. jaceae* (Linnaeus) *D. sonchi* (Linnaeus)
Caterpillar	*Prospalia capensis* (Guenèe)
Semi-looper	*Plusia orichalcea* (Fabricius)
Cabbage looper	*Trichoplusia ni* (Hübner)
Stem and flower head borer	*E. rivula* Moore
Lucerne caterpillar	*Spodoptera exiqua* (Hüner)
Gram pod borer	*Helicoverpa armigera* (Hüner)
Bihar hairy caterpillar	*Spilarctia* (*Spilosoma*) *obliqua* (Walker)
Gujhia weevil	*Tanymecus indicus* Faust
Fruitfly	*Acanthiophilus helianthi* (Rossi)
Stem - fly	*Melanagromyza obtusa* (Malloch)
Thrips	*Thrips carthami* singh *Taeniothrips frici* (Uzel) *T. flavidulus* (Bagnali)
Pod sucking bugs	*Dolycoris indicus* Stål *Creontiadas pallidifer* Walker *Cochlochila bullita* Stål

Cotton, ***Gossypium sp, Kapas***

(1)	(2)
American cotton	*G. hirsutum* Linnaeus
Desi cotton	*G. arboreum* Linnaeus
Desi cotton	*G. herbaceum* Linnaeus
Tetraploid species	*G. tomentosum* Nutt
Egyptian cotton	*G. barbadense* Linnaeus
Cotton Jassid	*Amrasca biguttula biguttula* (Ishida)
Whitefly	*Bemisia tabaci* (Grennadius)
Aphids	*Aphis gosypii* Glover
Spotted bollworms	*Earias insulana* Boisduval
	E. vittella Fabricius
Pink bollworm	*Pectinopora gossypiella* Saunders
American cotton bollworm	*Helicoverpa armigera* (Hübner)
Gray weevil	*Myllocerus maculosus* Desbrochers
Leaf roller	*Sylepta derogata* Fabricius
Stem weevil	*Pempherulus affinis* Faust
Stem borer	*Sphenoptera gossypii* Cotes
Red cotton bug	*Dysdercus lactus* Kirby
Dusky cotton bug	*Oxycatenus latus* Kirby
Thrips	*Thrips tabaci* Linermann
	Scirtothrips dorsalis Hood
Surface grashhopper	*Chrotogonus tranchypterus* Blanchard
	C. oxypterous Blanchard
Leaf eating caterpillar	*Spodoptera exigua* (Hübner)
	S. litooralis (Boisduval)
Mites	*Teranychus cinnabarinus* Biosduval
	Hemitasinemus latus Banks

Sugarcane, *Saccharum officinarum* Linnaeus, *Ganna, Paunda*

(1)	(2)
Termite	*Odontotermes obesus* (Rambur)
	O. assumthi Holmgren
	Microtermes andndi Holmgren
	Trinervitermes heimi Wasmann
Top borer	*Tryporyza nivelia* Fabricius
Early shoot borer	*Chilo infuscatellus* Snellen
Pink borer	*Sesamia infernes* Walker
Grenn borer	*Raphimatopus oblutella* Zeller
Guradas pur borer	*Rissetia steniellus* Hampson
Internode borer	*Procerus indicus* Kapur
Stalk (Tarai) borer	*Chilotracea auricilia* Dudgron
Stem (Pallasy) borer	*Chilo tumidicostalis* Hampson
Root borer	*Emmalocera depresella* Swinhoe
Leafhopper	*Pyrilla perpusilla* Walker
	P. aberrans Kirby
	P. pusana Distant
Whitefly	*Aleurolobus barodensis* Maskell
	Nemaskelia bergii Singnoret
Hispa	*Asamangulia cuspidata* Maulik
	Leptispa pygaea Baly
	Philodonta moderta Willemse
Black Bug	*Covelerius excavatus* Distant
Black leaf hopper	*Proutista moesta* Westwood
Cockehafar beetle	*Lachnosterna consanguinca* Blanchard
	Alssonotum inpressicolle Arrow
Grasshoppers	*Hieroglyphus banian* Fabricius
	H. nigrorepletus Boliver
	Choroedocus illustris Walker
Mealy bug	*Scchariococcus sacchari* Cockerell
	S. saccharifolii Green
	Ripersia sacchari Green
Scale insects	*Melanaspis glomeraia* Green
	Cercoplartes actiniforms Green
Aphids	*Ceatovacuna lanigera* Zehntner
	Longiungius sacchari Zehntner
	Rhoppalosiphum maidis Fitch
Red ants	*Dorylus orientalis* Weatwood
Mites	*Oligonychus indicus* Hirst
	Schizotetramychus andropogoni Hirst

Tabacco, *Nicatiana tabacum* Linnaeus *Tambaku*
N. rustica Linnaeus, Turkish or East Indian tobacco, *Vilaiti Tambacku*

(1)	(2)
Leaf eating caterpillar	*Spodoptera litura* Fabricius
Stem borer	*Scrobipalpa helicopa* Lower
Whitefly	*Bemisia tabaci* (Gennadius)
Ground beetle	*Mesomorphus villiger* Blanch
Aphids	*Myzus nicotianae* Blackman
	M. persicae Sulzer
Budworm/capsule borer	*Helicoverpa armigera* (Hübner)
Cigarette beetle	*Lasioderma serricorne* Fabricius

PEPPER Common Black Paper, ***Piper nigrum*** Linnaeus *Kali mirch, Gol mirch*

(1)	(2)
Flea (Pollu) beetle	*Longitarus nigripennis* Motschulsky
Top shoot borer	*Cydia hemidoxa* Meyr.
Marginal gall thrips	*Liothrips karnyl* Bagnail
Wild silk worm	*Cricula trifenestrata* H.
Pepper mussel scale	*Lepidosaphes piperis* Green
Soft scale	*Marisipococcus marsupiale* Green
Coconut scale	*Aspidiotus destructor* Sign.
White tailed mealy bug	*Ferrisia virgata* Cockerell
Whitefly	*Alcurocanthus piperis* Maskell
Leaf weevil	*Eugnathus curvus* Fst.
Berry gall midge	*Cecidomyia malabarensis* Felt.

Cardamom Small *Elettaria cardamomum* Maton, *Chhoti Elaichi*

(1)	(2)
Thrips	*Sciothrips cardamomi* (Ramk.)
Shoot, panicle and Early capsule borers	*Conogethes punctiferalis* (Guen.)
	Jamides sp. nr. *Electo*
	Jamides sp.
Beetle borer	*Onthophagus coorgensis*
	Onthophagus sp.
	Thammurigdes cardamomi
Hairy caterpillar	*Eupterote cardamomi* Renga
	E. canaica Moore
	E. fubia Cram
	E. testacea Walker
	Leuodera vittata Walker
	Euproctis lutifacia Hampsn.
	Alphaes biguttata Walker
	Pericallia ricini Fabricius
Shootfly	*Formosina flavipes* Malloch
Whitefly	*Dialeurodes cardamomi* David and Subr.
Aphid	*Pentalonia nigronervosa caladii* Vandor Goot
Spotted grasshopper	*Aulaches miliaris* Linnaeus
Banded spittle bug	*Cosmoscaria thoracica* Distant
Skipper	*Notocrypta feisthamelii* Boisduval
Looper	*Eumecella rosalia* Cramer
Cutworm	*Arcilasisa plagiata* M.
Root grub	*Basilepta fulvicorne* Jacoby
Root borer	*Hilarographa caminodes* Meyer
Rhizome weevil	*Prodioctes haematicus* Chev. Dames
Mites	
	Teranychus neocaledonicus and
	Brevipalpus delconi Pritchard & Baker
	Dolichotertranychus florifanus (Banks)
	D. elatariae Mohansundram
	Eriophyes elatariae Mohansundaram

Betelvine, *Piper beetle* Linnaeus Betal pepper, *Pan*

(1)	(2)
Betelvine	*Disphinctus politus* Walker
	D. maesayum Kirkaldy
Mealy bug	*Ferris virgata* Cockerell
	Lepidosaphes cornutus Green
	Pseudococcus spp.

Coconut, *Cocus nucifera* Linnaeus, *Nariyal*

(1)	(2)
Rhinoceros beetle	*Oryctes rhinoceras* Linnaeus
Red palm weevil	*Rhynchophorus ferrugineus* Fabricius
Black headed caterpillar	*Opisina arenosella*
	Nephantis serinopa Meyrick
Cockhafer beetle	*Leucopholis coneophora* Burmeister
Termites	*Odontotermes obesus* (Ram'ɔur)
Slug caterpillar	*Macroplectra nararia* Moore
	Latoia lepida Cramer
Coconut bug	*Stapanitis typicus* Distant
Leaf roller,	*Gangara thyrsis* Moore
	Suastuds gremius Fabricius
Coconut scales	*Aspidotus destructor* Signoret
	Aonidiella orientalis Newstead
	A. ficus Ashmead
Coconot mite	*Aceria guerreromis* Keifer

Pests of Vegetable Crops

Cauliflower, *Brassica oleracea* var. *botrytis* Linnaeus, *Phul Gobhi*

Cabbage, *Brassica oleracea* *var.* ***capitala*** Linnaeus, *Bandh Gobhi*

(1)	(2)
Diamondback moth	*Plutella xylostella* Linnaeus
Leaf webber	*Crocidolomia binotalis* Zetler
Cabbage	*Pteris brassicae* (Linnaeus)
	P. canidia (Linnaeus)
	P. napi (Linnaeus)
	P. rapae (Linnaeus)
Mustard sawfly	*Athalia lugens proxima* Klug
Cabbage aphid	*Brevicoryne brassicae* (Linnaeus)
Mustard aphid	*Lipaphis erysimi* (Kaltenbach)
Cotton aphid	*Aphis gossypii* Glover
Green peach aphid	*Myzus persicae* (Sulzer)
Cabbage aphid	*Siphocoryne indobrassicae* Das
Bugs	*Bagrada cruciferarum* Kirkaldy
	B. hilaris (Brumester)
Cabbage borer	*Hellula undalis* Fabricius
Termite	*Microterms anandi* Homlgren
Cutworm	*Agrotis ipsilon* (Hufnagel)
Groundnut earwig	*Euberllia annulipes* (Lucas)
Large brown criket	*Branchytrypes protentosus* (Lichienstein)
Grasshoppers	*Atractomorpha crenulata* Fabricius
Bombay locust	*Patanga succincta* (Linnaeus)
Thrips	*Callothrips indicus* (Bagnall)
	Thrips tabacci Lindemann
	Frankliniella schultzer Trybom
Flea hopper	*Phyllotreta cruciferae* Goeze
Leaf eating weevil	*Tanymecus circumdatus* (Wiedemann)
Leaf eating caterpillar	*Trichoplusia ni* Hübner

Okra, ***Abelmoschus esculentus*** Linnaeus Moench Lady's Finger, *Bhindi*

(1)	(2)
Leafhoppers	*Amrasca biguttula biguttula* (Ishida)
	Expoasca binotata Pruthi
	Corizus rubicundro (Distant)
Shoot and Fruit borer	*Earias vinella* (Fabricius)
	Helicoverpa armigera (Hübner)
	Spodoptera litura (Fabricius)
Leaf roller	*Sylepta derogata* Fabricius
Red cotton bug	*Dyddercus koenigii* (Fabricius)
Dusky cotton bug	*Oxycarenus* lactus Kirby
Cow bugs Treehopper	*Tricentrus bicolor* Distant
Lygeid bug	*Apilostethus pandrus* (Scopoli)
Thrips	*Microcephlothrips abdominalis* (Crawford)
	Frankliniella schulizei (Tryborn)
	Haplothrips gowdeyi (Franklin)
	Thrips tabaci Lindarnann
Semiloopers	*Anomis flava* Fabricius
	Xanthodes groellisi (Feisth)
	Acontia intersepta Guenèe
	A. malvae Esper
	A. tranversa Guenèe
	Tarache nitidula (Fabricius)
Cutworm	*Agrotis ipsilon* (Hufnagel)
	A. flammatra (Schiffer)
Stem borers	*Sphenoptera gossypii* Cotes
Cotton stem weevil	*Pempherulus affnis* Faust
	Alciodides affaber Glover
Grass hoppers	*Poekilocerus pictus* (Fabricius)
Aphid	*Aphid gossypii* Glover
Whitefly	*Bemisla tabaci* (Gennadius)
Mealy bugs	*Ferrisia virgata* (Cockerell)
	Nipaecoccus virdis (Newstead)
Scale insects	*Saissetia coffeae* (Walker)
	Parasaissetia nigra (Neitner)
	Ceroplastes floridensis Comstock

Leafminers	*Trachys herilla* Obenberger
Blossom beetle	*Oxycetonia albopunia* Fabricius
	O. veriscolor Fabricius
	Mylabris pustulata Thunberg
	M. phalerata Palias
	Podogria bowringi Baly
Leaf eating weevil	*Myllocerus undecimpustulatus* Desprochers
	M. variegatus Boheman
	M. virudanus (Fabricius)
	Ptochus ovulum Faust
Okra stem fly	*Melanagromyzza hibisci* spencer
Mites	
	Teranychus cinnabarinus (Boiduval)
	T. neocaledonicus Andre
	T. macrofarlanei Baker and Pritchard
	Oligonychus coffeae (Nietner)
	Phytoscius minuts (Narayanan, Kaur and Ghai)
	Amblyseius dethiensus (Narayanan and Kaur)

Eggplant, *Solanum melongena* Linnaeus, *Brinjal,* Egg plant, *Baigan*

(1)	(2)
Shoot and fruitborer	*Leucinodes oronalis* Guenèe
Stem borer	*Euzophera perticella* Ragonot
Leaf rollers	*Antoba olivacea* (Walker)
	Pterophoris lienigianus Zeller
Leaf feeding beetle	*Henospilachma viginitotopunctia* (Fabricius)
	Epilcacata Mulsant
	E. ocellata Redienbacher
	E. septima Dicke
	E. varvestis Mulsant
Leafhopper	*Amrasca biguttula biguttula* (Ishida)
	Empoasca binotata Pruthi
	E. parathea Pruthi
	E. punjabensis Pruthi
	Cestius phycitis (Distant)
Sap sucking bugs	*Anoplocnemis phasiana* (Fabricius)
	Tricentrust bicolor Distant
	Coptosoma nazirae Atkinson
	Haliticus minutus Reuter
Lace wing bugs	*Urentius hystricellus* (Richter)
Aphids	*Aphis gossypii* Glover
Termites	*Trienrvitermes biformis* Wasmann
	Microtermes Sp.
Grasshoppers	*Atractomorpha crenulata* Fabricius
	Orthacris simulans Boliver
	Oxya japonica (Thunberg)
Leaf roller	*Eublemma divaqa*
Ash weevil	*Myllocerus subfasicatus* G.
Coccids	*Coccidohystrix insolita* (Green)
	Aonideilla orientalis (Newstead)
	A. aurantii Maskell
	Quanaspidiatus destructor (Singnoret)
Lantana bug	*Orthezia insignis* Browne
	Sericothrips solanifolii Shumsher
	Thrips tabaci Lindemann
	Caliotrips indicus (Bagnall)

(1)	(2)
	Scitothrips dorsalis Hood
Bud worm	*Scrobipalpa blasigona* (Meyrick)
Leaf miner	*Scrobipalpa ergasima* (Meyrick)
	S. heliopa (Lower)
Eggplant Leaf webber	*Psara bipuntalis* (Fabricius)
Hairy caterpillar	*Selepa doclili* Butler
	S. celtis Moore
Ant Black	*Camponotus infuscus* Ford
Brown ant	*Solenpisis geminata* (Fabricius)

Chillies, *Capsicum annum* Linnaeus, *Red pepper, Lal mirch*

(1)	(2)
Thrips	*Scirtothrips dorsalis* Hood
	Caliothrips indicus (Bagnall)
	Frankliniella schultzer (Trybom)
Aphids	*Aphis gossypii* Glover
	Myzus persicae (Sulzer)
	Aphis cytisorum Hartig
Large brown cricket	*Bracytrypes pretentosus* (Litchtenstein)
Termites	*Odontotermes obesus* Rambur
Beetle grubs	*Anomala bengalensis* (Blanchard)
	Lachnosterna consanguinea (Blanchard)
	Holotrichia reynaudi (Blanchard)
Cutworm	*Agrotis ipsilon (Hufnagel)*
Sap sucking insect	*Ferrisia virgata* (Cockerell)
	Saisetia coffeae (Walker)
	Aspidotus destructor Signorer
	Lepdosaphes piperis (Green)
Lygaeid bug	*Spilostethus pandurns* (Scopoli)
Cow bug	*Tricentnis bicolor* Distant
Whitefly	*Bemisia tabaci* (Gennadius)
Leaf Defoliators	*Mythimna loreyi* (Duponchel)
Army worms	
Leaf roller	*Archips micaceanus* (Walker)
Lucerne caterpillar	*Spodoptera exigua* (Hübner)
Leaf eating beetle	*Monolepta signata* Oliveira
Fruit borer	*Longitarsus nigripennis* Motschulskey
Capsid bug	*Disphinctus politus* Walker
Tabacco caterpillar	*Spodoptera littoralis* (Fabricius)
Gram caterpillar	*Helicoverpa armigera* (Hübner)
Eggplant stem borer	*Euzophera pericella* Ragonot
Pepper top shoot borer	*Laspeyresia hemidosa* Meyrick
Fruit flies	*Bactrocera dorsalis* Hendel
	B. cucurbitae (Coquillett)

Fruit sucking moth	*Othreis* spp.
	Ophinusa coranata (Fabricius)
Cigrrette beetle	*Lasioderma serricorne* (Fabricius)
Storage Pests	
Drug store beetle	*Stegobium paniceum* Linnaeus
	Arthrodies sp.
Mites	*Polyphogotarsonemus latus* (Bankai)
	Tarsonemus tranlucens (Green)
	Tetranychus cinnabarinus Boisduval

TOMATO ***Lycopersicum esculentum*** Milliere *Tamatar*

(1)	(2)
Fruit borer	*Helicoverpa armigera* (Hübner)
	H. zea
	Heliothis assulata Guenee
Fruit - sucking moth	*Otheris fullonia* (Clerck)
	O. materna (Linnaeus)
Fruit flies	*Bactrocera cucurbitae* (Coquillett)
	Acriticheta excisa Thomson
Leaf eating caterpillar	*Spodoptera littoralis* (Fabricius)
	S. exigua (Hübner)
Green semilooper	*Trichoplusia ni* (Hübner)
Stem borer	*Euzophera perticella* Ragonot
Tomato pin worm	*Phthorimaca lycopersicella* Busck
	P. perculella (Zeller)
Tomato hornorm	*Protoparce quiquenaculata* (Haworth)
	P. sexia
Whitefly	*Bemisia tabaci* (Gennadius)
Hadda beetles	*Epilachna dodecastigma* Widemann
	Henospilachma vigintioctopuncitata Fabricius
Mealy bug	*Ferrisia virgata* (Cockerell)
Aphids	*Aphis gossypii* Glover
	Myzus persicae (Sulzer)
Leafthoppers	*Amrasca biguttula biguttula* (Ishids)
	Empoasca punjabensis (Pruthi)
Thrips	*Thrips tabacci* Lindermann
	Callothrips gangibaueri Schmutz
Blossom thrips	*Hoplothrips gangibaueri* Schumutz
	Franklintella schulizer (Tryborn)
	Megalurrothrips usitatus (Bagnall)
	Scirtothrips dorsalis Hood
Serpentine leaf - miner	*Liriomyza trifolii* Burgess

French Bean, *Phaseolus vulgaris* Linnaeus, *Vilaiti Sem*

(1)	(2)
Aphids	*Aphis craccivora* Koch.
	Acyrthosiphan pisum (Harris)
	Aphis fabae Scopoli
	A. gossypii Glover
	A. rumicis Linnaeus
	A. adusta Zehntner
	Cervaphis schoutedeniae van der Goot
	Therioaphis maculata Buckton
	Aphis medicaginis Koch.
	A. euonymi Fabricius
	A. lgycines Matsummura
	Macrosiphum euphorbiae (Thomas)
Sap sucking bugs	*Anchon philosum* Walker
	Tricentrus pilosum Distant
	Anoplocnemis phasiana Fabricius
	Chauliops nigrescene Distant
	Spilostethurs pandus (Scopoli)
	Aspongopus fanus Fabricius
	Coptosoma cribraria Fabricius
	C. nizirae Atkinsoni
	Cyclopelta siccifolia Westwood
Pod borers	*Helicoverpa armigera* (Hübner)
	Etiella zinckenella (Treitschke)
	Adisura atkinsoni Moore
	Maruca testulalis (Geyer)
	Sphenarches caffer (Zeller)
	Euchrysops cnejus (Fabricius)
	Lampides boeticus (Linnaeus)
Leaf miners	*Cosmopterys minetis* Meyrick
	C. phaeogatra Meyrick
	Cyphositicha coerulea Meyrick
Leaf eating caterpillar	*Amsacta* spp.
	Acherontia styx (Westwood)
	Spilarctia obliqua (Walker)
	Pericallia ricini (Fabricius)
	Esstigmene lactinea (Cramer)

	Spodoptera exigua (Hübner)
	S. littoralis (Fabricius)
Pod boring bug	*Riptortus fuscus* Fabricius
	R. linearis Fabricius
Stem fly	*Ophiomyia phasedi* Tryon
Bean beetle	*Epilachna varivestis*
	R. pedestris Fabricius
Thrips	*Ayyaria chaetophora* Kary
	Caloithrips indicus (Bagnall)
	Sericothrips ramaswamiahi Karny
Blossom thrips	*Frankliniella schulizei* (Trybom)
	Megalurothrips ustitatas (Bagnall)
	Taenithrips distalis Kary
	T. leyfroyi (Bagnall)
	T. longistylus Karny
Pest under storage Condition	*Callosobruchus affinis* Froll
	C. chinensis (Linnaeus)
	C. quadrimaculatus (Fabricius)

Bitter gourd *Momordica charantia* Linnaeus, *Karela*

(1)	(2)
Fruit flies	*Bactrocera cucurbitae* (Coquillett)
	B. cilliatus (Loew)
	B. diversus (Coquillett)
	B. latifrons (Hendel)
	B. tau (Walker)
	B. zonatus (Saunders)
	Myriopardlis pardalina (Bigot)
Pumpkin beetle	*Rhaphidopalpa foveicollis* (Lucas)
Blister beetle	*Mylabris pustulatus* Thunberg
Flea beetle	*Phyllotreta cruciferae* (Goezze)
Aphids	*Aphis gossypii* Glover
	Myzus persicae (Sulzer)
Leafthopper Jassids	*Eutettix phycitis* Distant
Stem gall fly	*Lasioptera falcata* Felt
Mites	*Tetranychus mecaledonicus* Andre
	T. angolensis Meyer
	T. cinnabarinus (Biosduval)
	T. urticae Koch.
	Bravipalpus californicus (Banks)
	B. phoenicis (Geijskes)
	Euretanychus anneckei Meyer
	E. maximae Nasser
	E. orientalis (Klein)

Mushrooms

(1)	(2)
White button mushroom	*Agaricus bisporus*
Oyster mushrooms	*Pleurotus* sp.
Straw mushrooms	*Volvariella., V. dipplasia., V. esculenta*
Shiitake mushrooms	*Lentinula edodes*
Black car mushrooms	*Auricularia rugosoanulata*
Giant mushrooms	*Stroppharia rugosoanulata*
	Flammulina velutipes
	Tremella fuciformis
	Hypsizygus marmoreus
	Pholiotata nameko
	Crifola frondosa
Insect pests	
Phorid fly	*Megaselia* sp.
	M. sandhuri
Sciarid fly	*Bradysia paupera*
	B. tritici

POTATO
Solanum tuberosum Linnaeus, *Alu*

(1)	(2)
Potato tuber moth	*Phthorimaea opeculella* (Zeller)
cutworms	*Agrotis ipsilon* (Hufnagel)
	A. segetum (Dems and Schiffermiller)
	A. intracta Walber
	A. flammantra Schiffer-miller
	A. spinifera (Hübner)
Aphids	*Aulacorthum solani* (Kaltenbach)
	Acyrthosiphon pisum (Harris)
	Myzus persicae (Sulzer)
	M. ornatus Laing
	Aphidula nasturtii (Kaltenbach)
	Aphis fabae Scopoli
	A. mepicaginus Koch
	A. rhamni Koch
	A. gossypii Glover
	A. nasturtii (Kaltenbach)
	Brevicoryne brassicae (Linnaeus)
	Lipaphis erysimi (Kaltenbach)
	Macrosiphum euphorbiae (Thomas)
	M. rosae (Linnaeus)
	Rhopalosiphoninus eatysiphen (Davidson)
	Rhopalosiphum rufiabdominalis (Sasabi)
Molecriket	*Gryyllotalpa africana* Palisot de Beauyoes
Termites	*Eremotermes* sp.
	Microtermes obesi Holmgress
	Odontotermes obsus (Rambur)
Ground beetles	*Gonocephalum hoffmanseggi* (Steven)
	Hapatrophalum seriatoporus Fairmare
Wireworm	*Drasterius* sp.
Red ants	*Dorylus oriendalis* Westwood
Capsid bug	*Creontiadus pallidifer* Walker
Green sting bug	*Nezara viridula* (Linnaeus)

Soybean bug	*Piezodorus hybneri* (Gmelen)
Leafhoppers	*Amrasca biguttulla biguttulla* (Ishida)
	Empoasca kerri motti Pruthi
	E. pujabensis Pruthi
	E. solanifolia Pruthi
	E. fobae Harris
Whiteflies	*Bemisia tabaci* (Gennadius)
	Trialeurodes vaporariorum (Westwood)
Thrips	*Selenothrips indicus* Bagnall
Lace-wing bug	*Recaredus* sp.
Beetle grubs	*Anomala* sp.
	Holotrichia conferta Sharp
	H. coriacea Hope
	H. longipennis Blanchard
	H. serrata Fabricius
	Xylotrupes gideon Linnaeus
Hadda beetles	*Epilachna dodecastigma* Mulsand
	E. ocellata Redtenbacher
	Henosepilachna vigintiodopunctata Fabricius
	Epitrix cucumeris Harran
	Epilachna philippinensis Dieke
	Epitrix tuberis Gent
Green leaf beetle	*chalaenosoma metallium* Fabricius
Leaf eating beetle weevils	*Atmetonychus peregrinus* Olivier
Leaf eating caterpillars	*Spilarctia (Spilosoma) obliqua* (Walker)
	Spodoptera exigua (Hübner)
Tussoch moth	*Dasychira mendosa* (Hübner)
Eggplant shoot borer	*Leucinodes orbonalis* Guenèe
Tuber mealy bug	*Niplaecoccus vastator* (Maskell)
Leaf weevils	*Myllocerus subfasicatus* Glover

SWEET POTATO

Ipomoea batatas Linnaeus, *Shakarkandi*

(1)	(2)
Sweet Potato weevils	*Cylas formicarius* (Fabricius)
	Alcidodes fabricii Fabricius
	A. dentipes
	A. erroneus
Weevils (leaf eating)	*Blosyrus asellus* Marshall
	Myllocerus sabulosus Marshall
	Phytoscaphus triangularis Olivier
	Protocylas coimbotorensis Subramaniam
Leaf eating caterpillars	
Sphinx caterpillar	*Agrius convolouli* (Linnaeus)
	Theretra oldenlandiae Fabricius
Bihar hairy caterpillars	*Spilarctia (Spilosoma) obliqua* (Walker)
Minor pest	
Blue pansy caterpillar	*Junonia orithya* (Butter)
	Euchromia polymena (Linnaeus)
	Creatonotus gangis Linnaeus
	Hyposidra succossaria Walker
Black hairy caterpillar	*E stigmene lactinea* (Cramer)
	Catephia inquita Walker
	Spodoptera littoralis (Fabricius)
Leaf-feeding beetle	*Aspidomorpha miliaris* (Fabricius)
	A. furcata Thunberg
	A. unelica
	A. sanctaecrucis (Fabricius)
	Cassida indicola Duvi vier
	Chiridia sexnotata (Fabricius)
	Glyphocassis trineata (Hope)
	Metriona circumdata (Herbest)
	M. varians (Herbest)
Sphingid caterpillar	*Herse convolvuli* Linnaeus
Brown hairy caterpillar	*Euchromia polymena*
	Oncocephala tuberculata Olivier
	Leucophalis coneophora Burmeister

	Anomala dussumieri Blanchard
	Gonocephalum sp.
Gap sucking bug	*Exitianus indicus* Distant
	Riptortus linearis Fabricius
	Halticus minutes Reuter
	Graptosethus servus Fabricius
Mealy bug	*Geococcus coffeae* (Green)
Leaf rollers	*Brachmia arotracea* (Meyrich)
	B. convolvuli Walsingham
	B. effera (Meyrich)
	B. engraota (Meyrich)
	B. macroscopa Meyrich
	Helcystogramma lamprostona (Zeller)
Leaf miner	*Acrocercops proscata* Meyrick
Stem or Vine borer	*Omphisa anastomosalis* (Guenèe)
Tuber borer	*Metasia coniotalis* Hampson

GARLIC

Allium sativum Linnaeus

Insect Pests

(1)	(2)
Aphid	*Myzus persicae* Sulzer
Mite	*Rhizoglyphus echinopus*
	Spodoptera litura Fabricius
Thrips	*Thrips tabaci* Lind,
	Caliothrips vidicus (Bagnall)

ONION

Allium cepa Linnaeus *Pyaz*

(1)	(2)
Borer	*Helicoverpa armigera* (Hübner)
	Spodoptera litura Fabricius
Onion maggot	*Hylemia antiqua*
Thrips	*Thrips tabaci* Lind
	Caliothrips indicus (Bagnall)
Mite	*Rhizoglypus* sp.

AMARANTH

Amranthus bliteem Linnaeus var ***oleracea***	*Duthie Chaulai, Sadanatiya*
A. cruentus Linnaeus	*Ramadana, Chaular*
A. spinosus Linnaeus	*Kantewali chaular*
A. tricolor Linnaeus	*Bari chaular*

(1)	(2)
Stem weevil	*Hypolixus truncatulus* (Boheman)
Leaf eating caterpillar	*Eretmocera impactella* Walker
	Hymenia recurvalis (Fabricius)
	H. fasicialis (Cramer)
	H. perspectalis (Hübner)
Leaf roller	*Nacoleia indicus* (Fabricius)
	Dichocrocis puncti feralis (Guenee)
	Psara basalis (Moore)
	Junonia orithya (Linnaeus)
	Helicoverpa armigera (Hübner)
	Otheris fullonia (Clerck)
	Plusia eriosoma Doubloday
	Spodoptera exempta (Walker)
	S. exigua (Hübner)
	S. littoralis (Fabricius)
Grasshopper	*Atractomorpha crenulata* (Fabricius)
Sap sucking insect	*Aphis craccivora* Koch
	Lipaphis erysimi (Kaltenbach)
	Myzus persicae (Sulzer)
	Hyadaphis undobrassicae Das
Coreid bug	*Cletus pugnator* Fabricius
Lygaeid bug	*Nysius unconspicuous* Distant
Mealy bug	*Ferrisia virgata* (Cockerell)
Scale insect	*Coccus hesperidium* (Linnaeus)
	Pulvinaria durantae Targioni
Thrips	*Aeolethrips collaris* Priesner
	A. fulvicollis Bagnall
	Euroyapethrips crassus (Ramakrishna and Margabandhu)
	Haplothrips ceylonicus Schmutz
	Franbliniella intonsa Trybom
Leaf eating beetle	*Aspidomorpha exilis* Boheman
Leaf twisting weevil	*Apoderus tranquebaricus* (Fabricius)
	Cyratozemia cognata Marshall
	Ptochus ovulum Faust
Termites	*Microtermes anandi* Holmgren

SPINACH Beet (PALAK)

Beta vulgaris* var. *bengalensis	**Hortom spinach-beet** *Palak*
Spinacia deracea Linnaeus	*Vilayti Palak*
Basella alba Linnaeus	*Indian Spinach*
Ipomoea reptans (Linnaeus)	*Water Spinach*

(1)	(2)
Blue beetle	*Altica caerulescens* (Baly)
Aphids	*Lipaphis erysimi* (Kaltenbach)
	Myzus persicae (Sulzer)
	Hydaphis (Siphovoryne) indobrassicae (Das)
Grasshopper	*Atractomorpha crenulata* Fabricius
Leaf eating caterpillar	*Hymenia recurvalis* (Fabricius)

FENUGREEK

Trigonella foenum-graceum Linnaeus *Methi*

T. polycerate Linnaeus	**Wild fenugreek** *Kasuri methi*
T. corniculata, *Kasuri methi*	**Chanihon**

(1)	(2)
Aphids	*Aphis craccivora* Koch
	A. gossypii Glover
	Myzus persicae (Sulzer)
Leaf eating caterpillar	*Hymenia recurualis* (Fabricius)

LETTUCE

Lactuca indica Linnaeus	Indian lettuce
L. sativa Linnaeus	Lettuce salad
L. scariola Linnaeus	Wild lettuce, *Kahu*

(1)	(2)
Cotton aphids	*Aphis gossypii* Glover
Green peach aphids	*Myzus persicae* (Sulzer)
Mustard aphids	*Lipaphis orysimi* (Kaltenbach)
	Brevecoryne brassicae (Linnaeus)
Cabbage green semilooper	*Trichoplusia ni* (Hübner)
Semiloopers	*Plusia chalcites* Esper
	P. eriosoma Doubleday
	P. nigrisigna Walker
	P. signata

ASPARAGUS ***Asparagus officinalis*** Linnaeus *Halyun*

(1)	(2)
Asparagus beetle	*Crioceris asparagi*
Garden centipede	*Scutigerella immaculate*

RADISH ***Raphanus sativus*** Linnaeus *Mooli, Mul*

(1)	(2)
Aphids - cabbage	*Brevicoryne brassicae* (Linnaeus)
Mustard aphid	*Lipaphis erysimi* (Kaltenbach)
Peach green aphids	*Myzus persicae* (Sulzer)
Citrus black aphids	*Toxoptera aurantii* (Boyer de Fonscolombe)
Cutworms	*Agrotis ipsilon* (Hufnagel)
Mustard sawfly	*Athalia lugens proxima* (Klug)
Minor pests	
Grasshopper	*Atractomorpha crenulata* Fabricius
Cotton whitefly	*Bemisia tabaci* (Gennadius)
Pentomid bugs	*Bagrada cruciferarum* Kirkaldy
	Eurydema pulchrum Westwood
Thrip	*Thrips tabaci* Lindemann
Foliage caterpillars	*Crocidolomia binotalis* Zeller
	Hellula undalus Fabricius
	Spodoptera littoralis Fabricius
	Spilarctia obliqua (Walker)
Flea beetle	*Chaetocnema basalis* Baly
	Phyllotreta chotanica Dv viver
	P. downesi Baly
	P. vittata Fabricius
	Apion amplum Faust

TURNIP — ***Brassica campestris*** var. *rapa* Linnaeus
Shalgum

(1)	(2)
Mustard sawfly	*Athalia lugens proxima* (Klug)
Cabbage butterfly	*Pieris rapae* (Linnaeus)
	P. napi (Linnaeus)
	P. brassicae (Linnaeus)
	P. oleracea Harris
	P. protodice
Minor pests	
Mustard aphid	*Lipahpis erysimi* (Keltenbach)
Red ant	*Dorylus horticola* Westwood
Pea-leafminer	*Chromatomyia horticola* (Goureau)
Cutworm	*Agrotis ipsilon* (Hufnagel)
Cabbage borer	*Hellula undalis* (Fabricius)
Diamond back moth	*Plutella xylostella* (Linnaeus)
Flea beetles	*Chaetocnema basatis* Baly
	Phyllotreta vittata Fabricius
Termites	*Retienlitermes flavipes* Kollar
Cabbage maggot	*Delia brassicae* Bouche

BEET root	***Beta vulgaris*** Linnaeus *Chukandar*
Borer/Leaf eating caterpillar	*Plusia nigrisigna* Walker
	Spodoptera mauritia (Boisduvat)
Beet web worm	*Hymenia facialis* (Cramer)
Hairy caterpillar	*Estigmenne ncrea* (Drury)
Bihar hairy caterpillar	*Spliarctia obliqua* (Walker)
Blue beetle	*Altica cyanea* (Weber)
Leaf eating weevil	*Ptochus ovulum* Faust
Beet leaf miner	*Pegomyia* sp.
Semi-looper	*Plusia nigrisiana* Walker

CARROT	***Daucus carota*** Linnaeus *Gajjar* ***D. carota sativa*** *de Condolle*
(1)	(2)
Leafhopper	*Empoasca punjabensis* Pruthi
Pea-leaf miner	*Chromatomyia horticola* (Goureau)
Cutworm	*Agrotis ipsilon* (Hufnagel)
Red ant	*Dorylus orientalis* Westwood
Flea beetles	*Chaetonema basalis*
	Chlorophorus sp.
	Anthrenus coloratus Rottenburg
	A. jardanicus Picard
	A. oceanicus Favral
Carrot weevil	*Listronotus oregonensis* (Le Conte)
Carrot rust fly	*Psila rosae*

COLOCASIA (Taro)	***Colocasia esculenta*** Schott, Elephant ear, *Arvi*
(1)	**(2)**
Leaf feeder	
Banana aphids	*Pentalonia nigronervosa* Coquillett
Cotton aphids	*Aphis gossypii* Glover
Cotton whitefly	*Bemisia tabaci* (Genn.)
Taro leafhopper	*Tarophagus prosperpina* (Kirb.)
Taro hawk moth	*Hippotion celerio* (Linnaeus)
Root feeder	*Popuana* spp.
	P. huebneri (Fairm)
	P. intermis Proll
	P. woodlarkiana (Montr.)
Root borer	*Ligyrus ebenus* (Degeer)
	L. cuniculus (Fabricius)
Flea beetle	*Monolepta signanta* Olivier
Leaf eating caterpillar	*Pericallia ricini* Fabricius
	Agrius convolouli (Linnaeus)
	Theretra gnoma (Fabricius)
	T. pinastriana (Marytn)
	T. oldenlandiae (Fabricius)
	Spodoptera littoralis (Fabricius)
Grasshopper	*Gesonula punctifrons* (Stal)
Lace-wing bug	*Stephanitis typicus* Distant
Thrips	*Caliothrips indicus* (Bagnall)
	Helionothrips kadaliphilus (Ramakrishna and Margabandhu)
Greenhouse thrips	*Helicothrips haemorrhoidalis* (Bouche)
Cardomon thrips	*Scirtothrips cardamons* (Ramakrishna)
Mite	*Tetranuchus marianae* Mc Gregor

CASSAVA	***Manihot esulenta** Crants* **CSM.** ***M. utilissima*** **Bhl, Tapioca,** *Tikhoor*
(1)	**(2)**
Termites	*Odontotermes obesus*
Leaf beetle	*Apogonia rauca* Fletcher
Whitefly	*Bemisia tabaci* (Gennadius)
Cassava scale	*Aonidomytilus albus* Ckll.
Rose thrips	*Rotithrips syriacus* (Mayel)
Mealy bug	*Pseudococcus filamentosus* Ckll.
Stem borer	*Petrolphia melanura*
	Sybra praeusta
Cock chafter grub	*Leucopholis coneophora*
Millipede	*Haropurostreptus* spp.

ELEPHANT Foot	***Amorphophallus companulatus*** Blume *Jimikand, Sooran*
(1)	**(2)**
Leaf eating beetle	*Galerucida bicolor* (Hope)
Leaf eating caterpillar	*Pericallia ricini* Fabricius
	Spodoptera littoralis Fabricius
Sphingid moth	*Rhyncholaba acteus* (Cramer)
	Hippotion celerio (Linnaeus)
	Theretra gnoma (Fabricius)

YAMS	*Dioscorea* spp.	
	D. alata Linnaeus	greater yam
	D. bulbifera Linnaeus	potato yam
	D. esculenta (Lour) Burkill	Chinese, Asiatic yam
	D. rotundata Poir	white yam
	D. cayenesis Lam	yellow yam
	D. dumetorum (Kunth) Pax	bitter yam
	D. trifida Linnaeus	cush-cush yam
	D. opposita Thunb.	cinnamon yam
	D. japanica (Thunb.)	

(1)	(2)
Leaf eating beetle	*Galerucida bicolor* (Hope)
	Lema lacordairei Baly
	Crioceris impressa Fabricius
	Heteroligus meles (Billb)
	H. appius (Burmeister)
Mealy bugs	*Geococcus coffeae* Gran
	Planococcus citri (Risso)
Scale insects	*Aspidiella hartii* (Cockerell)
	Quadraspidiotus destructor (Signoret)
	Aspidiella hartii
Sawfly	*Senoclidia dioscorease* Rohwer
Leaf eating caterpillars	*Theretra oldenlandiae* Fabricius
	Rhyncholaba acteus (Cramer)
	Pericallia ricini Fabricius
	Spilarctia obliqua (Walker)
	Spodoptera littoralis (Fabricius)
Skipper butterfly	*Parnara* sp.

DRUMSTICKS	***Moringa deifera*** Lamarck *Sojna Sahinjna, Soanjna*
(1)	**(2)**
Leaf-eating caterpillar	*Eupterote mollifera* Walker
	E. geminata Walker
	Noorda blitealis Walker
	Pericallia ricini (Fabricius)
	Protrigonia zizanialis Swinhoe
	Actias selene Hübner
	Metanastria hyrtaca (Cramer)
	Streblote siva (Lefevre)
Bark borer	*Indarbela tetraonis* (Moore)
Longhorn beetles	*Batocera rubus* (Linnaeus)
	Monochamus versteegi
Aphids	*Aphis craccivora* Koch
Whitefly	*Trialeurodes raja* Shumsher
Sclae insect	*Ceroplastodes cajani* Maskell
	Diaspidiotus sp.
Thrips	*Scirothrips dorsalis* Hood
	Ramaswamiaheilla subnudula Karny
Termite	*Neotermes flecheri* Holmgren
Beetle grubs	*Holotrichia reynaudi* Blanchard
Bud borers (midge)	*Stictodiplosis moringae* Mani
Drosophil fly	*Gitona* sp.
	Noorda moringae Tams
Leaf eating beetles	*Myllocercus discolor variegatus* Boheman
	M. teniclavis inferior Marshall
	M. undecimpustulatus maculasus Desbrochers
	M. viridans Fabricius
	Ptochus ovulum Faust

BANANA	***Musa paradisiaca*** **Linnaeus** ***Kela***
(1)	**(2)**
Banana weevil	*Cosmopolites sordidus* (Germar)
Banana stem borer	*Odoiporus longicollis* (Oliv.)
Flea beetle	*Nodostoma subcastatum* Jacoby
	N. viridipennis Motschulsky
Banana stalk borer	*Castniomera humboldi*
Leaf eating caterpillar	
Bagworm	*Acanthopsyche-minima* Hampson
	Kophene cuprea Moore
Black hairy caterpillar	*Pericallia ricini* Fabricius
	Asura ila Moore
	Creatotus gangis Linnaeus
Bihar hairy caterpillar	*Spilarctia obliqua* (Walker)
	Dasychira mendosa Hübner
	Euproctis fraterna (Moore)
	Eupterote geminata Walker
Leaf roller	*Ericonota throx* Linnaeus
Slug caterpillar	*Belippa lateane* Moore
	Miresa decedesas Walker
	Spdoptera litura (Fabricius)
Leaf sucking bugs	
Lace wing bug	*Stepjanitis typicus* Distant
Spittle bug	*Phymatosetha deschampsi* Lethierry
Leaf-hopper	*Euybrachys tomentosus* Fabricius
	Prodromus sabviridis Distant
Coccoids	*Perissopnemon tomarindus* (Green)
	Ferrisia virgata (Cockerell)
	Abgrallaspis cyanphylli (Signoret)
	Aonidiella aurantii (Maskell)
	A. orientalis (Newstead)
	Chrysomphalus ficus Ashmead
	Hemiberlesia lataniae (Signoret)
	Coccus acutissimus (Green)
	C. discrepans (Green)
	C. hesperidum Linnaeus
	Saissetia depressum Cockerell
	Pseudococcus comstocki Kuw.

Banana aphids	*Pentalonia nigronervosa* Coquillett
	Aphis gossypii Glover
Thrips	*Heliothrips radaliphilus* (Ramakrishna and Morgabandhu)
	Panchaetothrips indicus Bagnall
	Thrips florum Schmutz
	Astrothrips parvilimbus Stannord and Mitri
	Chaetana phothrips signipennis Bagnall
	Hercinothrips bicinctus Bagnall
Fruit flies	
Orientalis fruit fly	*Bactrocera dorsalis* (Hendel)
Mediterranean fruit fly	*Ceralitis capitata* Wiedemann
Fruit scarring beetle	*Colapis hypochlora* Lefore
Fruit boring caterpillar	*Heteromicata latra*
Fruit sucking moth	*Anua coronata* (Fabricius)
	Otheris fullonia (Clerch)
Red spider mite	*Tetranychus lambi* Pritchard & Baber

JACK Fruit	***Artocarpus heterophyllus*** Lamarck *'Kathal', 'Kanthal'*
(1)	**(2)**
Shoot borer	*Diphanua caesalis* (Walker)
Bark borer	*Indarbela tetraonis* (Moore)
	Apriona rugicollis Chevrolet
	Bactocera rufomaculata (de Geer)
	B.rupus (Linnaeus)
	Epepeotes lascus Fabricius
Longicorn beetle	*Glenea belli* Gahan
Girdler beetle	*Sthenias grisator* Fabricius
Pinhole borer	*Platypes indicus* Strohmeyer
Spiitle bugs	*Cosmoscarta relata* Distant
	Clovia lineaticollis Motschulscky
Mealy bug	*Nipaecoccus virdis* (Newstead)
	Ferrisia virgata (Cockerell)
Leaf webbers	*Perina nuda* Fabricius
	Diaphania bivitralis (Guenèe)
Castor capsule borer	*Dichocrocis punctiferalis* (Guenèe)
Scale insect	*Icerya aegyptica* (Douglas)
	Hemiberlesia lataniae (Signoret)
	Parlaspis papillosa (Green)
	Pinnopsis aspidistrae (Signoret)
	Semelaspidus atrocarpt (Green)
	Ceroplastes rubens Maskell
	Chloropulvinaria psidii (Maskell)
	Coccus acutissimus (Green)
Aphids	*Greenidia artocarpi* (Westwood)
	Toxoptera aurantii (Fonscolmbe)
Whiteflies	
	Aleurotrachelus caerulosans Singh
	A. rachipora Singh
	Realius schimae Takahashi
Thrips	
	Pseudodendrothrips divivarna (Ramakrishna and Margabandhu)
	Dendothrips divivarna
Red ant	*Oecophylla smaragdina* (Fabricius)

PAPAYA	***Carica papaya*** **Linnaeus '*Papita*'**
(1)	**(2)**
Ak Grasshopper	*Poekilocerus pictus* Fabricius
Aphids	*Aphis gossypii* Glover
	A. medicaginis Koch
	Myzus persicae (Sulzer)
	Urdeucom sonchi (Linnaeus)
Whitefly	*Bemisia tabaci* (Gennadius)
Fruitflies	*Bactrocera diversus* (Coquillett)
	B. cucurbitae (Coquillett)
	B. persestris Bezzi
	Toxotrypma curvicauda (Gerstacker)
Mediterranean fruitflies	*Ceratitis capitata* (Wieldemann)
Coccoids	
	Drosicha mangiferae (Green)
	Planococcus citri (Risso)
	Aspidiotus destructor (Signoret)
	Pseudoparlateria ostriate Cockerell
Stem borer	*Cerostoma rugosella* (Stainbow)
Green weevils	*Myllocercus viridanus* Fabricius
Mites	*Tetranychus cinnabarinus* (Biosduval)
	Bravipalpus phoenicus (Geijskes)

APPLE, *Malus pumila* Mill, *Seb*
Small fruited crab apple *M. floribunda*

(1)	(2)
San Jose scale	*Quadraspidiotus perniciosus* (Comstock)
Mealy bugs	*Icerya purchasi* Maskell
	Drosicha contrahems (Walker)
	D. mangigerae (Green)
Wooly aphid	*Erisoma langierum* (Hausman)
Green aphid	*Aphis pomi* de Geer
Melon aphid	*Aphis gossypii* Glover
Peach aphid	*Myzus persicae* (Sulzer)
Root borer	*Dorysthenes hugeli* Radtenbacher
Stem borer	*Apriona cinera* Chevrolat
	Batocera rugomaculata (De Geer)
Girdler beetle	*Linda migrosculata* Fairmaire
	Sthenian grisator Fabricius
Leopard moths	*Zeuzera* spp.
Apple weevil defoliators	*Myllocerus discolor* Boheman
Lesser apple weevil	*Dereodus poliinosus* Redtenbacher
Brown weevil	*Tanymecus circumelatus* Wiedemann
Apricot weevil	*Emperotthinrus defolitor* Marshall
Leaf roller	*Archips termias* (Meyrick)
	Gracillaria zachrysa Meyrick
Leaf eating caterpillar	*Actias selene* Hübner
Hawk moth	*Langia zeuzeroides* Moore
Indian gypsy moth	*Lymantria obfuscata* Walker
Blossoms thrips	*Frankliniella dampfi* (Priesner)
	Thrips falvus (Schrank)
	T. hawaiiensis (Morgan)
Brown apple moth	*Epiphyas postvittana* (Walker)
	Cydia pomonella (Linnaeus)
Apple maggot fly	*Planotortrix octo*
	Rahgoletis pomonella (Walsingham)
Fruitflies	*Bactrocera ciliatus* (Loew)
Oriental fruitflies	*B. dorsalis* (Hendel)
	B. zonaus (Saunders)
Colding moth	*Laspercysia pomonella* (Linnaeus)
Cutworm	*Agrotis ipsilon* (Hufnagel)
Leaf blotch miner	*Lithocolletis ringoniella* Mats
Mite	*Tetranychus viennensis*
	Panonychus ulmi

MANGO ***Mangiera indica*** **Linnaeus** *Am*

(1)	(2)
Mango hoppers	*Idioscopus clypealis* Lethierry
	I. niveosparsus Lethienrry
	Amritodus atkinsni Letheinrry
Mango mealy bug	*Drosicha mangiferae* Green
Fruitfly	*Bactrocera dorsalis* (Hendel)
	B. correctus (Bezzi)
	B. diversus (Coquillett)
	B. tau (Walker)
	B. zontatus (Saunders)
Mango leaf cutting weevil	*Euganmptus marginellus* Fst.
Mango stem borer	*Batocera rubus* Linnaeus
	B. rufomacularia Walker
Bark eating caterpillar	*Indarbela quadrinotata* Walker
	I. tetranonis Moore
Mango shoot gall	*Apsylla cistetlata* Buckton
Leaf cutting weevil	*Deporaus marginatus* Pascal
Mango leaf gallfly	*Procontainia matteiana* Keiffer and Cecconi
Mango gall midges	*Erosomiyia indica* Grover
	Procystiphora indica Grover
	P. mangigerae Felt
	Dasyneura amaramanjarae Grover
Stone weevil	*Sternochetus mangiferae* Fabricius
	S. frigidus Fabricius
Red ants	*Odcophylla smaragdina* (Fabricius)
Termite	*Neotermes gardeni* Snyder
	Odontotermes obesus (Rambur)
Mango scale insects	*Aspidiotus destructor* Signoret
	Pulvinaria polygonata Cockerell
	P. psidii Maskell
	Parlatoria pergandii Comstock
	Lepidosaphes gloverlii Packard
Thrips	*Acolothrips collaris* (Priesner)
	Rhipiphorothips creuntatus (Hood)
	Scirothrips dorsalis (Hood)
	Haplothrips ganglbaueri Schmutz
	Neoheegeria mangiferae (Priesner)
Mites	*Aceria mangiferae* Sayed
Mango malformation	Complex of Fungal *Fusarium moniliforme subglutinans* and *Aceria mangiferae*

POMEGRANATE, *Punica granatum* Linnaeus *Anar*

(1)	(2)
Pomegranate butterfly	*Virachola isocrates* (Fabricius)
Bark eating caterpillar	*Indarbela tetraonis* (Moore)
	I. guadrinotata (Walker)
Red borer	*Zeuzera coffeae* Nietner
Stem boring beetle	*Coleosterna spinator* Fabricuis
Sap sucking insects	*Drosicha mangiferae* (Green)
	Icerya purchasi Maskell
	Planococcus lilacinus (Cockerell)
	Pinnaspis theae (Maskell)
Aphid	*Aphis punicae*
Whiteflies	*Sipnimus finitimus* Slvestri
	S. phillyreae (Haliday)
Thrips (Leaves)	*Retithrips syriacus* (Mayet)
	Rhipiphorothrips cruentatus (Hood)
	Franklinneilla damfi (Prisener)
Thrips (Flower)	*Anaphothrips oligochaetus* Karny
	Ramaswamiahiella subnudula Karny
	Scirtothrips dorsalis Hood
Leaf eating caterpillar	
Castor semilooper	*Achaea janata* (Linnaeus)
Ber hairy caterpillar	*Euproctis flava* (Bremer)
Plum hairy caterpillar	*E. fraterna* (Moore)
	Pothesia scientillans (Walker)
	Creatonotus gangis (Linnaeus)
Beetles	*Anomala dimidiata* Hope
	Mimastra cyanura Hope
	Myllocerus spp.
Fruitflies Mango	*Bactrocera dorsalis* (Hendel)
Fruitflies Peach	*B. zonatus* (Saunders)
Termite	*Odontotermes obesus* (Rambur)

Acid lime ***Citrus aurantifolia*** (Christm) *Swingle Nimbu*
Sweet orange ***Mosambi, Citrus sinenses*** Linnaeus Osbeck,

(1)	(2)
Termites	*Odonototermes obesus* (Rambur)
Chafer beetles	*Anomala tenella* blanchard
	Lachnosterna consanguinea (Blanchard)
	Holotrichia repetita Sharb
	H. rufoflava Brenske
	Acanthophorus serraticornis (Olivier)
	Belionata prasina Thunberg
Trunk borer	*Indarbela quadrinotata* (Walker)
	I. tetraonis (Moore)
	Anopalophora versteegi (Risema)
Stem boring beetle	*Chelidonium cinctum* (Guerin- Meneville)
	Chloridolum alcmene Thomson
	Stherias grisator (Fabricius)
Foliage Pest	
Grasshopper	*Poekilocerus pictus* (Fabricius)
Leaf eating caterpillar	*Papilio demoleus demleus* Linnaeus
	Parasa lepida (Cramer)
	Narosa propolia Hampson
	Chilades laius (Cramer)
	Thiacides postica (Walker)
	Euproctis fraternal Moore
	Tarucus theophrastus (Fabricius)
Leaf eating caterpillar	*Myllocerus denlifer* Fabricius
	Pellotrachelus pupes (Faust)
Citrus psylla	*Diaphorina citri* Kuwyama
Whiteflies	*Aluerocanthus citriperdus* (Quaintance and Baker)
	A. woglumi (Ashby)
	Aluerolobus morlatti (Quaintance)
	Dialeurodes elongata Dozier
	Bemisia giffardi (Kotinsky)
	Dialeurodes citri (Riley & howard)
	D. citrifolii (Morgan)
	Aleurotuberculatus citrifolii (Corbett)

Aphids	*Toxoptera aurantii* (Boyer de Fonscolome)
	T. citriciuds (Kirkaldy)
	Aphis citricola van de Goot
Melon aphid	*Toxoptera aurantii* (Boyer de Fonscolombe)
	T. citricidus (Kirkaldy)
	Aphis citricola van deer Goot
Cotton aphid	*A. gossypii* Glover
Green peach aphid	*Myzus persicae* (Sulzer)
Green apple aphid	*Aphis pomi* Degeer
Peach leaf curl aphid	*Branchycaudus helichrysi* Kaltenbach
Coccids	*Drosicha mangiferae* (Green)
	Planoccus sp.
	Icerya purchasi Maskell
	Piandoccus citri Risso
	Pseudicoccus adonidum (Linnaeus)
	Aonidiella aurantii (Maskell)
	Chryomphalus ficus Ashmead
	Parasiassetia nigra (Neither)
Thrips (foilage)	*Helithrips haemorrhoidalis* (Boucher)
	Thrips pandu Ramakrishna
	T. nilifiriensis (Ramakrishna)
	Frankliniella dampfi Priesner
	Ramawamiacheilla subnudala Karny
	Scirothrips dorsalis Hood
	Thrips flavas Schrank
	T. florum Schmutz
Thrips (fruits)	*Scirothrips aurantii* Faure
	S. citri (Moulton)
Leafminer	*Phyllocnistic citrella* Stainton
Flower feeders	
Blossom midge	*Dasineura citric* (Rao and Grover)
Flower moth	*Prays citri* (Milliere)

Fruit Pests

Fruit sucking bugs	*Cappaea taprobanensis* (Dallas)
	Rhynchocaris humeralis (Thunberg)
	Neazara viridula (Linnaeus)
	Lepioglossus australis (Fabricius)

Fruit sucking moths	*Otheris materna* (Linnaeus)
	O. fullonica (Linnaeus)
	O. ancilla Cramer
Fruit flies	*Bactrocera correctus* (Bezzi)
	B. diversus (Coquillett)
	B. tau (Walker)
	B. dorsalis (Hendel)
Fruit borers	*Prays endocarpa* Meyrick
	Cryptophlebia illepida (Butter)
Mites	*Entetranychus citri* (Mc - Gregor)
	Tenuipalponychus citri Channa Basavanna and Lakkundi
	Brevipalpis delconi Pritchard and Baker
Citrus bud mite	*Aceria sheldoni* (Ewing)
Citrus silver mite or broad mite	*Hemitarsonemus fatus* Banks
Carmine spider mite	*Tetranychus cinnabarinus* Biosd

Wine Grapes, *Vitis vinifera* Linnaeus, European grape, *Angur*
Fox Grapes, *V, labrusca* Linnaeus

(1)	(2)
Flea beetles	*Scelodonta strigicollis* (Motshulsky)
	Monolepta erthrycephala Baly
	M. signata Olivier
Leaf eating beetle	*Adoretus bengalensis* Blanchard
	A. brachypgus Burmeister
	A. horticola Arrow
	Anomola bengalensis Blanchard
	Dasus depressum Fabricius
	D. hoffmanngeggi Steven
Thrips	*Rhipiphorothrips cruentatus* Hood
	Retithrips syriacus (Mayet)
Blossom thrips	*Thrips glarum* Schmutz
	Frankliniella dampfi (Prienser)
	Scirtothrips dorsalis (Hood)
Coccids	*Drosicha mangiferae* (Green)
	Icerya aegyptiaca (Douglas)
	I. purchasi Maskell
	Ferrisia virgata (Cockerell)
	Nipaecoccus virdis (Newstead)
	Aspidiotus destructor Signoret
	Parlatoria camelliae Comstock
	P. oleae (Colvee)
Girdler beetle	*Stenias grisator* Fabricius
Stem boring beetle	*Sinoxylon anale* Lesne
Leaf roller	*Sylepta lunalis* Guenèe
Termite	*Odontotermes obesus* (Rambur)
Leafthopper	*Empoasca minor* Pruthi
	Arbaridia viniferata Sohi and Sandhu
	Erythroneura sp.
Grapes phylloxera	*Daktulosphaira vitifolia*
Grapes berry moth	*Lobesia botrana*
One banded grapes tortrieid	*Eupoecilia ambiguella*
Peach mealy aphid	*Hyalopterus pruni* (Geoffrey)
Whitefly	*Aleurocanthus spiniferus* (Quaintance)
Bagworm	*Clania crameri* (Westwood)
Hawk moth	*Agrius convolvuli* (Linnaeus)
Fruit borer	*Oxyptilus regulus* Meyrick
Hornet wasp	*Vespa orientalis* Linnaeus

BANANA, ***Musa paradisiaca*** **Linnaeus,** ***Kela***

(1)	(2)
Banana borer (beetle)	*Cosmopolites sordidus* (Germer)
Weevil stem borer	*Odoiporus longicollis* (Olivier)
Fruit scarring beetle	*Colaspis hypochlora* Lefevre0
Rust thrips	*Chaetanaphothrips signipennis* (Bagnall)
Silvering thrips	*Hercinothrips bicinetus* Bagnall
Flower thrips	*Thrips florum* Schmutz
Aphid	*Pentalonia nigronercosa* Coquillett
	Aphis gossypii Glover
	Macrosiphum euphorbiae (Thomas)
Mealy bug	*Psuedococcus comstocki* Kuw.
Scab moth	*Nacoleia octasema* (Meyer)
Stalk borer	*Castiniomera humboldti* (Biosd.)
	Castnia licoides Biosd.
Leaf roller	*Erionota thras* Linnaeus
Fruit scarring bee	*Trigona amalthea* Ol.
Leaf eating caterpillar	*Ceramidia virdis* (Druce)
Bagworm	*Acanthopsyche minima* Hampoon
	Kophene cuprea Moore
Black hairy caterpillar	*Pericollia ricini* Fabricius
Bihar hairy caterpillar	*Spilarctia obliqua* (Walker)
Lace wing bug	*Stephanitis typicus* Distant
Fruitflies	*Bactrocera dorsalis* (Hendel)
	B. musae (Tryon)
	B. tryoni Froggot
Mediterranean fruit fly	*Ceratitis capitata* (Wiedermann)
Red spider mite	*Tetranychus lambi* Pritchard & Baker

Chinese date *Ziziphus mauritiana* Lamark, *Ber*
Root stock, *Z. nummularia* (Burm) Wight & Arn, *Wild Jujube, Jharber*

(1)	(2)
Fruitflies	*Carpomya vesuviana* Costa
Oriental fruit (*ber*)	*Bactrocera dorsalis* (Hendel)
	B. correctus (Bezzi)
	B. zonatus (Saunders)
Mealy bug	*Perissopneunon tamarindus* (Green)
	Drosicha mangiferae (Green)
	Nipaecoccus viridis (Newstead)
Leaf defoliating beetles	*Holotrichia insularis* Brenske
	Adorestus nitidus Arrow
	Anona bengalensis Blanchard
Leaf eating weevils	*Amblyrrhinus pericollis* Boheman
	Sitones crinitus Olivier
	Tanymecus cirucmdatus Wildermann
	Xanthochelus superciliosus Gyllenhal
	Occassida obscura (Fabricius)
	Platypria andrewesi Weise
Leafhopper	*Eurybrachys tomentosa* Fabricius
	Zynidia pakistanica (Ahmed)
Scale insects	*Aonidia zizphi* Rahman and Ansari
	Aonidia aurantii (Maskell)
	A. orientalis (Newstead)
	Aspidootus desnictor Signoret
	Kerria indicola (Kapur)
	K. lacca (Kerr)
	K. pusana (Misra)
Thrips	*Dolichothrips indicus* (Hood)
	Sciriothrips dorsalis (Hood)
	Scolothrips asura Ramakrishna and Margabandhu
Bark eating caterpillar	*Indarbela quadrinotata* (Walker)
	I. tetranonis (Moore)
Leaf eating caterpillar	*Tarucus theophrastus* (Fabricius)
Ber butterfly	
Hairy caterpillar	*Thiacidas postica* Walker
	Euproctis fraterna (Moore)

	E. lunata (Walker)
	E. flava (Bremer)
	Beara dichromella Walker
	Streblote siva (Lefevre)
Wild silk moth	*Cricula trifenestrata* Helfer
Leaf roller and Webbers	*Anchli lutescens* Meyrick
	Psorosticha zizyphi (Stainton)
Stem boring beetle	*Synclera univoclius* (Walker)
Longicorn beetle	*Celosterna scabrator* Fabricius
	C. spinator Fabricius
Fruit borer *Bee* fruit borer	*Meridoechus scyrodes* Meyrick
Pomegranate butterfly	*Virachola isocrates* (Fabricius)

GUAVA, *Psidium guajava* Linnaeus, *Amrud*
Strawbery guava, *P. cattleianum* *Sabine*

(1)	(2)
Oriental Fruitflies	*Bactrocera dorsalis* (Hendel)
	B. diversus (Coquillett)
	B. cucurbitae (Coquillett)
	B. zonatus (Saunders)
Bark eating caterpillar	*Indarbela tetramonis* (Moore)
	I. quadrinotata (Walker)
Castor capsule borer	*Dichocrosis punctiferalis* (Guenee)
Coccids	*Drosicha mangiferae* (Green)
	Icerya purchasi (Maskell)
	Ferrisia virgata (Cockerell)
	Planoccocus lilacinus (Cockerell)
Scales	*Coccus hespridum* Linnaeus
	C. viride
	Saissetia coffeae (Walker)
Aphids	*Aphis gossypii* Glover
Whiteflies	*Aleurocanthus rugosa* Singh
	A. wolglumi Ashby
	Aleyrtoberculatus psidii (Singh)
	Pealius misrae Singh
Thrips	*Selenothrips rubrocinctus* (Giard)
Guava scab	*Helopetis antonii* Signoret